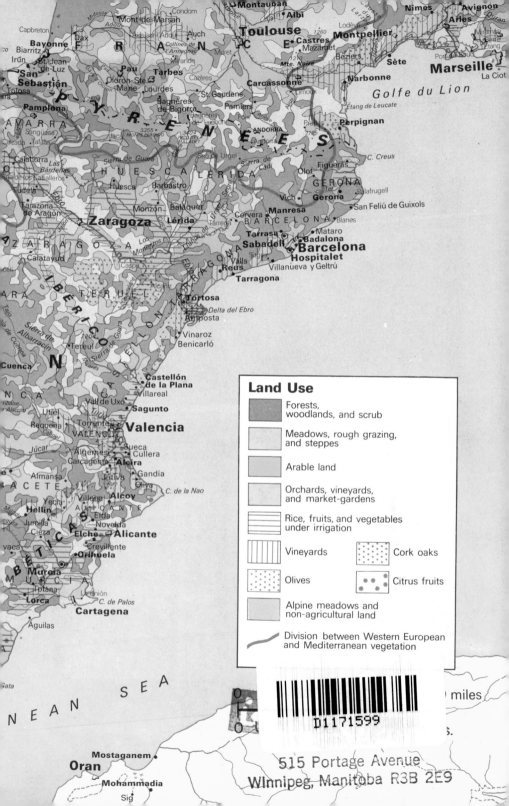

Land Use

- **Forests, woodlands, and scrub**
- **Meadows, rough grazing, and steppes**
- **Arable land**
- **Orchards, vineyards, and market-gardens**
- **Rice, fruits, and vegetables under irrigation**
- **Vineyards**
- **Cork oaks**
- **Olives**
- **Citrus fruits**
- **Alpine meadows and non-agricultural land**
- **Division between Western European and Mediterranean vegetation**

miles

Flowers of South-West Europe

Flowers of

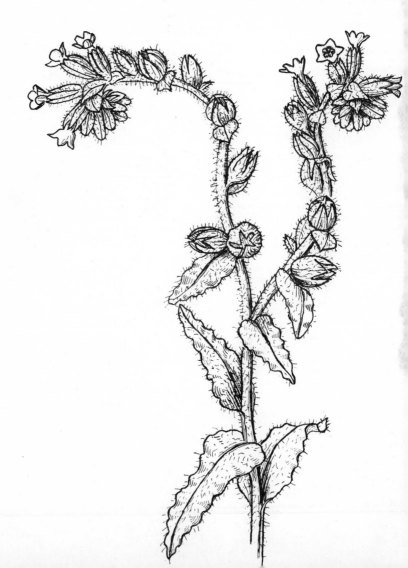

OLEG POLUNIN
B. E. SMYTHIES

South-West Europe
a field guide

With 80 pages of illustrations in colour
from photographs taken by the author and others,
27 pages of line drawings by Barbara Everard,
34 pages of line drawings by Jill Smythies
and maps by John Callow

London · *Oxford University Press* · New York · Toronto · 1973

Oxford University Press, Ely House, London W. 1

GLASGOW NEW YORK TORONTO MELBOURNE WELLINGTON CAPE TOWN IBADAN
NAIROBI DAR ES SALAAM LUSAKA ADDIS ABABA DELHI BOMBAY CALCUTTA
MADRAS KARACHI LAHORE DACCA KUALA LUMPUR SINGAPORE HONG KONG
TOKYO

ISBN 0 19 217625 0

© Oxford University Press 1973

Set by Filmtype Services Limited, Scarborough and printed
in Great Britain by Jarrold and Sons Ltd., Norwich

To Charterhouse

NOTE

The publication of volume 3 of Flora Europaea *since the proofs of this book were corrected necessitates the following small changes:*

p.42, line 29, p.50, line 32, p.101, line 4: *for* subsp. *pedunculata read* subsp. *lusitanica*

p.57, line 14: *delete* endemic

p.59, line 8 fr. foot, p.109, line 26: *for Echium lycopsis read Echium plantagineum*

p.153, caption 12, plate 40: *for Onosma arenaria read Onosma fastigiata*

p.298, line 5: *after* Cantabrian mountains *insert* Central Sierras

p.298, line 8: *for* N. and S. Spain *read* N., C. and S. Spain

p.309, line 23: *for O. arenaria* Waldst. & Kit *read O. fastigiata* (Br.-Bl.) Lacaita

p.309, line 27: *delete* S. C. and E. Spain

p.313, line 5: *for* Subsp. *Dunense* Sennen (including *T. Vincentinum*) *read* Subsp. *vincentinum* (Rouy) D. Wood

p.325, line 28: *delete* S. and E. Spain

The following further corrections should also be noted:

p.45, caption 12: *for* 1355a *read* 1335a

p.77, caption 12: *for* 518 *read* 518b

p.87, line 13 fr. foot: *for Arenaria capillipes read Arenaria lithops*

p.93, line 12: *delete Thymus cephalotos*

p.110, caption 9: *for* 1201i *read* 1201d

p.381, line 25: *after* (Algarve) *insert* S. Spain (near Marbella)

index: Aaron's Rod and *Abies* are listed at the foot of p.450 and the top of p.451 respectively

Contents

List of Terrain Colour plates

List of Maps

Preface

This book is designed for the traveller in Portugal, Spain and south-western France (west of the Rhône and south of the Loire) as a readily portable guide to the most botanically interesting and most beautiful regions of south-western Europe. The text is in two main sections: the first part describes in outline the twenty-three botanically richest regions in the south-west. The most interesting or characteristic species are named, and indications given of where these plants may be found. The double-page drawings done by Jill Smythies from collections and observations in the field give, for many of these regions, a region-by-region selection of some of these plants and should, in particular, help the beginner to name many species. The second part of the guide is concerned with the identification of species by means of keys and short diagnostic descriptions.

This field guide, though it can stand alone, is planned to be used in conjunction with the 'mother' volume *Flowers of Europe*, which describes and illustrates most of Europe's widespread species. The majority of these are not re-described in this volume, although some which are particularly characteristic of the south-west may be described again more briefly. However, in chapter 3 all plants illustrated in *Flowers of Europe* are indicated by a ** in this regional volume. The publication of three subsequent regional volumes will give a unique collection of plant photographs taken in the field in all parts of Europe. There will, of course, be some duplication in the colour illustrations: some characteristic species of each region will be illustrated again by new photographs taken in that region, despite their appearance in *Flowers of Europe*.

The selection and identification of species

A probable total of well over 6,000 species of seed plants is to be found wild in the area covered by this volume. As the aim of this book is to be a true working field guide, there has been, once again, the difficult problem of selection to keep the book to a reasonable size. In consequence, certain families—often 'difficult' families with insignificant flowers—where some botanical training and experience is needed to arrive at the correct identification of species, have been largely omitted or treated summarily. In particular, the *Polygonaceae, Chenopodiaceae, Euphorbiaceae, Umbelliferae, Cyperaceae* and *Gramineae* have been given very brief treatment, with descriptions of only a very small proportion of species—those that are conspicuous or in other ways notable.

For similar reasons, many of the largest genera have been considerably curtailed, and only those species which have been encountered by the authors on their travels, or are well represented in recent collections, have been included. As a high proportion of excluded species are endemic Iberian species, or restricted to limited areas, their absence will not, it is hoped, cause dismay except to the most diligent of plant-hunters. Other species of the larger genera, which are more widely distributed elsewhere in Europe, and will qualify for entry into other regional volumes, have also been excluded. Examples of genera receiving such limited treatment are: *Arenaria, Saxifraga, Genista, Astragalus, Trifolium, Limonium,* and *Centaurea.*

The experienced European botanist will surely understand the need for this selectivity and will appreciate that the requirements of the amateur are in all probability better served by keeping the volume within these bounds. To find any such excluded species the botanist must turn to the local (and often outdated) flora, or to *Flora Europaea,* the first two volumes of which have been published at the time of writing, out of a total of five.

A great many genera have been treated in full. And in these all southwestern species are included either in the form of keys or by short individual descriptions, or both. In the larger genera in which only a proportion of the species are described, the total number of species found in our area is given; in genera with full treatment the total number is self-evident.

For reasons of space the keys to families and to genera, and the brief diagnoses of the genera are not given; these will be found in *Flowers of Europe.* The numbering of species may initially seem complex, but the numbers have been designed as additions to those already used in *Flowers of Europe*: a continuous numbering system will be used in all the regional volumes. This greatly facilitates not only reference between volumes, but also between the illustrations and the text. For example, in the genus *Saxifraga*: 404, (404), 404a, 404b, are similar species in some ways and have the same basic number: the first two occur in *Flowers of Europe* (and in this volume), the second two occur only in this volume.

Conservation

It has been estimated by a reliable authority, who is in the process of preparing a Red Data book of rare and threatened plants, that about ten per cent of all species of flowering plants—something like 20,000 species

—are in danger of extinction in the world. The danger comes from many sources. In the French Mediterranean region, for example, the most serious threats are urbanization, tourist development, hydro-electric schemes, rice culture, wine-growing extensions, use of herbicides, pollution, airfields, conifer plantations, over-grazing, horticultural vandalism, and depredation by collectors. Only the last two are the immediate concern of naturalists in the field; and with the great increase of mobility, leisure, and communications, these particular threats are (like the others) undoubtedly on the increase. In the majority of circumstances the individual can do little to counteract these dangers to plants or to vegetation in a foreign country, except, it is suggested, to pass on information to the appropriate authority of the country concerned. International European co-operation is not yet sufficiently developed for a central dossier of threatened plants to be kept; yet international pressures on the countries concerned may well be the ultimate means by which active steps for the conservation of a species or community of plants and animals are undertaken. The authors suggest that individuals should give information about threatened plants in the area covered by this volume to one of the following societies:

The Portuguese League for the Protection of Nature
Faculdade de Ciencias
Lisbon 2
Portugal

Agrupación Española de Amigos de la Naturaleza
Plaza de Santo Domingo 16, 4°
Madrid
Spain

Fédération Française des Sociétés de Protection de la Nature
Musée Nationale d'Histoire
57 rue Cuvier
Paris Ve
France

The most vulnerable areas in Europe are undoubtedly the Iberian Peninsula, the Mediterranean Islands, the Alps, and the Balkan Peninsula. The first and last have the richest floras in Europe and it is possible that something like 200–300 species are in imminent danger of extinction in these areas alone.

It is inevitable in a volume of this nature that the whereabouts of a considerable number of rare species is revealed. While this information

xi

has, in almost every case, been previously published, mainly in inaccessible or obscure journals, it cannot be denied that it is, for the first time, being made available to a much wider public. The authors take the view that a knowledge of the locality of a rarity may, in the long run, focus attention on its preservation, and in consequence rarities have been marked by an appropriate symbol: †. Species marked in this way should not be disturbed or collected, unless they occur in considerable numbers in the particular locality. Other species which, though not threatened in Europe as a whole, may be in danger in a particular country or region, have also been marked in this way.

To those who have benefited from the information given in this volume, we would urge that they lend their full support to any national or international organization which is furthering the work of conserving the extremely rich and unique heritage of wild life in this continent.

The authors recommend the following guide lines to the botanist and plant-hunter in foreign countries.

1. Photography is one of the most rewarding and least damaging methods of plant collecting, but care must be taken to ensure that the damage to the surrounding vegetation by trampling and 'gardening' does not harm the environment or reveal the plant's whereabouts to unscrupulous people.

2. Critical species cannot be identified by photographs alone, and the only sure method is to collect and dry specimens. These retain almost all their botanical characters, except of course colour, and can be identified later at leisure. Collecting specimens must be done with care and forethought to the future of the particular colony of plants from which they are taken. Do not remove any whole plant or an unusual species from an area unless there are a considerable number of the same species in the vicinity.

   The removal of ripening fruits and the collection of seeds is certainly permissible, and carefully taken cuttings need not damage a plant. Dried fruits and seeds can be brought into the British Isles without any restrictions. Living plants or parts of living plants, including bulbs and corms, require a permit for entry. There are a few species which it is prohibited to import.

3. To avoid suspicion always try to inform local people what you are about. Collecting of supposedly 'medicinal' plants always causes great interest, and often results in the revelation of interesting local 'uses' of plants, as well as ready help with your search.

4. Do not collect in Nature Reserves.

Acknowledgements

The authors must, in the first instance, acknowledge their great indebtedness to the authors of *Flora Europaea*, for not only have they been privileged to use Volume III of this authoritative work in manuscript, but they have received help and guidance from them throughout the preparation of this volume. In addition, many of the keys and diagnoses have been freely adapted from *Flora Europaea*. Collections of specimens made during the authors' quite extensive survey of South-Western Europe, which are often voucher specimens for photographs illustrating this work, have been deposited in the herbaria at Leicester University and at the Royal Botanic Gardens, Edinburgh. A special debt of gratitude is due to Professor T. G. Tutin and A. O. Chater, not only for their prompt identification of specimens, but also for the loan of large numbers of herbarium specimens which have been invaluable for the preparation of the line drawings and for the diagnostic descriptions of species. To the staff of our national herbaria at Kew and the British Museum thanks are also due; the fact that help is so readily available is of enormous value to those working outside the usual academic institutions. J. E. Dandy has kindly checked the names of species not covered by Volumes I–III of *Flora Europaea*, and we should also like to thank Miss K. A. McInerny, Dr R. Melville, and Dr J. M. Moreira for advice and help.

The list of popular names is always fraught with difficulties and we have been fortunate in having the collaboration of Professor E. F. Galiano and Professor C. N. Tavares with the Spanish and Portuguese names respectively. Dr A. R. Pinto da Silva has kindly checked the distribution of Portuguese species and has provided a list of species to be protected in Portugal; he has also checked the Portuguese regional accounts. Professor A. Fernandes at Coimbra kindly identified some specimens of *Narcissi*. Mrs I. K. Ferguson has made many corrections and additions to the account of the Balearic Islands, and E. Chennery and C. E. Wuerpel have helped with the Algarve account. Miss A. M. Adeley continues to be an indispensable member of the team working on the volume, and despite the irregularity of the authors' production she has kept everything under control and has helped to iron out many of the wrinkles. We must also thank her for preparing the index. The maps have been prepared by J. Callow and J. Dawson, and we would like

ACKNOWLEDGEMENTS

to thank them, in particular, for preparing the four-colour Vegetation and Land-use maps of our area. These are in many ways new, and will, it is hoped, give the reader a clear picture of the present day vegetational cover of the land surface.

Illustrations are a very important part of a work designed for the identification of plants in the field. Once again we must thank Mrs Barbara Everard for preparing such a fine set of botanical drawings, and for her enthusiasm and co-operation in the work.

An unusual feature are the double-page spreads of drawings of plants characteristic of the majority of the plant-hunting regions we have described. Jill Smythies' work has been a real labour of love involving many hours of hard work and many miles of travel; to her we are greatly indebted.

To collect a set of first-class colour photographs of unusual plants growing in their natural surroundings is no easy matter, and we have covered many thousands of miles to obtain them. There always remain gaps, and these have most generously been filled by the loan of colour transparencies by the following people, some of whom have gone out of their way to obtain them for us. We would like to thank L. E. Perrins in particular for the loan of thirty-five superb transparencies, numbers: 349a, 347b, 374a, 511, (517), (528), 519, (527), 528d, (547), 694a, 736a, 763a, 781, 788, 793, 793b, 789, 960a, 972, (976), 1047a, 1110a, 1196, 1123, 1122b, 1361a, 1460b, 1495a, 1500a, 1501b, 1414c, 1611, 1639a, 1902; also A. W. Taylor, numbers: 231a, 1623, 1665j, 1665n, 1676; A. J. Huxley, numbers: 720, 1588a, 1668f; Mrs I. K. Ferguson, numbers: (199), 631b, 1234a; J. C. Leedal, numbers: (399), 1268; D. N. Paton, numbers: (404), 769a; Mrs J. Bowden, numbers: 1642b, 1668d; C. E. Wuerpel, number (121).

We would also like to thank Dr D. S. Ranwell for the loan of the photograph of the Coto de Doñana facing page 49, and A. J. Huxley for that of the Montes Universales that follows page 128. Finally we would like to extend our gratitude to the Publishers; their patience and thoughtful advice has been of great value during the lengthy preparation of this book, and we are particularly indebted to Carol Buckroyd for her work on the later stages of the manuscript.

O. Polunin. B. E. Smythies. 1972

Signs and Abbreviations

agg.	an aggregate of two or more closely related species
ann.	annual
bienn.	biennial
c. (*circa*)	about, used in measurements
fl. fls.	flower, flowering, flowers
fr. frs.	fruit, fruiting, fruits
incl.	including
introd.	introduced
lv. lvs.	leaf, leaves
Med.	Mediterranean
perenn.	perennial
Pl.	plate
sp. sps.	species
subsp. subsps.	subspecies
var.	variety
×	(in text) indicates hybrid; (on colour plates) indicates degree of magnification. For example, $\times \frac{1}{3} = \frac{1}{3}$ life size
**	Illustrated in *Flowers of Europe*
*	Illustrated in *Flowers of South-West Europe*
†	Plants to be conserved

Authorities
The names of the authorities first describing the species are often abbreviated. For the abbreviated forms used see *Flora Europaea*, vols. I and II.
auct. = various authorities
Genera
The number of species in each genera found in our area, whether native, naturalized or commonly cultivated is given, except where all the species of the genus are described or keyed.

1. Landform Climate and Vegetation

Landform and Geological Regions

The Iberian peninsula is largely composed of a hard crystalline core of ancient, partly metamorphosed rocks. This remains exposed to the present day only in the north-west of Spain and in the French Massif Central. Around the ancient core there have been successive periods of mountain upheaval, land subsidence and rising, folding and faulting, which has built up the present landscape and determined the main geographical regions.

By the end of the Miocene era the basic 'grand plan' of Iberia had been created; the peninsula extended further north and west than it does now, but the main ranges of the highlands (Montes de Leon, Cantabrians, Pyrenees, Central Sierras, Iberian mountains, Catalan hills, Sierra Morena, and Betic Cordillera) were already in evidence, separated by depressions which filled gradually with lacustrine or occasionally marine sediments. During the succeeding years, from late Tertiary times onwards, the Iberian peninsula gradually changed its shape to a pattern strongly resembling that of today. The Balearic Islands, a continuation, in part, of the Betic Cordillera, were still joined to the mainland; but the great expanse of land to the north and west of Spain and Portugal was gradually submerged beneath a proto-Atlantic. . . . Along the coasts of Galicia floundering created drowned river valleys or 'rias', while the coasts of the south-east were likewise partially submerged, thus opening the Strait of Gibraltar and producing conditions for the formation of alluvial plains.[1]

In consequence, we can today recognize the following main geographical regions. The ancient *crystalline massifs* of Galicia and the Massif Central of France, at approximately the north-western and north-eastern confines of our area. The *central plains* or *mesetas* where more recent Tertiary deposits overlay the ancient core in Iberia. The *old-fold mountains* which include the Central Sierras, the Iberian Mountains, the Sierra Morena and others formed from the ancient core. The *new-fold mountains* or marginal ranges, where, as a result of the alpine upheaval against the resistant core, the highest mountain ranges of our

1

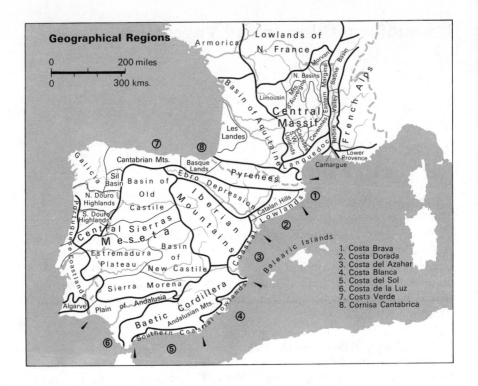

area, the Cantabrians, the Pyrenees and the Sierra Nevada, have been formed. The *depressions*, where low-lying basins are drained by the great rivers, in particular the Tagus, Guadalquivir, Ebro, and Garonne. The *coastal plains*, narrow strips of low-lying land, often dissected by coastal mountains and presenting a scenic variety that is such a feature of our area.

The crystalline massifs are formed of the most ancient rocks of granite, gneiss, and some Palaeozoic sediments. Today they form a grey landscape with rounded undulating hills, and fairly steep valley slopes. The hard rocks have resisted denudation despite the heavy rainfall, particularly in north-western Spain, and the general level is still high, with mountains up to 1500 m or more, and with steep V-shaped valleys with fast rivers which have cut into the ancient plateaux. Where these rocks form the present-day coastline, steep cliffs alternating with flooded valleys form a deeply indented coastline with fiord-like *rias* typical of north-western Spain, where the sea penetrates many miles inland. Despite the great age of this landscape in Spain, it is not of great importance either as a refuge for ancient plant species or as a site for new specia-

2

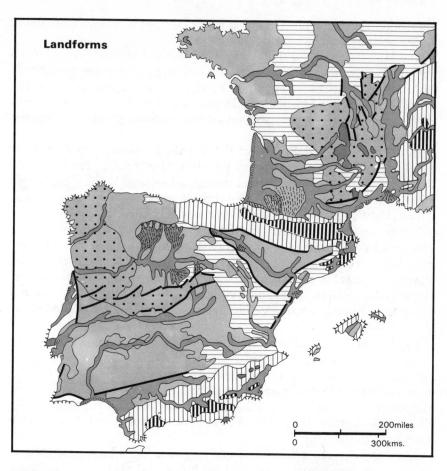

Landforms

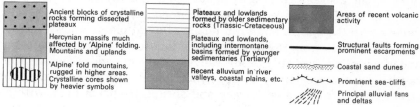

(dotted)	Ancient blocks of crystalline rocks forming dissected plateaux
(grey)	Hercynian massifs much affected by 'Alpine' folding. Mountains and uplands
(vertical lines)	'Alpine' fold mountains, rugged in higher areas. Crystalline cores shown by heavier symbols
(horizontal lines)	Plateaux and lowlands formed by older sedimentary rocks (Triassic-Cretaceous)
(light grey)	Plateaux and lowlands, including intermontane basins formed by younger sedimentaries (Tertiary)
(dark grey)	Recent alluvium in river valleys, coastal plains, etc.
(solid)	Areas of recent volcanic activity
——	Structural faults forming prominent escarpments
(dashed)	Coastal sand dunes
(hatched)	Prominent sea-cliffs
(lines)	Principal alluvial fans and deltas

0   200miles
0   300kms.

3

tion, and in consequence the north-west of the peninsula does not figure in the accounts of the botanical regions.

The Massif Central, which is twice the size of Switzerland, is of much greater importance in this respect. Not only does it comprise a relatively high mountain area with subalpine conditions, which acts as a 'stepping stone' between the Alps and the Pyrenees, but it also has an extensive recent volcanic area, an important site for endemic plants, of considerable botanical interest.

In the formation of the Massif Central the ancient crystalline rocks have acted as a rigid mass against which the softer rocks, formed on the floor of the Tethys Sea, have buckled and folded during the alpine earth movements. As a result of these enormous pressures the old rocks were tilted slightly in a north and north-westerly direction, and intense volcanic activity followed, continuing almost into historical times. Pinnacle-like volcanic plugs, piles of ash, streams of black lava, clusters of volcanic cones, give the country an almost lunar landscape. The Plomb de Cantal is the remnant of a large volcano thirty miles across. Outside the volcanic areas, high rounded hills with thin acid soils and deep, often heavily wooded valleys, produce a typically 'old' landscape.

The central plains, or mesetas are the vast, often extremely flat tablelands which cover nearly half of the interior of Spain. They are composed of Tertiary sedimentary rocks which overlay the ancient core and are largely formed of clays, marls, limestones, and sandstones. The Central Sierras traverse the Central Plains diagonally and divide them into two distinctive regions; those of Old Castile to the north and those of New Castile to the south.

Old Castile is about half the size of England. It is a huge upland basin lying at the average altitude of about 800 m and drained by the tributaries of the Douro river. These cut broad, shallow, but steep-sided valleys into the *meseta* and form fertile alluvial plains known as *campinas*. But away from the few rivers the landscape is flat, monotonous and parched, relieved only by low ranges of hills known as *cuestas*. They result from the erosion of an upper horizontal stratum, revealing the strata below; they are very steep and abrupt, 20–100 m high, with an eroded, serrated front. Seen from below they stretch across the horizon in tawny serried ranks of low hills.

Elsewhere, flat isolated tablelands, known as *paramos*, with steep-sided cliffs stand out from the plains; they are arid and uncultivable and fit only for goats and sheep. Bleak frosty winters and hot dry summers have made this a 'monotonous treeless plateau, falling at long intervals into a bluff-edged alluvial valley, repeated again and again'. Man and his animals have, in all probability, been largely responsible for the present-day aridity of this region. Remains of pine and oak woods testify to a once extensive tree cover where today only a low, sparse scrubland or *matorral* covers vast areas, whilst in the driest regions only steppe vegetation can survive.

The *meseta* of New Castile falls into two contrasting regions. To the west is the region of Estremadura where the ancient rocks lie close to the surface resulting in a more varied, undulating landscape, and where the rocks break through they form the low rounded hills of the Montes de Toledo and Sierra de Guadalupe. To the east, there is the same monotonous featureless plain stretching for mile after mile that occurs in Old Castile.

South-east of Madrid is a distinctive region known as La Mancha, where surface water is very scarce and where there is an accumulation of salt in the surface soil layers as a result of evaporation, forming the typical salt-steppes of semi-desert regions. The Spanish salt-steppes are unique in Europe and have a unique flora. A number of plants found there occur nowhere else in Europe, the nearest members of the same species growing as far away as North Africa, eastern Turkey, or even the Caspian Sea (which all have similar salt-steppe areas).

The old-fold mountains were probably formed during Hercynian folding and are composed of rocks of the ancient crystalline core. Characterized by the presence of granite, they have remained as relatively high summits owing to the very slow rate of denudation. The two most important old-fold ranges are the Central Sierras—the Serra da Estrêla-Sierra de Guadarrama group—which are such a dominant feature of the Iberian peninsula, and the Iberian mountains with two main highland areas centred around Soria in the north, and Teruel in the south. Both ranges are of great interest botanically as they act as refuges for montane and alpine plants, and as centres of diversification of recent species, being considerably isolated from other comparable montane regions. Of lesser importance are the other old-fold mountains, which include the Sierra Morena, the Montes de Toledo, and other lower ranges, where there are fewer endemic species. In general, these old-fold mountains have rounded summits often strewn with granite blocks and deep rocky glens, alternating with gently sloping hillsides.

The new-fold mountains, or marginal ranges, include the Cantabrian mountains, the Pyrenees, and the Andalusian or Baetic mountains. They are all younger and composed largely of sedimentary rocks which have been thrust upwards against the hard crystalline core during the alpine upheavals 50–100 million years ago. They are the highest mountains of our area and are truly alpine in character. They have jagged peaks, steep ridges, deep valleys, and show all the features of recent weathering, and the effect of snow and ice. U-shaped glacial valleys, moraines, and glacial lakes, are found throughout the higher regions and although there are no glaciers of any size, patches of snow persist throughout the year in some ranges.

The Cantabrian mountains range along the northern coast of Spain and include the Picos de Europa, the highest group. With their very heavy rainfall from the Bay of Biscay, they put their mark on the whole character of this northern pluviose region of Iberia. The country is difficult, heavily wooded, and deeply dissected by spectacular gorges, or *desfiladeros*, which have been carved out by tumultuous rivers which carry off the heavy rainfall in their relatively short passage to the sea. The Spanish Basque lands form a complex of lower mountains of great diversity between the Cantabrian mountains and the Pyrenees.

The Pyrenees are certainly the most impressive and important of the new-fold mountains of our area and their general character and importance is described in the plant-hunting regional account (p. 136 below).

The Andalusian or Baetic mountains in the south-west of Spain run discontinuously from Gibraltar to Cabo de la Nao, and continue under the Mediterranean to the Balearic Islands. A great fault, running in the same general direction as these ranges, has dissected the region even more. To the north lie the more recent Sierra de Cazorla and Sierra de Segura; while to the south are the Sierra Nevada and the Serranía de Ronda. In general, their contours are more rounded than those of the other new-fold mountains, and even the highest peak of the Sierra Nevada, Mulhacen, 3478 m, shows relatively little evidence of glacial activity, while the micaceous schist of which the summits of the sierra are composed makes for a barren and forbidding landscape, very like that of the Atlas mountains of North Africa. The Sierra Nevada seems a rocky wilderness with rounded peaks, with the northern slopes so gentle that a road goes to the summit. The aridity of the whole range is coupled with occasional torrential rains which have carved out deep river gorges, with but a trickle of a stream for the greater part of the year in their slaty bottoms. Elsewhere there is typical bad-land scenery of the severest kind with steeply eroded gullies running down through the soft tawny layers of more recent geological deposits.

The other important ranges of the Andalusian mountains are the Serranía de Ronda and Sierra de Grazalema, the Sierras of Cazorla and Segura, which are all largely composed of limestone. All act as botanical refuges for old and retreating species and are consequently of great interest, and they are briefly described in the accounts of plant-hunting regions (p. 67ff below). But there are many other smaller sierras in the south which are less well known, and which will undoubtedly repay detailed botanical exploration.

The depressions   Four large depressions occur in our area, where subsidence has taken place against the hardcore of the ancient rocks during the alpine upheaval. They are flooded with Tertiary deposits and form low plains.

In the west are the Tagus and Guadalquivir depressions drained by the rivers of the same name, where flat usually fertile alluvial land runs deep into the heart of the peninsula. The Tagus depression is relatively unimportant as far as plant life is concerned, but it does tend to form a dividing line between a vegetation which is predominantly Mediterranean to the south, and one containing a greater proportion of central European and Atlantic species to the north.

The Guadalquivir depression is more important. It extends about 320 km inland, and is composed mainly of Miocene and Pliocene marine deposits of marl and limestone. To the north is the eroded edge of the crystalline massif and the Sierra Morena, while to the south are more recent deposits, some washed down probably from the Sierra Nevada. In places, fertile *terra rossa* occurs as a result of limestone weathering, and with the proximity of the Atlantic moisture, provides rich, well cultivated soils. Elsewhere on siliceous and gypsum soils bad-land landscapes occur. At the seaward end of the depression lie the *marismas*. These extensive low-lying marshlands, fringed by sand dunes along the coast, give a very distinctive character to south-west Spain.

The Ebro depression, in the east of Spain, is another special feature of the peninsula. It lies south of the Pyrenees, is roughly the same length as the Guadalquivir depression, and is a great hollow, once a lake cut off from the sea by the Catalan hills. The Ebro river long since cut itself a narrow channel through the coastal hills to reach the sea, and the lake dried up, leaving the most extensive steppe area in Spain. It is a flat, monotonous country, and because of the very low rainfall and high salinity, much of it is still uncultivated. The Ebro river and its tributaries have cut deeply into the sedimentary rocks, dissecting the plain into barren isolated tablelands reminiscent of the *meseta*.

The Garonne depression, the country of Aquitaine, is once again quite different. 'The shadow of the sea seems to dominate the whole of Aquitaine and the breath of the ocean permeates every quarter of the land.' It is also a low monotonous country with none of the brilliance of light of the Mediterranean, but a country with great fertility. Humidity is high, and the great rivers draining largely into the Garonne bring down enormous quantities of water from both the Pyrenean heights to the south and the Massif Central to the east. At the same time alluvium and other debris brought down from the Pyrenees have spread over the plain of Aquitaine, bringing further uniformity to the landscape. Along the seaward side of the depression is the sandy coast of the Landes, stretching 300 km or more. This is an infertile area of acid soils, with shallow lakes, dunes and marshes, and extensive pine woods and heaths. But inland

the south-west of France is a region of rich and varied cultivation, and a surprising number of crops have been grown at different times. Thanks to its transitional position between temperate and warm temperate latitudes it will grow cereals, fruits and vegetables of many kinds, and its agriculture has been called polyculture.... It specialises in dye-crops—saffron and woad. Contacts with the New World brought maize, tobacco, beans and later tomatoes. Of its many fruits the vine and the plum are of ancient repute.[2]

The coastlands are probably the greatest of all the attractions that the Iberian peninsula has to offer to its visitors. There are over 3000 km, or 1800 miles, of coastline in our area. The powerful erosive action of the Atlantic rollers, the rises and falls in the land surface, and the very different erosive forces of the Mediterranean have resulted in every imaginable variation of scenery. Steep cliffs rise precipitously to snow-covered mountains, rocky headlands border sunken valleys, sandy bays alternate with rocks and cliffs, golden sands stretch for scores or even hundreds of miles, and rolling dunes, sandy spits, salt marshes, and lagoons are all part of the rich variety of probably the finest coastlands in Europe.

The Atlantic coastland includes two quite distinctive regions: the low-lying Biscayan coast of France, and the predominantly rocky Atlantic coastlines of Spain and Portugal. The western coast of France is one uninterrupted sandy shore backed by sand dunes, running for more than 300 km, from La Rochelle to Biarritz near the Spanish border, while further north to the mouth of the Loire the coastline continues to be low, with shallow bays and extensive salt marshes and salines.

By contrast the north Atlantic coast of Spain, sometimes known as the Costa Verde or the Cornisa Cantábrica, is steep, rocky, and without any coastal plain. Fast rivers rush down narrow valleys from the Cantabrian mountains and enter the sea through narrow estuaries between steep headlands, often at the head of sunken valleys. In the north-west of Spain there is a fiord-like coastline where the sea has flooded inland for many miles as a result of land subsidence. The steep hills on each side of the valleys (known as *rias*) and the rocky coastal islands represent all that remains above water of the summits of mountains. The *rias* are in two main areas, one in the extreme north-west known as the Rias Altas and one further south and more extensive, the Rias Bajas. Tiny rivers, disproportionate to the *rias*, flow in the valleys, with the result that the country is difficult of access.

The rugged rocky coast continues southwards through Portugal to Oporto, after which it becomes lower and less rocky, and long sandy coastlines with sand dunes and marshy lagoons inland become characteristic. Occasionally headlands such as the Cabo da Roca, the most westerly point of mainland Europe, and the Serra Arrábida break the otherwise low-lying coastland on its sweep down towards the Algarve. The wild, windy headland of Cape St Vincent with its perpendicular cliffs of hard limestone is a distinctive feature; but once again the southern Atlantic coast of Portugal and Spain (sometimes called the Costa de la Luz) is mainly low-lying and often sandy or marshy with small alluvial plains inland, though it becomes progressively steeper and more rocky towards the Straits of Gibraltar.

The southernmost coast of Spain, the Costa del Sol, is a narrow strip of plain intersected by mountains which come down to the sea in high cliffs: luxuriant green alluvial plains alternating with dry brown or dull green hills of the coastal ranges. Torrential rains occur very suddenly in these hills and create swollen, turbulent rivers within a few hours, which flood down to the tideless Mediterranean carrying enormous quantities of debris. The alluvial plains and deltas thus formed, where they can be regularly irrigated by water from the hills, provide some of the richest agricultural lands of Spain, the *vegas*, where a great variety of fruit and vegetables is grown.

The Costa Blanca lies further east, stretching approximately from Almería to Valencia. Between the Cabo de Gata and the Cabo de Palos is one of the remotest and least developed coastal regions of Spain, with barren cliffs and arid hinterland. North and east of Cabo de Palos the coastline becomes lower, with frequent lagoons, salt marshes, and sand-spits, while much of the fertile irrigated areas lies inland, and here such semi-tropical crops as date palm, bananas, rice, sugar cane and cotton are grown.

The Levantine coast from Cabo de la Nao northwards to the Ebro delta is one continuous series of *huertas* where the main cultivation of citrus fruits occurs. In general, there is a succession of cultivation as one progresses inland, with rice fields near the coast, vegetable *huertas* and orange groves further inland, and vineyards climbing the foothills of the interior.

The Costa Brava, or Catalan coast, is by contrast a predominantly rocky coast of considerable diversity. The different rocks from which it is carved—which range from Palaeozoic schists, sandstones, crystalline rocks, limestones to conglomerates—make for an attractive variety of coastal scenery, with small headlands, rocky bays, wide sandy beaches, and small deltas. Unfortunately the rapid development of tourism has spoiled much of the scenic beauty of this coastline.

The French Mediterranean coast from the foot of the Pyrenees at Port Vendres to the delta of the Rhône is low-lying, flat, and marshy with long sand spits backed by a series of salty lagoons with only occasional rocks of limestone or lava breaking the monotony of the plain. It is currently the scene of a huge tourist development programme.

The Camargue—the flat wind-swept country at the mouth of the Rhône delta—culminates in some of the most extensive salt marshes and marshland of the whole of our area. Intensive rice cultivation is largely destroying the natural vegetation and restricting its botanical interest.

Climate and Weather

Three main types of climate occur in our area. A predominantly oceanic climate affects the west and north coasts of the Iberian peninsula, the Atlantic coast and mountains of central France and the north Pyrenees, with a moderate equable temperature and high rainfall—a southern variant of the typically oceanic climate found in north-western Europe. A Mediterranean climate affects not only the Mediterranean coast, but also the Atlantic coasts of southern Portugal and south-western Spain, with hot dry summers and mild moist winters. A continental type of climate affects the high plateau of the Iberian peninsula, which experiences extremes of temperature both daily and seasonally, and very low and unreliable rainfall.

The Iberian peninsula is such an extensive land mass that it generates weather systems of its own, giving it a climate unlike that of any other region in Europe. It lies between the warmish moist air streams of the north Atlantic, and the warm dry area of North Africa and the Sahara. It also lies in the path of the Atlantic depressions which during a large part of the year are deflected either to the north or to the south, so that for months at a time the ameliorating effect of this moisture-laden air coming in from the Atlantic makes very little impression on the climate of central Spain. In spring and autumn, however, these depressions can penetrate deeply into the peninsula and produce very different kinds of weather. But in general, the climate of the central plateau is continental in character with hot summers and cold winters, while the peripheral regions have largely an Atlantic or Mediterranean type climate which is more equable and humid.

The central plateau heats up rapidly in summer under clear skies and long days of sunshine, causing inblowing winds from the Atlantic and Mediterranean. Because of the heat this incoming moist air soon dries

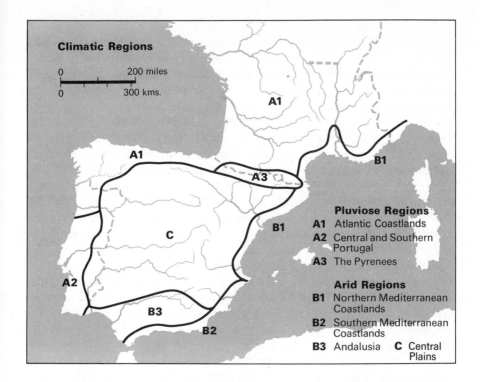

Climatic Regions

0   200 miles

0   300 kms.

A1

A1

A3

B1

Pluviose Regions
A1 Atlantic Coastlands
A2 Central and Southern Portugal
A3 The Pyrenees

B1

C

A2

Arid Regions
B1 Northern Mediterranean Coastlands
B2 Southern Mediterranean Coastlands
B3 Andalusia   **C** Central Plains

B3

B2

and little or no rain falls in the interior. However, there may be intermittent short and very violent thunderstorms with torrential rain, which may result in disastrous floods in the coastal regions.

In the winter, by contrast, freezing air settles on the *mesetas* and heavy snow falls in the mountains. This air may spill down into the Mediterranean coastal strip causing cold spells sometimes late into the spring with serious results for fruit crops along the coast. Springs are earlier on the Costa del Sol where the Andalusian mountains protect the coast from the cold air inland, but they may be surprisingly late on the Costa Brava because of this cold air-mass.

The complexity of the weather is further increased by the position of the new-fold mountain ranges. The Cantabrian mountains, ranging along the north coast of Spain, successfully catch much of the moisture from the northern and western winds, while immediately to the south of this range the *meseta* lies in the rain-shadow and receives a very low rainfall. The Pyrenees likewise act as an almost continuous rain-barrier, so that the Ebro depression, to the south of the Pyrenees, is one of the lowest rainfall areas in Spain; and the Andalusian mountains in the south form a barrier to the south-westerly Atlantic winds. It is only on

11

the southern and central Portuguese coast that no high mountains stand in the line of incoming moist air streams, which can there penetrate deeper into the peninsula, making the dividing line between the Atlantic and continental climates approximately along the Portuguese-Spanish border. Only in the extreme north-west of the peninsula can the Atlantic weather hold sway throughout the year.

The Massif Central has a similarly dominating influence on the French part of our area. These highlands act as a rain-barrier and catch a great deal of the moisture coming in from the Atlantic, while the low-lying plain of Aquitaine allows its passage inland unimpeded. The dividing line between the predominantly Atlantic climate to the north and the Mediterranean climate of south-western France is roughly along a line drawn through the Carcassonne gap and the Cévennes. To the south there is the familiar pattern of warm dry summers and mild winters, with rainfall occurring largely in the winter half of the year.

These are the general summer-winter climatic patterns; but quite different conditions prevail at other times of the year producing very variable weather. In spring and autumn the normal pressure systems are disrupted; there may be long periods of instability. Secondary depressions may occur, particularly in autumn along the south-eastern coast of Spain, and the weather may become cloudy and rainy with sudden cloudbursts of great severity.

Secondary depressions to the north of the peninsula in winter and spring may produce a different type of weather still. Colder air is diverted southwards, producing at first mild weather along the north-eastern coast, followed by cold blustery weather and strong winds—the *pargorii* of the Basque coast, and the *tramontana* of the Catalan coast. In winter secondary depressions are formed in the Bay of Biscay causing gales and heavy rain over the Basque country and the western Pyrenees. A further type of weather is produced by large Atlantic depressions which pass over the north-west of the peninsula bringing mist, drizzle, and fog over much of the coast and at times deep into the *meseta* in winter. In summer, north Portugal and Galicia continue to be deeply affected by these Atlantic depressions.

Rainfall

The Iberian peninsula shows extreme variation both in the amount of rain and in its seasonal occurrence. Annual rainfalls as low as 300 mm (12 inches), with half this amount in some years, occur in parts of the central plateau and the south-east; drought may occur for over seven months of the year; Zaragoza in the Ebro depression has four months of winter drought and three months of absolute summer drought (see map, page 15). By contrast, the highlands of Iberia and the Massif Central are

PLACE	Total Annual Rainfall mm	% of Summer Rainfall	Number of Days of Rain	Number of Cloudless Days	Days of Drought (Gaussen)	Mean Annual Temperature °C	Mean January Temperature °C	Mean July Temperature °C	Absolute Minimum Temperature °C
Barcelona	526	14	72	120	6	15·2	9	23	−9·6
Cabo de Gata	122	0	—	—	365	18·6	13	26·5	—
Cartagena	205	·5	14	223	215	17	12·2	27·2	−1·2
Lagos	516	6	76	—	160	17·9	12·4	22·4	+·4
Lisbon	726	4	112	—	95	15·3	10·4	21·1	−1·5
Madrid	419	8	94	130	100	14	4·3	24·3	−12·5
Málaga	600	6	49	177	145	18·6	11·8	25	−·9
Porto	1291	9	113	—	60	15·3	10·2	21·8	−3·6
San Sebastian	1396	20	175	56	0	14·1	9·9	22·3	−8·2
Santiago de Compostella	1655	13	176	63	0	12·9	9·7	19·9	−5·1
Serra da Estrêla	2365	15	—	—	0	8·2	3·5	17	—
Sevilla	500	4	67	218	140	19·8	10·3	29	−2·7
Sierra de Guadarrama	780	12	—	—	45	8·7	2	18	−16·2
Valencia	472	9	56	108	135	16·9	10	25	−8·2
Valladolid	308	9	71	70	95	11·8	4·9	26	−21
Zaragoza	295	20	66	178	97	14·7	6·9	23·6	−16·6

CLIMATES OF THE IBERIAN PENINSULA[3]

extremely wet and have over 2000 mm a year (80 inches). The highest total annual rainfall occurs in the Serra da Estrêla in Portugal with 2825 mm (113 inches). In winter, much of it is precipitated as snow with average depths of 1½–2 m, and snow lies late into spring and early summer.

Rainfall has a very marked effect on the distribution of plant life, more so than any other environmental factor, and consequently it is convenient to divide the area into two rainfall regions: pluviose and arid. In general, the pluviose areas can be considered as having over 600 mm (24 inches) of rain (see map, page 11).

The following pluviose rainfall regions can be distinguished:

A.1. Dominated by Atlantic weather, with high rainfall and humidity, mild winters, and comparatively cool summers. Heavy rainfalls become heavier towards the east, the Basque provinces receiving the heaviest;

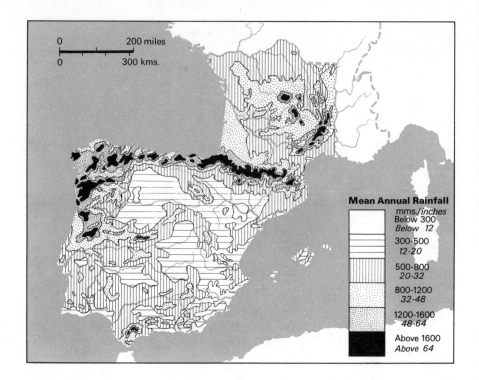

Mean Annual Rainfall

mms./inches
Below 300 / Below 12
300-500 / 12-20
500-800 / 20-32
800-1200 / 32-48
1200-1600 / 48-64
Above 1600 / Above 64

while the presence of the Azores high pressure system causes summer droughts in Galicia. Much of the moisture-laden Atlantic air crosses the low plain of Aquitaine to the centre of France and is precipitated as rain and snow on the Massif Central.

A.2. The Atlantic coast of central and southern Portugal has a climate which is largely Mediterranean, with rainfall mainly in the winter half of the year and increasing summer drought as one moves southwards. However, the total amount of rain fluctuates widely from year to year and with the proximity of the ocean, the temperature is kept below that of the true Mediterranean coast.

A.3. Pyrenean weather. In spring this weather penetrates only as far south as Pamplona, but it extends along the whole northern flank of the Pyrenees becoming progressively less rainy from west to east. The full effect of the moisture-laden winds from the Atlantic gives heavy rainfall and snowfall with peaks in spring and autumn.

The arid areas generally have a rainfall below 600 mm and are divided into:

B.1. The northern Mediterranean zone from the Cabo de la Nao to the Rhône delta. It receives most of its rain from the south-easterly winds,

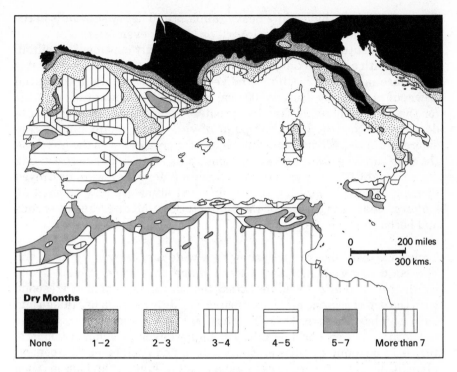

Dry Months

| None | 1-2 | 2-3 | 3-4 | 4-5 | 5-7 | More than 7 |

Dry month = PJ/T < 10; where P = monthly precipitation in mm, J = number of rain-days, T = mean monthly temperature (Birot, 'Sur une nouvelle fonction d'aridité en Portugal', *Ann. Fac. Sciences*, Oporto, xxx, 1945, pp. 90–101).

which follow in the rear of depressions passing through the Carcassonne gap.

B.2. From Cabo de la Nao westwards to the Straits of Gibraltar lies a much more arid area. Cabo de Gata, in the centre of this area, has the lowest rainfall in Spain and continuous drought for over seven months in the year. The climate is more characteristic of North Africa than of Europe and it is in this minimal rainfall area of Spain that most of the unique steppe plants occur. Towards the west, in coastal Andalusia, the rainfall becomes progressively heavier owing to the influence of the Atlantic.

B.3. The region of Andalusia inland from the coast is also affected by the Atlantic and has a somewhat higher rainfall than that of the Mediterranean coastlands. It becomes progressively drier further inland, however, and here the highest mean average temperatures of Iberia occur. The mountains of this region, such as the Sierra Nevada and

15

Sierra de Cazorla, have quite heavy rainfalls and heavy snowfalls in the winter, and snow persists until April or May, or even later.

C.1. The central plateau of Iberia has a markedly continental rainfall pattern, with a low maximum in spring and another in autumn, but it is very unreliable and irregular, and as often as not occurs in torrential downpours. It rarely exceeds 500 mm, except in the mountains; and it is on the central plateau that the greatest steppe areas of Castile are developed. From the botanical point of view this climate may be considered akin to a Mediterranean climate, for crops such as the olive and the vine can be grown, and at one time sparse forests of evergreen oak were thought to cover much of the terrain before they were destroyed by man. In the place of forest, uncultivated areas are now covered by *matorral* containing many of the characteristic Mediterranean shrubs and herbaceous plants.

Temperatures, like rainfall, show very wide ranges over our area, and very contrasting summer and winter averages. In summer, the south and south-east of Iberia have maximum temperatures, which correspond with the driest regions of the peninsula. The July maxima of Málaga and Jaén may be as high as 30°C (86°F) and 35°C (95°F) respectively, but the fall at night is very considerable. The highest average summer temperature is to be found in parts of Estremadura and Andalusia. The central plateau, despite its altitude, is extremely hot in summer, and has maxima of over 38°C (100°F), but again a great fall at night—seldom less than 10°C. By contrast, in the north and west, the influence of the ocean produces a much more equable temperature with lower summer maxima of 21–27°C and a night drop of only 3–6°C.

In winter conditions are in many respects reversed. The coastal regions show the highest temperatures, and from Galicia to the Ebro the winter average is over 8°C, with an average minimum temperature in the colder months of about 4°C. Consequently frosts are extremely rare for the whole of this coastline; it is only in the Basque provinces that they become more prevalent. The south coast of Spain and Portugal remains relatively warm throughout the winter with average winter temperatures of above 10°C (50°F). The central plateau by contrast has very much lower temperatures with the numbers of days of frost rising steadily from south to north, with Valladolid having up to 70 days of frost during some winters. In the region of Madrid the average minimum temperature of the coldest month (January) is 0°C while in Zaragoza the average winter minimum is as low as −7°C. It is here that one of the lowest temperatures of the peninsula has been recorded, −16°C.

In the main mountain ranges of the Iberian peninsula temperatures are much lower. On average there is a fall of about $\frac{1}{2}$°C for every 100 m

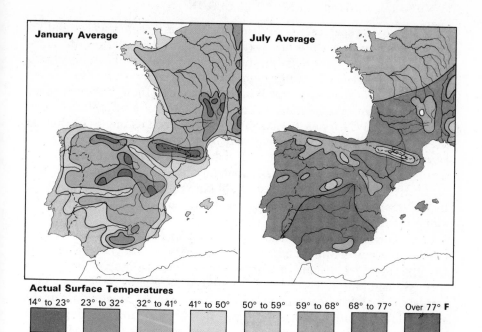

Actual Surface Temperatures

14° to 23°	23° to 32°	32° to 41°	41° to 50°	50° to 59°	59° to 68°	68° to 77°	Over 77° **F**
−10° to −5°	−5° to 0°	0° to 5°	5° to 10°	10° to 15°	15° to 20°	20° to 25°	Over 25° **C**

rise in altitude, and in consequence in the highest mountains and above 2000 m there may be a reduction of 12–18°C, giving continuous temperatures below freezing for four to five months in the year.

In the south-west of France, the warmest regions in the winter occur along the Atlantic and Mediterranean coasts, where the mean winter temperatures lie between 4–7°C. Moving inland to the Massif Central they fall to between 2–4°C.

In summer the coolest regions lie along the Atlantic coast with the July mean at 21°C; further south and inland it increases to 24°C and more on the Mediterranean coast.

The Vegetation

The vegetation covering the earth's surface is, in the broadest analysis, the result of the interplay of the climate and the soil with the animal and plant world. Today we find a mosaic-like covering of plant communities composed not so much of independent plants but of a delicately balanced assemblage of species which are interdependent and rely on each other for their survival. If the forest covering is removed from an area, for example, many woodland species quickly disappear to be replaced by more sun-loving species. Each community in this mosaic may be relatively stable or be in a state of development or of degradation. Primeval forests and some alpine vegetations may be in a relatively stable climax stage and are least affected by man; but the majority of plant communities can be termed 'semi-natural' in the sense that, though they are composed of wild species, man has in course of time profoundly altered their structure. The Cork oak forests of Iberia are good examples of this. At the other extreme, the plant cover may be quite artificial and maintained as vineyards, arable, orchards, and such like; but even here the 'weeds', though natural and unwanted companions, are characteristic and distinctive and are often of considerable interest. The roadside weeds in Iberia have a rich variety which cannot but impress the traveller along the open road.

Plant communities, like those briefly outlined below, have a life of their own: a mature forest may take tens or hundreds of years to develop. During this development a number of distinctive stages are passed through and the stages have a regular progression or 'succession'. Grassland may be succeeded by scrub, and scrub by forest, while at the same time the soil is deepening and developing, and the dependent species are continuously changing as new conditions arise. Ultimately a relatively stable balance is obtained between the plant community and the environment, and this is known as a 'climax community'. Evergreen oak forests, montane pine and fir forests, beech forests are all examples of these 'climax communities': so also are certain steppes and *matorral* shrub communities.

Left undisturbed the vegetation would develop a monotonous uniformity over vast tracts of country in the Iberian peninsula. We must largely thank man and his animals for the rich variety of communities that we find at the present day, though we have also to thank him for the degeneration that results from overgrazing, soil deterioration, and activities such as felling, firing, and cultivation, so that where there was once forest, as on the *meseta*, there is now a poor degraded scrub or grass steppe. In low rainfall areas, the time taken to develop a climax community may be more a matter of hundreds of years than of decades. Mediterranean vegetation is particularly vulnerable—once destroyed it may never fully regenerate.

If the traveller can comprehend the underlying vegetational systems or plant associations, and can learn to pick out the various stages of development and degradation in each association, his progress through the countryside will be much more interesting. Similarly a knowledge of geology and landform can illuminate his comprehension of village patterns and local variations in husbandry.

Willkomm estimated in 1896 that there were 5,660 species of flowering plants, cone-bearing plants, and ferns native in the Iberian peninsula, and that 1,465 of these were exclusive, or endemic, to the peninsula. These numbers may have to be modified somewhat, particularly in the light of more recent and detailed exploration of North Africa, and also because of a fuller understanding of the status of the species. But for want of a more up-to-date appraisal, they give a good indication of the richness and interest of this unique Iberian flora. In Europe, only the Balkan peninsula can match these numbers. Turrill, after an exhaustive analysis of the Balkan flora, estimated in 1929 a total of 6,530 species of flowering and cone-bearing plants, of which 1,754 are endemic Balkan species. The south-western and south-eastern extremities of the European continent are by far the richest in plant species.

The reasons for such a rich diversity in the flora of Iberia are many. In the first instance the absence of any permanent ice cap in the south-west during the last ice age has resulted in the survival of a large number of ancient Tertiary species, which have been eliminated elsewhere in Europe. Many of the 'old' and most distinctive endemic species belong to this category. Secondly, the proximity of a rich North African flora has helped to swell the numbers. Though the Straits of Gibraltar may have been broached up to ten million years ago, it has not acted as a complete barrier to the northern spread of plant species and many have penetrated into the southern hotter regions of Iberia from North Africa. Thirdly, the relative geographical isolation of the Iberian peninsula has resulted in the separation of its own breeding species from those of adjacent breeding stock, with the result that cross-breeding has been reduced and gradual evolution of new species has occurred, and is still occurring. This accounts for a large proportion of the Iberian endemic species and particularly those which are relatively 'recent', which may still have closely related species in adjacent countries. Fourthly, the climate of Iberia has as varied a range of conditions as anywhere in Europe. This has helped to produce 'niches' where plants of different tolerances can survive: notably the unique steppe regions and the isolated high mountain ranges with climates unlike that of any other part of Europe.

In the Iberian peninsula there are today a number of genera which are in all probability actively evolving. They produce a high proportion of new species which are able to survive in the great variety of habitats and climates which the Iberian peninsula offers. The most noteworthy are: *Centaurea* with 90 species including about 50 endemic species; *Linaria* with 52 species including 36 endemic species; *Thymus* with 31 species including 24 endemic species; *Genista* with 33 species including 22 endemic species; *Cytisus* with 15 species including 9 endemic species;

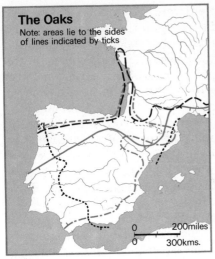

The Oaks

Note: areas lie to the sides of lines indicated by ticks

0 ——— 200miles
0 ——— 300kms.

- – – – – *Q. ilex & Q. rotundifolia* - Holm oak
- –·–·–·– *Q. pyrenaica* - Pyrenean oak
- ·········· *Q. suber* - Cork oak
- ——— *Q. coccifera* - Kermes oak

The Pines

- ·········· *P. halepensis* - Aleppo pine
- – – – – – *P. sylvestris* - Scots pine
- ——— *P. nigra subsp. salzmannii* - Black pine
- –··–··– *P. pinea* - Stone pine

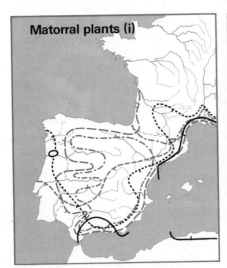

Matorral plants (i)

- ·········· *Nerium oleander* - Oleander
- –·–·–·– *Erica arborea* - Tree heather
- – – – – – *Arbutus unedo* - Strawberry tree
- –––––––– *Rosmarinus officinalis* - Rosemary
- ——— *Vitex agnus-castus* - Chaste tree

Matorral plants (ii)

- ——— *Anthyllis barba-jovis* - Jupiter's beard
- ·········· *Lavandula stoechas* - French lavender
- – – – – – *Cistus monspeliensis* - Narrow-leaved cistus
- –··–··– *Teucrium polium* - Felty germander
- – – – – – *Globularia alypum* - Shrubby globularia

Distribution of

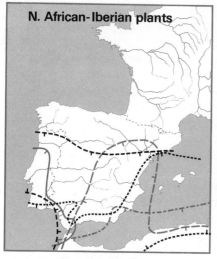

N. African-Iberian plants

.... Chamaerops humilis
------- Lygos sphaerocarpa
———— Drosophyllum lusitanicum
-·-·-·- Ziziphus lotus
- - - - Lygeum spartum

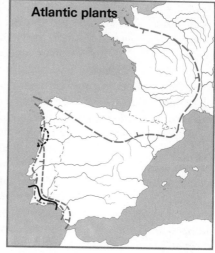

Atlantic plants

-·-··-·- Corema album
- - - - Erica vagans
.... Narcissus cyclamineus
------ Leucojum trichophyllum
———— Lavandula viridis

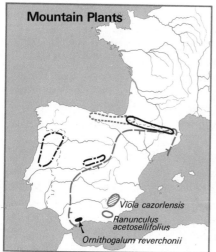

Mountain Plants

Viola cazorlensis
Ranunculus acetosellifolius
Ornithogalum reverchonii

- - - - Erinacea anthyllis
-·-·-·. Crocus carpetanus
———— Gentiana burseri
.... Gentiana occidentalis

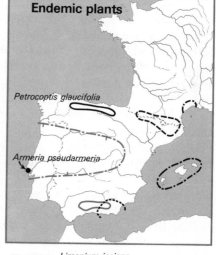

Endemic plants

Petrocoptis glaucifolia

Armeria pseudarmeria

.... Limonium insigne
-·-··-·- Cyclamen balearicum
-·-·-·- Digitalis thapsi
———— Echium albicans
------ Ramonda myconi

important plants

Ononis with 41 species including 17 endemic species; *Armeria* with 30 species including 23 endemic species; *Teucrium* with 35 species including 20 endemic species; *Narcissus* with 45 species including 22 endemic species. It should, however, be stressed that the exact numbers depend upon the interpretation botanists put on the term 'species'; and that not all endemic species are evolutionarily 'recent'.

The main floristic elements of our area can be summarized as follows. The *Western Mediterranean* element comprises about 2,350 species, none of which occurs further east than the Cévennes. They include not only the 1,465 Iberian endemic species, but also 480 North African-Iberian species. It is this element that is so very important in Iberia; it accounts for two-fifths of the total and gives the flora a very distinctive character quite unlike any other part of Europe. The *Circum-Mediterranean* element accounts for another one-fifth of the total. These are plants which occur throughout the whole Mediterranean region. Many of these plants penetrate into the arid regions of the interior of Spain and into the *meseta*, and give it a peculiarly Mediterranean aspect. Portugal also possesses a high percentage of Mediterranean species despite its Atlantic position. In the coastal regions and evergreen forests of the south and centre of Portugal the proportion of Mediterranean species may be as high as fifty to sixty per cent, while the Atlantic element here may be as low as twelve per cent. The *Central European* and *Alpine* elements account for much of the remaining two-fifths of the flora. There are about 1,630 species that have a wide distribution in Europe and which occur in the Iberian peninsula mostly in the pluviose regions and the montane and sub-alpine regions. Many of these do not penetrate as far west as Portugal and this is one of the reasons why the flora of Portugal is relatively poor in species and contains a total of less than 3,000. Also included in this category are the 180 or so Pyrenean endemic species.

Other less important but equally interesting floristic elements include the *Western Atlantic* or *Lusitanian* element which occupies the Atlantic seaboard of western Europe and adjacent inland areas. It comprises many species in genera such as *Erica, Ulex, Genista, Halimium*, as well as *Corema, Drosophyllum, Daboecia,* and *Lygos*. The *North African-Iberian* element is also of great interest. It consists of about 480 species which have their main distribution in North Africa, but which spread into the arid regions of south and south-eastern Spain and into the mountains of the south.

One of the smallest elements of all are the 48 or so species which are found in the steppe regions of Spain. They occur nowhere else in Europe or in North Africa, and their nearest living relatives are to be found in the steppes of Asia Minor and Central Asia.

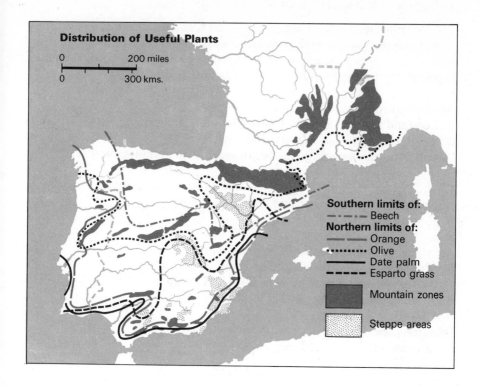

Distribution of Useful Plants

0   200 miles

0   300 kms.

Southern limits of:
– · – · – Beech
Northern limits of:
Orange
········· Olive
Date palm
– – – – Esparto grass

Mountain zones

Steppe areas

To these natural elements of the flora must be added a further category —that of plants introduced by man during historical times and earlier. There are many examples, such as vines and olives from the Middle East, citrus fruits from the Far East, *Eucalyptus* and *Acacia* species from Australia, and many sub-tropical species, all of which can flourish and often become naturalized in the warm climates of Southern Iberia. Other examples are weeds like the Bermuda Buttercup and the Hottentot Fig from South Africa, which have become naturalized and are now well established as an integral part of our contemporary flora.

Rather than considering species individually, it is more revealing to consider the types of plant community that exist, for they largely determine the character of the landscape. Each plant community has its own characteristic assemblage of species, and the least experienced planthunter knows that some species, often the less common ones, can only be found in a certain type of community and that it is useless to search for them elsewhere. The Central European school of plant-sociologists (Zurich-Montpellier School) has, during the last half century, devised a nomenclature for the main types of such communities found in Europe. Dr Rivas Goday is one of the main Spanish workers in this field, and he

23

has described many different types of forest, shrub, and steppe communities in the peninsula, each with their own distinctive floristic composition. As the system is complex, a much simplified analysis of the main types of vegetation found in our area is outlined as follows:

Mediterranean evergreen plant communities
1. Lowland and hill forests. Evergreen oaks and pine forests.
2. Scrub communities. *Matorral* and pseudo-steppes.
3. Sub-montane forests. Semi-deciduous and deciduous forests.

Central European and Atlantic deciduous plant communities
4. Deciduous forests. Oak and chestnut forests.
5. Heathlands and grasslands.

Montane, sub-alpine, and alpine plant communities
6. Montane forests. Beech, pines, silver fir forests.
7. Sub-alpine communities. Meadows and scrub.
8. Alpine communities.

Maritime and salt-tolerant plant communities
9. Dunes, sands, and salt marshes.
10. Steppes.

Freshwater, wetland communities

Each of these plant communities has one or more dominant species by which it can easily be recognized, and which has a very profound effect on all the associated species by helping to form micro-climates in which the associated species can survive. Once the dominant species of each community is known it is not a difficult matter to identify broadly the type of plant community that confronts one. Each community has also a number of indicator species which will give a clue to the type of community under investigation, even if the dominant species is absent. The presence of the Lentisc, *Pistacia lentiscus*, for example, immediately identifies the Mediterranean evergreen complex of plant communities.

Mediterranean evergreen plant communities

1. *Lowland and hill forests*
Evergreen oak forests are potentially the climax communities in the Mediterranean region, but man in particular has had a devastating effect on them, clearing great areas of forest for agriculture and subjecting the remainder to intensive cutting, grazing, and firing. The Iberian

peninsula is now poorly forested: Portugal has a twenty-five per cent forest cover and Spain only ten per cent, and most of the existing forests occur not in the Mediterranean region but in mountains and in the pluviose regions.

Forest of Holm oak probably at one time covered two-thirds of the peninsula, with such characteristic associated tall shrubs as *Viburnum tinus* and *Arbutus unedo, Phillyrea* and the climbers *Smilax aspera*, and *Lonicera* species. In Portugal and south Spain, in areas of lower rainfall (around 650 mm) a closely related species, *Quercus rotundifolia*, takes the place of the Holm oak. Good examples of Holm oak forest still persist in the Catalan hills, but most of it is reduced to a maquis-like thicket containing only scattered oaks and with an abundance of the Kermes oak, *Quercus coccifera*, which is invasive in conditions where the original cover has been destroyed.

Indicator species of this climax community in the Iberian peninsula are: *Quercus coccifera, Juniperus phoenicea, Juniperus oxycedrus, Pistacia terebinthus, Lonicera etrusca, Lonicera implexa, Phillyrea* species, *Jasminum fruticans, Arbutus unedo, Ruscus aculeatus, Cistus salvifolius, Daphne gnidium, Asparagus acutifolius, Bupleurum fruticosum, Coronilla juncea, Teucrium fruticans*, and others.

So extensive has been the destruction of the 'typical' Mediterranean forest that its interest is largely academic; a much more characteristic feature of the landscape is provided by the maquis which has replaced it so widely . . . and which represents a degeneration of the evergreen forest caused by man with his fire and domestic animals. Its appearance is infinitely varied: in some places it is a stunted woodland dotted with conifers and oaks, survivors of a nobler vegetation; in others an immense shrubbery or a tangled thicket so laced with briars as to be almost impenetrable; it may be contiguous or patchy, but everywhere woody evergreen bushes are the most important element. . . . The woody shrubs . . . are principally those that constituted the brushwood of the evergreen forests and given a sufficiently long undisturbed period the original vegetation would no doubt reassert itself. Unfortunately this rarely happens. Quite often maquis is allowed to colonize exhausted land and act as a kind of long-term fallow, only to be burned off again, perhaps after a few years, perhaps after a generation and be subjected once more to man's needs.[4]

All stages of regeneration and degeneration of this forest can be found and the stages are largely responsible for the present day appearance of much of the Mediterranean landscape. In general, the stages are as follows:

Degeneration

Evergreen forest⇌Maquis⇌Garigue⇌pseudo-steppe

Regeneration

2. *Scrub communities*

These communities can be divided roughly into *maquis*, where the ever-green scrub is up to a man's height or to twice that height, typically so dense that it is difficult to force a way through and with a very sparse ground flora, and *garigue* (or *garrigue*), which is a more or less open community of small shrubs, often knee high, and with very aromatic foliage. There are all gradations between these two and the Spanish term *matorral* includes both these and seems more appropriate and will be used as a general term in the account that follows.

Matorral now covers enormous areas in the Mediterranean zone and the central regions of the peninsula. In its most developed form, it may occur as dense thickets 2–4 m high, composed largely of a mixture of evergreen shrubs, the most widespread of which are *Arbutus unedo*, *Erica arborea*, and *Cistus monspeliensis*; while other common and characteristic species are *Olea europaea*, *Phillyrea* species, *Pistacia* species, *Viburnum tinus*, *Calicotome* species, *Spartium junceum*, *Juniperus phoenicea* and *Juniperus oxycedrus*, *Quercus ilex* and *Quercus coccifera*, *Rosmarinus officinalis*, *Erica multiflora*, *Cytisus villosus*, *Teline monspessulana*, *Smilax aspera*, and *Asparagus* species. In differ-ent regions certain species may dominate large areas, thus giving many distinctive local variations to the landscape. Usually the presence of *Ulex*, *Cytisus*, *Genista*, *Erica*, and *Cistus* species is characteristic of siliceous, more acid soils; while the practice of burning the scrub to improve grazing potentialities favours fire-resistant species such as *Cistus* species, *Arbutus unedo*, and *Pistacia lentiscus*. Less developed *matorral* in the Iberian peninsula consists largely of evergreen shrubs, with small leathery leaves, usually 1–1½ m high and often spiny. These shrubs may form a dense, almost impenetrable growth, or they are often more scattered and have considerable patches of bare ground between the bushes. The dominating families in this low *matorral* are undoubtedly the *Leguminosae*, *Cistaceae*, *Ericaceae*, and *Labiatae*, while families such as *Compositae*, *Ranunculaceae*, *Liliaceae* provide many of the spring- and autumn-flowering annuals and herbaceous perennials which are such an interesting feature of Mediterranean scrub in general.

There are many types of *matorral* in the peninsula, many distinguished by vernacular names:

brezal	—heaths with *Erica* and *Calluna* species predominating;
jaral	—*Cistus* and *Halimium* species predominating;
jaro	—tree-heather scrub;
butjedal	—box scrub
goscojal	—scrub with Kermes oak predominating;
tomillar	—*Salvia* and *Thymus* species predominating;

esplegara —*Lavandula* among other species common;
retamar —broom-heath, dominated by *Lygos* and *Cytisus* species.

Rivas Goday lists the indicator plants of the peninsula low *matorrals* as: *Rosmarinus officinalis, Cistus albidus, Halimium atriplicifolium, Teucrium polium* subsp. *capitatum, Helichrysum stoechas, Stipa lagascae, Lithospermum fruticosum, Cistus monspeliensis, Dorycnium pentaphyllum, Thymus vulgaris, Thymus zygis,* and *Ruta angustifolia,* while on acid soils the following are characteristic: *Cistus crispus, Halimium umbellatum, Lavandula stoechas, Erica scoparia, Erica umbellata, Cistus ladanifer,* and *Genista hirsuta.*

Matorrals are particularly widespread on shallow soils, and in lower rainfall areas with high evaporation. They are typical of the sunny hills of the south and the stony plateau of the centre of Iberia. One of the most widespread is the *tomillar,* with *Thymus* species and many *Labiatae* predominating, and it is in this type of community that a number of the Iberian endemic species occur. *Jaral,* dominated by *Cistus* species, in particular *C. ladanifer,* is very widely distributed and covers vast areas in southern Portugal and south-western Spain. Another distinctive type is that in which the Dwarf Palm, *Chamaerops humilis* is abundant and it extends over considerable areas of the drier hills of the south and east of the peninsula.

Heavily over-grazed, or continuously fired or cut *matorral* will eventually revert to the most unproductive of all vegetation, where the soil cover is lost and where only extremely resistant species can survive. These communities are sometimes called pseudo-steppes or Asphodel ste.·ne, because of the predominance of this latter species, and they consist largely of plants which can not only survive such extremely difficult conditions, but which are also, for some reason or another, unpalatable to grazing animals. The spread of such plants as the mulleins, sea squill, asphodel, and some of the excessively spiny plants, is no doubt very greatly facilitated by their resistance to grazing. These pseudo-steppes are thus biologically determined and in this differ from true steppes which are primarily determined by climatic and soil conditions. These pseudo-steppes often have a large number of annual species of families such as the *Cistaceae, Leguminosae, Compositae,* and *Graminae,* which can make a bright show of colour for a short time during the spring. There are also a number of bulbous and rhizomatous plants which flower briefly at this time or in the autumn.

Though forests, reduced to *matorral* or eliminated by agriculture, comprise such a small percentage of the present day plant cover in our area, yet they nevertheless, potentially, form the climax community in many environments. The Holm oak forests are the characteristic forest to develop in a typical Mediterranean climate and over wide areas and on a great variety of bedrocks in both acid and alkaline conditions. However variations of the typical Mediterranean climate, either drier or wetter, or hotter or colder, result in other potential forest dominants, and their inter-relationships are outlined in the diagram below.

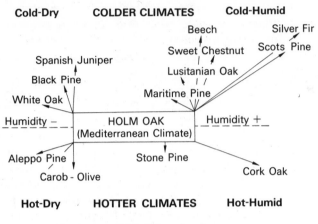

The Inter-relationships of some
Iberian forest communities[5]

The Cork oak, *Quercus suber*, is extensively grown in the Alentejo in Portugal and in this region alone one-third of the world's cork supply is produced. It is also grown in south-west Spain and to a lesser extent in Catalonia in east Spain. The Cork oak woods are grown in open, often park-like forests. They may be regularly cultivated and kept clear of ground vegetation, or a dense scrub may be allowed to develop among the trees, mainly of *Cistus ladanifer* and *Erica arborea* with many other species of *Cistaceae, Ericaceae, Leguminosae,* and *Labiatae* in particular. Many of these great Cork oak woods have been developed during the last century by large estates or *latifundia*, from waste land which was previously covered by *matorral*.

A sample of a Cork oak wood *matorral* in Andalusia contained the following woody species:

Quercus coccifera	*Daphne gnidium*
Calicotome villosa	*Cistus ladanifer*
Chamaespartium tridentatum	*C. crispus*
Stauracanthus boivinii	*C. monspeliensis*
Adenocarpus telonensis	*C. salvifolius*
Pistacia lentiscus	*Halimium halimifolium*
Thymelaea lanuginosa	*Myrtus communis*

Erica australis   *Phillyrea angustifolia*
E. scoparia   *Lithospermum diffusum*
Arbutus unedo   *Lavandula stoechas*
Olea europaea   *Teucrium fruticans*
              Chamaerops humilis

The Olive is another important tree, which though not truly native in the Iberian peninsula, has been under cultivation in the west for at least two thousand years. Regular intensive cultivation is required to maintain olive groves in productivity; however a wild form of the olive does occur regularly in the hotter and drier *matorrals* of the peninsula. For example, in Catalonia a *matorral* with the characteristic Dwarf Fan Palm *Chamaerops humilis* is associated with the following: *Pinus halepensis, Juniperus oxycedrus, Quercus coccifera, Euphorbia characias, Pistacia lentiscus, Pistacia terebinthus, Rhamnus lycioides, Rubia peregrina,* and the wild olive *Olea europaea* var. *oleaster.*

The Aleppo pine, *Pinus halepensis,* is another important and widespread tree in the Mediterranean region and it forms open forests on shallow denuded soils which at one time may have supported the Holm oak. Like the Kermes oak, it is an invasive species on poor dry soils, particularly on limestone. The Aleppo pine often supports beneath it a typical *matorral* of Mediterranean evergreen shrubs as well as some bulbous and rhizomatous plants. It is particularly well developed on dry coastal hills and is largely responsible for the wooded appearance of much of the Mediterranean coastal landscape which still remains unspoiled by development. In south Portugal it has been planted on coastal hills from which it has spread.

The Maritime pine, *Pinus pinaster,* dominates the more humid coastal hills particularly on acid siliceous soils, as well as the sands of much of the littoral. Extensive planting in Portugal has increased the area of Maritime pine five times during the last century. Maritime pine woods also occur inland on the sandy soils of Old Castile. These pinewoods, native or planted, are characterized by a *matorral* of *Cytisus, Genista, Ulex* and other shrubby *Leguminosae,* and *Cistus, Halimium, Erica* species in particular. *Eucalyptus* species are often planted with the pines, particularly in Portugal.

The Stone or Umbrella pine, *Pinus pinea,* has a limited distribution in Iberia, being restricted to light sandy soils on the littoral. It forms natural forests on fixed dunes of the Coto de Doñana and the Camargue.

3. Sub-montane forests

In cooler, more humid climates in the hills with rainfall between 700–900 mm, the semi-evergreen Lusitanian oak, *Quercus faginea,* replaces the Holm oak. It forms extensive forests along the western Atlantic hills, and in the south-west of Spain, between the coastal pine forests and the deciduous montane forests. An example of this forest in the Mondego valley of central Portugal has the following assemblage of species:

Laurus nobilis	*Vinca difformis*
Rosa sempervirens	*Origanum virens*
Euphorbia characias	*Centaurea sempervirens*
Rhamnus alaternus	*Asparagus aphyllus*
Daphne gnidium	*Ruscus aculeatus*
Hedera helix	*Smilax aspera*
Phillyrea species	*Iris foetidissima*

The White oak, *Quercus pubescens*, may also form an open climax forest in the hills of the Mediterranean region where the temperature is a little lower and rainfall higher; it is sometimes called the montane sub-mediterranean climax. It occurs in the Cantabrian mountains between the Pedunculate oak, *Q. robur*, forests and the beech forests, while in the Pyrenees it is found between the Lusitanian oak, *Q. faginea*, forests and the beech forest. With it, the lime and Montpellier maple, *Acer monspessulanus*, may often be found, and the associated flora is rich in species.

The deciduous Pyrenean oak, *Quercus pyrenaica*, is a characteristic forest-former in the Central Sierras and the Iberian Mountains, at altitudes ranging from 1200–1400 m, particularly in Burgos and Logroño provinces. But over much of the peninsula these forests have been reduced to *matorral* which is now dominated by *Cistus laurifolius* and *Adenocarpus* species, and often with these shrubs *Paeonia broteroi* is found. In the Sierra de Cazorla the latter occurs with the Corsican pine, *Pinus nigra* subsp. *salzmannii*. This tree forms important forests in hills of central and eastern Spain, the Pyrenees, south-west France, and the Cévennes.

The Scots pine, *Pinus sylvestris*, forms extensive forests in the Iberian mountains and the Central Sierras but occurs only as a sparse outlier further south in the Sierra Nevada. In the Central Sierras, it is well developed at altitudes of 1400–1800 m with an under-scrub of *Juniperus communis, Genista florida, Adenocarpus telonensis, Erica arborea,* etc., while above 1800 m they are replaced by a spiny hedgehog scrub of *Cistus* and *Genista* species. On the wetter slopes and in the valleys of the Northern Iberian mountains there are magnificent pine forests, perhaps some of the best in Spain, while the south Iberian mountains have largely been deforested of their oak and pine, and heathlands have taken their place. Between Cuenca and Teruel on the Serranía de Cuenca there are extensive pine woods and *Juniperus* scrub.

Central European and Atlantic deciduous plant communities

The forests, heathlands, and meadowlands of the pluviose regions of northern, north-western Spain, and western France are Central European in type, and the majority of plant communities and species belong to this element of the flora. The climate which favours the development of these communities has a rainfall of between 1000–2000 mm per annum with rain falling at all times of the year, and with winter snow in the mountains. The result is a typically Atlantic climate with cool humid summers and mild wet winters. In addition, most of the soils are siliceous, and owing to leaching by heavy rainfall, become strongly acidic, particularly in the west.

4. *Deciduous forests*

Oak forests dominated by the Pedunculate oak, *Quercus robur*, develop over wide areas if undisturbed by man. With it often occurs the Pyrenean oak, *Quercus pyrenaica*, the Sessile oak, *Quercus petraea*, while widely planted and rapidly self-seeding is the Sweet Chestnut, *Castanea sativa*. Oak and chestnut forests are well developed in the Cantabrian mountains (see description of Parque Nacional de la Montaña de Covadonga, p. 129 below), especially in the Basque provinces and Massif Central. When the oak in particular is cleared, the beech may take its place, especially in the more humid regions.

5. *Heathlands and grasslands*

More complete forest clearance may result in characteristic western heathland with *Erica, Calluna, Daboecia, Ulex, Genista* and *Cytisus* species becoming abundant and dominant. The heaths are directly comparable to the western heathlands of north-western France and Great Britain, though the latter are, in general, not so rich in species.

There are many variants of these heathland communities. *Arbutus unedo* may be dominant; another is dominated by *Quercus pyrenaica* and *Genista florida*. In the extreme west there are also all graduations between the Holm oak communities of southern Portugal and the heathlands of north-west Portugal and Spain. Thus in Galicia, *Cistus* and *Lavandula* heaths with *Erica umbellata* and *Genista triacanthos* become more abundant in drier areas.

Meadowlands are, nearly always, man-made by clearing and firing, and maintained by mowing or by grazing. Montane meadows are often very rich in herbaceous species and so also are the verges where they border on woodlands or heathlands; this is where many of the 'hedgerow' species can be found.

Most of south-western France, the French Pyrenean foothills, and the region north of the Mediterranean zone, is situated in this deciduous forest zone, but on the dry acid soils of Les Landes, the Maritime pine, *Pinus pinaster*, has long been planted and is widely naturalized. In Cantabria, Galicia, and west Portugal, *Eucalyptus* species and *Pinus radiata* are sometimes widely planted forest trees. The extensive Sweet Chestnut forests of Cantabria and south-western France were also originally planted by man, but are now naturalized.

Montane, sub-alpine, and alpine plant communities

A convenient division of altitude zones, given by Lázaro e Ibiza, in our area is as follows:

Littoral zone	0–100 m
Lower (hill) zone	100–600 m
Sub-montane or middle zone	600–800 m
Montane zone	800–1600 m
Sub-alpine zone	1600–2000 m
Alpine zone	2000–3500 m

In the Cantabrian mountains, Pyrenees, and Massif Central the four highest zones commence approximately 300 m lower, while in the south and south-east they are elevated 300 m or more.

6. *Montane forests*

The beech is the dominant forest tree in the Cantabrian mountains, the Pyrenees, and the Massif Central. It has a wide altitude range from about 600–1900 m and forms very extensive, dense and uniform forests, particularly on steep valley sides where it is not so easily exploited by man. Growing with the beech, other trees, such as ash, elm, and sycamore and the evergreens, box and yew, may be frequent, but they are all subsidiary to the dominance of beech. The field layer of the Iberian beechwoods is, surprisingly enough, directly comparable with those of southern England and the Rhodope mountains of southern Bulgaria—to indicate the extent of its range. In all of these forests *Mercurialis perennis*, *Sanicula europaea*, *Anemone nemorosa*, *Rubus* species, and the saprophytes such as *Neottia nidus-avis* and *Monotropa hypopitys* occur. The similarity is such that if one were brought blindfolded to a beechwood in any of these three areas, even a botanist might well take some time before he could distinguish which part of Europe he was in when he opened his eyes!

Beech favours the warmer, drier, south-facing slopes of the mountains in general, but in the Cantabrian mountains, on the boundary of the pluviose regions, the dry stony southern slopes are scattered with Spanish Juniper, *Juniperus thurifera*, and *Quercus ilex*, or a montane-Mediterranean type of *matorral* (particularly *tomillares*), while the northern slopes are clothed in heavy beech forests.

The Silver fir, *Abies alba*, forms the highest forest zone in the more humid north-facing valleys of the Sierra del Montseny, the Pyrenees, and Massif Central, where mists collect and are slow to disperse. By contrast, in the drier mountain climates of the arid region of Iberia, where rainfall may still be considerable, the silver fir is missing and the highest forests are those of Scots pine, *Pinus sylvestris*, particularly in the Central Sierras and Iberian mountains, with *P. nigra* subsp. *salzmannii* and *P. uncinata*. A few relict stands of the Spanish fir, *Abies pinsapo*, are now all that remain of once probably extensive montane forests in Andalusia.

7. Sub-alpine communities

At approximately 1600–2000 m is a very distinctive vegetational zone, particularly characteristic of the mountains of the arid regions. This is, in popular parlance, the 'hedgehog' zone and it replaces a scrub of *Rhododendron*, *Vaccinium*, *Arctostaphylos*, and Junipers, that occurs above the tree-line in the higher mountains of central Europe.

The 'hedgehog' zone is composed of low, rounded, intricately branched, spiny shrubs, often domed in shape. They are able to tolerate the extremely severe conditions that prevail and must withstand heavy snow cover, short growing periods, dry arid summers, and often strong winds. Reduction in leaf size, and the general development of woodiness, often associated with spines, are adaptations which help them to survive, while a short-lived flowering period, very intense when it comes, is characteristic. Consequently in the space of a few days the dry stony mountainsides may suddenly burst into flower and as quickly fade into drabness. A number of families take on the 'hedgehog' form, but the *Leguminosae* predominate with *Astragalus, Genista, Echinospartum*, and in particular *Erinacea anthyllis. Berberis hispanica, Vella spinosa, Ptilotrichum spinosum, Bupleurum spinosum*, are other examples from widely differing familes, which develop this very distinctive plant form. This 'hedgehog' zone is well developed on the drier, southern slopes of the Pyrenees, and on the highest Sierras of central and south Spain, and in particular on the Sierra Nevada, as well as on the Atlas mountains of Morocco. The unique climatic conditions coupled with the isolation of these high mountains have helped to evolve communities of plants which are rich in endemic species (see Sierra Nevada, page 74). Each isolated mountain-top may show an interesting collection of species, many with varied and often disjointed types of distribution. For example, *Prunus prostrata* occurs only on the high mountains of south Spain, North Africa and on the 'stepping-stones' of Corsica and Sardinia, and is then more widespread further east, from Yugoslavia to the Lebanon. By contrast, *Erinacea anthyllis* occurs only in the mountains of the arid zone of Iberia (including the Pyrenees) and in the high mountains of North Africa.

In sub-alpine climates with heavier rainfalls, particularly during the summer season (as for instance on the Serra da Estrêla in central Portugal, the northern slopes of the high Pyrenees, and the Massif Central) mountain grasslands develop on the deeper soils. Mountain pastures everywhere in Iberia are important summer grazing areas: there is regular transhumance from the plains to the mountains and in consequence the climax dwarf shrub communities, largely of *Juniperus communis* subsp. *nana*, which develop naturally above the tree-line, have been greatly modified by firing and grazing.

8. *Alpine communities*
True alpine communities are not developed much below 2600 m, and in our area only the Pyrenees and Sierra Nevada achieve altitudes above this. As the descriptions of these two ranges will show, they both possess a percentage of central European alpine species, with a distribution that spreads from the Sierra Nevada to the Carpathians and beyond. At the same time, because of their isolation, they also possess a high proportion of endemic species, both recent and ancient in origin. The Pyrenees has 180 or more endemic species, and the Sierra Nevada has about 40 and it is for this reason that these mountains are of particular interest to the botanist and plant-lover.

In general, the alpine vegetation is sparse and often confined to favoured habitats such as cliffs and rock crevices, screes, glacial moraines, gullies, moist areas, glacial lakes, and places where there is sufficient stability for soil to accumulate. Each habitat has its own assemblage of a few specialized species which are able to survive successfully in these difficult conditions. The majority of species are low cushion-forming perennials, often densely hairy, sub-shrubs creeping over the rocks, herbaceous perennials, rosette plants, and a few bulbous and corm-forming plants; annuals are rare. Protection by snow in winter, and protection from excessive sun and exposure in summer are essential for survival. Many of the distinctive plants inhabiting these areas are listed in the area accounts that follow.

Maritime and salt-tolerant plant communities

9. *Dunes, sands, and salt marshes*
Much of the central and southern coast of Portugal, the Biscayan coast of France, the Mediterranean coast of Languedoc, and the deltas of the

large rivers of our area, are sandy, often with attendant dunes. The dominant trees are the widespread Maritime pine, which forms extensive forests, often preserved and planted by man, and the Stone or Umbrella pine, which has a more restricted distribution. These pine forests quickly develop on consolidated sand behind the mobile dunes, and may cover very extensive areas as in Les Landes and the Pinhal de Leiria. With the pines is often a typical western heathland, rich in species of *Erica, Cistus, Halimium*, and with many *Leguminosae*. In the south-west of Iberia *retamales*, heathlands dominated by *Lygos* species, are very distinctive.

On the sandy strands and young dunes, a number of widespread western Atlantic strand plants can be regularly found, such as the Marram grass, *Ammophila arenaria*, Sea Holly, *Eryngium maritimum*, Cottonweed, *Diotis maritima, Euphorbia paralias*, and the more restricted *Corema album* and *Crucianella maritima*. In addition, a number of local species of the Atlantic strands may be found, including species of *Echium, Ononis, Silene*, and *Matthiola*, as well as some adventitive species from other regions, like *Cryptostemma calendulaceum* and *Carpobrotus edulis* from South Africa.

Salt-marsh communities are well developed along the Mediterranean coast, particularly in the Gulf of Lions and the Rhône delta, and also in the *marismas* of the Coto de Doñana and Tagus estuary. They are characterized by succulent plants of the family *Chenopodiaceae*, which are able to withstand periodic flooding by salt water. The south of Spain is particularly rich in members of this family (see p. 176) and some species have very interesting disjointed distributions, indicating survival from much wider distribution in the past.

Most families of plants are completely eliminated by these saline conditions, but there are others that have a few genera or species which are salt-tolerant. The *Compositae*, for example, has *Aster tripolium, Inula crithmoides*, and some *Artemisia* species, while genera such as *Frankenia, Tamarix, Spergularia, Mesembryanthemum* are characteristic. The *Plumbaginaceae* is a family that is predominantly salt-tolerant; the genera *Armeria* and *Limonium* are particularly rich in the south-west. *Armeria* has 15 species and *Limonium* 26 endemic species. Otherwise these salt marsh communities are poor in species of flowering plants, except among the grasses, rushes and sedges, and are consequently of limited interest to the layman. By contrast, they provided a unique and highly specialized environment of the greatest interest to the ornithologist and zoologist, though many areas are now very severely threatened by drainage and agriculture.

10. *Steppes*

These occur where rainfall is below 300 mm (12 inches), where there are five to seven months or more of drought, and over 200 days of sunshine in the year. Added to this, temperature ranges are extreme and both daily and seasonal fluctuations excessive; at the same time excess of evaporation over precipitation coupled with poor drainage results in soil with a high salt content. It is easy to see that these add up to some of the most difficult conditions for plant life in any part of Europe.

The steppes of the Iberian peninsula occupy an area nearly equal to that of Scotland, but the exact area is not always easy to define as there are gradual transitions to a sparse *matorral*. They have also been reduced by methods of temporary dry cultivation, while recently developed irrigation schemes have further restricted these otherwise unproductive areas. The four main steppe areas in Spain are in Murcia and Valencia; La Mancha, New Castile; Andalusia; and the Ebro basin. These steppes are surprisingly quite rich, and something like 200 species are restricted to this habitat. About one-third of these are North African steppe plants (most of which occur only in these Iberian steppes in Europe). About thirty per cent of all these plants are very short-lived annuals, which only grow after periods of rain and are able to flower and seed in a matter of a few weeks, but the majority of species are perennials and show special adaptations to the harsh environment. Succulence, woodiness, absence or early loss of leaves, cactus-like forms are common modifications, while many species roll up the margins of their leaves, or are covered with waxes, blooms, or scales, or are hairy, or excrete aromatic oils.

The main types of steppe are *salt steppes*, *grass* (or *esparto*) *steppes*, and *stony steppes*; they are described in outline on pages 92–3.

The steppes of the Ebro basin have a climate of extremes. In the summer the temperature may rise well above 100°C, while in the winter the average of the coldest month is as low as − 5°C; there are very cold snaps during the spring, added to which frequent and very strong north-westerly winds occur, which bowl along the dead spiny bushes of *Salsola*.

The grass steppes, as elsewhere, are dominated by tufts of the Albardine, *Lygeum spartum*, and the steppe grasses *Stipa barbata*, *S. lagascae* and *S. parviflora*. Old dry cultivated areas are invaded by *Salsola vermiculata*, *Eruca vesicaria*, and *Artemisia herba-alba* which are usually grazed down to the ground, while fallow plants, largely members of the Poppy family, *Papaveraceae*, include *Hypecoum imberbe*, *H. pendulum*, *Roemeria hybrida*, *Glaucium corniculatum*, and *Platycapnos spicata*. *Asphodelus fistulosus* is abundant and brightens the steppe in spring, which is at other times of the year grey and monotonous.

Around the villages, where the sheep are folded and the ground is enriched by their droppings, *Peganum harmala, Sisymbrium irio,* and *Raphanus raphanistrum* subsp. *landra,* and the massive thistles *Onopordum illyricum* and *Silybum eburnum* occur commonly.

In depressions, and around pools of brackish water are such succulent plants as *Arthrocnemum glaucum, Suaeda brevifolia, Salsola vermiculata, Atriplex halimus, Limonium* species, and the tall graceful shrub *Tamarix africana.*

On gypsum soils, often of startling whiteness owing to the accumulation of calcium salts, the following are characteristic:

Herniaria fruticosa	*Helianthemum squamatum*
Gypsophila hispanica	*H. lavandulifolium*
Lepidium subulatum	*Lithospermum fruticosum*
Ononis tridentata	*Thymus zygis*

So barren is the region around Zaragoza that it is like a 'corner of Africa lost in Europe'.

The La Mancha and Andalusia areas also show the three main types of steppes, grass, salt and stony, and are similar to those already described. In addition, the dry valleys or *ramblas* of the New Castile steppes may be mentioned, where a small yellow-flowered shrub, *Securinega buxifolia,* occurs characteristically. Another feature of these arid areas is the abundance of tall and spectacular thistles lining the roadsides. They include species of *Carlina, Carduus, Carthamus, Scolymus,* and

Onopordum nervosum	*Cirsium echinatum*
O. illyricum	*C. flavispina*
O. macracanthum	*C. acarna*
O. acanthium	*C. ferox*
Cynara humilis	*Centaurea ornata*
C. cardunculus	*C. solstitialis*
	C. toletana

Freshwater, wetland communities

Plant communities in these wetlands are surprisingly uniform throughout temperate Europe. The majority of species of pond, lake, marsh, bog, and river are very widespread and differ little over the greater part of Europe. As most of these freshwater wetland communities occur in the pluviose region of our area and as they have few species of special interest, a detailed summary of these communities will not be given in this volume.

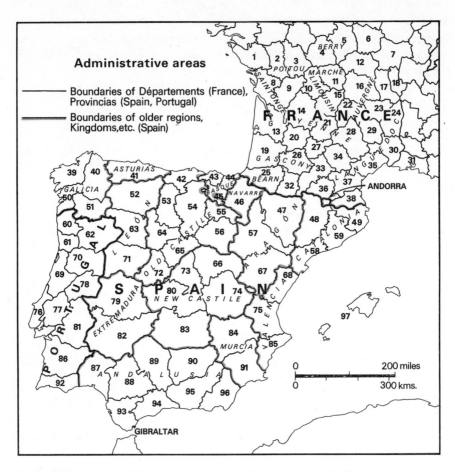

Administrative areas

———— Boundaries of Départements (France), Provincias (Spain, Portugal)

━━━━ Boundaries of older regions, Kingdoms,etc. (Spain)

ANDORRA

0   200 miles
0   300 kms.

GIBRALTAR

France (Départements)	26 Gers	49 Gerona	84 Albacete
1 Vendée	27 Tarn-et-Garonne	50 Pontevedra	85 Alicante
2 Deux-Sèvres	28 Aveyron	51 Orense	87 Huelva
3 Vienne	29 Lozère	52 León	88 Sevilla
4 Indre	30 Gard	53 Palencia	89 Córdoba
5 Cher	31 Bouches-du-Rhône	54 Burgos	90 Jaén
6 Nièvre	32 Hautes-Pyrénées	55 Logrono	91 Murcia
7 Saône-et-Loire	33 Haute-Garonne	56 Soria	93 Cádiz
8 Charente-Maritime	34 Tarn	57 Zaragoza	94 Málaga
9 Charente	35 Hérault	58 Tarragoña	95 Granada
10 Haut-Vienne	36 Ariège	59 Barcelona	96 Almería
11 Creuse	37 Aude	63 Zamora	97 I. Baleares
12 Allier	38 Pyrénées-Orientales	64 Valladolid	
13 Gironde		65 Segovia	
14 Dordogne		66 Guadalajara	Portugal (Provincias)
15 Corrèze	Spain (Provincias)	67 Teruel	60 Minho
16 Puy-de-Dôme	39 La Coruña	68 Castellón	61 Douro Litoral
17 Loire	40 Lugo	71 Salamanca	62 Trás-os-Montes e Alto Douro
18 Rhône	41 Oviedo	72 Avila	69 Beira Litoral
19 Landes	42 Santander	73 Madrid	70 Beira Alta
20 Lot-et-Garonne	43 Vizcaya	74 Cuenca	76 Estremadura
21 Lot	44 Guipúzcoa	75 Valencia	77 Ribatejo
22 Cantal	45 Alava	79 Cáceres	78 Beira Baixa
23 Haute-Loire	46 Navarra	80 Toledo	81 Alto Alentejo
24 Ardèche	47 Huesca	82 Badajoz	86 Baixo Alentejo
25 Basses-Pyrénées	48 Lérida	83 Ciudad Real	92 Algarve

38

REFERENCES TO CHAPTER 1

[1] Way, *A Geography of Spain and Portugal* (London, 1962), p. 2.

[2] Estyn Evans, *France* (London, 1966), p. 145.

[3] After Delvosalle et Dubigneaud, *Itineraires Botaniques en Espagne et en Portugal* (Bruxelles, 1962), p. 11.

[4] Walker, *The Mediterranean Lands* (London, 1964), p. 41.

[5] After Paviari, 'Fundamentos ecologie, e tecnicos, de silvaculture nos paises mediterraneon', *Estudos e Informacão Direcc. Genal das Servicos Flor. e Aquae*, 85, Feb. 1958, pp 1–28.

2. The Plant-hunting Regions

Our area, the Iberian Peninsula and south-western France, has perhaps the greatest variety of habitats in Europe. They range from coastal dunes and salt marshes, almost Asiatic steppes, montane forests and pastures, to high alpine floras. There is much ground to explore, and ever changing vistas for the traveller. Many regions are little known and will undoubtedly repay diligent searching by the naturalist, while others are well known for their great richness and variety of plant life and are frequently visited.

The authors have selected twenty-three regions which they consider to be of the greatest interest to the botanist and plant-lover and which are likely to be most rewarding during short visits. A thumb-nail sketch of each region has been given, drawn from the authors' own experience and from the botanical literature available. The accounts have to be brief and can only name those species which are of particular interest for one reason or another. The selected regions vary considerably in importance and extent; there has, perhaps, been an over-emphasis of high mountain areas because of their richness in endemic species, but this seems unavoidable. Only regions which are reasonably accessible have been chosen. It must be emphasized that much remains for the energetic traveller to explore. The map on the facing page shows the selected regions.

1. Algarve

This, the most south-westerly region of Europe, has many distinctive features. Triassic and Jurassic rocks dominate; in places they rise to promontories or fall away, in the more sheltered easterly part, to sandy coasts with dunes and lagoons, while in the exposed west they end abruptly in rugged dolomitic cliffs. To the north of the region the underlying carboniferous rocks come to the surface in a low range of mountains which protect the southern coastline from the cold north winds. The Algarve is a sunny region, tempered even in the summer drought by

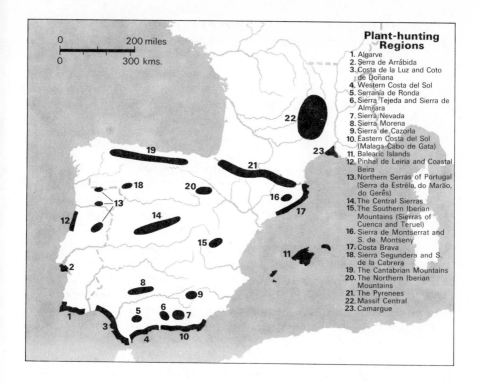

Plant-hunting Regions

1. Algarve
2. Serra de Arrábida
3. Costa de la Luz and Coto de Doñana
4. Western Costa del Sol
5. Serranía de Ronda
6. Sierra Tejeda and Sierra de Almijara
7. Sierra Nevada
8. Sierra Morena
9. Sierra de Cazorla
10. Eastern Costa del Sol (Malaga-Cabo de Gata)
11. Balearic Islands
12. Pinhal de Leiria and Coastal Beira
13. Northern Serras of Portugal (Serra da Estrêla, do Marão, do Gerês)
14. The Central Sierras
15. The Southern Iberian Mountains (Sierras of Cuenca and Teruel)
16. Sierra de Montserrat and S. de Montseny
17. Costa Brava
18. Sierra Segundera and S. de la Cabrera
19. The Cantabrian Mountains
20. The Northern Iberian Mountains
21. The Pyrenees
22. Massif Central
23. Camargue

sea breezes, while by contrast the extreme westerly promontories of Sagres and Cape St Vincent receive the full force of the south-west winds and are covered with treeless heathlands.

The motorist in the Algarve cannot fail to notice the work of the road maintenance men, the *cantoneiros* of the Ministry of Public Works, and the neat and tidy appearance of their road verges, which are often colourfully planted with ornamental acacias, quince trees with large white blossoms faintly tinged with pink in March and April, irises (mainly *I. albicans* with some *I. germanica*), *Salvia officinalis*, pelargoniums, and so on. Nature also contributes a generous quota of roadside flowers, mostly widespread Mediterranean species, and these include *Narcissus obesus* which is very abundant; *N. papyraceus*; the asphodels *A. morisianus* (*lusitanicus*), *A. aestivus* (the latter commencing to flower as the former is ripening its fruits) and *A. fistulosus*; *Gladiolus segetum* abundant in the corn fields and the smaller *G. illyricus* in scrub, and *Iris sisyrinchium*. The various species of *Erica* and *Cistus*, especially *C. ladanifer*, are much in evidence, with *Lithospermum diffusum* subsp. *lusitanica* and other common associates of *cistus* scrub. Later flowering are the yellow composites *Chrysanthemum coronarium*, usually var. *discolor* with the bi-coloured flowers, *C. myconis* and its twin *C. macrotus* turning whole fields yellow, and in high summer the yellow thistles *Scolymus hispanicus* and *S. maculatus.* Tall yellow umbellifers of the genus *Thapsia* are seen everywhere in April and May.

Other plants have a more limited distribution, and are restricted to one or other of the three contrasting zones which distinguish the Algarve and run parallel to the southern coast, namely, the coastal zone, the limestone zone (the *Barrocal*), and the Serras.

The coastal zone There is a marked difference between the *sotavento* or leeward coast east of Faro, with sand dunes and salt lagoons, and the *barlavento* to the west, a more humid area of mainly rocky coast exposed to the Atlantic winds. Around Monte Gordo in the *sotavento* there are extensive woods of Stone pine with an undergrowth of the graceful slender white-flowered broom, *Lygos monosperma*, a wonderful sight when in full bloom in February with a carpet of *Narcissus bulbocodium* beneath. Salt marsh vegetation can be studied along much of this coast, especially around Santa Luzia, where the robust deep yellow-flowered parasite *Cistanche phelypaea* is the most spectacular.

Between the *sotavento* and the *barlavento* there are more extensive pine woods on Pliocene sands, extending from Faro towards Albufeira. The undergrowth here is colourful when in full bloom between February and April. White-flowering shrubs include *Cistus ladanifer, *C. libanotis, *C. salvifolius*; the many yellow-flowered species include *Halimium commutatum, Spartium junceum, *Genista hirsuta* and *G. triacanthos, *Stauracanthus boivinii, *Anagyris foetida* and *Tuberaria major†*, endemic to south Portugal, *T. guttata*, and *T. bupleurifolia*. Shades of blue and purple are contributed by *Cistus crispus, *Erica umbellata, *Anchusa calcarea, *Anagallis monelli*, and *Lavandula stoechas* subsp. *pedunculata; Calluna vulgaris* is also abundant but flowers at the end of the year.

The Tertiary and Quaternary deposits behind the sandy soils of the littoral are intensely cultivated in small plots and most of the native vegetation has long since disappeared. Driving through this area along the Vila Real-Faro road one gets the impression that every possible nook is lovingly and tidily cultivated.

The *barlavento* begins a little to the east of Portimão and extends to the west coast. West of Lagos the hilly coastal strip becomes increasingly less fertile, the population decreases, and the traveller arrives at the dramatic arid wind-swept headlands of Sagres and Cape St Vincent. They are composed of hard calcareous dolomitic rocks, weathered and battered by the elements into a karst-like plateau which ends abruptly in overhanging cliffs 80 m high, with the blue sea below. Although heavily grazed by cattle and sheep, most of this headland is relatively undisturbed by man and there is a most interesting and distinctive flora.

The dominant plant is *Cistus palhinhae*† which forms a low compact spreading bush of semi-prostrate habit, growing gregariously, with extremely shiny sticky leaves, and with pure white flowers the size of *Cistus ladanifer*, to which it is closely related. The other common cistus species here are *C. salvifolius* and *C. albidus*, and the smaller white-flowered *C. monspeliensis* and pink *C. crispus*, and associated with them many of the common *matorral* species such as:

Shrubs	Herbs
Juniperus phoenicea	*Anemone palmata*
Dorycnium hirsutum	*Silene colorata*
Pistacia lentiscus	*Iberis procumbens*
Malva hispanica	*Onobrychis peduncularis*
Daphne gnidium	Lithospermum apulum
Halimium commutatum	*Salvia verbenaca* (very
Helianthemum origanifolium	dark blue)
Corema album	*Campanula lusitanica*
Lithospermum diffusum	C. rapunculus subsp.
subsp. *lusitanica*	verruculosa
Lavandula stoechas	*Calendula suffruticosa*
Rosmarinus officinalis	*Asteriscus maritimus*
Thymus camphoratus†	*Dipcadi serotinum*
	Narcissus obesus

In April the Sea Pink *Armeria pungens*† is a fine sight, the vegetable hedgehog *Astragalus massiliensis*† is starting to open its white flowers, and so also is the shrubby Calendula suffruticosa with its orange flowers. *Antirrhinum majus* subsp. *linkianum* has developed the curious habit of pushing its way up through bushes and supporting itself by twining petioles and young shoots around twigs. A dwarf form of *Cerinthe major* subsp. *gymnandra* with bristly leaves and pale whitish flowers, with yellow and red markings at the base of the corolla, occurs here and there. Other plants of special interest include: *Biscutella vincentina*†, distinguished like all members of the genus by the paired disks of the fruit, *Viola arborescens*†, partial to calcareous soil on rocky headlands (it also grows near Cape Trafalgar), *Linaria algarviana*, *Teucrium polium* subsp. *dunense*†, and *Scilla vincentina*† with paired bracts below each flower. The rare *Ionopsidium acaule*† forms its little circular mats here and there and flowers in February or even January, and a few plants of *Bellevalia hackelii*† may be seen on the Sagres promontory. On more sheltered ground the brilliant blue of *Anagallis monelli* contrasts with

ALGARVE

1. *Anagyris foetida* 495   2. *Lygos monosperma* 514   3. *Astragalus massiliensis* (528)
4. *Asphodelus fistulosus* 1592   5. *Iberis procumbens* 347b   6. *Armeria pungens* (976)
7. *Biscutella vincentina* 349a   8. *Chrysanthemum myconis* (1424)

9. *Gladiolus segetum* 1695   **10.** *Anchusa undulata* (1057)   **11.** *Genista hirsuta* 511a
12. *Campanula lusitanica* 1355a   **13.** *Linaria algarviana* 1201e   **14.** *Lavandula viridis* 1110a   **15.** *Bellevalia hackelii* 1644a   **16.** *Orchis morio* 1885

the tiny white flowers of the sweet-scented *Lobularia maritima*; two docks with showy purple inflorescences are *Rumex intermedius* and *R. papillaris*. *Nepeta tuberosa* is spectacular by the road.

A similar flora, less exposed and dwarfed, can be seen under the protection of *Pinus pinea* on giant sand dunes on the west coast, off the Lisbon road about 2 km west of Bordeira, and on the headland of Pontal.

The limestone zone (the *Barrocal*) is a curved lens-shaped area, commencing at Cape St Vincent in the west, with its greatest width of about 20 km in the centre of the Algarve, and tapering off and disappearing north of Tavira in the east. It is formed of Jurassic limestones, dolomites, and marls, usually white in colour but sometimes orange when freshly fractured, and weathering to a grim grey. These rocks form a series of crests running roughly parallel to the coast, which are covered by scrub, and they separate the fertile valleys. It is here that the characteristic crops of the Algarve are grown. Rich orchards of almond, olive, fig, and carob spread from crest to crest; natural water resources are sufficient for heavy crops. The display of almond blossom, from mid-January to late February dependent on the arrival of spring, is a spectacle for which the Algarve is famous. Beneath the clouds of pale pink, the ubiquitous weed *Oxalis pes-caprae* spreads its yellow carpet.

The heavy clay *terra rossa* soil is cultivated wherever possible, often after much laborious clearing of surface rocks, but the hill slopes remain clad in a scrub which is rich in Mediterranean species. In addition to those already listed for Cape St Vincent it includes:

Quercus coccifera	*Phlomis purpurea*
Q. ilex	*Scilla peruviana*†
Olea europaea	*Asphodelus morisianus*
Jasminum fruticans	*A. aestivus*
	Chamaerops humilis

Plants of special interest include the dwarf yellow narcissi *N. gaditanus*†, and *N. willkommii*†, in small colonies here and there and a white form of *Anemone palmata* which not uncommonly grows with the normal yellow form. North of the line Portimão-Lagos there are large colonies of *Iris planifolia*†, a rare species which reappears in Spain east of Niebla where it is very common, and *Bellevalia hackelii*† which is locally abundant. Many of the common Mediterranean orchids can be found in this zone.

The Serras, the hills that make up the northern two-thirds of the Algarve, are built of schists of the carboniferous period (except for the Monchique syenite and foyaite), and are separated from the limestone

zone to the south by a narrow band of Tertiary sandstones and conglomerates. These hills consist of the Serra do Caldeirão in the east, S. do
Malhão in the centre, and S. de Monchique in the west, separated by the
low land of the São Marcos depression. Their acid soils support a quite
different vegetation from that developed on the limestone further south.

The original vegetation of the Serra do Caldeirão consisted of oaks
and *cistus* undergrowth but much of it has disappeared. Some cork oaks
remain, but the general impression is of bare rounded hills, mostly cultivated or fallow, with occasional trees or bushes. Here and there remnants
of the *cistus* scrub persist, dominated by *C. ladanifer*, with *C. monspeliensis* and a little *C. populifolius* at the higher elevations. The
commonest heathers are *Erica australis*, *E. lusitanica*, *E. scoparia*,
while the graceful pagoda-like inflorescences of *Genista hirsuta* are
commonly seen throughout the hills, often associated with *G. triacanthos*, which is easily distinguished by the deeper and more orange tint
of its very numerous small flowers, and by its inflorescence which ends
in a leafy tassel. Both are extremely prickly. Other conspicuous shrubs
are: *Lavandula stoechas*, *L. viridis* with small white flowers set in a
pale green head, with a pale green top-knot, *Chamaespartium tridentatum*, and *Ulex argenteus*. The most richly wooded part of the Serra is
along the road north from São Braz do Alportel, but this gives a mislead-
ing impression of the area as a whole.

The Serra de Monchique is the highest part of the range, with the twin
peaks of Foia, 902 m, and Picota, 744 m, both formed of foyaite and
syenite, surrounded by schists. The western end of the Serra has the
highest annual rainfall of the Algarve, and the abundant water available
from perennial springs has led to the building of many rich irrigated
terraces. Between the two peaks nestles the village of Monchique at
460 m, where fine camellia trees are grown, while lower down near the
foot of the hills on the south side, lies the spa of Caldas de Monchique
at 200 m, which was known to the Romans.

This is one of the classical botanical localities of the Algarve, but
unfortunately most of the original oak forests of *Q. canariensis†* and *Q.
faginea* that once clothed the hills have long since disappeared, though
some chestnut coppice is still maintained on the slopes of Picota. Today
these hills appear well wooded owing to the extensive plantations of
Eucalyptus and *Pinus pinaster*, while olive groves and cork woods are
grown up to altitudes of 550 m. There are also small plantations of the
creamy-white-flowered *Acacia melanoxylon*, and the more ornamental
yellow-flowered species have been planted along the roadsides. Of the
two peaks, Foia is accessible by a good road to the summit, rather bare
and uninteresting and now much cluttered up with television masts, but

the lower slopes are more rewarding; Picota is wilder and more interesting than Foia; one has to walk up most of the way from Monchique. The more conspicuous plants include *Paeonia broteroi*, which flowers before the end of April, and *Rhododendron ponticum*† flowering at the same time or a little later in gullies, and the two tall white heathers *Erica lusitanica* and *E. arborea* which may start flowering in February. The former seems to be the commoner of the two, but *E. arborea* is dominant on the upper slopes of Picota, and in sheltered places grows to 4 m in height. *Arbutus unedo* grows luxuriantly, and a type of brandy, *medronheira*, is distilled from the berries in Monchique. The primrose, *P. vulgaris* is found here and there, notably in the chestnut coppice on the lower slopes of Picota. Of the bulbs, *Scilla monophyllos* is conspicuous everywhere in April; *Romulea bulbocodium* is also abundant, notably on the summit of Foia; and the Spanish bluebell, *Endymion hispanicus*, is locally common.

As one climbs up Picota one passes for two-thirds of the way through farms and chestnut coppice, where the most conspicuous plants in spring, in addition to those already mentioned, are *Saxifraga granulata* on walls, the two tall spurges *E. characias* and *E. amygdaloides*, *Anchusa undulata*, and *Asphodelus morisianus*. On the rocky summit slopes, above the limit of cultivation, the most colourful shrubs are the golden-yellow *Halimium commutatum*, the purple *Erica australis*, as well as pink *Cistus crispus*.

The rare tree *Myrica faya*† grows here and there, usually by streams, on Picota, but we have only seen one specimen on Foia. It resembles the Strawberry Tree, *Arbutus unedo*, but is easily distinguished from it by its darker green and narrower leaves. Other more unusual plants of this area are *Campanula primulifolia*†, *Centaurea longifolia*†, *Senecio grandiflorus*†, the handsome robust, white-flowered *Astragalus lusitanicus*, and the very rare *Leucojum longifolium*† which may now have become extinct. *Erica ciliaris* has recently been found on Foia.

---

1 **Algarve** Dolomitic cliffs near Cape St Vincent with *Armeria pungens*, *Calendula suffruticosa*, *Astragalus massiliensis*, *Lobularia maritima*, and *Juniperus phoenicea* growing to the cliff edge. Inland is a heath of the endemic Portuguese *Cistus*, *C. palhinhae*.

2 **Serra da Arrábida** Limestone mountains rising to 500 m above a wild unspoiled coastline. Though Atlantic in position, the scrub or *goscojales* of Kermes Oak has many typical Mediterranean species mixed with western bulbous species, such as *Scilla monophyllos*, *Fritillaria lusitanica*, *Ornithogalum concinnum*, *Endymion hispanicus*.

2. Serra da Arrábida

The Serra da Arrábida is an isolated outcrop of limestone—a rounded whale-back of a mountain about 55 km long, running parallel to the coast and falling in steep scrub-covered cliffs to the sea. Inland to the north are fine views across the low alluvial lands, around the estuary of the Tagus, to the granite outcrops of the Serra da Sintra beyond. The heights of Arrábida rise to no more than 500 m and are eroded and weathered into pitted, fluted, and flaked rocks, with pockets of red *terra rossa* soil between: the terrain known in Spain as the *campos de lanar*.

To the west the range falls away gradually to the high cliffs of Cabo de Espichel, a remote western promontory with windmills and small white houses tucked into the folds of the landscape, and with the wide horizon of the Atlantic stretching away on every side. Marble quarries, donkey carts, and small stone-walled fields of beans and corn, and the ruined monastery of Nossa Senhora do Cabo complete the picture. A good road from the port of Setubal crosses the seaward flank of the Serra da Arrábida and in places runs along the crest of the mountain itself, making access to the interesting flora easy. The terrain is characteristically covered by a garigue, the *carrascais*, dominated by the spiny-leaved Kermes oak, together with many typical Mediterranean shrubs such as:

Juniperus phoenicea	*C. monspeliensis*
Genista tournefortii	*C. salvifolius*
Astragalus lusitanicus	Arbutus unedo
Coronilla valentina	Jasminum fruticans
subsp. *glauca*	Lonicera implexa
Pistacia lentiscus	Rosmarinus officinalis
Phillyrea angustifolia	*Lavandula multifida*
Daphne gnidium	*Phlomis purpurea*
Cistus albidus	

More interesting are the many herbaceous plants which grow under the protection of this scrub and in the numerous rock crevices. The following are in flower towards the end of March:

Anemone palmata   Ranunculus paludosus

---

3 **Coto de Doñana** Inland from the coastal dunes are fixed dune slacks colonized by Stone Pine 'savannah' with rushes such as *Juncus acutus* and *Scirpus holoschoenus*.

4 **Costa del Sol** Between Gibraltar and Estepona, looking eastwards towards the Sierra Bermeja. Often a dense maquis of cistuses, brooms, gorses, ericas, Kermes Oak, Dwarf Fan Palm, and asphodels, and many other species cover the steeper rocky slopes where cultivation is not possible.

*Lobularia maritima
*Saxifraga granulata
*Anagallis monelli
*Omphalodes linifolia†
Valeriana tuberosa
*Allium roseum
*Fritillaria lusitanica
  subsp. stenophylla†
*Tulipa australis†

*Scilla monophyllos
Ornithogalum concinnum
*Narcissus obesus
*Ophrys lutea†
*O. speculum†
*O. fusca†
Orchis italica†
*O. morio†
O. papilionacea†

In sheltered valleys on the seaward slopes, and in particular where the road goes down to the little village of Portinho da Arrábida, there are dense forests of *Quercus faginea* the dominant tree, under the shade of which grow the Spanish bluebell *Endymion hispanicus*, as well as the coral-pink *Paeonia broteroi*, while through the thickets scrambles the climbing *Antirrhinum majus* subsp. *linkianum*, which with its prehensile lateral shoots can climb 2 m or more up through the scrub to the light. Other noteworthy species are:

*Matthiola fruticulosa
*Cistus populifolius
*Erica arborea
*Convolvulus althaeoides

Echium tuberculatum
Asphodelus morisianus
*Gladiolus illyricus
*Arisarum vulgare

Luisier records over 1,000 species in the Serra da Arrábida and the surrounding lowlands; there is much of interest to be searched for.

The Pinhal d'Aroeira, which lies on the south flank of the Tagus estuary, though not more than 20 km away, has an entirely different, acid-loving flora. In the recent deposits of loose sand, the Maritime pine has been extensively planted, and below the trees has developed a more or less closed shrub layer of *Halimium alyssoides* and *Halimium commutatum* with *Ulex parviflorus* subsp. *jussiaei*. Other distinctive shrubs are:

*Corema album†
*Daphne gnidium
*Lithospermum diffusum

*Lavandula stoechas
  subsp. pedunculata
*Erica umbellata
Juniperus oxycedrus

Undoubtedly the most attractive plant is the bulbous *Leucojum trichophyllum† with pendulous white or flushed-pink bells, often growing in company with the delicate blue-flowered *Scilla monophyllos. The scarlet and orange parasite Cytinus hypocistis grows under many of the bushes of *Halimium alyssoides, bursting unexpectedly through the fine, loose sand.

3. Costa de la Luz and Coto de Doñana

The Atlantic coast of south-west Spain is cooler than the Mediterranean Costa del Sol, more exposed to strong westerly winds, and except for the region round Cadiz bay much less populated and developed. Two areas on this coast are of particular interest to the naturalist: the rocky coast from Tarifa to Cape Trafalgar in the south, and the sandy coast, which includes the Biological Reserve of the Coto de Doñana, between Huelva and the mouth of the Guadalquivir river, to the north.

Between Tarifa and Cape Trafalgar the coastline is one of the wildest stretches of the entire southern coast of Spain. It is a naturalist's paradise. The main stream of bird migration from Africa to Europe, and vice versa, crosses the Straits of Gibraltar and passes along this coast. Parties of storks, cranes, kites, honey buzzards, avocets, and many other species, sometimes in spectacular numbers, may be seen moving northwards in the spring and southwards in the autumn. Centrally situated in this area between Barbate and Tarifa is the Hotel Cortijo de la Plata near Zahara de los Atunes, regularly visited by parties of both ornithologists and botanists. In spring, in particular, the flowers make a fine show of colour and many species typical of Mediterranean coasts can be seen here, such as:

*Silene littorea†	Eryngium maritimum
S. nicaeensis	*Limonium sinuatum
S. obtusifolia	Solanum sodomeum
Cakile maritima	Otanthus maritimus
Ononis natrix	*Centaurea sphaerocephala
*Lotus creticus	*Pancratium maritimum
Medicago marina	

The rocky sierras, high sand dunes, and sandy bays between Zahara and Tarifa are exciting to explore. Access to the coast is restricted by military activities, but a reasonable road leads to the old Roman harbour of Bolonia and another to the lighthouse, Faro de las Palomas. Sand dune plants include: *Lygos monosperma, and *Corema album†, *Malcolmia littorea, *Anagallis monelli†, *Orobanche ramosa, *Ophrys speculum†. Both the yellow-flowered Ononis variegata and Linaria pedunculata are often half-buried in the sand, while the prostrate silvery-leaved *Lotus creticus is often parasitized by Orobanche densiflora or O. sanguinea, or sometimes by both species at the same time. Juniperus phoenicea is abundant on the headlands. The rocky sierras inland are covered with cistus scrub, with *C. populifolius dominant at the higher levels.

COSTA DE LA LUZ AND COTO DE DOÑANA

1. *Pancratium maritimum* 1673 **2.** *Chrysanthemum coronarium* 1425 **3.** *Viola arborescens* 781 **4.** *Hedysarum coronarium* 634 **5.** *Drosophyllum lusitanicum* (378) **6.** *Atractylis cancellata* 1470 **7.** *Anagallis monelli* (967) **8.** *Narcissus bulbocodium* 1666 **9.** *Triguera ambrosiaca* 1185a.

10. *Gladiolus illyricus* 1697 **11.** *Halimium commutatum* 796 **12.** *Ornithogalum arabicum* 1642a **13.** *Cotula coronopifolia* 1433 **14.** *Ophrys scolopax* 1876
15. *Malcolmia lacera* 297c **16.** *Romulea gaditana* 1681b **17.** *Orobanche ramosa* 1270
18. *Narcissus gaditanus* 1668a

On the low ground round Zahara de los Atunes the most prominent species in the riot of colour that appears at the end of April include:

*Polygonum equisetiforme†   *Lavatera trimestris
Spergularia fimbriata†   *Thapsia villosa
*Vicia villosa   *Convolvulus tricolor
Scorpiurus vermiculatus   Stachys germanica
*Hedysarum coronarium   *S. ocymastrum
*Erodium laciniatum   *Chrysanthemum coronarium
Malope malacoides   *Gladiolus illyricus

and *Cotula coronopifolia in wet ditches beside the road. The handsome solanaceous annual *Triguera ambrosiaca† with large deep violet, bell-shaped flowers with a darker centre and conspicuous yellow stamens, grows in sandy places—a North African plant which has its only European station in south-western Spain. The rare yellow-flowered *Allium stramineum† grows on rock ledges locally and one of the finest of all the hardheads *Centaurea polyacantha† with very large pink flower heads up to 7 cm across is common along the coast.

There is an interesting flora in the pine woods south of Vejer with Crocus clusii flowering in November, and here also can be found *Cistus libanotis† which is endemic to Cadiz province, Ononis subspicata†, and *Omphalodes linifolia†. The South American shrub *Solanum bonariense grows in hedges by the main road.

Between the Barbate road and Cape Trafalgar, on a tableland of calcareous Miocene sandstone and conglomerates, are extensive pine woods extending to the edge of the cliffs. Here there are fine displays of Armeria macrophylla† and *Anagallis monelli†. In other areas there is much *Halimium commutatum and Lotus angustissimus, with some of the tiny yellow Linaria haenseleriit and the richly coloured *Romulea gaditana†. Along the cliff edge the rare *Viola arborescens† is common, growing up through the scrub and flowering best in November. Six species of orchid, including the Woodcock orchid, *Ophrys scolopax†, grow under the pines. Further south, just before reaching Tarifa, there is a stretch of sandy ground between the main road and the sea, with stunted pines, where *Romulea clusiana† is a beautiful sight in February, and *Pancratium maritimum flowers in high summer.

From the Straits of Gibraltar thickly wooded sierras stretch inland for 50 km, to the peak of Algibe at 1092 m. They are composed of Tertiary sandstones overlying Jurassic limestone, and ericaceous plants flourish in the acid soils. The forests are mostly Cork oak, with an undergrowth of scrub formed by *Erica arborea, *E. scoparia, *E. erigena, *E. umbellata, *E. australis, *E. ciliaris, and Calluna vulgaris mixed with many of the shrubs typical of Cistus matorral, and in addition *Halimium *lasianthum*†, and *Rhododendron ponticum*† growing in gullies. The yellow-flowered leguminous shrubs *Teline linifolia*†, *T. monspessulana*, and *Cytisus villosus* are common in places. *Scilla monophyllos*† flowers in January-February; *Sedum brevifolium* grows on bare rock.

Amongst the higher rocks grow splendid specimens of *Laurus nobilis* and *Viburnum tinus*; only in these natural conditions can their real beauty be appreciated, there being small resemblance to their mutilated and grimy cousins so often seen at home. However, even more outstanding were the groups of strawberry trees,*Arbutus unedo*, laden with pendant fruit growing out of and on the edge of the escarpments. Descending on dry rocky slopes there are various other interesting plants to be seen in their season, the insectivorous *Drosophyllum lusitanicum*†, whose yellow cistaceous-looking flowers are not to be despised, an unusual blue milkwort *Polygala microphylla*, and if one looks hard in February one may find the reddish-brown bells of *Fritillaria lusitanica* growing up through the scrub on foot-high stems.[1]

As Dwight Ripley says, 'Wherever a spring occurs underneath the oaks, edging its way gingerly through wads of pallid sphagnum, will be *Sibthorpia europaea*, *Wahlenbergia hederacea*, and *Anagallis crassifolia*†.' *Arisarum proboscidium*† with its extremely long spadix grows here in two localities but nowhere else nearer than Italy.

The highest parts of these sierras are mostly covered with a dense growth of *Quercus fruticosa* and *Cistus populifolius*, the latter normally found above 300 m though it does occasionally occur lower down. *Narcissus bulbocodium* flowers in winter, usually on the ledges of large isolated outcrops of rock, an unusual habitat for this species; yellow and citron colour forms occur and so does the smaller white-flowered *N. cantabricus*†.

The low-lying ground of the Campo de Gibraltar is rapidly becoming industrialized, and some of the terrain that Wolley-Dod used to explore on foot and on horseback is now built over or wired off. One of the interesting plants that is becoming increasingly rare, as a result of these disturbances, is the green-flowered autumnal, *Narcissus viridiflorus*†, a North African species which is only found in this small corner of Spain. But spectacular plants like *Salvia bicolor*†, one of the finest of the genus in Europe, and the 3½ m tall thistle *Cirsium scabrum* may still be seen by the roadside.

The rock of Gibraltar, which is a limestone mountain, has an interesting flora with more than one North African species which grows on the Rock and nowhere else on the mainland of Europe as, for example, *Iberis gibraltarica*†, which is locally common in bushy ground. It is a robust plant with a woody stock and large showy, flat-topped clusters of pink flowers which appear in early spring. The climate is mild, though humid and even at Christmas time there is quite a sprinkling of plants in flower, such as: *Aristolochia baetica, *Clematis cirrhosa, *Lobularia maritima, and *L. libyca†, Vinca difformis, *Lavandula dentata, *L. multifida, Ruscus hypophyllum†, *Narcissus papyraceus, *Arisarum vulgare. Later, in March, the more interesting plants in flower include:

Matthiola tricuspidata	*A. fistulosus
Aeonium arboreum	*Allium roseum
Coronilla valentina	*A. triquetrum
Theligonum cynocrambe	Scilla peruviana†
*Limonium sinuatum	*Dipcadi serotinum
L. spathulatum	*Romulea clusiana†
Solanum sodomeum	*Iris sisyrinchium
*Calendula suffruticosa†	*Ophrys bombyliflora†
Asphodelus aestivus	*O. speculum†
*Gennaria diphylla†	

Late spring and early summer flowering plants include:

Glaucium flavum	*Antirrhinum majus
Saxifraga globulifera†	Linaria tristis†
*Teline linifolia†	Acanthus mollis
Lavatera arborea	Lonicera implexa
Ferula tingitana†	Campanula velutina
*Convolvulus althaeoides	Helichrysum rupestre
*Echium boissieri†	*Iris filifolia†
*Phlomis purpurea	Gladiolus communis

In June-July and August the following interesting plants can be found:

Dianthus caryophyllus†	Datura stramonium
Delphinium pentagynum†	Nicotiana glauca
D. gracile†	Verbascum sinuatum
*Daphne gnidium	Carlina corymbosa
Ecballium elaterium	Senecio cineraria
Bupleurum fruticosum†	*Pancratium maritimum

Autumn flowerers appearing on the arrival of rain in October, if not before, are: *Ranunculus bullatus, Inula viscosa, *Colchicum lusitanum, *Urginea maritima, Crocus salzmannii†, Spiranthes spiralis.

The best way to approach the Coto de Doñana and adjacent coast, particularly if coming from Portugal, is to by-pass Huelva by the recently opened bridges across the Odiel and Tinto rivers, and make for the monastery of La Rabida, where Christopher Columbus planned his voyage to the west, and where many relics and models of his ships can be seen.

Early March is the best time, for in the pine wood surrounding the monastery grow sheets of *Narcissus bulbocodium*, while across the tidal creek in front of the monastery, between the oil refinery and the Rio Tinto, on pure sand *Halimium commutatum* grows in abundance, making a lovely display of yellow. Between the bushes, the sand is carpeted with the small mauve flowers of *Malcolmia lacera†*, the white bells of *Leucojum trichophyllum†*, and the delicate pink flowers of *Ornithopus sativus*. *Halimium halimifolium* flowers a month later, and is not nearly as abundant here as further east and in addition *Corrigiola litoralis*, *Lavandula stoechas*, and a few blue lupins are also present. *Narcissus gaditanus†*, a rare species with up to eight flowers per stem, seems to grow only in a small area near the refinery. The yellow *Linaria viscosa* flourishes round the cement pipe factory, and is abundant all through the area. A month later in the tidal creeks the rare endemic *Cistanche phelypaea†* may be seen thrusting its brilliant yellow flowers through the tangle of *Salicornia* on which it is parasitic. On bare sand *Echium gaditanum†* flowers in May.

From the mouth of the Rio Tinto to the mouth of the Guadalquivir the coast consists of a long and almost straight sandy beach, backed in the eastern part by a formidable line of sand dunes, the Arenas Gordas, which rise as high as 60 m in some parts. They are partly covered with marram grass, *Ammophila arenaria*, while on the landward side there are scattered shrubs of *Juniperus oxycedrus*, *J. phoenicea*, *Corema album†* and a mixture of *Cistus* and *Genista* species.

An excellent road runs to Mazagon, and thence a rough forest road (negotiable by car) leads to the newly opened resort of Matalascañas. It passes through stunted forest of Umbrella pines which were probably planted during the eighteenth or nineteenth century, replacing the former thickets of *Halimium* and park-like sprinkling of Cork oaks. Under the pines a fairly dense scrub layer of junipers occurs with *Cistus salvifolius*, *Halimium commutatum*, and *Rosmarinus officinalis*. *Halimium halimifolium* forms extensive thickets in places, and there is some *Astragalus lusitanicus* and *Erica erigena*, with *Pistacia lentiscus* in moist hollows.

East of the El Rocio–Matalascañas road lies the famous Biological Reserve of the Coto de Doñana. This, and the Camargue in the Rhône delta, are the two most famous wildfowl areas in south-west Europe, and it has been very well described in the writings of Buck, Chapman, Yeates, and Mountfort. Formerly a private shooting estate, it was recently purchased by the World Wild Life Fund and set up as a permanent nature reserve. Permission to visit the reserve must be obtained by writing in advance to The Director, Estación Biológica de Doñana, Paraguay 1, Seville. It is usually granted only for specific research projects, to avoid undue disturbance of feeding and nesting birds. The botanist, however, can see all the interesting plants of this area outside the reserve. In general the Coto de Doñana consists mainly of heaths of *Halimium halimifolium*, with some **Erica scoparia*, **Atractylis cancellata*, **Iris xiphium*† and others, and some rather scattered oaks and pines. Much of the area of the reserve consists of *marismas*, which are usually flooded during and after the winter rains, but which quickly dry out and are left baked and cracked for most of the year. The *marismas* support a typical salt-marsh vegetation, consisting of *Salicornia europaea*, *Arthrocnemum fruticosum*, *A. glaucum*, and a little **Limoniastrum monopetalum*. Dotted over the whole extent of the *marismas* are little eminences which, when the area is flooded, form islands, called locally *vetas*. They are in general of small area, surrounded by a band of pure clay covered with *Suaeda*; and on the crest of each tiny island is a dense growth of grasses, camomile, and such thistles as *Silybum marianum* and *Cirsium arvense*, contrasting with the surrounding extensive salt marsh. The traveller can also enter this area by the Ruta de las Marismas which runs southwards through Almonte and El Rocio where it crosses a freshwater stream, the Madre de las Marismas. Here typical swamp and marshland plants make a welcome relief from the heathlands, with *Phragmites*, *Typha latifolia*, *Iris pseudacorus*, *Scirpus lacustris*, and **Cotula coronopifolia*. In fresh water lagoons and pools in the area can be found the unusual floating yellow composite with hollow stems and leaves, *Scorzonera fistulosa*† which grows in company with water crowfoots. Along the side of the road as one travels southwards there are fine groups of **Leucojum trichophyllum*† and brilliant blue patches of **Anagallis monelli*† and **Lithospermum diffusum*, while in damper patches in the *Halimium-Erica* heath grows a magnificent thrift, *Armeria gaditana*†, with very large pink globular flower heads 8 cm across, borne on stems up to 1 m high. Other interesting species are: **Reseda media*, **Lygos monosperma*†, *Lathyrus palustris* subsp. *nudicaulis*, *Armeria baetica*†, and **Pterocephalus intermedius*.

4. Western Costa del Sol (Sotogrande-Málaga)

So rich in flowering plants is the Costa del Sol and the coastal mountains that this area has been divided into two contrasting regions: a western region where a closer proximity to the Atlantic results in heavier rainfall, and an eastern region including the Cabo de Gata which is the driest area of all Spain.

Gibraltar is often shrouded in the mists of the Atlantic and receives 560 mm of rain per year, while Málaga has an average of 400 mm. There is a marked difference between the rich and extensive Cork oak woods covering the Sierras west of Estepona and the bare hills about Torremolinos. The rains usually start towards the end of October, with the highest monthly falls in January and February, and they virtually cease in May, followed by a drought of five to six months. A few plants come into flower in the autumn, but the majority start their growing period after the arrival of the first rains, and flower in the spring and early summer. Visitors to the Costa del Sol at the height of summer, when everything is brown and parched, must find it difficult to appreciate the true richness of the flora. Limited space makes it possible to indicate only some of the more showy and interesting plants of this area. Though the best time is undoubtedly between March and May when the great majority of species are flowering, at most other times, except in the height of summer, plants are in flower. There are a number of late summer and early autumn flowerers, such as: *Kickxia spuria, Heliotropium europaeum, *Solanum bonariense, Inula viscosa, *Smilax aspera*, and in the sierras *Arbutus unedo, *Trachelium caeruleum*. Autumn-flowering bulbs include *Scilla autumnalis, *Leucojum autumnale, *Colchicum lusitanum†, *Urginea maritima, *Narcissus serotinus*, all of which appear in open patches amongst *cistus* scrub. The tiny, rather rare, yellow-flowered lily *Tapeinanthus humilis†* grows in waste grassy places. *Ranunculus bullatus* is often conspicuous locally in November and so are the large violet flowers of the Autumn Mandrake, *Mandragora autumnalis*. *Bidens aurea*, an American native, is now extensively naturalized up the Guadiaro valley and around Sotogrande and it also is in full flower in November. *Lonicera implexa* may be seen flowering at Christmas time and soon afterwards *Fedia cornucopiae, *Chamaemelum fuscatum, Calendula arvensis, *Asteriscus maritimus*. Other species continue all the winter including *Aristolochia baetica†, *Lobularia maritima, *Salvia verbenaca, *Calamintha nepeta, *Narcissus papyraceus*, while the ubiquitous weed, the Bermuda buttercup, *Oxalis pescaprae* forms sheets of welcome colour by the roadsides from early December to April.

Spring arrives any time between February and late March, depending on the coastal weather. The most distinctive herbaceous species are:

Hedysarum coronarium	*Cerinthe major*
Erodium cicutarium	*Echium lycopsis*
E. guttatum (local)	*Teucrium fruticans*
Malva hispanica	*Linaria hirta*
Lavatera cretica	Scrophularia sambucifolia†
Convolvulus althaeoides	*Allium triquetrum*
C. tricolor	*Scilla ramburei†
Borago officinalis	*Iris sisyrinchium*
Anchusa azurea	Asphodelus morisianus

WESTERN COSTA DEL SOL (SPRING)

1. *Allium triquetrum* 1610 2. *Borago officinalis* 1054 3. *Teucrium fruticans* 1101
4. *Erodium guttatum* 652d 5. *Iris sisyrinchium* 1684 6. *Ophrys bombyliflora* 1879
7. *Scilla ramburei* 1633a 8. *Linaria hirta* 1201f

9. *Anchusa azurea* 1056   10. *Malva hispanica* 736a   11. *Asphodelus morisianus* 1591b
12. *Convolvulus tricolor* 1035   13. *Serapias lingua* 1902   14. *Leucojum trichophyllum*
(1662)   15. *Saxifraga biternata* 411e   16. *Erodium cicutarium* 653

WESTERN COSTA DEL SOL (AUTUMN AND WINTER)

1. *Solanum bonariense* 1183a 2. *Bidens aurea* 1406a 3. *Urginea maritima* 1630
4. *Aristolochia baetica* 77a 5. *Orchis saccata* 1899a 6. *Calamintha nepeta* 1156
7. *Asteriscus maritimus* 1398 8. *Ranunculus bullatus* 234a

9. *Arbutus unedo* 924   **10.** *Smilax aspera* 1659   **11.** *Chamaemelum fuscatum* 1413a
12. *Salvia verbenaca* (1148)   **13.** *Leucojum autumnale* 1662a   **14.** *Romulea bulbocodium*
1681   **15.** *Oxalis pes-caprae* 639   **16.** *Narcissus serotinus* 1672   **17.** *Narcissus papyraceus* (1667)

After April the tall and vigorous thistles and umbellifers tend to dominate the roadsides, notably *Silybum marianum, Galactites tomentosa, Notobasis syriaca, *Cynara humilis,* and the robust *Scolymus hispanicus* which is a fine sight when it covers whole fields. The more striking umbels are yellow-flowered and include *Cachrys sicula, Thapsia garganica* and *T. villosa,* and *Elaeoselinum foetidum.*

Behind the littoral the ground rises steeply to a succession of tall sierras, which include behind Estepona one of the largest outcrops in the world of the intrusive volcanic rock called Peridotite. This produces a fine-textured soil, with a characteristic red colour due to the abundance of iron oxide; it is known locally as *tierra colorada*. These red sierras of Peridotite contrast markedly with the limestone sierras such as the Sierra Blanca behind Marbella to the east and the Serranía de Ronda to the north.

The red sierras have a distinctive vegetation and were once covered with forests of *Pinus pinaster*, but there has been much destruction by fire in recent years. To salvage the remaining timber a forest extraction road has been built up to the Puerto de Peñas Blancas, to 900 m, thus making the 1450 m summit of Reales more accessible.

The highest part of the Sierra Bermeja is mostly covered with open pine forest with an undergrowth of *Alyssum serpyllifolium, *Cistus populifolius, *Halimium atriplicifolium,* and *Ulex,* with some *Erinacea anthyllis* and *Armeria colorata* on the summit rocks. The tiny slender white-flowered annual *Arenaria capillipes*† grows in open patches among the scrub; it is endemic in these mountains. The north face has stands of *Abies pinsapo*† and it was here that Boissier in 1837 first discovered this unique Spanish fir. Under the trees there is dense shade with ground nearly bare apart from a little *Daphne laureola, Berberis hispanica,* and the herbs *Mucizonia hispida* and *Saxifraga gemmulosa*†.

On the slopes above the Puerto de Peñas Blancas, *Crocus nevadensis*† flowers in January. In May in addition to *Orchis pallens*† and *O. mascula* are other interesting species such as *Chaenorhinum rubrifolium,* *Centaurea haenseleri*† with attractive yellow flowers, and *Anthericum

---

5 **Sierra Nevada** The 'hedgehog' zone at 1700–2000 m on the Sierra Nevada with *Vella spinosa, Erinacea anthyllis, Astragalus granatensis, Astragalus sempervirens*. This zone of very spiny cushion-like shrubs is characteristic of the highest mountains of Southern Europe and North Africa.

6 **Serranía de Ronda** The limestone mountains of the Sierra del Pinar above the ancient town of Grazalema are the home of some interesting endemic plants, such as *Papaver rupifragum, Centaurea clementei,* and *Biscutella frutescens*. Cork Oak woods rich in *Cistus, Halimium, Cytisus* and *Erica* species occur in the foothills. *(overleaf)*

baeticum†. Below the Puerto *Ulex parviflorus and *Genista hirsuta turn the hillsides yellow at this time; and growing amongst them there are fine patches of *Arenaria montana, Alyssum minus, *Linum narbonense, *Coris monspeliensis, *Tulipa australis and *Aphyllanthes monspeliensis. One of the most interesting species is Scorzonera baetica† with solitary large pale yellow flower-heads borne on tall stems; it is an endemic plant of Málaga province. Staehelina baetica another endemic abounds at 400–800 m. By streams can be found a few bushes of the rare *Erica terminalis† which does not flower until July, also Coriaria myrtifolia, Antirrhinum majus, Saxifraga gemmulosa, and Linaria tristis. Lower down on the mountain *Erica umbellata becomes abundant under the pines, and amongst it can be found the extraordinary insectivorous plant *Drosophyllum lusitanicum†, here at the easternmost limit of its range. At about 200 m there is an outcrop of marble, marked by *Cistus albidus, *Coronilla juncea, and Anthyllis polycephala†.

The Sierra de Mijas is another mountain of marble and dolomite, with a rich flora on the north-east slopes, particularly between Churriana and Alhaurin de la Torre where the road skirts the foot of the hill. Here the pretty little pink woodruff Asperula hirsuta† is everywhere, and at one place there is the finest display of *Iberis linifolia† we have ever seen, with large cushions of purple flowers; other striking plants include *Coris monspeliensis, *Cleonia lusitanica, the deep yellow *Centaurea prolongi†, the lemon-yellow Serratula flavescens†, and the tall bigfruited *Asphodelus ramosus. The spidery spreading crucifer Crambe filiformis†, with very long slender branches and numerous small white flowers, is as common here as anywhere. On the slopes above Alhaurin el Grande there are fine displays of *Iris filifolia† up the fire breaks, with *Ranunculus gramineus; towards the top of the ridge *Matthiola fruticulosa and Saxifraga globulifera† are fairly plentiful.

The old spa of Carratraca, north-west of Málaga, is another good centre for plant hunting. The Sierra de Alcaparain opposite the village is of Jurassic limestone with a very different flora from the Sierra Bermeja. In the Aleppo pine woods, Anthony Huxley has described his

---

7 **Sierra Nevada** The summit of Mulhacen at 3400 m is formed of barren-looking schists, but in the alpine zone alone over 40 endemic species are found. Snow still lies in August, but there are no true glaciers.

8 **Sierra de Cazorla** Miles of rough roads through the pine forests of Cazorla help to make this area one of the most rewarding plant-hunting terrains in Spain. It has a rich flora and is a refuge for ancient species which have survived the last ice-age: notably a shrubby violet and alyssum, and a butterwort.

discovery of 'the most incredible collection of orchids I have ever seen', which included amongst others the Moroccan species *Ophrys atlantica†. Plants of interest at the lower altitudes include *Cytinus ruber† on *Cistus albidus*, *Saxifraga globulifera†*, *Echium albicans†, the most beautiful of all buglosses in Spain, *Thymus granatensis†, *Chamaepeuce hispanica*, *Jurinea pinnata*, and *Serratula pinnatifida†. In shady places there is the big-leaved umbellifer *Magydaris panacifolia* with cream-coloured flowers, and the rare *Salvia candelabrum†. At the highest point of the road, between Carratraca and Alora, there is a fine display of *Iberis linifolia* varying from pale to deep purple, with *Centaurea carratracensis†*, *Serratula boetica†*, some *Omphalodes linifolia*, and abundant *Cerastium boissieri*.

Another mountain worth visiting from Málaga is El Torcal de Antequera (about 1300 m high), where the limestone plateau has been deeply eroded into fantastic shapes resembling a ruined city with narrow streets, caves, and tunnels. Similar *torcales* are found on the Jurassic limestone of the Sierra de Endrinal near Grazalema. In May *Iris subbiflora†* which we have seen nowhere else in Spain, is in flower with peonies on the plateau, and the endemic *Saxifraga biternata† is abundant on the rocks of the *torcal*. *Linaria anticaria†* was originally described from this area, though it is not confined to it, and the bright yellow-flowered *Viola demetria† grows here and there among the rocks. *Rhamnus alaternus* grows freely all through the *torcal* and one sees again some of the typical Ronda species, like *Endymion hispanicus*, *Arabis verna*, *Geranium lucidum*, and *Omphalodes linifolia*.

The whole area is rich in orchid species and at least 30 species have been recorded including the following:

*Ophrys lutea	*Serapias lingua †
*O. bombyliflora†	*S. cordigera †
O. apifera	S. parviflora
*O. fusca subsp. *atlantica †	*Aceras anthropophorum
O. speculum	Neotinea intacta
O. scolopax†	Himantoglossum
*O. tenthredinifera †	longibracteatum †
*Orchis coriophora†	*H. hircinum†
*O. morio	Gennaria diphylla†
O. palustris	*Epipactis atrorubens
O. mascula	E. helleborine†
O. papilionacea†	Cephalanthera longifolia
O. tridentata	*Limodorum abortivum
O. italica	Spiranthes spiralis
*O. saccata†	S. aestivalis
O. laxiflora	

5. Serranía de Ronda

The little town of Ronda lies at 723 m in the mountains and it is visited by many people for its Moorish associations; but the journey is equally worth while for its wonderful mountain views, for Ronda is at the centre of some thousand square kilometres of rugged mountains, rising in places to nearly 2000 m, which can be seen on every hand from the ancient town. The mountains are largely of hard Triassic and Jurassic limestones, overlain in places by Tertiary rocks, and the whole area is botanically one of the richest and most interesting in Spain. Five roads converge on Ronda, but probably the most usual approach is from San Pedro de Alcántara to the south.

The road soon begins a spectacular climb for over 1000 m up through volcanic rocks of Peridotite, at first through woods of Cork oak where the handsome white *Linum suffruticosum* is conspicuous. Higher up it climbs through forests of *Pinus pinaster*, where several colonies of the lovely white hoop-petticoat daffodil *Narcissus cantabricus*† flower in March on the face of steep rock cuttings. Here, too, is *Saxifraga boissieri*†, an endemic species with small white flowers with a yellowish-green eye in the centre. In May the handsome 1½ m tall Mullein, *Verbascum rotundifolium* subsp. *haenseleri*, is in flower as well as the lovely rust-red foxglove *Digitalis obscura* subsp. *laciniata*†, while the undergrowth is dominated by *Halimium atriplicifolium*†, finest of this genus, with greyish leaves and large pale yellow flowers, and *Cistus populifolius*. The slopes to the west at the top of the pass are covered with *Ulex parviflorus*, a gorse that starts flowering as early as Christmas time. *Gagea* species, *Alyssum serpyllifolium*, and *Saxifraga dichotoma* may be found here. About five kilometres beyond the pass, the Puerto del Alijar, there is an abrupt change from the reddish peridotite rocks to the barren and forbidding greyish-white dolomite, and here the pine forest ends equally abruptly. Here and there will be found colonies of the small yellow daffodil *Narcissus requienii*, flowering in April.

Near km 13 a private motor road turns off to the right, and one can enjoy a magnificent walk by following this road on foot. For the first hour it runs more or less level, through cork woods, where the peonies *P. broteroi* and *P. coriacea* bloom in May. On the slopes on the right-hand side can be seen a clump of the endemic Spanish silver fir *Abies pinsapo*†, one of the few remnants of the extensive forests of this species that once covered much of the Serranía. Turning to the left at a farm, the road continues for another hour of zig-zag climbing up scrub-covered slopes, past a notice marking the boundary of the Coto Nacional (a protection for the Spanish ibex, rather than the flora of these mountains), to end at another farm. On this last stretch of road grow *Iberis*

SERRANÍA DE RONDA

1. *Moricandia moricandioides* 356a **2.** *Linaria platycalyx* 1205a **3.** *Cynoglossum cheirifolium* (1049) **4.** *Ornithogalum reverchonii* 1642b **5.** *Omphalodes brassicifolia* 1048c **6.** *Iris planifolia* 1684a **7.** *Viola demetria* 784a **8.** *Saxifraga granulata* 404 **9.** *Narcissus requienii* 1668

10. *Hesperis laciniata* (296)   **11.** *Linum suffruticosum* (664)   **12.** *Iberis crenata* 348a
13. *Endymion hispanicum* (1638)   **14.** *Ophrys lutea* 1874   **15.** *Prolongoa pectinata*
1428e   **16.** *Narcissus cantabricus* (1666)   **17.** *Arabis verna* 319   **18.** *Digitalis*
obscura (1233)

saxatilis and *I. pruitii, Aethionema saxatile, and masses of Helleborus foetidus, and *Helianthemum apenninum. A robust thistle with strong yellow spines and leaves with white-felted undersides, *Chamaepeuce hispanica, is conspicuous throughout these mountains. Above the last farm is another clump of Abies pinsapo†, and climbing through this one can reach the sub-alpine zone at about 1500 m. Narcissus hispanicus abounds in places below 1400 m.

Blue and purple hummocks of the vegetable hedgehog *Erinacea anthyllis, Genista triacanthos, and *Lavandula lanata† are found on the edge of water-courses, and in a natural rock garden at the summit there were delightful combinations of *Erodium daucoides† with monkey face flowers—white buttonheads of Armeria villosa, *Arenaria montana, pink stonecrops and that pretty pest *Cerastium boissieri.[2]

Clinging close to the rocks are windswept examples of *Prunus prostrata† and the tiny white-flowered Myosotis minutiflora. From here it is easy to reach the summit of Torrecilla, 1919 m, the highest peak in these mountains; but botanically it is disappointing and there is little to see on the top apart from juniper scrub and a few yew trees.

Continuing on the main road to Ronda one may find a number of orchids, including:

*Ophrys fusca	*O. morio
*O. lutea	O. papilionacea
*O. scolopax†	O. tridentata subsp. lactea†
*O. speculum†	*Aceras anthropophorum
*O. tenthredinifera†	Neotinea intacta
Orchis italica	*Himantoglossum hircinum†
O. mascula	H. longibracteatum†
*O. pallens†	

But they are not all to be found in one locality; the most abundant are *Ophrys fusca and *O. lutea, which are typically roadside plants.

At Christmas time one of the sights of Ronda is the beautiful violet-flowered *Iris planifolia†, which grows abundantly around the town, and on both sides of the road just above Atajate on the road to Jimena. It is the only member of the section Juno which grows in Europe and can be distinguished by the three small standards which spread outwards at the base of the flower. The fine big buttercup growing round Ronda is *Ranunculus rupestris†; and on waste land near the town grows the handsome mauve-flowered *Moricandia moricandioides. *Iberis crenata† rapidly colonizes disturbed sandy soil and in a little pine wood north of the town *Linaria amethystea† and *Prolongoa pectinata carpet the floor.

The road to the Cueva de la Pileta, with its neolithic cave paintings, passes through some remarkable rock formations just before Montejaque. The interesting endemic *Ornithogalum reverchonii†, like a white bluebell, grows out of crevices on the cliffs. There are large tufts of *Silene pseudovelutina*; and the minute rosette-forming *Arabis verna, and *Linaria platycalyx† grow on these rocks. *Cynoglossum cheirifolium, the loveliest of the Hound's-tongues, with its soft silvery white leaves and small deep red flowers, may be found near the cave.

Grazalema, once a centre of the wool trade, is a famous botanical locality of the mountains. Three of the more spectacular species are described by Dwight Ripley:[3]

*Biscutella frutescens† . . . How superb it is in the cliffs of Grazalema, with its showers of gold and thick rosettes of scalloped velvet, sometimes as many as fifty of them forming plants a yard in diameter! . .

Papaver rupifragum† . . . its poppies of delicate brick prefering to conceal themselves in the shadier and damper corners, where its narrow, distinctive leaves may rest comfortably on a bed of moss. . .

*Centaurea clementei†, a vast coarse thing, chooses the hottest exposure it can find . . . huge white leaves and gold sweet-sultans. . .

The poppy grows nowhere else, but the *Centaurea* is found also in the Sierra de Yunquera. In April, the rocks above the village are covered with *Saxifraga globulifera, with some *S. boissieri† and *S. haenseleri†. Growing out of crevices in the rocks is the little yellow *Narcissus rupicola* subsp. *pedunculatus* (there is still some disagreement about its correct name), while another species *N. jonquilla* is found on heavy clay in wet meadows beside the road to Ubrique. *Ornithogalum reverchonii† is abundant on the rocks above Grazalema and flowers in May. Other plants of the rock fissures are the handsome cut-leaved Dame's violet *Hesperis laciniata, *Omphalodes brassicifolia†, and the Spanish bluebell *Endymion hispanicus. The red-berried mistletoe *Viscum cruciatum† is a North African species which occurs rarely in south-west Spain; its fruits become become conspicuous in December. Its favourite host is the wild olive but it is sometimes found on cultivated olive trees which have been neglected and it occurs less commonly on ivy, hawthorn, and other shrubby species of the *matorral*. In the Cork oak woods on the way to Grazalema the tiny *Parentucellia latifolia* grows in patches by the roadside.

A magnificent mountain excursion is to climb El Cristobal which takes about two hours from Grazalema, and then to traverse along the ridge to El Pinar (1654 m), the second highest summit in these ranges, so conspicuous from Ronda but rarely visited. Allow six to eight hours out and back. On the north-east slopes there is a large block of *pinsapo* forest, a few trees even reaching the crest of the ridge, and here a few trees of *Sorbus aria* and *Acer granatense*† grow. In the more open parts of this forest the following shrubs are found:

Berberis hispanica	*Ulex parviflorus* subsp.
Ptilotrichum spinosum	*funkii*
Crataegus species	*Daphne laureola*
Erinacea anthyllis	*Hedera helix*
Viburnum tinus	*Bupleurum spinosum*
Lonicera etrusca	

For some reason *Adenocarpus decorticans*† does not grow on this mountain but on the Sierra de Endrinal opposite.

Senecio minutus and the small bright yellow *Viola demetria*† are among the plants that grow along the summit ridge, which is a favourite haunt of the Red-billed Chough.

Some of the more interesting herbaceous plants to be found in the Serranía de Grazalema are:

Arenaria aggregata	*Ionopsidium prolongoi*†
Silene cretica	*Reseda undata*†
Dianthus hispanicus	*Ononis reuteri*†
Rupicapnos africana†	*O. saxicola*†
Erysimum grandiflorum	*Viola parvula*
Draba hispanica†	*Senecio petraeus*†

The Sierra Blanca is another summit of this same limestone massif which is much nearer the coast and towers above Marbella. Access is via the Marbella-Ojén road, where *Putoria calabrica* is common on the rocks. Shortly after crossing the Puerto de Ojén (580 m) there is a turning off to the left, leading up to the Refugio. The three most striking plants of the area, which grow near this turning are the lovely silvery-leaved *Echium albicans*†, the tall wand-like *Reseda suffruticosa*†, and the tall *Linaria clementei*†.

Another oddity is *Biarum carratracense*† which contrives to wedge its ovoid tubers into impossible positions in the crannies; the leaves are short and spathu-late, and the ephemeral pink spadix appears only after the foliage has died down, this then vanishes and in due course a group of red berries slowly emerges from the soil.[4]

This plant flowers in the autumn. Up near the Refugio there is a group of *Ononis speciosa*† and the charming little *Senecio minutus* is common, with purple undersides to the yellow petals. In rock crevices, the dark green leaves with a pale central band of *Lapiedra martinezii*† can be found; late in the summer it produces a cluster of white flowers. This is the southern end of the Coto Nacional, and there is a good chance of seeing ibex within walking distance of the Refugio.

6. Sierra Tejeda and Sierra de Almijara

These mountains lie roughly mid-way between Málaga and the Sierra Nevada, and dominate the coast road between Torre del Mar and Nerja. Viewed from the Sierra Nevada they appear as a long range of mountains to the west, rising to 2134 m, snow-capped in parts until early May. The name Tejeda is said to be derived from *Tajo*, the yew tree, which once clothed the higher parts of the range, but according to Willkomm these had been reduced to a few gnarled remnants as long ago as the 1840s. The road from Torre del Mar to Granada, via Ventas de Zafarraya, provides access to the Sierra Tejeda, and onwards via Jatar, to the western part of the Sierra de Almijara. The latter can also be reached from the coast road by driving up from Mezquitilla to the Canillas de Albaida, and from here a mule track leads across the range via the Puerto Blanquilla, 1463 m, to Jatar.

It is a fine walk through pine woods and steep limestone country with magnificent views of the sierras. May is the best time for the lower slopes, and June and July for the summits. Near the start of the climb the distinctive tufted silvery-grey *Anthyllis* species, with pale yellow flowers, is abundant, with *Cerastium boissieri*. But the majority of interesting plants will be found on the south-facing slopes at 1100–1200 m. Here the yellow-flowered *Chaenorhinum rubrifolium* subsp. *raveyi*† is common, with *Silene psammitis*†, *Biscutella frutescens*†, *Coris monspeliensis*, *Alkanna tinctoria*, *Echium albicans*†, *Lavandula lanata*†, *Verbascum rotundifolium*†, *Linaria saturejoides*†, and the lovely red-purple *Linaria amoi*†. Higher up *Alyssum montanum*, *Thymelaea tartonraira* forming bushes to ½ m, and *Rhamnus myrtifolius*†, much dwarfed and growing prostrate on the rocks, may be seen. The rare *Teucrium fragile*† forms a dense low mat in rock crevices; it has small purple flowers and small crowded serrate leaves and is so fragile that it is difficult to gather intact. The track down from the pass to Jatar runs at first through pines with a fine display of *Adenocarpus decorticans*† in the undergrowth; lower down the slopes are covered with rosemary and *Cistus clusii* scrub.

At the Ventas de Zafarraya the limestone cliffs have a flora similar to that of the Ronda mountains, with *Silene pseudovelutina*, *Hesperis laciniata*, *Biscutella frutescens*†, and so on. The limestone gorge on the road to the Balneario at Alhama de Granada is worth a visit; here *Antirrhinum barrelieri*† and *Linaria anticaria*† are abundant.

The road from Almuñecar to Granada climbs to 1300 m over the eastern part of the Sierra de Almijara, through some spectacular scenery. *Dianthus anticarius†* flowers here in May, and the unusual *Thymus longiflorus†* in early June.

The whole area was exhaustively explored by Dr Laza, a well-known chemist from Málaga, in the 1930s, and his collections are in the botany department of the University of Granada.

7. Sierra Nevada

The Sierra Nevada is the highest mountain range in south-west Europe; its highest peak, Mulhacén, 3478 m (11,420 ft), just beats Pico de Aneto, 3404 m, the highest in the Pyrenees. The sierra is formed of a central dome of rocks, composed largely of mica schist, 100 km or so in length and with rounded, relatively smooth contours at higher altitudes. Its flanks on the other hand, are of younger Triassic limestones and sandstones, and have become deeply dissected by rivers into steep gorges.

The close proximity of the Mediterranean and the great height of the Sierra Nevada have combined to create quite exceptionally wide variations in climate at different altitudes and aspects. Snow lies for as long as nine months in the year above 2500 m, and hollows in the highest peaks may remain snow-filled throughout the year; but there are no true glaciers. Yet the action of earlier glaciers—there was probably a small isolated ice-cap over the Sierra Nevada during the last ice age—has left characteristic, if rather weak, alpine features above about 2000 m. There are glacial moraines—one is about 2½ km in length—and there are small corries and glacial lakes.

The northern valleys are petrified by snow and ice; the southern ones roasted by the heat of the sun. In the farmhouses around Granada people crouch beside an open fire, and keep the doors and windows well shut; in those of the Alpujarra, they sleep in the moonlight, on piles of straw, in the open air. The farmers of Guejar or Monacil store away acorns and cherries; those of the Contraviesa put by almonds and oranges. The former keep potatoes under the snow; the latter dry figs in the sun. In the south there are wide rolling hills, light blue in colour; in the north menacing peaks rise up, shadowed and black.[5]

Such a climatic range must inevitably result in a rich diversity of types of vegetation and the following zones can be distinguished:

The Mediterranean zone may reach 1200 m, and has a characteristic array of shrubs such as *Lygos sphaerocarpa, Cytisus* species, *Genista umbellata, Helichrysum stoechas*, and many herbaceous plants, in particular *Anthyllis vulneraria* subsp. *argyrophylla*†, a densely silvery-haired plant with red or purple flowers which is restricted to south Spain. Even the olive can be grown to altitudes of 1300 m and the orange to over 1100 m. Towards the upper limit spiny thickets of deciduous shrubs occur such as *Rosa micrantha*, *Crataegus monogyna* subsp. *brevispina*, *Berberis hispanica*, *Ulex parviflorus* subsp. *funkii* and *Rubus* species.

The Mediterranean-montane zone Altitudes of 1200–1700 m once supported natural oak woods which have long since disappeared. But despite this, some plants characteristic of the woodlands still persist, such as *Helleborus foetidus*, *Paeonia coriacea*†, *Geum heterocarpum*, *Digitalis obscura*†, as well as such members of the Mint family as *Salvia lavandulifolia*, *Phlomis crinita*†, *Lavandula lanata*†, and *Stachys lusitanica*.

The 'vegetable hedgehog' zone ranges from 1700–2000 m, covered by an assortment of dense, low, dome-shaped shrubs widely scattered over the stony ground and beset with tough needle-sharp spines. An unlikely collection of families has contributed to this distinctive type of vegetation, which is characteristic of similar mountain zones in North Africa and Mediterranean Europe. Thus there is *Vella spinosa*†, a spiny member of the *Cruciferae*, a family which only very occasionally takes on this extreme xeromorphic form, while *Bupleurum spinosum*† is a member of the *Umbelliferae*, and the *Leguminosae* is represented by *Erinacea anthyllis*, *Astragalus sempervirens* subsp. *nevadensis*† and *A. granatensis* Lam.†.

The pre-alpine zone Between 2300 and 2600 m the spiny shrubs are replaced by the dwarf Junipers, *J. communis* subsp. *nana* and *J. sabina* subsp. *humilis*.

The alpine zone From 2600 to 3500 m herbaceous perennials predominate, plants that can tolerate many months under snow. The combination of high altitudes and the extreme geographical isolation of the Sierra Nevada has naturally resulted in a unique assemblage of something like 200 species of plants with the most varied affinities. About 70 species are common to the Sierra Nevada and the Alps, including *Ranunculus glacialis*, *Saxifraga oppositifolia*, *Draba aizoides*, *Gentiana verna*, *Vitaliana primuliflora*, and *Erigeron alpinus*. Another 20 are found in the Atlas Mountains of Morocco as well as the Sierra Nevada, and a further dozen occur in other mountains of southern Spain in addition to the Sierra Nevada. However, the greatest interest is afforded by the 40 or so species that are truly endemic to the Sierra Nevada and occur on

SIERRA NEVADA

1. *Narcissus nevadensis* 1665c   **2.** *Geum heterocarpum* 444b   **3.** *Erinacea anthyllis* (515)   **4.** *Andryala agardhii* 1533c   **5.** *Helianthemum apenninum* 803   **6.** *Prunus prostata* 476b   **7.** *Saxifraga oppositifolia* 398   **8.** *Crocus nevadensis* 1678a
9. *Chaenorhinum macropodum* 1211d

10. *Polygala boissieri* 699a **11.** *Vella spinosa* 365a **12.** *Adenocarpus decorticans* 518
13. *Ptilotrichum spinosum* 325b **14.** *Lavandula lanata* 1111a **15.** *Silene boryi* 170d
16. *Ranunculus acetosellifolius* 248b **17.** *Colchicum triphyllum* 1588b

this massif only. The commonest of the endemic species are probably: *Ranunculus acetosellifolius†, *Saxifraga nevadensis†, Meum nevadense†, Eryngium glaciale†, and *Linaria glacialis†.

The northern slopes, particularly in the upper regions, are surprisingly gentle and the summit of Veleta, at 3392 m, can be reached by car by a road from Granada, in July and August when clear of snow. Hotels at 2500 m, ski-lifts, and a projected cable car are rapidly opening up the area to sportsmen and tourists, during both summer and winter seasons. By contrast, the south side of the sierra, the region known as the Alpujarras (so well portrayed by Gerald Brenan in South from Granada) is still remote and wild. The roads are very rough, and small terraces are cultivated on the steep hillsides, irrigated by the ample water which flows down from the melting snow in spring, and led by artificial channels to the cultivations in summer. Sweet chestnuts, mulberries, and walnuts are grown at upper levels, and lower down, grapes and olives. But the hot dry valleys are barren wastelands, eroded into steep cliffs and gullies, devoid of soil, and capable of growing nothing of use to man, except perhaps the Esparto grasses. Only along the narrow, almost waterless river valleys is cultivation possible.

The flowering season on the Sierra begins in April, and with increasing altitude successive zones of vegetation come into flower; the alpine zone is not at its best until early August. Let us imagine a series of trips out from Granada up the Veleta road at periodic intervals. In late March or early April, on the slopes at the 1000 m level, appear *Saxifraga erioblasta†, Gagea species, Narcissus concolor† in plenty, and Orchis mascula var. olbiensis in a wide range of colours, from white to deep pink. Higher up, near the melting snow, *Crocus nevadensis† appears in thousands, with colours ranging from lilac veined with white, to clear blue; and often associated with it are a few plants of *Colchicum triphyllum† buried up to the neck in wet earth, with delicate shell-pink petals surrounding chocolate-coloured anthers. *Narcissus nevadensis† grows only in boggy patches trickling with water, up to 1900 m; it is very similar in appearance to our own native daffodil, but the flowers are usually in pairs; it is nearly over by mid-May.

From May to June there is much more to be seen. Many of the spiny hemispherical bushes of the 'vegetable hedgehog' zone are flowering. The Blue Gorse, *Erinacea anthyllis, the yellow Echinospartum boissieri, the pale yellow *Vella spinosa†, the purple *Astragalus sempervirens subsp. nevadensis† and the Barberry, Berberis vulgaris are also common. Later flowering is Bupleurum spinosum†. Growing amongst the bushes may be seen *Paeonia broteroi, Saxifraga granulata, *Polygala boissieri, *Fritillaria hispanica†, and much *Tulipa australis.

Prunus prostrata† grows here and there, flattened against the rocks, and is covered in small pink flowers; *Draba hispanica*† grows in rock crevices. Common roadside plants are the white-flowered *Helianthemum apenninum*, the tufted *Chaenorhinum macropodum*† and the grey-felted *Andryala agardhii*. Round the Hotel Sierra Nevada, the white-flowered *Ranunculus acetosellifolius*†, with mauve sepals and curious halberd-shaped leaves with undulate margins, carpets acres of ground in mid-May. *Gentiana brachyphylla*, rich blue with a white eye, grows in moist short grass, while in wet flushes the oxlip, *Primula elatior* is fairly common, usually with *Myosotis alpestris* and the small mauve bog violet, *V. palustris*.

Other plants that grow above 2000 m include *Dianthus* species in tight bluish cushions with masses of rather small, deep pink blooms. *Ptilotrichum spinosum*, another vegetable hedgehog with off-white flowers and covered with the dead persistent remains of last year's inflorescences, is extremely abundant on the rocks. *Arenaria nevadensis*† forms large cushions 30 cm across amongst prostrate junipers; while *Androsace argentea* has large compact silvery cushions studded with pearl-like, round-petalled flowers. The stemless *Gentiana alpina*† grows in moist short grass.

In the screes above the Parador the yellow-flowered daisy, *Tanacetum radicans*, with a tufted creeping habit, grows in association with *Arenaria armerina*, *Plantago maritima*, *Jasione amethystina*, and *Artemisia granatensis*†; *Sempervivum nevadense*† grows on rocks.

In July above 2500 m Dwight Ripley describes 'massive screes of slate, stretching endlessly, seem dark and fobidding even in the sun. Here and there long stripes of unmelted snow relieve the dismal scene with a fugitive brilliance, occasionally such zebraed slopes drop headlong into the waters of a lake.' But despite its barren appearance there are many interesting plants to be found nestling in among the slates, notably the pink-flowered *Ptilotrichum purpureum*† which forms quite prostrate growths, the pale yellow *Sideritis glacialis*†, a deep rosy-purple form of *Anthyllis vulneraria* subsp. *atlantis*†, the pink-flowered cushions of *Dianthus subacaulis* subsp. *brachyanthus*, and *Senecio boissieri*†. Round the summit of Veleta the following are common: *Viola crassiuscula*† with large violet to white flowers borne above tiny rounded leaves; *Plantago nivalis*† with a silvery rosette of leaves and black flower clusters; *Erigeron frigidus*†, *Jasione amethystina*, *Ranunculus demissus*, *Linaria glacialis*†, and *L. aeruginea* var. *nevadensis*, *Lotus glareosus*†, *Arenaria tetraquetra*, and the yellow-flowered *Biscutella glacialis*†. *Saxifraga globulifera*† and *Silene boryi*† grow on the cliffs below the summit.

We camped at 9,000 feet by the Laguna de Yeguas, a small tarn in a glaciated valley, which was reached by a steep winding track, quite safe if taken slowly. It is a delightful spot, a green oasis surrounded by the gaunt and barren slopes of the Valeta, which sun and frost have turned into a wilderness of shattered rocks and scree. There were still many snowdrifts, and small streams gurgled cheerfully from them down into the lake, their banks sprouting azure trumpets of *Gentiana alpina†, while pale blue mats of *Veronica repens* grew half-submerged in the running water. Near our tent vivid tufts of *Gentiana brachyphylla* were interspersed with the grey stars of *Plantago nivalis† and pink *Armeria australis†*. In stonier ground there were *Chrysanthemum radicans*, *Ranunculus acetosellifolius†*, *Eryngium glaciale†*, and everywhere the tight hard mats of *Arenaria tetraquetra† studded with little white flowers. On some damp ledges was a deep blue harebell, *Campanula herminii*.[6]

The comparatively small limestone outcrop of Cerro Trevenque is the home of three interesting species:

Helianthemum pannosum†, bearing rounded leaves of the purest silver and abundant yellow rock-roses; *Santolina elegans†*, most elegant indeed with its glistening, narrowly pinnate foliage; . . . and *Scabiosa pulsatilloides†*, a plant of exceptional charm . . . The leaves are deeply shredded and covered with long hair, and the heads of flowers, rising on the briefest of peduncles from Trevenque's arid screes, are of a quite disproportionate size and beauty.[7]

Perhaps the finest plant of this area, however, is the little cushion-forming *Convolvulus*, *C. boissieri†*, described by Farrer as:

One of the loveliest of all. . . . This treasure forms enormous wide and perfectly tight masses in the dolomitic fissures, dry stony places, and barrens in the sub-alpine and alpine regions of the Sierra Nevada between 6,000–7,000 feet (as on the summits of Dornajo, Trevenque, and Aquilones), but is nowhere common; the cushion is built of ovate, blunt, folded little leaves, marked with nerves, and gleaming brilliantly with a plating of the finest silver sheen, soft and silky. Upon this spring stems so short as to be no stems at all, each carrying from one to four large and ravishing cups of rosy white, seeming to be scattered upon the surface of the mat.[8]

The most spectacular shrub growing on the south side of the Sierra is undoubtedly the tall silvery-leaved bush *Adenocarpus decorticans†*

---

9 **Majorca** In the eroded limestone precipices of the 'cloudy regions' above 500 m in Majorca grow most of the 50 or so of its unique plants which are known in no other part of the world. On these cliffs below Puig Mayor are to be found *Helichrysum lamarckii*, *Hippocrepis balearica*, *Cyclamen balearicum*, *Helleborus lividus*, and *Hypericum balearicum*.

with orange-yellow flowers, which forms thickets near Trevélez. This is the highest village in the Alpujarras and it is possible to climb from here in five or six hours, via Las Siete Lagunas, up Mulhacén, the highest summit. The climb is steep and most of the plants already mentioned on the northern slopes can be seen again, but the ascent is well worth while on account of the magnificent scenery. Above 2500 m *Sisymbrium laxiflorum*†, a small pink and yellow crucifer is found, with the white-flowered, extremely hairy, dwarfed *Potentilla caulescens*.

Round the Lagunas there is a green sward with masses of gentians, including *Gentiana pneumonanthe*, which grows only a few centimetres high. Below the string of lakes a stream cascades downwards and here are tall clumps of *Aconitum nevadense*† and the minute pink-flowered *Leontodon microcephalus*†.

8. Sierra Morena

The 'Dark Mountains' are but the eroded edge of the *meseta* of central Spain, where they come against the great fault of the Guadalquivir. The range is about 500 km long and nowhere exceeds 1323 m, becoming pro-gressively lower in the west where it ends in the hills of Monchique in the Algarve, and the cliffs of Cape St Vincent. Viewed from the north the sierras appear as sinuous swellings rising above a rolling plain, but viewed from the south the effect is very different. They present a formid-able range of steep rounded mountains with their slopes deeply dissected into spectacular valleys by streams running down to the Guadalquivir river. Along the streams are narrow strips of typically central European vegetation with alder, sedges, and grasses and many widespread herba-ceous species which are not normally found in this otherwise dry region. The main ridge of the Sierra Morena is composed largely of slates and greywacke, and most of it is covered with an evergreen mantle of *cistus* scrub. If one stands on one of the higher summits especially in the eastern or central part, one sees on all sides perhaps the greatest ex-panse of *matorral* in the Iberian peninsula, with dark green mountains stretching away to the horizon. The *matorral* has a characteristic, rather

---

10 **Cabo de Gata** The date palm has become naturalized in this, the driest region of Spain, with seven rainless months. Tufted grasses such as the False Esparto Grass and the Alfa form a characteristic steppe. In salt-rich soils a number of unusual Saltworts, Sea Blights, and Sea Lavenders have their only station in Europe.

11 **Serra da Estrêla** The granite tors forming the summits of the highest mountains in Portugal (2000 m) recall Dartmoor. In the moist Atlantic climate five species of heather, several *Cytisus* species, and *Halimium alyssoides* form low bushy growths.

uniform collection of species similar in composition to that described in the Costa de la Luz region, while in more sheltered areas there are mixed oak woods of *Q. faginea, Cork and Holm oaks with a typical undergrowth of cistus and leguminous shrubs. What has been described as 'the great dreary sweep of the Morena' has little attraction for the tourist, but the Desfiladero de Despeñaperros, the only important break in the wall of the escarpment, where the road and railway between Granada and Madrid pass through an exciting gorge, is well known.

A scramble up the steep slopes above the Despeñaperros gorge, in the second half of May, is worth while if only to see the magnificent display of white-flowered purple-blotched *Cistus ladanifer covering whole hillsides. Amongst the rocks will be found here and there *Iris xiphium†, and growing in crevices the little Jasione crispa subsp. mariana†. The endemic foxglove, Digitalis purpureus subsp. mariana†, with soft silky-white leaves, is best seen in June, as also are two other rock-fissure plants, members of the Caryophyllaceae, *Dianthus lusitanus and Bufonia willkommiana†, both of which are woody and shrubby at the base. The Sedum family is well represented by Sedum tenuifolium, Umbilicus horizontalis, U. rupestris, and *Mucizonia hispida†, an attractive grey, fleshy-leaved plant with pink flowers. Other interesting species are:

Paronychia argentea	Anthyllis lotoides
Dianthus crassipes†	A. cornicina
Brassica barrelieri†	Ornithopus compressus†
Biserrula pelecinus	*Onobrychis peduncularis†
*Saxifraga granulata	Sideritis lacaitae†
var. glaucescens	*Phlomis lychnitis
Trifolium cherleri	*Linaria aeruginea
T. hirtum	*Senecio minutus
	*Iris sisyrinchium

In the mixed oakwoods are:

*Anemone palmata	Smyrnium perfoliatum
*Paeonia broteroi	Magydaris panacifolia
Sedum fosteranum	Origanum virens
*Geum sylvaticum	Doronicum plantagineum
Sanguisorba hybrida	Centaurea citricolor†
Genista tournefortii†	*Endymion hispanicum
*Thapsia villosa	*Tulipa australis
*T. maxima	

Widespread bulbous plants are *Asphodelus ramosus, *Urginea maritima, *Dipcadi serotinum, *Ornithogalum umbellatum, and *Muscari comosum.

Another locality in the Sierra Morena worth a visit is Virgen de la Cabeza near Andujar. A few days' stay here offers an excellent opportunity to explore the range. On the way up to the village grows the lovely tall *Iberis linifolia†*, with deep pink flowers. Round the Parador there is a cream-coloured population of *Digitalis purpurea* subsp. *mariana†*, which is usually purple-flowered, and on bare baked rocks *Sedum caespitosum* manages to thrive. Some miles up the road running northwards from the Parador there is a good population of the bug orchid, *Orchis coriophora*, which is uncommon in these parts.

9. Sierra de Cazorla

This area, famous for its silver and lead mines, was known to the Romans as Mons Argentarius, but today its wealth lies in the timber and resin obtained from the extensive pine forests, probably the most important in southern Spain.

The Sierra de Cazorla consists mostly of hard Jurassic and Cretaceous limestones. A band of Triassic clays and red sands is exposed beneath this cap by the relatively young Guadalquivir river, which rises in the Sierra and has cut deep valleys through the heart of the mountains. As one drives up into the mountains from the town of Cazorla, one leaves behind wide horizons, and undulating brown plains covered with neatly planted rows of olive trees, and enters over the rim an entirely different landscape of jagged mountains and forests. Ahead lies a wild and untamed region, for long inaccessible; the haunt of eagles, game, and trout. But a recently constructed network of rough forest roads totalling 250 km or so in length, have made many of the wildest spots accessible.

The rainfall is exceptionally heavy, rising to 2000 mm (80 inches) per annum in some places; it is more or less continuous in the winter, sometimes tempestuous in spring and autumn, while the summers are dry and often very hot.

The highest peaks are Cerros de las Empanadas, 2107 m, and Pico de Cabañas, 2036 m, where snow may lie at least to the end of May. The clear waters of the Guadalquivir, rising at Canada de las Fuentes and running down through steep gorge country into a long reservoir known as the Embalse de El Tranco del Beas, adds much to the beauty of the scenery.

It is a country of deep ravines, sparkling streams, magnificent gnarled trees, enormous bare weather-worn cliffs and bluffs, and fantastic views. A fine Parador built in the heart of this country helps to make it one of the most worthwhile regions for the visiting naturalist and sportsman.

The sierra lies between the Sierra Nevada to the south-west and the Montes Universales to the north-east, and in consequence it has acted

SIERRA DE CAZORLA

1. *Vicia onobrychioides* (544)   **2.** *Tulipa australis* 1625   **3.** *Linum narbonense* 661
4. *Linaria aeruginea* 1204b   **5.** *Pterocephalus spathulatus* 1324a   **6.** *Erodium
daucoides* 653f   **7.** *Sarcocapnos crassifolia* 279b   **8.** *Narcissus hedraeanthus* 1666b
9. *Armeria filicaulis* 975a

10. *Narcissus longispathus* 1665a   **11.** *Cleonia lusitanica* 1121a   **12.** *Phlomis herba-venti* 1122   **13.** *Cistus laurifolius* 794   **14.** *Pinguicula vallisneriifolia* 1280b
15. *Erysimum linifolium* 295a   **16.** *Linum tenue* 658b   **17.** *Viola cazorlensis* 781a
18. *Tanacetum pallidum* 1428a

as a refuge for a number of montane species, which had in all probability previously undergone migrations and recessions culminating in the last ice age.

The most interesting of the Tertiary relicts is *Viola cazorlensis†, a shrubby, rock-loving species with deep crimson-carmine flowers with very long slender spurs. It is confined to the Sierras de Cazorla, Magina (Jaén province), and Castril (Granada province), but a closely related shrubby species V. kosaninii grows in Montenegro, and another V. delphinantha on Mt. Olympus in Greece. V. cazorlensis is generally found in crevices or more or less shaded rock faces, but occasionally in screes at high altitudes; it flowers abundantly in May along the road below the Parador. *Pinguicula vallisneriifolia† is another relict, with long pale green, strap-shaped insect-catching leaves and violet flowers. It was discovered as long ago as 1851 and then not seen again for more than half a century. It grows in several places, but only under towering, overhanging limestone cliffs where there is a continual drip of water, usually north-facing so that it is rarely touched by the direct rays of the sun; it is closely related to *P. longifolia of the Pyrenees.

Another old species is Ptilotrichum reverchonii†, a white-flowered, Alyssum-like shrublet that clings to dry vertical limestone cliffs. It is closely related to the celebrated P. pyrenaicum, which grows in one locality only in the eastern Pyrenees. *Ptilotrichum spinosum, so abundant in the Sierra Nevada, is rare here and confined to the Sierra de la Cabrilla, but P. longicaule†, with frail stems carrying showers of white blossom from July to August, is more frequent. The white-flowered plant with silvery leaflets that is everywhere on the face of most cliffs is Potentilla caulescens.

Two interesting endemic Narcissus species occur here: *N. hedraeanthus†, an attractive sulphur-yellow miniature hoop-petticoat daffodil which grows near melting snow at the highest altitudes, comes into flower early in May. *N. longispathus† is a giant by contrast; flowering in April, it grows up to 1½ m high, and has a very long spathe. An endemic species of columbine, Aquilegia cazorlensis†, is known only from shady limestone slopes around the summit of the Pico de Cabañas, where it flowers early in June.

Driving along through the pine forests one sees little in the way of massed colours by the roadside, due partly to the scarcity of yellow-flowered leguminous shrubs, heathers, or cistus species. Here and there are flashes of vivid blue where *Linum narbonense grows in quantity. Also in July the Cupidone *Catananche caerulea is very common; while on sheltered damp banks grows *Trachelium caeruleum with its numerous tiny tubular flowers forming dense broad, flat-topped, blue clusters. Occasionally *Silene colorata makes a carpet of pale pink in clearings in the pine forests. There are a number of fearsome spiny thistle-like plants such as *Carduus granatensis†, *Carduncellus monspelliensium, *Cynara cardunculus, *C. humilis, and *Chamaepuce hispanica†.

The vegetation can be divided into three main altitudinal zones:

The lower zone, from 700 to 1300–1400 m is characterized by the Aleppo pine, *Pinus halepensis*, which has its strongest development in this zone, with *Juniperus phoenicea, Pinus pinaster, Quercus ilex* (usually as scrub), and *Pistacia lentiscus*. In particularly humid enclaves, there is a marked development of *Quercus ilex* scrub with *Pistacia, Phillyrea*, and *Arbutus unedo*, known as the *maleza*. The olive is frequently cultivated in this zone.

In damp places the slender pale violet *Lysimachia ephemerum*, the yellow-flowered *Linum tenue*, the large-flowered pink *Centaurium erythraea* subsp. *grandiflorum*, *Lythrum junceum*, and the distinctive hairy-stemmed St John's Worts, *Hypericum caprifolium†* and *H. tomentosum* can be found.

The middle zone, from 1300–1400 to 1700–1800 m is distinguished by the maximum development of the Black pine, *Pinus nigra* subsp. *salzmannii*, which forms extensive forests with *Quercus faginea*, and some *Quercus ilex, Pinus pinaster, *Acer granatense†*, and *A. monspessulanum*. The rock-loving plants in this zone are very rich and include the following characteristic species: *Viola cazorlensis†, Ptilotrichum reverchonii†, P. longicaule†, Anthyllis ramburii, *Teucrium rotundifolium†, *Hypericum ericoides†, Erinus alpinus, *Sarcocapnos crassifolia†*.

The widespread rock-rose *Helianthemum croceum* often forms extensive sheets of yellow, while other interesting, largely Iberian, species are:

Arenaria capillipes†	*Nepeta reticulata†*
Iberis linifolia†	*Sideritis incana*
Pistorinia hispanica†	*Anarrhinum laxiflorum*
Sedum sediforme	*Digitalis obscura†*
Thymelaea sanamunda	*Linaria aeruginea*
Cistus laurifolius	*L. lilacina†*
Eryngium dilatatum†	*Campanula specularioides*
Marrubium supinum	*C. lusitanica†*
Cleonia lusitanica	*C. mollis†*
Phlomis herba-venti	*Santolina rosmarinifolia†*
Salvia blancoana†	*Centaurea granatensis†*
Ballota hirsuta†	*Allium polyanthum*
	Tulipa australis

In the course of walks in the vicinity of the Parador the following may be seen:

*Erysimum myriophyllum†	Hippocrepis glauca
Iberis pruitii	*Armeria filicaulis
Adonis flammea	*Cynoglossum cheirifolium
*Geum sylvaticum	*Lithospermum fruticosum
*Vicia onobrychioides	Parentucellia latifolia
*Orchis patens†	

The upper zone, from 1700–1800 to 2100 m is characterized by dwarf junipers—*Juniperus sabina* and *J. communis*—and dwarf spiny shrubs such as *Erinacea anthyllis, *Astragalus sempervirens, *Ptilotrichum spinosum*, and *Echinospartum boissieri. Pinus nigra* subsp. *salzmannii* still persists to some extent but with less vigour, while the shrub layer is better developed.

In this zone up among the rocks above the tree-line, accessible in half-an-hour's walk along a mule track from the Parador, are:

*Berberis hispanica	Buxus sempervirens
Saxifraga carpetana†	*Globularia spinosa†
S. corbariensis†	*Tanacetum pallidum
*Erinacea anthyllis	subsp. spathulifolium
*Erodium daucoides†	*Prolongoa pectinata
	*Narcissus rupicola

Elsewhere in cracks and fissures and exposed summit screes of calcareous rocks the scabious-like *Pterocephalus spathulatus† forms extensive silvery-white mats with almost stemless dark or pale pink flowers; it may cover the limestone screes for several yards to the exclusion of all other plants, or it may grow in association with the beautiful silvery-leaved, pink-flowered *Convolvulus boissieri†, and *Fumana procumbens. *Teucrium rotundifolium† is another unusual plant with velvety leaves and tiny milky-white flowers fading to pale purple, or reddish-purple with a cream-coloured lip. *Saxifraga rigoi†* with numerous summer dormant auxillary buds, and *S. camposii†* are two more Iberian saxifrages with limited distributions which can be found here.

Two peonies, *P. broteroi* and *P. officinalis* subsp. *humilis*; the purple-flowered *Erysimum linifolium†, and the primrose come into flower in May, while later follows the Burning Bush, *Dictamnus albus, and Verbascum hervieri†, the most spectacular of the five species of Mullein recorded in the area, which grows up to 3 m high. It has a very large rosette of silky leaves and quite hairless shining mahogany flowering stems.

Space does not allow an exhaustive list of the Sierra de Cazorla plants, but the great majority of those mentioned were found by the authors during short visits to the region.

10. Eastern Costa del Sol (Málaga-Cabo de Gata)

As one proceeds eastwards from Málaga the climate becomes progressively drier until one reaches Almeria where one enters the driest and hottest part of the whole peninsula. The Cabo de Gata and surrounding regions have an average rainfall of only 122 mm (under 5 inches) with over seven months without measurable rain each year, and in consequence possess a vegetation which is unlike anything else in Europe.

On the Málaga-Almeria coast road the first plant to catch the eye is *Echium sabulicola* with red-purple flowers growing all along the roadsides in March (or even February in some years).

Around Nerja the Balearic box, *Buxus balearica†*, has one of its few stations on the mainland of Europe, close to the famous cave, and here also on the limestone outcrop grows the interesting dwarf evergreen shrub *Cneorum tricoccon†*. It is a member of the tiny family *Cneoraceae* which has only two genera, and its nearest relative is to be found in Cuba. From Nerja to Adra the road is a continuous corniche except for a few kilometres across the Motril plain where sugar cane is grown in the *vegas*. Characteristic plants include *Genista spartioides†*, *G. umbellata†*, *Psoralea bituminosa*, *Maytenus senegalensis†*, *Withania frutescens†*, **W. somnifera†*, and **Launaea spinosa†*. Some way inland up the Motril-Granada road **Ononis speciosa†* displays its handsome yellow candles in April or May. On rocky headlands near Calahonda **Lavatera maritima* grows, and a prostrate form of **Rosmarinus eriocalix†* clings closely to limestone rocks. Past Calahonda an attractive stock, **Matthiola lunata†*, flowers beside the road in March, and a yellow composite with a dark purple centre, **Reichardia tingitana†*, is conspicuous in April.

Along the last few kilometres before entering Almeria, on another corniche road, the slopes are ablaze with yellow **Coronilla juncea* and **Launaea spinosa†* in spring, and in a few places by the roadside may be seen the tall narrow violet spikes of the rare and unusual plant **Lafuentea rotundifolia†*, a member of the *Scrophulariaceae* whose nearest relative is found in the eastern Sahara. The Sierra de Gador rising to 2236 m, which dominates this area, is well worth exploration, though somewhat inaccessible. Several endemic plants grow here, as well as such rock plants as **Polygala rupestris* and **Chaenorhinum villosum*.

The province of Almeria has in all probability a richer flora than any other part of Spain. It has been estimated that about 2,500 species of seed plants grow in the province, which is getting on for half the total number of species in the whole Spanish flora. About 200 of these species are con-

EASTERN COSTA DEL SOL

1. *Limonium insigne* 972c   2. *Lavatera maritima* 743   3. *Rosmarinus eriocalix* 1105a
4. *Euzomodendron bourgaeanum* 356b   5. *Sideritis romana* 1114   6. *Polygala rupestris* 697a   7. *Chaenorhinum villosum* 1211c   8. *Asphodelus tenuifolius* 1592a   9. *Coris monspeliensis* subsp. *fontqueri* 970

10. *Launaea spinosa* 1538a   **11.** *Ononis speciosa* 571b   **12.** *Cistanche phelypaea* 1269a
13. *Lavandula dentata* (1110)   **14.** *Helianthemum almeriense* 803b   **15.** *Lycium intricatum* 1173b   **16.** *Androcymbium gramineum* 1588c   **17.** *Fagonia cretica* 655b

fined to the arid areas of Almeria and Murcia provinces and more than half of these are endemic to this region. Others occur in North Africa as well, while a few have no near relatives at all, as for example the monogeneric Crucifers *Guiraoa arvensis*†, and *Euzomodendron bourgaeanum*†, or *Lapiedra martinezii*† of the daffodil family. About 30 per cent of the total are short-lived annuals which only appear after the rains and are quite absent during dry periods.

The Cabo de Gata and the hills inland form a desolate, sparsely populated region 'a rose-pink landscape with the wild desolate decor of a western film'. Ravines, badlands, interminable steppes, all showing the bare bones of the underlying rocks and the results of the erosion that follows occasional torrential rains, have left their permanent scars over the landscape. Only a very sparse covering of widely spaced steppe plants relieves the monotony. Grey leafless plants, wire-netting-like shrubs, switch plants, succulents, and woolly-leaved perennials all testify to the difficulties of survival in an almost waterless climate. There are occasional oases of great luxuriance, with Date Palms, and rich *huertas* of sugar-cane, banana, oranges, and rice wherever there is sufficient irrigation. The whole area is reminiscent of North Africa and in general the following types of vegetation can be distinguished: salt steppes, esparto steppes, and stony steppes.

The salt steppes cover large areas along the coast and inland to altitudes of up to 700 m, and they are characterized by the predominance of often succulent members of the Goosefoot family, *Chenopodiaceae*. Some, such as *Atriplex glauca* and *Salsola vermiculata* are common in similar localities throughout Spain, but others like *Anabasis articulata* occur only in south Spain as well as in the North African salt steppes; while *Salsola papillosa* and *S. genistoides* are endemic to south Spain.

These *Chenopodiaceae* are not infrequently parasitized by a strange leafless plant *Cynomorium coccineum*†, which produces above ground a single fleshy, dark reddish-black, club-shaped flower-spike arising directly from the soil. Another parasite growing on members of this family, is the striking yellow-flowered *Cistanche phelypaea*† which flowers in early spring.

Mesembryanthemum crystallinum, with fleshy leaves covered with sparkling crystal-like hairs; *Frankenia corymbosa*†; the Sea Lavenders *Limonium furfuracea, L. cymuliferum*, and in particular the lovely pink-flowered endemic *L. insigne*†, all occur in these salt steppes.

The esparto steppes are characterized by the tufted False Esparto grass, or Albardine, *Lygeum spartum*, which is readily distinguished by the terminal pale boat-shaped sheaths which enclose the inflorescences.

The wiry stems of this grass are one of the few useful products of the steppes and it is harvested regularly. Another taller tufted grass with stems 1 m or more high is the Alfa, *Stipa tenacissima*, also characteristic of this region.

The stony steppes are often scattered with extremely spiny and intricately branched hummocks of *Ziziphus lotus*, which appear quite dead until the leaves begin to sprout in April, but much of this land has been cleared and extensive plantations of sisal *Agave sisalana* have taken their place. The curious crucifer *Succowia balearica*, the attractive mauve-flowered *Fagonia cretica*, *Asphodelus fistulosus*, and its dwarf relative *A. tenuifolius* are all abundant. Other species include: *Peganum harmala*, *Thymus cephalotos*†, *Zygophyllum fabago*, *Haplophyllum linifolium*†, and *Plantago albicans*. Another unusual plant is *Androcymbium gramineum*†, rather like an *Ornithogalum* with a stalkless cluster of white flowers finely streaked with mauve, nestling among the narrow leaves; it flowers in mid-winter.

An example of this stony steppe can be seen along the main road (N 340) out of Almeria between Rioja and Tabernas. It runs through typical *Euzomodendron* steppe country, with deeply eroded arid ravines sparsely covered by an interesting vegetation with the *Chenopodiaceae* again well represented by *Salsola genistoides, S. webbii, S. verticillata, Anabasis articulata*, and the parasite *Cistanche phelypaea*†, growing on *Salsola genistoides*. Here too the very lovely *Limonium insigne*† is fairly plentiful, flowering from the end of March onwards; its rival in grace and beauty *L. caesium*† grows in Murcia province, especially round Hellin.

A rarity of this area is *Senecio auricula* var. *major*†, known only from one other locality in Spain; a fine plant with fleshy entire basal leaves and large yellow flower heads.

The arid hills of the Sierra de Gata are covered with a low scrub of *Thymus vulgaris, Atractylis humilis*, *Lavandula multifida*, and *L. dentata, Asparagus stipularis, Helianthemum* species, and the tufted spiny bush with fleshy leaves *Lycium intricatum*. The rare endemic *Antirrhinum charidemi*† grows only in this sierra, on volcanic rocks, while on sandy soil near the sea may be seen the pale blue calyces of *Limonium thouinii*, an annual with broadly winged stems. The long twin horns and dark purple flowers of *Periploca laevigata*†—growing here as a dense shrub and not as a climber—are a feature of the hills round San José, while on the beach *Tamarix boveana*†, a North African species rare in Europe, produces its lovely pink four-petalled flowers in early April. Along the Nijar-Carboneras road there is a very lovely tall erect form of *Rosmarinus eriocalix† (looking very different from the prostrate form of Calahonda), and a tall candelabra form of *Coris monspeliensis (named by some subsp. fontqueri).

A similar desert landscape of badlands covers extensive areas round Puerto Lumbreras and Velez-Rubio, where the friable red sandstone and grey mica schists have been heavily eroded into a patchwork of colours and which bears a very sparse plant cover. Among the species that are somehow able to survive the severe conditions, and the goats, are:

Capparis ovata	Teucrium gnaphalodes
*Ononis fruticosa	T. saxatile†
*Anthyllis cytisoides	*Lavandula dentata
*Polygala rupestris	*Marrubium supinum†
Rhamnus alaternus	*Chaenorhinum villosum
*Helianthemum almeriense†	Phagnalon rupestre
Fumana thymifolia	*Scorzonera graminifolia
Chamaerops humilis	

Echium humile† is a North African species found only in these dry areas of Spain.

11. Balearic Islands

The Balearic Islands are the eastern extension of the Andalusian mountains, or Baetic Cordillera, of southern Spain. They are new-fold mountains of the alpine upheaval, running in a north-east and south-west direction. Of the three main islands Majorca is the largest and the most varied geologically, and in consequence of the most interest botanically.

Majorca is about 100 km long and is dominated by its main mountain range which runs the whole length of the north-western coast for 80 km, rising to 1440 m in Puig Mayor in the centre of the range. The mountains are of folded Triassic and Jurassic limestones, with several ridges running parallel to each other and enclosing inter-montane basins. Steep gorges, lofty ridges, and spectacular scenery, together with precipitous cliffs 300–400 m high and wide vistas across the Mediterranean, are characteristic. This range acts as a natural barrier, both by protecting the low-lying plains in the centre of the island from the prevailing north-westerly winds or tramontana, and also by catching the moisture and rain coming in from the sea. The rainfall may be as high as 1500 mm (60 inches) in the mountains, while inland the lower hills may have as little as 800 mm. Palma, the capital, has 73 rainy days in the year.

The main low-lying central part of the island is formed of Miocene limestones, overlaid in places by recent sands and gravel, and with its generous covering of *terra rossa* soil is the most highly populated and richly cultivated part of the island. Almonds, figs, grapes, mixed with cornfields, potatoes, and other vegetables and salad crops give the typical poly-culture of the Mediterranean, while in the north-east, bordering the Bay of Alcudia, is a low-lying coastal region with lagoons and marshes. Along the south-eastern coast are low-lying limestone hills with low but steep cliffs and narrow inlets indicating flooding of sunken river valleys, or miniature *rias*. These limestones are largely covered with the characteristic *matorral* of the islands, which now takes the place of the oak forests which probably once covered these lower hills. Forests of Aleppo pine occur on the coastal hills and in the mountains up to about 700 m, while above this they are generally replaced by forests of evergreen oak. Some olives, sweet oranges, and cherries are grown in the foothills.

The *matorral* of the Balearic Islands is very similar to that of the mainland. It is dominated by *Pistacia lentiscus*, *Cistus albidus*, *C. monspeliensis*, *Rosmarinus officinalis*, *Olea europaea* var. *oleaster*, with other common species such as

Osyris alba	*Lonicera implexa*
Psoralea bituminosa	*Helichrysum stoechas*
Dorycnium pentaphyllum	*Phagnalon saxatile*
Ruta chalepensis	*Asphodelus aestivus*
Daphne gnidium	*Asparagus albus*
Thymelaea hirsuta	*A. stipularis*
Fumana procumbens	*Smilax aspera*
Teucrium polium	

By contrast, the *matorral* of the hills have the following dominant species: *Rosmarinus officinalis*, *Erica multiflora*, and the robust grass *Ampelodesma mauritanica*, together with the endemic shrubby *Hypericum balearicum†*, *Asphodelus albus*, *Rhamnus alaternus*, with such species as *Clematis cirrhosa*, *Astragalus balearicus†*, *Dorycnium hirsutum*, *Cneorum tricoccon*, *Euphorbia dendroides*, *Teucrium subspinosum†*, and the endemic *Genista lucida†* in the hills of the west and east.

With 1,280 native species of seed plants, the flora of the Balearic Islands is not very rich by comparison with the adjacent mainland areas or with much larger islands such as Corsica and Sardinia; but like these islands it has a considerable number of exclusive species and subspecies, which grow nowhere else in the world. Consequently it is imperative that prompt and forthright action should be taken to preserve this unique collection of plants. Goats, sheep, and horses are probably the major threat, particularly in the mountains where the majority of endemic species live, but there is also an ever increasing threat from the explosive building development which has taken place in the island during the last decade. Fortunately there is still time to take protective measures.

H. Knoche in *Flora Balearica* has estimated that nearly five per cent of the flora is endemic, a total of over 60 species and subspecies, but more detailed study and comparison with mainland floras has reduced this number considerably, and Majorca probably has an endemic flora of something like three per cent, a little lower than other islands with comparable isolation. Endemic species may be either very old relicts which have survived since the Tertiary period, or of medium age and evolved sometime during the interglacial periods of the Quarternary; or relatively recent species, which have evolved, often as a result of geographical isolation from their nearest relatives, over periods of tens of thousands of years. Of particular interest are those species which have no relatives in any part of the world and which must therefore be considered to be of very ancient stock. They include the evergreen shrubby St John's Wort *Hypericum balearicum†*, with very distinctive crimped leaves with warty lumps on the underside and large solitary yellow flowers, and *Daphne rodrigueziï†*, a dwarf evergreen shrub of Minorca with clusters of fragrant purple flowers tinged with yellow. Other attractive endemic species occurring on Majorca with relatives existing on the mainland (and consequently younger in origin) are: *Paeonia cambessedesiï†*, with large deep rose flowers, a local plant of the east of the island; **Helleborus lividus†*, also a local plant often associated with the peony and growing at the base of cliffs in the eastern and mountain areas, and distinguished by its large three-lobed leaflets with whitish veins and its usually pink flowers flushed with purple; and *Crocus cambessedesiï†* a widely distributed, autumn-flowering plant with grass-like leaves and tiny lilac or white flowers feathered with purple and with brilliant scarlet stigmas. Other endemics include *Senecio rodrigueziï†*, with powder blue foliage and pink flowers; the dense, stiff-leaved evergreen shrub **Rhamnus ludovici-salvatoris†*, widespread in mountain woods and valley sides; the hedgehog-plant **Teucrium subspinosum†*, growing abundantly on the coastal limestone, in woodlands and on mountain tops; *Genista lucida†*, an erect spiny shrub with simple leaves and spike-like clusters of yellow flowers; also *Cymbalaria aequitriloba†*, and *Lotus tetraphyllus†*. **Cyclamen balearicum†* was considered to be an endemic species, but is now known to occur in the south of France. It is a widespread plant in the islands and enjoys shady and rocky places. It has beautifully mottled leaves and small delicate white flowers, and is the most westerly growing species of this genus.

Though some of the endemic 'balearic' species grow in the plains and foothills, the majority grow in the 'cloudy region' in the mountains of the north-west coast where there is a unique combination of climate and environment with heavy rainfall and high humidity, together with sheltered limestone gorges, exposed limestone cliffs, fantastic weathered rocks, and screes, that together create one of the finest natural rock gardens to be found anywhere in the western Mediterranean. The best approach to this region is from the Corniche road, particularly in the region of Puig Mayor where the road tunnels through the mountain, or from the brilliantly engineered road which drops down from the main ridge to the minute harbour of La Calobra. It is from about 1000 m upwards that most of the interesting species are to be found, but some may be carried down to lower levels in screes and in the gravels formed by the perennial water torrents, for example *Crepis triassii*†; the sweet-scented, golden-yellow **Hippocrepis balearica*†, really a plant of woodlands and *matorral* but surviving against predators on the cliffs; the silvery rosette-forming **Helichrysum lamarckii*†, and the huge, fetid, shiny-leaved endemic parsnip, *Pastinaca lucida*†.

Two interesting evergreen shrubs occur in the lower 'cloudy region'; they are **Rhamnus ludovici-salvatoris*†, with rounded leathery leaves with silvery undersides and red berries, whose nearest relative is in California, and *Buxus balearica*†, a box found also in south and east Spain, which is very like the common box, but with larger leaves (2½–4 cm), and larger flower clusters (1 cm across). Other endemics include: *Aristolochia bianorii*†, with brownish-yellow striped flowers and a reddish lip, which can also be found almost at sea level, as at the lighthouse of Puerto Pollensa where it is abundant; *Sibthorpia africana*†, which occurs under shady rocks; *Teucrium lancifolium*† commonly found in the screes between 1000 and 1200 m; *Phlomis italica*† with woolly leaves and purple flowers, most abundant at about 900 m; the uncommon *Scutellaria balearica*† of shady slopes; **Brassica balearica*† with bright yellow flowers, and a small rosette of fleshy leaves; and the evergreen shrubby *Bupleurum barceloi*†, with narrow lance-shaped leaves.

If one climbs the highest summits one meets other species. On the summit of Ternellas '*Calamintha rouyana*†' is abundant on the rocks. But it is Puig Major, thanks to its altitude, which is richest in species. In the first place, one meets on the escarpment *Ligusticum lucidum* subsp. *huteri*. In shady hollows on the northern slopes, particularly where snow has remained for a long time and in consequence has conserved more moisture, the rare *Ranunculus weyleri*† carpets the soil; and a few metres higher up, there is the white-flowered form of *Primula acaulis* [*P. vulgaris* subsp. *balearica*]. At the foot of these escarpments in deep soil, *Arenaria grandiflora* var. *incrassata* grows very abundantly. Certain endemic species do not find a sufficiently humid atmosphere in these mountains. They take refuge in a deep gorge which descends from the Gorch Blau towards the sea. In the shade *Erodium chamaedryoides†* [*E. reichardii*] spreads over the soil. On the rocks one sees the shining leaves of *Viola jaubertiana†*, a plant which is found in certain stations in the high sierra. At the bottom of the gorge grows *Hypericum cambessedesii†* [*H. hircinum* but possibly an endemic subspecies].[9]

Up to the summit of Puig Mayor *Teucrium subspinosum†*, and *Astragalus balearicus†*, two very spiny 'hedgehog' plants, are dominant; while *Rosmarinus officinalis*, *Asphodelus aestivus*, *Euphorbia characias*, *Digitalis dubia†*, *Evax pygmaea*, *Bellium bellidioides*, and *Santolina chamaecyparissus* also occur round the summit. In shady places where snow lies late are the following: *Cyclamen balearicum†*, *Arenaria balearica†*, *Sibthorpia africana†*, *Chaenorhinum origanifolium*, *Cymbalaria aequitriloba†*, and others.

Undoubtedly the best time to see many of these unique species in Majorca is during May, for by the end of July they have mostly finished flowering. Nevertheless, autumn, with the arrival of the rains, is a time when other interesting species come into flower. The glories of autumn are provided by the blue flowers of the shrubby *Globularia alypum*, and the pink *Erica multiflora*, while bulbous and rhizomatous plants, such as the Autumn saffron, *Merendera filifolia†*, with pinkish-purple flowers, may occur locally in quantity, often with the leafless, white-flowered narcissus *N. serotinus*, and the blue Autumn squill, *Scilla autumnalis*. The striped brown and purple flowers of the Friar's Cowl, *Arisarum vulgare* occur in shady spots, and *Viola arborescens†*, a woody plant with pale violet flowers, grows by the wayside; it also flowers in spring. The Saffron crocus, *C. sativus*, flowers in the autumn in the orchards, and its scarlet stigmas, highly prized as a drug, are collected at this time of year; while in the *matorral* the tiny crocus, *C. cambessedesii†* comes into flower at this season, and the dark purplish-red spathes of *Arum pictum†* are occasionally seen.

Minorca, the second largest island, is only 50 km long, and though it is more varied geologically than Majorca with the inclusion of some much older rocks, in all probability fragments of the ancient land mass, it is much less spectacular scenically and possesses a more meagre flora. The hilly country lies in the north-east of the island, rising to the summit of Monte Toro, 357 m, and is composed of old siliceous rocks of the Devonian and Jurassic times; while southwards the land gently falls in an undulating plateau of younger Miocene limestones. The landscape is monotonous, with innumerable walled fields of corn and barren exposed heathlands relieved only by the many gorges and ravines which cut into the central limestone plateau. These gorges give protection from the strong and often violent northerly winds, or *tramontana*, which may blow continuously for days on end, and for up to 165 days in the year. In these gorges the vegetation is rich and varied, and conditions may be sufficiently humid for the truly 'balearic' species to survive. These include: *Paeonia cambessedesii†, Lotus tetraphyllus†, Pastinaca lucida†, Teucrium lancifolium†, and Digitalis dubia†, while in one ravine, growing in full sun, is the one locality of the endemic Lysimachia minoricensis†. In the most shady and dampest ravines, where water oozes from the slopes—often shown by the presence of Samolus valerandi— more 'balearic' species can be found, such as: Silene mollissima†, *Hippocrepis balearica†, Scabiosa cretica†, Crepis triasii†, and rarely Euphorbia maresii†, and Cymbalaria aequitriloba subsp. fragilis†.

The coastal cliffs, particularly those of the limestone to the south, are rich in species and are well worth exploring, for here again can be found some more of the 'balearic' species. Interesting plants of these cliffs include:

Capparis spinosa	*Bellium bellidioides
Cheiranthus cheiri	Inula viscosa
*Sedum sediforme	*Asteriscus maritimus
Thymelaea hirsuta	Artemisia arborescens
*Teucrium subspinosum†	*Asparagus stipularis
*Smilax aspera	

The siliceous cliffs of the north-east coast, though more humid than the calcareous cliffs, have a very similar flora but it is here that the endemic Daphne rodriguezii† can be found, and it only grows on Minorca.

Ibiza is the third island of any size. It is 45 km long and is composed of Secondary limestones which have been folded and eroded into very hilly country, culminating in the summit of Atalayasa, 475 m, in the west of the island.

By comparison with Minorca it is a much more wooded island with orchards of almonds, carob, and citrus fruits and in particular figs, while higher up there are extensive young forests of Aleppo pine, Pinus halepensis. Water is short and much of the vegetation is xeromorphic. Cistus scrub is characteristic with Juniperus phoenicea unusually common, and Helichrysum stoechas and Ononis natrix conspicuous. In the dried up river beds the oleander, Nerium oleander, and the myrtle, Myrtus communis are a feature of the island, while Vitex agnus-castus and Inula viscosa are characteristic where the rivers enter the sea. Localities where 'balearic' species can survive are generally lacking on the island, and it is only at the Cala Aubarca where anything like the right conditions prevail. Here can be found: Silene mollissima†, Sedum rubens, Micromeria filiformis, *Chaenorhinum origanifolium, and Sibthorpia africana†. Other plants of interest include:

Silene littorea	*Lavandula dentata†
Biscutella frutescens	Orobanche latisquama
Cneorum tricoccon	*Ballota hirsuta
*Hypericum balearicum†	*Iris sisyrinchium
Cistus clusii	Chamaerops humilis
Helianthemum origanifolium	Ophrys bertolonii†

and a beautiful form of *Chaenorhinum origanifolium*† which is unique to this island and to Formentera, and which grows on coastal sand dunes.

Urginea maritima and *Tamarix africana* are also plants of the coastal hills and sands.

12. Pinhal de Leiria and Coastal Beira

The coastal region of central Portugal forms a transitional area between the Mediterranean flora to the south and the truly Atlantic flora to the north. On the drier warmer calcareous soils up to 56 per cent of the plants are Mediterranean species; on the colder siliceous soils it is only about 36 per cent, while in the mountains in the interior the percentage is lower still.

The Pinhal de Leiria, and the salt marshes and lagoons of Aveiro—Portugal's Little Holland—occupy low-lying sandy soils along the western seaboard. As early as the thirteenth century, during the reign of King Dinis the Labourer, plantations of Maritime pine have been grown in an attempt to keep the sand dunes at bay, and for timber. Today after six centuries of continuous afforestation bare sandy wastes still occur. These *landes* are similar in general appearance to those of western France but contain many exclusively western European plants, particularly among the brooms, ericas, cistuses, and grasses. Driving on the narrow forest roads through this extensive area one sees uniform stands of Maritime pine of different ages, and evidence of intensive afforestation, which affords valuable timber and turpentine. Strips of rye, potatoes, and maize cultivation occur here and there; and in places stands of *Eucalyptus globulus*, now grown to an enormous size, and *Acacia melanoxylon* have been planted. It is a poor country which will support little permanent agriculture. Often the pine woods have a rich and well developed scrub layer which includes:

Cytisus grandiflorus	*C. salvifolius
Ulex europaeus	Myrtus communis
U. minor	Halimium commutatum
*Daphne gnidium	Calluna vulgaris
*Cistus crispus	*Erica scoparia
*C. psilosepalus	*E. umbellata

Corema album   *Lonicera periclymenum*
Lithospermum diffusum   *Helichrysum serotinum*
Lavandula stoechas   *Scilla monophyllos*
    and subsp. *pedunculata*   *Ruscus aculeatus*

In the more mature vegetation grow the taller *Arbutus unedo*, *Viburnum tinus*, *Quercus coccifera*, often with *Phillyrea angustifolia*, *Pistacia lentiscus*, and less commonly *Rhamnus alaternus*, and with the climbers *Smilax aspera* and *Rubia peregrina* growing up through the shrubs.

On the sand dunes near Praia da Vieira the following zones of vegetation occur as one passes from the coastal dunes to the stabilized forest inland.

On the seaward side of the partly consolidated dunes, growing in blown sand are *Otanthus maritimus*, *Ammophila arenaria*, *Euphorbia paralias*, *Eryngium maritimum*, and *Crucianella maritima*.

On the consolidated crests of the dunes, between the Marram grass, are mats of *Crucianella maritima*, bushes of *Corema album*, and the pink-flowered annual *Silene littorea*, also *Malcolmia littorea*, and the Sea daffodil *Pancratium maritimum*. On the landward side of the dunes *Halimium commutatum* soon makes its appearance, while the dune slacks themselves are colonized by dark thickets of *Pinus pinaster* only 2 m or so high. In the stabilized areas an interesting assemblage of both western Atlantic and Mediterranean species occur including:

Matthiola sinuata   *Medicago marina*
Iberis procumbens   *Lotus creticus*
Cytisus striatus   *Frankenia intermedia*
   Ulex europaeus subsp.   *Crithmum maritimum*
     latebracteatus   *Anchusa calcarea*
Chamaespartium tridentatum   *Armeria welwitschii*
Stauracanthus genistoides

as well as *Carpobrotus edulis* and *Cryptostemma calendula* both native plants of the Cape of Good Hope and now widely naturalized on the littoral of the south-west.

13. The Northern Serras of Portugal (Serra da Estrêla, do Marão, do Gerês)

The Serra da Estrêla is the highest range in Portugal; the mountains just fail to reach 2000 m, and form a high barrier across the centre of the country from south-west to north-east. They are old-fold mountains, formed of old crystalline and palaeozoic rocks, and are the westernmost

NORTHERN SERRAS OF PORTUGAL

1. *Amelanchier ovalis* 471   **2.** *Chamaespartium tridentatum* 513   **3.** *Cistus psilosepalus*
(790)   **4.** *Adenocarpus complicatus* 518   **5.** *Erigeron karvinskianus* 1372
6. *Phalacrocarpum anomalum* 1428f   **7.** *Narcissus bulbocodium* var *nivalis* 1666
8. *Silene foetida* 165a   **9.** *Arenaria montana* 126

10. *Cytisus multiflorus* 503 **11.** *Omphalodes nitida* 1047a **12.** *Iris boissieri* 1685c
13. *Prunella grandiflora* (1120) **14.** *Halimium alyssoides* 797b **15.** *Tuberaria globularifolia* 799a **16.** *Narcissus asturiensis* 1665h **17.** *Polygala microphylla* 696b
18. *Simethis mattiazzii* (1596)

range of the Central Sierras of the peninsula. Their summits are rounded and grassy. Huge granite tors lie scattered over their surface, recalling Dartmoor or parts of Scotland; and deep glacial valleys, like that of the Zêzere, give striking evidence of the importance of ice and snow during the last ice age. There are glacial lakes, and extensive moraines occurring as low as 700 m.

Today, snow lies on the heights from November to April, and being relatively near the Atlantic, a rainfall of over 2500 mm (100 inches) a year may occur. In consequence, a great variety of climates and vegetational types is found. Olives and vines, mixed with forests of Maritime pine, may grow up to altitudes of 800 m in warm pockets, and the vegetation may be predominantly of Mediterranean type, with *Cistus ladanifer, *C. crispus, *C. salvifolius, and the handsome *Lavandula stoechas making fine splashes of purple. Above this, from 800–1500 m, there is a montane zone, with forests of Pedunculate oak, chestnuts and pines, and an undergrowth mainly of ericaceous shrubs such as *Erica australis, *E. arborea, and a little *E. umbellata, *E. lusitanica, and Calluna vulgaris. At lower levels in this zone the common broom, Cytisus scoparius is a magnificent sight in early June, with tall bushes 2 m or more high covered in masses of large golden-yellow flowers; while higher up *C. striatus, *C. purgans, and the white-flowered *C. multiflorus are common. Also present are *Arenaria montana, Ranunculus nigrescens, Genista cinerea, Centaurea paniculata, Luzula lactea, and Stipa gigantea. Natural forests are rare, but near Poço do Inferno (The Well of Hell) at 1000 m there are forests of the Pyrenean oak, *Quercus pyrenaica, with such species as:

*Cytisus multiflorus	Galium vernum
C. grandiflorus	*Linaria triornithophora
*Erica arborea	Jasione montana

In the montane zone, rye may be cultivated up to 1600 m in temporary fields for the period of a single year, after which they revert to secondary heath composed of *Cytisus multiflorus, *Halimium alyssoides, and bracken, often covering vast areas. The uncultivated heathlands have a richer flora.

The alpine zone is characterized by the dwarf juniper, which occurs above 1500 m. In such a harsh climate, with up to thirty-five days of snow each year, the absence of trees is not surprising, although they occur elsewhere at comparable heights on other mountains of the peninsula.

The summits are covered with a heavy sward of Mat grass, Nardus stricta, very similar to some of our own mountain pastures. Low bushes of Juniperus communis subsp. nana less than ½ m high form scattered clumps. The associated species are few and include: Rumex *acetosella, Arenaria aggregata, Cerastium gracile, Paronychia polygoni-folia, Sedum arenarium, Galium saxatile, Pedicularis sylvatica, Plantago radicata, Gentiana pneumonanthe,* and **Narcissus bulbocodium* var. *nivalis.*

On the rocky parts of the Serra can be found:

*Silene ciliata†	*Armeria alliacea*
*S. foetida†	*Teucrium salviastrum*
*Sedum brevifolium	*Campanula herminii*†
*S. hirsutum	*Jasione humilis*†
*Ornithogalum concinnum	

and **Saxifraga spathularis*† in wet flushes, and *Leontodon hispidus*† in its only locality in Portugal.

The best centre for exploring the Serra da Estrêla is the magnificently sited Pousada de São Lourenço, high above the small spa of Manteigas. Mid-April to mid-May is the best time to make a visit for the bulbous species, while June and July are better for other plants, and even later on the summits.

Round the Pousada **Chamaespartium tridentatum* turns whole hill-sides into gold, and **Erica australis* makes splashes of purple. The small purple **Linaria elegans* may be seen occasionally by the roadside, but it is not nearly as common here as on the Sierra de Gredos in Spain. **Saxifraga spathularis*† and *S. continentalis*† like wet places by streams and waterfalls, while **S. granulata* is more widespread. The attractive white daisy *Phalacrocarpum amomalum*† and the sticky **Silene foetida*† may be seen on steep shady rocks, especially at the Poço do Inferno, a waterfall accessible by a pleasant forest road 7 km above Manteigas.

To reach the higher ground, take the road to Penhas Douradas, a collection of holiday bungalows about 5 km from the Pousada. From here, a magnificent mountain walk of three to three and a half hours over easy ground, up the gently sloping main ridge, will lead to the summit of Torre, 1991 m. At the end of April the bulbous plants are at their best. The charming little **Narcissus bulbocodium* var. *nivalis* and **N. asturi-ensis*† may be seen in their thousands, covering wide stretches of brown turf, while **Crocus carpetanus*, in all shades from deep purple to almost white†, grows in patches near melting snow. **Narcissus rupicola*† is less common and is found growing in rock crevices, or in the shade of huge rocks. In ponds and streams grows an aquatic *Ranunculus*, *R. ololeucos*†, with particularly large white flowers $2\frac{1}{2}$ cm across.

Six weeks later the dwarf shrubs are coming into flower; mainly **Erica australis* and **Halimium alyssoides*, with a little **Erica umbellata*, while **Echinospartum lusitanicum*† grows at the higher levels. Another crocus,

C. asturicus†, flowers here in the autumn. The yellow-flowered *Viola langeana* may be seen in May beside the road.

The Serra do Marão lies between the Douro and Tamega valleys and rises to a height of 1415 m. These mountains are composed of Silurian slates and have steep, very rocky slopes. The lower slopes, at least on the Douro side, are covered with vineyards up to altitudes of 600 m, where the vines are grown on trees and trellises. Above the cultivations, *Erica* heaths are very extensive and the vegetation has much in common with the mountain heathlands of Leon and Asturias, with largely central European species predominating; few interesting Iberian species occur as far north-west as this. However, a few days spent at the Pousada São Gonçalo would be an ideal way of exploring this area, with fine walks in all directions.

The hillsides around the Pousada are covered with a heath of *Cytisus scoparius*, **Cytisus multiflorus*, **Chamaespartium tridentatum*, **Halimium alyssoides*, **Erica umbellata*, with some **Erica arborea*, while **E. australis* appears only near the highest point. In the undergrowth **Simethis mattiazzii* is abundant, and the white-flowered *Halimium umbellatum* is common between 1100–1200 m. Below the Pousada **Silene foetida*†, and **Tuberaria guttata* make splashes of pink and yellow along the side of the road; **Saxifraga spathularis*† grows in damp spots and a few plants of one of the finest of the Linarias, **L. triornithophora* may be seen.

The Serra do Gerês is composed of granite mountains in the north-west of Portugal, and though not the highest in the country, they are in many ways the most spectacular. Seen from the Chaves-Braga road (N 103) they rise up boldly from deep gorges, the lower slopes largely forested with pines, the upper treeless with much bare granite. The Pousada de São Bento lies just off this road to the north, and is magnificently sited with enormous views across the reservoirs deep down in the valley, and to the ranges beyond, rising above the little spa of Caldas do Gerês. A number of forest roads, unmarked on the maps, radiating from the spa, make it possible to explore much of the area by car.

A road with seventeen U-bends rises rapidly to Leonte, the watershed between the Gerês and Homem rivers, where **Narcissus triandrus* is abundant in April. It then drops down through woods of Pyrenean oak, **Q. pyrenaica*, and birch to Albergaria and the Ponte de S. Miguel over the river Homem. A side road leads off from the bridge up to a disused wolfram mine, called Carris, to near the top of the Serra, Altar de Cabroes. Though not negotiable by car the whole way up, this is a lovely walk and takes one up into a highland glen ablaze with colour, mainly of *Erica arborea*, *E. australis*, *E. umbellata*, *Chamaespartium tridentatum*, and *Halimium alyssoides*. *Polygala microphylla*, the best of the milkworts, is common here at the northernmost limit of its range, while *Thymelaea broteriana†*, an attractive shrublet endemic to Portugal with numerous tiny yellow flowers and contrasting bright green heather-like leaves, also grows here. There is also some *Fritillaria lusitanica*, *Tulipa australis*, *Endymion hispanicum*, *Narcissus bulbocodium*, and the white daisy *Phalacrocarpum amomalum†*.

At the higher levels the Dog's Tooth violet, *Erythronium dens-canis†*, grows at the most western point of its range in Europe, and is in flower from March to April, with *Crocus carpetanus*. The most interesting plant in this region is *Iris boissieri†*, which is endemic to these mountains. It can be found from 600 m upwards, but seems to be rare at lower elevations and more abundant in the neighbourhood of Carris at about 1450 m. It resembles *I. filifolia* with the same very long and narrow leaves and the same habit of growing singly up through the scrub, but the flowers are somewhat smaller and bluer, and the falls are bearded.

The road from Albergaria runs down the valley of the Homem, through lovely forests of *Quercus robur*, with an undergrowth of a mixture of *Prunus lusitanica†*, *Ilex aquifolium*, *Acer pseudoplatanus*, *Arbutus unedo*, and *Amelanchier ovalis*. Mossy banks dripping with water are colonized by *Saxifraga spathularis†*, and the lovely blue-flowered *Omphalodes nitida†*, and the early flowering *Anemone trifolia* subsp. *albida†*. *Saxifraga clusii* is less common, but it sometimes forms loose cushions 2 m across, covered with a cloud of small white flowers. *Lilium martagon†* flowers here in July.

The return to Gerês is up to the belevedere of Junceda overlooking the Gerês gorge, and back down through the woods of *Pinus pinaster*, where *Simethis mattiazzii* and *Lithospermum diffusum* are to be found. At Gerês itself the most noticeable plant is the pink and white daisy *Erigeron karvinskianus*, a native of Central America now widely naturalized in Europe, which has run wild on the walls. *Anarrhinum bellidifolium* and *Prunella grandiflora* grow beside the road, and the autumn-flowering bulbs of this area are *Crocus clusii* and *Merendera montana*.

The most conspicuous families of plants of this region are the *Leguminosae*, *Cistaceae*, and *Ericaceae* and they include a number of northwestern species:

Cytisus multiflorus	*Adenocarpus complicatus*
C. scoparius	*Cistus psilosepalus*
C. striatus	Halimium umbellatum
Genista micrantha	*H. alyssoides*
G. florida	*Tuberaria guttata*
Ulex minor	*T. globularifolia*

Daboecia cantabrica	*E. ciliaris*
Calluna vulgaris	E. tetralix
Erica cinerea	Vaccinium myrtillus

In October 1970 the Peneda-Gerês National Park was established by decree as Portugal's contribution to the International Conservation Year, and an office opened in Caldas do Gerês where a brochure and map of the park can be obtained. The construction of many kilometres of forest road, marked on the National Park map but not as yet on the usual motoring maps, has made it possible to explore much of the area by car— an area of 68,000 hectares of some of the most beautiful country in Portugal, with magnificent displays of heather in June. One of the most attractive plants of the granite areas of northern Portugal, the lemon-scented *Thymus caespititius*, forms low pink mats from early June here and there between the granite rocks.

14. The Central Sierras of Spain

The Central Sierras, or *Cordillera Central*, are a high range of mountains, 400 km in length, arranged *en echelon* in a south-west/north-east direction right across the centre of the Iberian Peninsula. Starting from the west in Portugal the Serra da Estrêla, the Sierra de Gata, S. de Peña de Francia, S. de Gredos, S. de Guadarrama, S. de Somosierra, S. de Ayllon lead eastwards, to end in the high tableland of Soria. They rise to heights of over 2500 m and form a natural barrier separating the *meseta* plains of Old Castile in the north from those of New Castile.

The rock scenery is very varied, from the sombre slates of Somosierra to the high gneiss summits of the Guadarrama, reaching 8,100 ft. in the high peak of Peña-lara, and the more uniformly granitic and more intensely glaciated slopes of the Gredos. After the S. Nevada and Pyrenees, the S. de Gredos comprise the highest mountains of Spain. . . .

The mountains are a climatic divide between the harsher continental conditions to the north, and the milder climate to the south. Thus in the northern valleys, the holm oak climbs alone, but on the southern flanks, it mingles with the cork oak and gum cistus, the latter filling the air with its distinctive scent. In the sheltered southern basins, there are mimosa, magnolias, tamarisks and eucalyptus, flora quite different from the Central European species to the north. As a climatic and botanic island, the central Cordillera has been the refuge of plant migrations since the cold Quaternary phases; this explains the curious occurrence of the beechwoods in the Somosierra valley of Montejo. There is however a transition from east to west. In the Somosierra and Guadarrama mountains, the Scots pine rising from 4,000 to *c.* 6,000 ft. has successfully ousted the deciduous oak [*Q. pyrenaica*]. But in the Gredos where the rainfall is higher and the winter temperatures less severe the *Quercus [pyrenaica]* mingled with *Pinus pinaster* is still the prevailing cover of woodland.[10]

Above about 1700 m lies the 'vegetable hedgehog' zone, dominated by *Cytisus purgans*, with *Echinospartum lusitanicum* and *Genista hystrix* higher up, and for short periods in the summer they cover vast expanses of the upper slopes with a carpet of golden-yellow, and the air is scented with vanilla. Such zones are characteristic of the high mountains of North Africa, the Sierra Nevada, and Asia Minor, and a similar zone occurs in the Pyrenees. Heavy winter snow, extremes of temperature, violent winds, and short growing periods have resulted in the evolution of these low densely branched, domed, and usually spiny shrubby growths, and their geographical isolation has often led to the evolution of distinctive endemic species.

The Sierra de Gredos appears from the south as a formidable mountain barrier rising steeply from the plain to the highest summit, the 2592 m of Pico Almanzor. Arenas de San Pedro and Candelada in the south, or Hoyos del Espino and the Parador de Gredos in the north, are good centres for exploring this massif; there is quite a good network of forest roads into the interior.

Towards the end of May a number of interesting plants may be seen in the pinewoods around Arenas de San Pedro, such as *Lavandula stoechas* subsp. *pedunculata*, which is ubiquitous; and where the road crosses the Rio Pelayo *Dictamnus albus*, *Geranium sanguineum*, and *Melittis melissophyllum* are abundant. *Arenaria montana* and *Linaria triornithophora* also grow here in addition to the widespread buglosses *Echium lycopsis* and *E. creticum*. A species endemic to west-central Spain, *E. lusitanicum* subsp. *salmanticum†* is common by roadsides and in open spaces in the pinewoods. It has rather small pale blue flowers, softly white-haired stems and leaves, and numerous tall and gracefully upcurved stems, arising from the large basal rosette. The glandular-leaved foxglove *Digitalis thapsi* is abundant on rocky sections of the road, and common orchids include *Orchis laxiflora* and *Epipactis helleborine*, while the broomrape *Orobanche rapum-genistae* fattens on the leguminous shrubs. The *Cistus-Erica* scrub includes:

Genista florida	*C. ladanifer*
Polygala microphylla	*Halimium ocymoides*
Cistus psilosepalus	*Erica arborea*
C. populifolius	*E. umbellata*

The endemic lupin *L. hispanicus†*, with a short spike of three or four whorls of purple flowers, is common by the wayside.

The motor road to Avila from the south zigzags up to the 1395 m pass,

THE CENTRAL SIERRAS OF SPAIN

1. *Dictamnus albus* 688   **2.** *Cytisus purgans* 502   **3.** *Malva tournefortiana* 734a
4. *Lupinus hispanicus* 519a   **5.** *Crocus carpetanus* 1678b   **6.** *Saxifraga pentadactylis*
404c   **7.** *Tanacetum pallidum* 1428a   **8.** *Allium stramineum* 1606b   **9.** *Linaria*
elegans 1201i   **10.** *Hispidella hispanica* 1511b

11. *Linaria triornithophora* (1204) **12.** *Halimium ocymoides* 797d **13.** *Echinospartum lusitanicum* 513d **14.** *Lavandula stoechas* subsp. *pedunculata* 1110 **15.** *Narcissus rupicola* 1668c **16.** *Armeria juniperifolia* 975b **17.** *Ranunculus abnormis* 245a **18.** *Digitalis thapsi* 1234b

Puerto del Pico, crossing and recrossing the old Roman road. Here the white-flowered *Cytisus multiflorus*, and *C. striatus* with big yellow flowers and snow-white pods, are the dominant shrubs, while the pretty little mauve *Linaria elegans* lines the roadside. A few Spanish bluebells *Endymion hispanicum*, and the purple *Erysimum linifolium* grow near the top of the pass. The tall *Echium flavum*† with small flesh-coloured flowers is fairly common and an abundant roadside plant is *Tanacetum pallidum*, a daisy with white and pale yellow colour forms and tiny silvery leaves.

In the wood of Scots pine, near the village of Hoyocasero there is a magnificent display in May of the massed yellow flowers of *Pulsatilla alpina* subsp. *apiifolia*, covering acres of north-facing slopes with out-sized plants up to 70 cm tall, and with flowers up to 7½ cm across. Other species flowering in late May are *Convallaria majalis*, *Primula elatior*, *Geum sylvaticum* and two peonies coming out a little later. This wood is also famous for the first discovery of *Centaurea rhaponticoides*† in central Spain; it grows 1–1½ m tall and its purple flowers emerge at the end of June from a massive involucre looking like a pine cone. *Lilium martagon* and *Anthericum liliago* also flower at this time, and the yellow-flowered *Centaurea ornata* is notable for the long spines on the flower heads. Several species of *Dianthus* including *D. laricifolius*†, an endemic of central Spain, occur here.

Below the Parador Nacional de Gredos, there is another wood of Scots pine with a rather different flora. At the end of May there is little to be seen but violets and buttercups, *Ajuga pyramidalis*, and *Caltha palustris* in marshy spots. But by the end of June the enchanting little endemic composite *Hispidella hispanica*†, with yellow flower heads with purple centres, flowers abundantly and also the endemic *Erodium carvifolium*†, with attractive bipinnate leaves and flowers in largish umbels. Other woodland species include *Fritillaria hispanica*, *Campanula patula*, *Doronicum carpetanum*, and *Aquilegia dichroa*†. *Linaria elegans* and *Ornithogalum concinnum* are abundant round the Parador

---

12 **Sierra de Gredos** The long range of high mountains spreading across Central Spain form a climatic divide between the colder north and the milder south: they also act as a refuge for plants. The 'hedgehog' zone comprising low bushes of broom, gorse, furze and allied species lies above the forest zone of Scots Pine, which occupies many of the valleys.

13 **Sierra de Montserrat** The fantastic pinnacles or *monserratinos* recall those of the Meteora in Central Greece and like the latter harbour an early Christian monastery. The higher slopes are covered with forests of Holm Oak and numerous shrubby species, while such Pyrenean outliers as *Ramonda myconi* and *Saxifraga callosa* grow on the perpendicular slopes above.

at the end of May, and the hedgehog shrub *Echinospartum lusitanicum* flowers at the end of June.

From Hoyos del Espino a road runs southwards for 12 km into the heart of the range, ending up some distance below the Refugio de Cazadores and the Spanish Alpine Club hut at 1892 m, from which Pico Almanzor the highest point can be reached in three to four hours. Around these huts on the flat moorland in May *Narcissus bulbocodium* var. *nivalis* grows in sheets, while *N. rupicola* is abundant near rocks, and down by the stream there are some clumps of *Narcissus* very like *N. pseudonarcissus* but deeper yellow and with other small differences. (This is probably Pugsley's species *N. confusus* which may prove to be only a form of *N. hispanicus.*) Near patches of melting snow *Crocus carpetanus* grows in groups, and the Spanish bluebell *Endymion hispanicus* grows among the rocks. *Ranunculus abnormis*† grows abundantly at higher altitudes; it has narrow basal *Scilla*-like leaves and solitary yellow flowers. In marshy meadows the pale blue *Scilla verna*, and *Saxifraga latepetiolata*† grow in profusion.

A month later, above the huts, the endemic *Antirrhinum grossii*† can be found in flower on perpendicular rock faces; it has large white flowers with faint purple stripes on the lip, and is a very brittle plant. The other interesting endemic is a curious plant of the rocky slopes, *Reseda gredensis*†, known only from the Sierra de Gredos and S. da Estrêla. On the highest slopes both *Cytisus purgans* and *Echinospartum lusitanicum* make a fine show by early July, and other plants of interest are *Linaria saxatilis*†, *Jasione amethystina*†, *Campanula lusitanica*, and *Senecio adonidifolius* with attractive leaves. In general, however, the heights have few of the true high alpine plants characteristic of either the Pyrenees or the Sierra Nevada.

The Sierra de Guadarrama has a less rugged appearance than the Sierra de Gredos although it is almost as high; by mid-June most of the snow has disappeared except for a few patches high up on the northern slopes. Puerto de Navacerrada, the pass on the road north to Segovia, is probably the best point from which to explore the highest mountain, Pico Peñalara (2430 m).

The Sierra de Guadarrama is much more heavily forested than the Sierra de Gredos, and extensive forests of *Pinus sylvestris* clothe both sides of the range. Venerable, weather-beaten giants, with their dis-

---

14 **Picos de Europa** The Desfiladero de la Hermida is but one of several dramatic gorges carved through the Cantabrian mountains by fast running rivers that drain into the Bay of Biscay. On cliffs such as these, species of an interesting genus *Petrocoptis* of the Pink family can be found.

tinctive rust-coloured bark on the upper half of the trunk, climb the steep mountainsides to 2000 m or more in places. There is a shrub undergrowth of *Juniperus communis*, **Genista florida*, *Adenocarpus telonensis*, and **Erica arborea*, with herbaceous species such as *Linaria nivea†*, **L. triornithophora* and *L. incarnata*, *Galium verum*, *Senecio adonidifolius*, and *Luzula lactea*. In shady places are *Valeriana tuberosa*, *Galium broterianum*, *Ranunculus bulbosus* subsp. *gallecicus*, and **Endymion hispanicus*. Above the forests lies a dense belt of dwarf juniper and **Cytisus purgans*, with very little else.

From the Puerto de Navacerrada there is a pleasant walk to the west, up to Siete Picos (2138 m). This area is one of the main strongholds of **Narcissus rupicola*, which for some reason grows more freely here than on the Sierra de Gredos or Serra da Estrêla. Both the charming dwarf pink thrift **Armeria juniperifolia*, and *Ranunculus gregarius* are indeed gregarious, especially near the summit, as well as a yellow *Gagea* species. **Tanacetum pallidum* is frequent and attractive; **Saxifraga pentadactylis* grows in rock crevices, and the small *Biscutella intermedia* occurs sparsely. **Crocus carpetanus* appears in these mountains as soon as the snow melts, and *C. asturicus* flowers in the autumn.

The Sierra de Peña de Francia lies west-north-west of the Sierra de Gredos and seen from the east it appears as an imposing pinnacle of rock, with a monastery on its summit. This is accessible by car and it is worth a visit not only for its extensive views, but for the magnificent display of floral colour at the end of May. A patchwork quilt of purple and gold, composed of the white-flowered **Erica arborea*, the purple-flowered **Erica australis*, and several golden-yellow leguminous shrubs, notably tall bushes of the common broom *Cytisus scoparius*, a low-growing thicket of **Halimium alyssoides*, and higher up on the mountain **Chamaespartium tridentatum*. A few scattered examples of the very handsome shrub **Adenocarpus complicatus* with tall inflorescences of rich gold occur on the mountain and comes into flower at the beginning of June. The fine-grained quartzite of the summit supports little plant life, except **Narcissus rupicola* and a *Gagea* species at the end of May.

15. The Southern Iberian Mountains (Sierras of Cuenca and Teruel)

These mountains comprising the Montes Universales, Sierra de Albarracin, Sierra de Javalambre and many others, form the southern part of the old-fold Iberian mountain system. Lying north of the Madrid–Valencia road about 150 km east of Madrid, they are formed largely of Jurassic and Cretaceous limestones but other outcrops such as sandstones, shales, and Triassic schists occur in places as, for example, around Bronchales. Both calcicole and calcifuge species are consequently found in parts of the range and the flora is rich and diversified. Early summer is the time to see the plants at their best, though towards the end of July may also be rewarding for the later flowerers.

These mountains may reach altitudes of over 2000 m; but as they rise from a plateau nearly 1000 m high, no great exertion is required to reach the highest points. There are no prominent or sharp peaks; instead long series of rising and falling ridges, following one after the other, make it difficult to decide exactly where the highest points lie. There are some spectacular gorges, steep hill slopes, and rocky precipices, and, surprisingly in limestone country, broad glaciated U-shaped valleys, as at Tragacete. The scenery is thus very different from that of the Central Sierras.

From Cuenca a road leads northwards to Beteta, the last few kilometres running through the celebrated 'Hoz de Beteta', or gorge, of the Rio Guadiela. Just before the start of this gorge a turning on the right leads to the Balneario de Solan de Cabras, at the entrance of another gorge, that of the Rio Cuervo, up which there is no motor road, only a goat track running for 10 kilometres through superb scenery with a rich and varied flora. Both gorges are well forested with pines and an undergrowth of calcicole shrubs such as *Genista scorpius*, *Cistus laurifolius*, *Amelanchier ovalis*, rosemary, box, thyme, *Lavandula latifolia*, and *Lithospermum fruticosum*, and small trees such as *Sorbus aria*, *Acer monspessulanum*, and *Viburnum lantana*. Interesting herbaceous species are: *Ononis rotundifolia*, *Lathyrus filiformis*, the white and blue flaxes *Linum suffruticosum* and *L. narbonense*, *Geranium sanguineum*, *Allium moly*, *Anthericum liliago*, and *Gladiolus illyricus*. Two of the typical rock plants of this region, *Antirrhinum pulverulentum*† and *Sarcocapnos enneaphylla* can be seen near the Balneario. *Asphodelus ramosus* is abundant. Here and there the fine *Orobanche macrolepis* may be seen growing on rosemary. Somewhere in the Hoz de Beteta *Pinguicula vallisneriifolia*† was discovered far from its other colonies in the Sierra de Cazorla.

From Beteta one can drive via Cañamares and Poyatos direct to Tragacete on roads not to be found on small-scale motoring maps, or one can drive round via Cuenca and thence to Tragacete, passing on the way, near Uña, Ciudad Encantada—the Enchanted City—where the soft limestone has been eroded into such weird shapes and fantastic formations that it has become a major tourist attraction. Forest roads, though not marked on the usual road maps, are open to the public and make exploration of the hills possible; one of these is a short cut from Tragacete over the watershed to the Montes Universales, Sierra de Albarracin and Teruel. The whole region is well forested with *Pinus*

THE SOUTHERN IBERIAN MOUNTAINS

1. *Salvia lavandulifolia* 1143c **2.** *Fritillaria hispanica* 1623b **3.** *Thalictrum tuberosum* 255b **4.** *Consolida orientalis* (213) **5.** *Limodorum abortivum* 1921 **6.** *Antirrhinum pulverulentum* 1197f **7.** *Hepatica nobilis* 219 **8.** *Ranunculus gramineus* 245

9. *Arctostaphylos uva-ursi* 925   **10.** *Matthiola fruticulosa* (301)   **11.** *Cephalanthera rubra* 1920   **12.** *Berberis hispanica* 261a   **13.** *Orchis patens* 1894b   **14.** *Hedysarum humile* 634b   **15.** *Sarcocapnos enneaphylla* 279a   **16.** *Geum sylvaticum* 442a

nigra and a little Maritime pine at lower altitudes, and Scots pine above. Thickets of box, *Buxus sempervirens*, and small trees of the Snowy Mespilus, *Amelanchier vulgaris*, are characteristic.

Among the colourful plants seen along the Cuenca-Tragacete road are the handsome shrubby pink *Ononis fruticosa*; the lovely blue flowers of *Vicia onobrychioides*; the flaxes, *Linum narbonense* and *L. suffruticosum*; *Consolida orientalis*, and *Salvia phlomoides*. There are several species of salvias, including the striking woolly-leaved, white-flowered *S. aethiopis* and the widespread *S. verticillata* and *S. pratensis*, while the tall, handsome, lavender-flowered *S. lavandulifolia* forms low bushes.

Other attractive species are:

Thalictrum tuberosum	*Cistus laurifolius*
Lathyrus filiformis	*Helianthemum croceum*
Euphorbia serrulata	H. cinereum
E. nicaeensis	Echium italicum
Dictamnus albus	Centaurea triumfetti

Various interesting plants grow on the rock and pinnacles of the Ciudad Encantada: *Sarcocapnos enneaphylla*, with yellow and white fumitory-like flowers, clings to impossibly dry-looking rock faces, while the white flowers of *Antirrhinum pulverulentum*† peer out of the rock crevices, with *Saxifraga corbariensis*† and *Draba dedeana*†. *Tulipa australis* and *Tanacetum pallidum* may be seen on the flat tops of the rocks.

A walk up the hill above Tragacete into the subalpine zone near the top of the ridge, is a memorable experience on a fine day. Halfway up, the gently sloping grassy meadows, though grazed by large flocks of sheep, are covered with the massed flowers of *Adonis annua*, *Ranunculus gramineus*, *Geum sylvaticum*, *Muscari neglectum*, *Fritillaria* species, the small white-flowered *Saxifraga carpetana*† in abundance, and three orchids *Orchis morio*, *O. mascula* and a few *O. patens*†. Morels grow under the pines, and a hatful makes a pleasant addition to the evening meal. The crest of the ridge is reached after passing through a scrub of box in which *Hepatica nobilis* grows, and on the crest are *Arenaria aggregata*, *Biscutella valentina*, *Tanacetum pallidum*, *Polygala calcarea*, and on the rocks *Draba dedeana*†.

In June a number of conspicuous yellow-flowered species are at their best; among them, the shrubs *Berberis hispanica* and *B. vulgaris*. *Genista scorpius*, and *Ononis aragonensis*†, and the herbaceous species *Allium moly*, which favours roadside ditches, *Coronilla minima*, forming a lovely yellow carpet under the pines at the higher elevations, and *Tetragonolobus maritimus*. The beautiful white-flowered *Thalic-*

trum tuberosum, the best of all European Meadow Rues, is commonly seen, while *Convolvulus lineatus*, with white flowers and silvery leaves, forms low mats on the road verge.

This part of Spain is rich in species of orchids and the following have been recorded in this region:

Ophrys scolopax†	Anacamptis pyramidalis
*O. sphegodes	Coeloglossum viride
*Orchis coriophora	*Epipactis atrorubens
O. mascula	E. helleborine
*O. patens†	E. leptochila
*O. ustulata	E. palustris
Dactylorhiza elata	Cephalanthera damasonium
D. incarnata	C. longifolia
	*C. rubra
	*Limodorum abortivum

A new subspecies possibly of *Dactylorhiza romana*†, with distinctive orange-marked flowers, was discovered here as recently as 1966.

The Teruel side has a rather different flora. On the slopes round the headwaters of the Tagus, *Juniperus thurifera* reaches a height of 10 m and a girth of 4 m in woods that are probably the finest in Europe. The crown is broad and almost globular, and dark green, so that the hillsides seen from a distance appear as if flecked with black. Other junipers *Juniperus oxycedrus*, *J. phoenicea*, and *J. sabina* are common in the undergrowth of the forests, together with much *Cistus laurifolius*.

Plants of interest round Albarracin include *Colchicum triphyllum*†, abundant on calcareous soils round the town and in the surrounding mountains, and flowering from November to February; *Artemisia assoana*, common on walls and stony waste ground; *Saxifraga carpetana*†, the Spanish endemic with white flowers in a rather compact cluster; *Astragalus turolensis*†, a stemless, yellow-flowered perennial of stony ground, a North African species whose only occurrence in Europe is in this part of Spain; and *Draba dedeana*† in rock crevices. On the rocks below Albarracin grow *Hypecoum imberbe*, *Nepeta nepetella*, and large tufts of *Ptilotrichum spinosum*.

Near the highest point of the road to Bronchales in the Sierra de Albarracin, the charming *Pulsatilla rubra*, which is always dark violet in Spain, pushes its way through a carpet of *Arctostaphylos uva-ursi*. *Ranunculus gregarius* grows in sub-alpine meadows with the prostrate *Rosa pimpinellifolia*, and *Narcissus triandrus* is abundant locally in the pine woods south of Orihuela. The Sierra Alta, at 1855 m the highest part of the Sierra de Albarracin, is easily reached from Bronchales and is worth exploring. Round a fire look-out tower on the crest of the ridge overlooking the Tagus at about 1750 m, grows, on bare limestone scree, the finest display of *Erodium daucoides† we have ever seen.

Roadside plants of the Cuenca-Teruel road include *Coronilla scorpioides*, with a projection like a scorpion's sting at the end of the pod; *Matthiola fruticulosa* a form with narrow wavy-edged brown petals; the pink-flowered *Teucrium gnaphalodes*; *Thesium divaricatum*; and *Vicia pannonica* subsp. *striata* with dirty purple flowers.

Also accessible from Teruel are the Sierra de Javalambre, 1958 m, to the south, and the Sierra de Gúdar, 2024 m, to the east. The rocks are mainly Triassic and Cretaceous limestone, with some siliceous outcrops. These mountains have been studied exhaustively by Rivas Goday and Carbonell, who have described the plant associations and provided a list of species.

16. Sierra de Montserrat and Sierra del Montseny

Many tourists to Catalonia visit the famous monastery of Montserrat founded in 880 A.D., where medieval legend placed the Holy Grail and which inspired Wagner's Parsifal. The monastery is situated high up amongst the towering masses of rock in a remarkable situation with magnificent views to the Pyrenees and the sea. These summit rocks are Oligocene conglomerates, eroded into fantastic shapes of rounded smooth-walled pinnacles and buttresses. The highest point is San Jerónimo, 1238 m; the whole ridge is rather festooned with cable cars and funiculars.

Forests of pine and evergreen oak cover the lower and middle slopes and a dense shrub undergrowth consists of *Juniperus phoenicea*, *J. communis*, *Coronilla emerus*, *C. minima*, *Amelanchier ovalis*, *Arbutus unedo*, *Erica arborea*, *Buxus sempervirens*, *Bupleurum fruticosum*, and *Viburnum tinus*. The field layer of these often dense forests include *Hepatica nobilis*, *Lilium martagon*, *Orchis mascula*, and *Ruscus aculeatus*.

On the more exposed gullies and ridges towards the rocky summits and on the cliff faces are: *Genista hispanica*, *Astragalus monspessulanus*, *Dorycnium pentaphyllum*, *Onobrychis saxatilis*, *Lithospermum fruticosum*, *Globularia valentina†*, *Aphyllanthes monspeliensis*, and on the vertical summit cliffs *Ramonda myconi*. Orchids flowering in June include *Ophrys fusca*, *Dactylorhiza fuchsii*, *Aceras anthropophorum*, and *Limodorum abortivum*.

On the drier south-facing slopes the vegetation has a more Mediterranean aspect with fewer trees and more open scrub of *Cistus*, *Rosmarinus*, and *Quercus coccifera*; and in open places *Ranunculus gramineus*, the beautiful *Thalictrum tuberosum*, *Astragalus monspessulanus*,

*Convolvulus lanuginosus, *Phlomis lychnitis, Thymus vulgaris, Antir-
rhinum majus subsp., Orobanche species, and Muscari neglectum can be
found.

The most interesting plant in these mountains is Saxifraga callosa
subsp. catalaunica† which has beautiful silvery, lime-encrusted rosettes
and large wand-like inflorescences of white flowers. It is similar to S.
longifolia but has clusters of rosettes some of which produce flowers
every year. It grows on the high cliffs here, and also in the hills near
Marseilles, the only two places it is to be found.

The Sierra del Montseny is quite different, being composed of granite
and gneiss, and has in consequence a less interesting flora; it rises to
1712 m. Quiet and secluded in its own woods, the Hostal San Bernat is an
ideal centre for exploration, with some fine walks through the woods
and up the mountains in splendid solitude (except at weekends).

The walk up to Matagalls, about 1½ hours each way, leads first through
woods of Quercus ilex, with a little holly, while higher up through
extensive beech woods the Col Pregon, 1600 m, can be reached. On the
shady side of the summit grows the endemic saxifrage, S. vayredana†,
with strongly aromatic resinous leaves and small white flowers with
petals 4–6 mm; and also S. granulata, Sempervivum species, Viola lutea
subsp. sudetica, Arenaria and Potentilla species, Plantago carinata,
Centaurea montana, with leaves decurrent to the leaf below and a black
border to the involucral bracts, and near the hotel, C. pectinata. In the
fields grows C. cyanus, and fine examples of *Limodorum abortivum grow
in the woods near the hotel. *Lavandula stoechas is abundant round San
Marcal and the Col de Sariera.

17. Costa Brava

The Catalan coast from the French frontier to Blanes is commonly
known as the Costa Brava. Its wide variety of rocks—ranging from
Palaeozoic schists and sandstones to crystalline rocks, and from Cretaceous and Jurassic limestones to conglomerates—gives it a great diversity of attractive scenery.

Thanks to the higher rainfall (often 800 m a year) and shorter summer
drought, the Costa Brava is well wooded compared with the equally
famous Costa del Sol. The coastal Catalan Mountains are covered by
extensive forests consisting mainly of Cork oak and the pines, P. pinea
and P. halepensis, and much of the country's beauty is due to these
forests coming steeply down to the coast, fringing a series of lovely blue
water bays and inlets. This is only broken by the low-lying estuaries of
the Rio Ter and Rio Fluviá which flow into the Golfo de Rosas.

The undergrowth of these forests is composed of the characteristic

Cistus-Erica community. On treeless slopes there is a high maquis, 3 m or more, of *Arbutus unedo*, **Erica arborea*, **Genista scorpius*, *Calicotome spinosa*, *Cistus salvifolius*, and *Ulex* species; and with only the two parasites **Cytinus hypocistis* and **Limodorum abortivum* under this dense impenetrable growth. In more open places are: *Clematis flammula*, *Coriaria myrtifolia*, **Daphne gnidium*, **Cistus monspeliensis*, *Calluna vulgaris*, **Lavandula stoechas,* and *Viburnum tinus.* But one has only to cross the crest of the coastal sierra and pass over to the landward slope, out of the direct influence of the Mediterranean, to arrive quite suddenly into coppiced chestnut woods and a vegetation which is almost purely central European with hazel, ivy, holly, and such herbaceous shade-loving species as:

Stellaria holostea	*Sanicula europaea*
Ranunculus ficaria	*Primula veris*
Geranium robertianum	*Symphytum tuberosum*
Mercurialis perennis	*Pulmonaria longifolia*
Euphorbia amygdaloides	*Teucrium scorodonia*
Viola riviniana	*Arum italicum*

The coastal plains are much cultivated but interesting species occur here, and on grassy banks grow **Lathyrus tingitanus*† a bright magenta, large-flowered robust annual pea, *Plantago sempervirens* a shrubby species, *Vicia villosa*, *Ononis natrix*, **O. fruticosa*, *Lactuca perennis*, and **Allium roseum*. On the sea cliffs in particular are *Carpobrotus edulis*, **Teline linifolia*, *Helichrysum stoechas*, and *Inula viscosa*.

18. Sierra Segundera and Sierra de la Cabrera

These mountains lie in the north-west corner of the province of Zamora, about 40 km south of Ponferrada, and are connected by the Montes de Leon with the Cantabrian Mountains to the north. They are most easily approached from the south, from Puebla de Sanabria. There is also an access road to the Peña Trevinca from the N120 between El Barco and Ponferrada.

The mountains are of Cambrian and Silurian rocks, with much granite; and the scenery, with heather dominating the vegetation, is reminiscent of the Highlands of Scotland. Several summits exceed 2000 m; Peña Trevinca and Vizcodillo in the Peña Negra are the highest at 2124 m and on them snow lies throughout much of the year.

The flora has affinities with that of the Central Sierras, for example *Centaurea raponticoides*† reappears here; and a number of central European species like *Streptopus amplexifolius* here reach their western-most limit.

Plants of interest round the lake near Ribadelago which is about 10 km north-west of Puebla, include *Hispidella hispanica†, which is as abundant by the roadsides as *Saxifraga granulata is on the rocks; *Echinospartum lusitanicum growing at unusually low elevation; a Serratula; Phalacrocarpum sericeum† a white composite endemic to this area. The hillsides above Ribadelago look rather bare, but there are pockets of sweet chestnuts and oaks, particularly *Q. pyrenaica, and much of the area is covered with the usual Erica-Cytisus heathland of *Erica umbellata, *E. australis, *E. arborea, *Cytisus purgans, *C. multiflorus, and *Chamaespartium tridentatum.

One of the best localities for plants is the Barranco de Fornillo some distance up the Rio Tera. Geranium bohemicum† has been collected here at 1300 m, its only known locality in Spain. Other species of interest include: Aconitum lamarckii, *Paeonia officinalis subsp. humilis, Saxifraga stellaris and S. cuneifolia, Lathyrus niger, Vicia orobus which is abundant in the oak wood, Geranium sanguineum, G. pyrenaicum, G. columbinum, Eryngium duriaei†, Verbascum lychnitis (here probably at its western limit), and Lilium martagon.

To the north the mountains tower up above Ribadelago in a steep escarpment. A two-hour climb brings one to the crest, where an extensive rolling plateau is unexpectedly revealed with several small lochs of glacial origin and a number of reservoirs built by the hydro-electric company. This plateau can also be reached by car up a side road that turns off the N 525 near the Zamora–Orense provincial boundary. This is all magnificent hill-walking and camping country, though with the limited flora associated with granite mountains. Some plants of this plateau region are Halimium umbellatum, *Narcissus rupicola, Ajuga pyramidalis; in wet patches, Viola lutea, Pedicularis sylvatica, Veronica serpyllifolia subsp. apennina; and in rock crevices Saxifraga continentalis with dark purple stems. A good list of plants found in this region is given by M. Losa España.

19. The Cantabrian Mountains

The high ridge of the Cantabrian mountains extends for 240 km running parallel to the north coast of Spain, from the Peña Labra and the Picos de Europa south of Santander to the Sierra del Caurel in Galicia in the west. The general altitude of the ridge is between 1600 and 2000 m, with summits of up to 2500 m; it seldom falls below 1500 m. The summits are often flat and desolate with steep cliffs and ravines at lower altitudes, particularly to the north, while to the south they fall gradually to the meseta plateau of North Castile. The eastern part of these mountains is largely composed of secondary limestones, the central region of primary slates

THE CANTABRIAN MOUNTAINS

1. *Narcissus nobilis* 1665k   **2.** *Dianthus furcatus* 191a   **3.** *Daboecia cantabrica* 922
4. *Chaenorhinum origanifolium* 1211b   **5.** *Lathyrus filiformis* 565c   **6.** *Androsace villosa* (954)   **7.** *Saxifraga conifera* 411b   **8.** *Ranunculus amplexicaulis* 248a
9. *Aceras anthropophorum* 1905

10. *Pedicularis foliosa* 1252   **11.** *Geum rivale* 443   **12.** *Petrocoptis glaucifolia* 163b
13. *Narcissus triandrus* 1668d   **14.** *Teesdaliopsis conferta* 348e   **15.** *Saxifraga*
aretioides 410b   **16.** *Linaria supina* 1206   **17.** *Himantoglossum hircinum* 1907

and limestones, and the western part of slates and quartzites; the mountains become progressively older towards the west.

One of the few roads running through the heart of the Cantabrian mountains passes south from Oviedo over the Puerto de Pajares, 1379 m. The parador here affords a convenient base for exploring the main watershed. On the northern slopes there are extensive forests of beech and oak with only a limited flora. But on the south side above the parador heathlands with heather and gorse and, higher up, the pastures are rich in species which in addition to common saxifrages and gentians include:

*Arenaria montana	*Daboecia cantabrica
Anemone nemorosa	Primula vulgaris
Ranunculus ficaria	Ajuga pyramidalis
R. aconitifolius	Pedicularis verticillata
*Teesdaliopsis conferta†	*Pinguicula grandiflora
Cardamine pratensis	*Erythronium dens-canis
*Chamaespartium tridentatum	*Fritillaria hispanica
Viola bubanii†	*Narcissus triandrus

To the east of the pass, flowering near melting snow at 2000 m are:

*Ranunculus amplexicaulis†	Phalacrocarpum victoriae?
Thymelaea calycina†	*Narcissus asturiensis†

Further east there used to be the village of Riaño in one of the most beautiful valleys in the whole of Spain, but a high dam across the Esla a few kilometres below the village is nearing completion at the time of writing, and the valley will soon be engulfed by a lake some 20 km long, drowning the existing roads and villages; new roads will presumably be cut to maintain communications from the south to the three passes across the watershed north of Riaño.

The Puerto de Tarna, on the N 635 to Orense and the nearby Puerto de las Señales† produce several species of narcissus, including *N. nobilis†, *N. bulbocodium, and N. hispanicus amongst an abundance of:

Helleborus viridis	*Daphne laureola
Anemone nemorosa	Primula elatior
*Ranunculus amplexicaulis†	*Lithospermum diffusum
Thymelaea coridifolia†	*Asphodelus albus
*Erythronium dens-canis	

Lower down there are fields of bluebells, Endymion non-scriptus.

The Puerto del Pontón on the N 637 to Cangas de Onis is approached by marshy meadows full of orchids and the orchid-like showy Pedicularis verticillata. Around the pass, or on the heights above it, plants listed by

Dresser include: *Saxifraga cotyledon, S. canaliculata†, *Cytisus purgans, Genista obtusiramea†* and **Prunella grandiflora*.

From the pass the road drops down through beech woods with an undergrowth of *Vaccinium myrtillus* until at an altitude of some 700 m begins the gorge of the Rio Sella, or Desfiladero de los Beyos—one of the finest of all limestone gorges. The river in places runs through a deep cleft with walls only a metre or two apart, and the rocks rise in great precipices to the sky. The two most striking plants of the rock walls are a white form of *Petrocoptis glaucifolia†* and *Antirrhinum braun-blanquetii†* which has a large cream corolla with yellow palate and a few purple lines on the upper lip. Other common plants include *Aquilegia* species, *Saponaria ocymoides, Geranium sanguineum, *Hypericum nummularium, Erinus alpinus, Centranthus ruber, Leucanthemum vulgare* and *Reseda glauca†* with long linear leaves.

Further east, dividing the western mountains from the central block of the Picos de Europa, is the similar gorge of the Rio Cares, but this has to be explored on foot either from Posada de Valdeón on the south or from Camarmeña on the north side. It forms the eastern boundary of the National Park of Covadonga, and the official guide describes the inhabitants of the isolated village of Caín as 'los Sherpas de la Peña', who exist on their sheep and cattle and by hunting. It must be a fascinating place to explore.

Further east again is the Puerto de San Glorio, 1609 m, which leads over to Potes and a third great gorge the Desfiladero de la Hermida. From the pass are tremendous views of the white limestone crags of the Picos de Europa to the north, and the rounded summits of the Peña Prieta to the south. Here grows *Narcissus nobilis†*, the finest of the daffodils; with it other interesting species such as: *Genista florida, G. obtusiramea†, Hypericum richeri, *Fritillaria hispanica*, and *Iris xiphioides†*. On the way down *Eryngium bourgatii†, *Linaria triornithophora*, and *Erica vagans* all grow in open spaces, and *Digitalis parviflorus†* becomes common near Potes on shale.

The Picos consist of highly folded Lower Carboniferous rocks, are dominated by limestone, and reach an altitude of over 2600 m. Vertical crags of resistant rock, with large areas of loose and fixed screes beneath them, often occur near the summits of the hills. A large flat area at 1700 m, shows signs of glaciation in the form of huge moraine deposits, and now forms an extensive pasture. This is the site of the well-known Refugio Aliva. Below are steep-sided valleys, which on the south-east of the range lead down to the lowlands of Liebana. The tree line is at about 1300 m, but varying by as much as 200 m, according to local conditions.[11]

Ideally situated for exploring the Picos is the Parador Fuente De, all on its own in a peaceful pasture where cow bells continually sound, set back from the base of the huge cliffs that form the southern buttresses of the Peña Vieja. From here a cable car takes one 800 m up to the top of the cliffs and drops one among rough mountain pastures and extensive limestone screes. A very rough road leads to the Refugio de Aliva which can also be reached direct from Espinama by another rough, very steep road.

In the pastures and screes near the parador in mid-June there is much of interest including:

*Linum suffruticosum
*Androsace villosa
*Teucrium pyrenaicum
Linaria alpina
*Chaenorhinum origanifolium
Erinus alpinus

*Globularia repens
G. nudicaulis
*Aceras anthropophorum
Gymnadenia conopsea
*Ophrys insectifera
*O. sphegodes
*Orchis ustulata

and a month later, Coeloglossum viride, Nigritella nigra, and *Epipactis atrorubens.

Between the upper cable car station and the Refugio de Aliva in mid-June *Ranunculus amplexicaulis† carpets the turf like daisies and one can also find:

*Draba dedeana†
*Matthiola fruticulosa
*Saxifraga aretioides†
*S. conifera†

*Daphne laureola
Gentiana verna
*G. acaulis
*Narcissus asturiensis†

*Linaria supina and eight other species of Saxifraga are listed by Dresser.

In mid-July high up amongst the splintered limestone and rock crevices the following are in flower:

*Anemone pavoniana†
Arenaria purpurascens†
Silene acaulis
Iberis procumbens?
Saxifraga canaliculata†

*S. conifera†
Potentilla nivalis
*Erodium petraeum subsp.
   glandulosum†
*Daphne laureola subsp.
   philippi†

---

15 **Sierra de Urbión**   The Laguna Negra is a glacial lake lying under the summit of Urbión, which rises to 2252 m. These mountains contain both acidic and calcareous rock and in consequence have an interesting flora worthy of further exploration.

16 **Montes Universales**   A Cuenca river gorge in the foothills of the Montes Universales. On cliffs such as this may be found Sarcocapnos enneaphylla, Antirrhinum pulverulentum, Saxifraga corbariensis and Draba dedeana. The surrounding country is rich in orchid species. (*overleaf*)

Helianthemum croceum	*Chaenorhinum origanifolium*
Gentiana occidentalis†	Erinus alpinus
G. lutea	*Globularia repens*
Lithospermum diffusum	Scilla verna
Myosotis alpestris	*Narcissus asturiensis*†
	Viola cornuta?

Sempervivum cantabricum† grows here, with deep red petals and large rosettes of dark green leaves with red apices. Another endemic *S. giuseppii*† grows on a nearby mountain, Peña Espiguette (south of Peña Prieta) in North Palencia; it has tiny rosettes which are very hairy and the petals are red with a narrow white margin. *Merendera montana* begins to flower in late July; it is locally plentiful.

Situated on the north-west flank of the Picos de Europa is the Parque Nacional de la Montaña de Covadonga (fully described in a publication of the Servicio Nacional de Pesca Fluvial y Caza by G. M. Goyanes, with an excellent map and a comprehensive list of the flora with both Latin and Spanish names). The park covers 16925 hectares and extends from Covadonga to the Rio Cares; its highest peak is the Peña Santa de Castilla, 2596 m. The only motor road entering the park runs through Covadonga, and it was in this narrow valley that Pelayo fought the famous battle of 718 A.D. against the Moors and was afterwards crowned king of the Asturias; an obelisk marks the spot. The park boundary lies just below this point, though no notice has been erected, here or elsewhere, to indicate that one is entering a National Park.

According to Muñoz, the original vegetation of the park below the alpine level consisted mainly of extensive forests of the Durmast oak *Quercus petraea*, and *Q. pyrenaica*, with alders, poplars, and willows along the margins of the streams and in the more humid areas. But centuries of uncontrolled timber and firewood extraction have so altered the forests that today beech, *Fagus sylvatica*, is the principal species with *Quercus petraea* and *Castanea sativa* less frequent. Similarly,

---

17 **Basque Lands** A typical agricultural pattern of the North-East of Spain, in a country of sea mist and heavy rainfall. In the valleys are apple orchards, cornfields, and meadows, while on the steep hillsides are forests of Oak, Chestnut, Ash and Beech; the bleaker summits are covered with western heaths.

18 **Spanish Pyrenees** On the Spanish side of the Pyrenees is the National Park of Ordesa, enclosing magnificent canyon-like country penetrating the heart of the high Pyrenees. Almost primaeval forests of Beech, Pine, and Silver Fir clothe the lower slopes while there is a rich Pyrenean alpine flora at higher altitudes.

centuries of intensive grazing, burning, and mowing have greatly modified the ground vegetation. Lack of humus has caused a progressive acidification of the soil, resulting in the development of heaths. *Daboecia cantabrica*, an indicator plant of leached acid soils, is now widespread on areas that formerly carried oak forest.

The sweet chestnut *Castanea sativa* is abundant on north-facing slopes; and associated with it is the hazel, *Corylus avellana*, especially in the gorges. Other trees occurring in the park include: ash, *Fraxinus excelsior*; lime, *Tilia* species; sycamore, *Acer pseudoplatanus*; walnut, *Juglans regia*; elm, *Ulmus glabra*; and holly, *Ilex aquifolium*; while the yew, *Taxus baccata* is found mostly as a stunted shrub.

In the shrub layer may be mentioned *Juniperus communis* subsp. *nana* mainly on north- and east-facing slopes, *Sorbus aucuparia*, and *S. aria*, *Genista hispanica* and *G. hystrix* mainly on south-facing slopes. The heaths are composed of *Erica cinerea*, *E. arborea*, *E. vagans*, *Vaccinium myrtillus*, *Crataegus monogyna*, *Genista florida*, and *Rubus idaeus*. *Calluna vulgaris* flourishes in areas where the beech has been destroyed by fire.

In the undergrowth of the beech forest the following occur:

Stellaria holostea	Euphorbia amygdaloides
Helleborus viridis	E. hyberna
H. foetidus	*Daphne laureola
Anemone nemorosa	Cornus sanguinea
*Hepatica nobilis	Hedera helix
Ulex species	*Daboecia cantabrica
Geranium robertianum	Primula elatior
Mercurialis perennis	Orchis species

From Covadonga a narrow steep road leads up to the two lakes of Enol and Ercina at 1100 m. On a ridge between them is a comfortable Refugio, which makes an ideal base for exploring the alpine zone for those who do not wish to camp out by the lakes. As it rains on average for 193 days in the year, good tents are necessary if camping is contemplated; the abundance of *Pinguicula grandiflora* by the roadside is a good indication of the general humidity.

The area round the lakes is heavily grazed by cattle and sheep, and consists of a lawn-like turf with scattered, closely cropped bushes of *Genista hystrix* and *G. hispanica*. On the cliffs near the Refugio grow *Gentiana occidentalis*†, *Globularia nudicaulis*†, and *Scilla verna*. Dresser recorded the following species among others around the lakes:

Aconitum vulparia	*Hepatica nobilis
A. napellus	*Cardamine raphanifolia

Reseda glauca	**Linum viscosum*
Saxifraga hirsuta	**Eryngium bourgatii†*
S. trifurcata†	**Campanula glomerata*
*Amelanchier ovalis	*C. pusilla*

The eastern end of the road from Covadonga to Panes passes through attractive scenery, and in places it follows a fine salmon river much frequented by fishermen. With luck one may see *Lilium pyrenaicum†* nodding its yellow head from the top of the roadside bank. From Panes the road turns south to Potes running up through the Desfiladero de la Hermida, another fine gorge but broader and less spectacular than the gorge of the Rio Sella. The flora is similar, except that **Petrocoptis glaucifolia†* growing on the rock walls here is pink and not white; at one place **Erigeron karvinskianus* grows in abundance; and a flax with with large pink flowers, **Linum viscosum*, and *Dianthus monspessulanus* can also be seen.

From Potes the road up to the Puerto de Piedras Luengas, 1329 m, is notable mainly for the abundance of lizard orchids, **Himantoglossum hircinum*, while just over the pass on the south side once again there are fields of that king of daffodils **Narcissus nobilis†*, adding a final touch to a splendid last view of the Picos de Europa.

20. The Northern Iberian Mountains

The mountains of Old Castile lie in the provinces of Burgos, Logroño, and Soria and they include the Sierra del Moncayo, the Sierra de Urbión, and the Sierra de la Demanda. They are well off the tourist track and can most easily be explored from Soria, though simple accommodation in the small forest villages is a good alternative.

A favourite excursion in the Sierra de Urbión, north-west of Soria, is to the Laguna Negra, a glacial trout lake at 1700 m, which is reached by a well signposted forest road from Vinuesa. At one point on the way up there is a massed display of *Aquilegia vulgaris*, common in these mountains but nowhere as spectacular as here. The forest round the lake is a mixture of beech and Scots pine with an undergrowth of *Erica* and *Vaccinium*. In the vicinity of the lake **Ranunculus amplexicaulis†*, *R. gregarius*, *Viola tricolor*, a little *Ajuga pyramidalis*, *Gagea* species, **Tulipa australis*, **Erythronium dens-canis*, **Scilla liliohyacinthus*, *Streptopus amplexifolius*, *Convallaria majalis*, and **Narcissus bulbocodium* are some of the species to be found.

From the lake one can climb to the summit of Urbión (2228 m) in one and a half to two hours, but this peak can also be reached by car along

THE NORTHERN IBERIAN MOUNTAINS

1. *Erodium carvifolium* 653c **2.** *Xeranthemum inapertum* (1464) **3.** *Anthericum liliago* 1596 **4.** *Linaria repens* 1202 **5.** *Iris spuria* 1688 **6.** *Astragalus sempervirens* 528a **7.** *Tanacetum alpinum* 1428 **8.** *Dianthus subacaulis* 194b **9.** *Orchis militaris* 1891

10. *Malva moschata* 734   **11.** *Dianthus lusitanus* 194a   **12.** *Genista hispanica* (511)
13. *Orchis coriophora* 1886   **14.** *Saxifraga cuneata* 404h   **15.** *Erythronium dens-canis*
1628   **16.** *Scilla liliohyacinthus* 1632a   **17.** *Sedum brevifolium* 391b

a forest road that turns off from Covaleda. The summit slopes are covered with ling, *Calluna vulgaris*, in which *Viola montcaunica*† is abundant; and *Narcissus bulbocodium* and *Tulipa australis* are still in flower in mid-June. Near the summit there is some *Tanacetum alpinum*, and a form of what is probably *Sempervivum cantabricum*†. *Erodium paui*† is endemic to this mountain; it is one of the *E. acaule* 'group' and distinguished from *E. carvifolium*,† another endemic in Spain, by the whitish stipules and bracts, the glandular-hairy leaves, and glandular pits of the fruits.

Instead of turning off to the Laguna Negra one can follow the forest road over the Puerto de Santa Inés, 1753 m, to Montenegro de Cameros, returning to Soria over the Puerto de Piqueras, 1710 m. The road up to the former pass runs through pine forests with an undergrowth of *Cytisus scoparius*, *Erica arborea*, and *E. australis*, with some *Halimium alyssoides* and *H. umbellatum*; *Genista pilosa* grows prostrate on this road. At one damp spot there is a fine display of *Pinguicula grandiflora*. Over the pass the forest disappears; *Vicia pyrenaica*, *Acinos arvensis*, and *Cerastium arvense* are conspicuous by the road, and lower down the grassy fields are rich in orchid species. In the valley of the Iregua through which the main Logroño–Soria road runs there is much *Quercus pyrenaica* woodland, and as one climbs up to the Puerto de Piqueras beech takes its place. On the way down to Soria large hummocks of *Genista hispanica* stand out, and near the bottom are a few plants of *Erodium carvifolium*† with exceptionally large flowers up to 4 cm across; *Linaria aeruginea* with streaked brown and yellow flowers and *Lavandula stoechas* subsp. *pedunculata* also become abundant.

The Sierra de la Demanda (its highest summit San Lorenzo is 2262 m) is less accessible, but a good impression of it can be gained from the valley of the Rio Najerilla, and by following the road C 113 down to Anguiano. The scenery is attractive with broad-leaved scrub covering the shaley hillside and a trout stream in the valley below. *Genista florida* provides plenty of yellow, and thick tufts of *Dianthus lusitanus* grow out of the shale; *Linaria repens* grows to 1 m tall.

North of Soria, on the road to Arnedo and Calahorra past the ancient city of Numancia sacked by the Romans in 133 B.C., there is a climb to 1454 m at the Puerto de Oncala. *Iris spuria* grows in roadside ditches across the plain. On the way up a fine form of *Linum suffruticosum* grows densely tufted with tiny closely overlapping leaves and pink buds that open out into petals of the palest rose. With it is the bushy, prickly *Astragalus sempervirens*, the stemless form of the blue thistle *Carduncellus monspelliensium*. In the closely grazed turf above the pass *Dianthus subacaulis* is common, forming little hummocks usually with pale pink flowers but grading from white to darkish purple. Just over the pass there is a field beside the road with abundant *Pulsatilla rubra*, with *Pedicularis foliosa*, *Digitalis parviflora†*, and *Senecio doronicum*.

One of the most interesting areas lies along the main Soria–Burgos road, particularly between Cidones and Navaleño. For many kilometres *Erodium carvifolium†* and *Echium vulgare* grow by the roadside inter-mittently, and every now and then one passes large colonies of *Orchis militaris*, *O. coriophora*, *Ophrys sphegodes*, some *O. scolopax†*, and at one spot there is a fine display of *Pulsatilla rubra*, with some *Antheri-cum liliago* and *Iris xiphium*. Further west a turning off the main road leads to Santo Domingo de los Silos and runs through limestone country. Just past the village, on the road leading south, there is the spectacular gorge of the Rio Yecla, which has cut a narrow passage only a few metres wide down through hundreds of feet of vertical limestone. A concrete cat-walk runs through the length of the gorge, while the road goes through a tunnel. Here there is a very distinctive flora, with an abund-ance of *Silene boryi†*, *Dianthus furcatus†*, *Alyssum montanum*, *Ptilotrichum spinosum*, *Sedum acre*, *Saxifraga cuneata†*, *Erodium petraeum†*, *Xeranthemum inapertum*, and *Matthiola fruticulosa*.

The highest mountain in this mountain system is the Sierra del Moncayo, 2313 m, which lies east of Soria. The upper part of the sierra is covered with boulders and scree, bare and treeless, and because of its loose rocks is known as the 'moving mountain'. Viewed from the west it might be one of the Red Coolins of the Island of Skye, except for the belt of deciduous woodland just above Cueva de Agreda. On the other side it rises majestically from the Ebro valley and in early July the wooded slopes, rich in springs, and the remains of the summit snow tempt the traveller sweltering down on the plain below. From Agramonte a good road zig-zags up the mountain to El Santuario de Moncayo, perched high up on the eastern slopes, whence it is easy to climb to the summit. From here in clear weather the Pyrenees can be seen and one would expect to find many Pyrenean plants—but in fact the flora resembles more closely that of the Central Sierras.

There are three zones of vegetation: the lower oak, the middle beech, and the upper sub-alpine and alpine treeless zone. The oak zone is dominated by *Quercus pyrenaica* and the ground flora is mostly grass, with much *Erica vagans*, and the pretty switch plant *Genista micrantha*; herbaceous species include *Digitalis parviflora†*, *Paradisea liliastrum*, and the rare *Cypripedium calceolus†*. The beech zone from 1000–1650 m comprises one of the finest forests in Spain with a tightly closed canopy and little growing underneath in the deep shade. Above this, in the sub-alpine and alpine zones the most interesting plants are: *Saxifraga mon-cayensis†* in luxuriant cushions in rock crevices, endemic to this sierra, *Silene ciliata*, *Paronychia polygonifolia*, *Ranunculus gregarius*, *Biscu-

tella intermedia, **Oxytropis pyrenaica*, **Onopordum acaulon*, and *Viola montcaunica*† like a small **V. cornuta* with a shorter spur, growing amongst the usual high-level scrub of **Cytisus purgans*, **Erica australis*, *Vaccinium myrtillus*, *Juniperus communis* subsp. *nana* and **Erinacea anthyllis*.

21. The Pyrenees

The Pyrenees run from the Bay of Biscay to the Mediterranean, a distance of about 450 km. At their widest, they are about 130 km wide, and by any standards they constitute a formidable barrier to the movements of plants and animals. The highest summit, Pico de Aneto, is 3404 m (11,169 ft); but more significant is the fact that the central section of the range never falls below 1700 m. Like the Alps they contain an interior of hard crystalline rocks such as granite, gneisses and crystalline schists which weather slowly, and outer bands of little altered sedimentary rocks, generally of Mesozoic and Tertiary age, in which limestones predominate. But by contrast with the Alps they have no glacier of any size today, though there is abundant evidence of glaciation during the ice age, with glacial lakes, U-shaped valleys, and moraines among the high mountains.

In general the northern slopes, on the French side, are very steep and fall abruptly to the lower country of Aquitaine; whilst on the Spanish side there are a series of progressively lower parallel ridges before the flat plateau of the Ebro basin is reached. There are three main parallel ranges; the first south of the main watershed is composed of Mesozoic rocks with much limestone; the middle range is of Tertiary marls and conglomerates; and the southernmost, the Sub-Pyrenean range, is largely composed of Mesozoic rocks again. The rivers running down from the main range often have to flow parallel to the lower ranges before turning at right angles to cut a way through them, often in deep gorges, to the plateau and ultimately the Ebro river.

The Pyrenees also act as a climatic barrier and cause much of the humidity from the Atlantic to be precipitated on the French flank of the range. By contrast the corresponding valleys to the south of the watershed suffer from very low rainfall, and indeed are some of the driest areas in the whole of Spain. The vegetation of the montane and sub-alpine regions of the French and Spanish Pyrenees is thus very different. The high Pyrenees have quite hot summers with violent thunderstorms, occurring roughly once every four days in August or September, but otherwise they are dry. In winter, snow falls are heavy, and spring is relatively late.

The Spanish Pyrenees   Until recently the Spanish side of the Pyrenees was much less accessible to the tourist than the French side, because of the lack of good roads and accommodation. Even today there is nothing comparable with the Route des Pyrénées and travel from one Spanish valley to the next usually involves long detours southwards. However, the Spanish government is rapidly developing this wild and difficult area, both for winter sports and summer tourism, and building new roads, paradors, and hotels.

Starting in the west, on the Pamplona–Jaca road (N 240), the traveller crosses dry cultivated country. At Alto de Loiti the road rises to 729 m, and here some interesting plants can be seen such as *Helianthemum nummularium* subsp. *pyrenaicum* with unusually large rose-coloured flowers; *Thalictrum tuberosum* with white 'petals' like an anemone; and *Catananche caerulea*, a brilliant blue everlasting with silvery-brown involucral bracts. Near the head of an enormous artificial lake, the Embalse de Yesa, a road turns north up into the Valle del Roncal, which is well worth a visit. In the gorge sections are some of the finest displays of *Saxifraga longifolia*† to be seen in the Pyrenees as well as *Dianthus hispanicus*†, and the slender, glaucous, pink-headed valerian *Centranthus angustifolius*.

From Jaca the high Pyrenees can be reached via the Col de Somport (1631 m), the main pass across the western Pyrenees; but much more rewarding is the more difficult road east of Jaca, up the Valle de Tena to Sallent. This is a good lead into magnificent alpine country around some of the finest Pyrenean peaks such as the Pic du Midi d'Ossau (2473 m), Pico Balaitus (3144 m), and Pic de Vignemale (3298 m). Near Sallent is El Escalar, an interesting road up to the Spa of Panticosa with *Ramonda myconi*†, *Lonicera pyrenaica*†, and *Antirrhinum sempervirens*†. From Panticosa there is easy access to very fine alpine country, some of which will be described in more detail from the French side.

By taking the side road eastwards at Biescas and crossing the Puerto de Cotefablo one can travel up the next valley past the ancient village of Torla to the Parque Nacional de Ordesa (which lies due south of Gavarnie in France). The park comprises some of the finest canyon-like country of the whole of the Pyrenees, overshadowed by the summits of Marbore and Monte Perdido, at 3352 m the third highest of the range. Magnificent unspoiled forests of beech, pine, and silver fir cover the main valley of the Ordesa river which rises between these two mountains.

The most worthwhile expedition here is up past the waterfalls to the alpine meadows at Soaso (2100 m), and in July and August when there is less snow about one can climb up to the watershed by various paths. At the entrance to the broad U-shaped glaciated valley of Soaso one passes

LOWER PYRENEES

1. *Lathyrus laevigatus* 565a   **2.** *Dianthus hispanicus* 194c   **3.** *Paradisea liliastrum* 1595   **4.** *Chamaespartium saggitale* 513   **5.** *Antirrhinum sempervirens* 1197e   **6.** *Lathraea clandestina* (1269)   **7.** *Pinguicula grandiflora* (1280)   **8.** *Ramonda myconi* 1268

9. *Lonicera xylosteum* 1302 **10.** *Pedicularis mixta* 1255b **11.** *Campanula hispanica* 1339a **12.** *Catananche caerulea* 1511 **13.** *Erodium petraeum* 653d **14.** *Veronica austriaca* 1229 **15.** *Scutellaria alpina* 1107 **16.** *Euphrasia salisburgensis* 1244 **17.** *Asarina procumbens* 1200

the lovely *Hyacinthus amethystinus† stretching for half a mile, and the turf in the valley is dotted with streaked white or pale pink flowers of *Geranium cinereum†, here sitting close to the turf because of the heavy grazing; *Dactylorhiza majalis grows in wet flushes. The flora of the high mountains will be described from Gavarnie.

The next valley further east is the Valle de Pineta above Bielsa, giving access to the east side of Monte Perdido and some fine country. Plants recorded here include:

Ononis natrix	Campanula speciosa†
Hypericum maculatum	C. cochleariifolia
H. montanum	Leontopodium alpinum
Astrantia major	Senecio alpinus
*Eryngium bourgatii†	*Anthericum liliago
*Ramonda myconi†	Lilium martagon
*Iris xiphioides†	

Benasque at 1138 m, in the valley of the same name, gives access to the Baños de Benasque from which very fine country to the west of the Maladeta group of mountains can be explored. This group includes Pico de Aneto, 3404 m, the highest in the Pyrenees.

So it is shortly after passing the ruined Hospital de Benasque, an ancient hostel on the road to France, that the track climbs sharply into thin woodland and here, as far as the Pllan del Estañ, you come to a very flowery stretch. Among the rocky outcrops grew a pretty little treacle-mustard, the yellow *Erysimum helveticum (pumilum), Globularia cordifolia* subsp. *nana†, *Daphne cneorum, Lotus corniculatus, Helianthemum nummularium, Erinus alpinus*, and one of the yellow Pyrenean snapdragons, *Linaria pyrenaica (supina)* while the alpine forget-me-not, *Myosotis alpestris* and *Hyacinthus amethystinus† were in large drifts. . . . On the other side of the track a small marshy valley was full of fluffy white cotton-grass *Eriophorum alpinum, *Cardamine raphanifolia, Trollius europaeus* and *Primula farinosa*, some of which were the largest I have ever seen.[12]

The track to the Puerto de Benasque turns off to the left shortly above the Hospital de Benasque; here the following can be found:

Cerastium alpinum	Trifolium alpinum
Silene acaulis	Helianthemum canum
Arenaria ciliata	*Androsace carnea var.
*A. tetraquetra†	laggeri†
*Saxifraga aretioides†	*A. villosa
Vicia pyrenaica	Gentiana verna
Veronica fruticans	

Another fine exploratory walk can be made into the Valle de Astós and the Pic Perdiguero where *Pulsatilla alpina, *Ranunculus amplexicaulis†, Lilium pyrenaicum†, and *Iris xiphioides† are conspicuous.

Further east again is the Valle de Arán, a Spanish valley running into the heart of the Central Pyrenees, and which contains the headwaters of the mighty French river, the Garonne. The main road cuts here through the Viella tunnel, and through the main Pyrenean watershed; the Valle de Arán is consequently much greener and more fertile than most Spanish valleys, and in September the meadows are covered with *Viola cornuta, Carlina acanthifolia, *Merendera montana, and *Crocus nudiflorus.

Salardú is the best centre.

In the valley of the Rio Aiguamoch which joins the Garonne a mile above Salardú, grow Narcissi of the Ajax group [section pseudonarcissus] in profusion: N. bicolor†, *N. pallidiflorus†, and *N. abscissus† were all there, with intermediate forms impossible to name. N. × bernardii†, the rare natural hybrid between N. abscissus and N. poeticus occurred in small groups amongst the limitless white seas of N. poeticus. Higher up the same valley above the first jasse and not far from the deserted Bains de Tredos were *Ranunculus amplexicaulis† and *Pulsatilla alpina, not in small groups, but in great stretches, competing even with N. poeticus. Higher up still *Daphne cneorum made spreading patches of pink between the granite boulders, with early spikes of Lilium pyrenaicum†, and by the stream sides amongst the birch trees grow *Gentiana acaulis (kochiana). On the third jasse, those characteristic steps up these Pyrenean valleys, growing not far from the still deserted herdsmens' huts, was the Pyrenean oxlip, Primula intricata† with Gagea liotardi.[13]

North of Salardú, up the Rio Iñola some crystalline rocks occur, and here are to be found:

*Aquilegia pyrenaica†	*Viola cornuta
*Saxifraga media†	*Asphodelus albus
Geranium sanguineum	*Fritillaria pyrenaica†
G. pyrenaicum	*Hyacinthus amethystinus†
G. phaeum	Dactylorhiza sambucina

To reach the next valley to the east, that of the Rio Noguera Pallaresa, a rough mountain road re-crosses the main Pyrenean divide at Puerto de la Bonaigua, 2072 m, and here in mid-June amongst the snow patches are *Ranunculus pyrenaeus, Lychnis alpina, Vitaliana primuliflora, *Androsace carnea, and Crocus albiflorus; and later in the year:

*Arenaria tetraquetra†	Iberis sempervirens
Gypsophila repens	*Daphne cneorum
Saponaria caespitosa†	*Linaria repens
*Pulsatilla alpina	*Cirsium monspessulanum

The Parque Nacional de Aigues Tortes y Lago de San Mauricio is a magnificent tract of wild mountainous country with numerous lakes and very jagged mountains: it is best reached from Espot. There is a very rough road up to Lago San Mauricio, from which the famous view of the Enchanted Mountains, Los Encantats, is seen. This is country for trekkers and campers and requires plenty of time and energy. Losa and Montserrat give an annotated list of the plants of this area.

The Sierra del Cadi lies east of Andorra to the south of the main Pyrenean axis and ranges eastwards to the mountains of Nuria and the Puigmal, 2913 m. Here lie some of the finest and wildest tracts of country in the eastern Pyrenees, crossed only by one pass, the Puerto de Tossas (1800 m) above which lies the ski resort of La Molina.

Up here in these Pyrenean heights grows what is probably the richest alpine flora of the eastern range, surviving in spite of the sheep and goats. The turf is full of *Gentiana pyrenaica*†, frequent *Gentiana alpina*†, with *Erigeron uniflorus* and much *Daphne cneorum* var. *pygmeă*. On the rocks grow abundant *Saxifraga exarata*, *Globularia repens* (*nana*) and *Silene acaulis*, while the cooler damper sides show *Primula integrifolia*† and *P. viscosa*. The charming little *Vicia pyrenaica* grows on steeper stony ground with *Androsace villosa*. In the screes is *Papaver rhaeticum* (*alpinum* var. *pyrenaicum*), *Ranunculus parnassifolius*, and *Erysimum decumbens* (*ochroleucum*), while higher up on dry stone screes of frightening steepness and bareness are great clumps of *Adonis pyrenaica*† blazing gold against the dark rock.[14]

Castellar de Nuch is a convenient place for exploring the eastern part of the Sierra de Cadi and a mountain road connects with the Puerto de Tossas where *Ononis rotundifolia* and *Narcissus alpestris*† can be found. Within walking distance of Castellar two botanists collected 400 species of flowering plants (excluding grasses, sedges, and ferns) in two weeks in July.

Andorra This tiny country is tucked in the heart of the Pyrenees and consists of the head valleys of Gran Valira, a tributary of the Rio Segre. It is surrounded by mountains and nowhere falls below 900 m in altitude. The rocks are largely granite but alkaline rocks occur at Pic de Casamanyá in the centre and Arinsal and Ordino in the north-west, where the most varied flora can be found. Over 1,000 species of seed plants have been recorded for the country, and Losa and Montserrat have provided an annotated list of species. They state that

The flora met with is varied and rich in species; many plants that are rare and local in the Pyrenees are found in the valleys of Andorra, some of them quite widely distributed. The floristic variation is due primarily to the variety of habitats to be found: cultivation, meadows, woods, torrents, stony slopes, rocky summits, etc. provide ample space for accommodating flowers of widely different habitat requirements. The whole range of altitude from 900 to nearly 3000 m enables plants to grow that are montane, or alpine, or confined to the rock fissures of the highest summits.[15]

Andorra is thus one of the best centres for exploring the Pyrenean plant world. Andorra La Vella is a bustling tourist centre, but Soldeu, 1826 m, on the main road, gives access to the Val d'Incles and the high peaks on the border with France. From Soldeu the Cirque dels Pessons can be explored; the cirque itself is not particularly rich but the surrounding country is very rewarding. On the ridge above the Bosc da Soldeu are the following in considerable quantity:

Pulsatilla alpina	*Vitaliana primuliflora*
P. vernalis	Gentiana pyrenaica†
Ranunculus pyrenaeus	G. verna
*Daphne cneorum	*G. alpina†
Loiseleuria procumbens	Linaria alpina
Androsace carnea var.	*Erigeron uniflorus
laggeri†	

The Vall d'Incles can also be reached from Soldeu where the following can be found:

Thalictrum aquilegifolium	*Pinguicula grandiflora
*Anthyllis montana	*Asphodelus albus
Gentiana nivalis	*Paradisea liliastrum
*G. lutea	Lilium pyrenaicum†
*Iris xiphioides†	

while higher up are *Geum montanum, Androsace argentea†, Soldanella alpina, Primula integrifolia†*, and *Narcissus pallidiflorus†*.

At Canillo, further down the main valley towards the capital on a limestone outcrop are:

Gypsophila repens	*Erodium petraeum
*Saponaria ocymoides	subsp. *glandulosum†*
Ononis cristata	*Asarina procumbens†
O. natrix	Aster alpinus

A walk along the ridge to the peak of Casamanyá (8,780 ft.) revealed many plants not previously met with in Andorra. *Teucrium pyrenaicum† a nice small compact sh~ublet, endemic to the Pyrenees; *Paronychia capitata (nivea), Erinus alpinus, *Globularia repens (nana), G. cordifolia, a form of *Saxifraga oppositifolia . . . *Daphne cneorum form, Erysimum helvaticum (pumilum), Ranunculus parnassifolius, Thlaspi rotundifolium, Dryas octopetala. . . . At the summit grew fine cushions of Saxifraga bryoides and Silene acaulis, with Soldanella alpina, Primula integrifolia†, Linaria alpina (faucicola), *Androsace carnea var. laggeri† and various gentians including some white G. verna.[16]

Side valleys such as Arinsal and Ordino are also excellent valleys to explore. In the latter one can drive as far as El Serrat above which, towards the Port de Rat at 2000 m, the turf is carpeted with *Ranunculus pyrenaeus, Anemone narcissiflora, *Androsace carnea, *Erythronium dens-canis, and Gagea soleirolii.

Up the right-hand path (marked to the lago) are good colonies of *Narcissus pallidiflorus† on steep turf-covered rocks at several places. Above the village of Arinsal a path takes one to a lake and the Refugio at the base of Pic Coma Pedrosa –at 2946 m the highest in Andorra – where masses of the yellow *Pulsatilla alpina subsp. apiifolia are a feature.

From the high pass of Port d'Envalira, 2407 m, where the main road crosses into France, many of the best Pyrenean alpine plants can be found.

The French Pyrenees Starting from the east and retracing our steps on the other side, there is first of all the somewhat isolated mountain of gneiss, Mont Canigou, 2413 m. From Fillols a mountain road takes one to the Chalet-Hotel des Cortalets from which the summit can easily be reached. Interesting plants round the hotel include: *Ranunculus pyrenaeus, Gentiana verna, and *G. alpina†, and the rare *Senecio

---

19 **Massif Central** The Causse of the Cévennes are high limestone plateaux dissected by deep gorges. The rocky soil is very shallow but it supports a rich flora of many interesting shrubby and herbaceous species.

20 **French Pyrenees** Near the Pic du Midi d'Ossau, the centre of a fine, recently established National Park in the Pyrenees. The region has yet to be fully developed as a reserve, for many thousands of head of cattle graze in the upper pastures and holiday-makers can still make off with enormous bunches of the beautiful Spanish Iris. The area is rich in Pyrenean endemic plants.

leucophyllus†. Higher up grow *Primula integrifolia*† in the utmost profusion below the Pic de Joffre and also *Primula viscosa*. Other species include:

Dianthus deltoides	**Gentiana burseri*†
Saxifraga moschata	**Asarina procumbens*†
Pyrola rotundifolia	*Erigeron alpinus*
Moneses uniflora	**Paradisea liliastrum*
*Androsace carnea	*Lilium pyrenaicum*†
A. argentea	*Polygonatum verticillatum*

The Val d'Eyne, well-known to alpine gardeners, lies south-south-west of Mont-Louis and it leads over a high pass, the Col de Nuria, to Nuria in Spain. Here the following species are of particular interest:

Lychnis alpina	*Iberis spathulata*
Delphinium species	*Biscutella laevigata*
*Adonis pyrenaica†	**Linum narbonense*
Ranunculus parnassifolius	*Viola biflora*
*Aquilegia pyrenaica†	*Armeria alpina*
Papaver rhaeticum	*Aster alpinus*

Near here in Gorge de la Caranca the big pale-flowered *Antirrhinum latifolium* is abundant and higher up **Chamaespartium sagittale*.

Further west lies Pic Carlit, 2921 m, considerably north of the main Pyrenean range. It is best explored from the village of Porte lying under the Col de Puymorens. At the pass **Cytisus purgans* covers the mountain-sides and other plants of interest include *Saxifraga retusa*, *Lilium pyrenaicum*†, **Scilla liliohyacinthus*, **Narcissus pallidiflorus*†, while *Aster alpinus* and *Erigeron uniflorus* grow on Pic Carlit.

Moving further westwards on the French side, Luchon is probably the next worthwhile centre. From here, or Superbagnères, one can explore a number of valleys into the heart of the main range, and in particular the area around the Lac d'Oo and the Val d'Esquierry. Of especial interest in this region are:

Arenaria biflora	*Saxifraga clusii*
*Cardamine raphanifolia	**S. aquatica*†
Reseda glauca†	**Astragalus monspessulanus*

---

21 **Auvergne** Looking northwards from the summit of Puy Mary 1786 m, into the mountains of the Auvergne. Moist north-facing cliffs form natural rock gardens harbouring a number of mountain species which are able to bridge the gap between the Alps and the Pyrenees.

Hypericum richerii	**Teucrium pyrenaicum†*
*Eryngium bourgatii†	**Chaenorhinum origanifolium*
Asperula hirta†	*Lathraea clandestina*
Tozzia alpina	**Fritillaria pyrenaica†*

From Luchon westwards the Route des Pyrénées crosses magnificent country over three high places, the Col de Peyresourde, 1563 m, the Col d'Aspin and the Col du Tourmalet, 2114 m. From the latter a rough road takes the traveller to below the summit of Pic du Midi de Bigorre, 2865 m, where the following can be found growing by the roadside:

Arenaria ciliata subsp.	*Hutchinsia alpina*
moehringioides	*Trifolium thalii*
Ranunculus gouanii†	*Campanula cochleariifolia*
Reseda glauca†	*Phyteuma hemisphaericum*
*Sesamoides pygmaea	*Gnaphalium supinum*

To the south of the Col du Tourmalet, up in the mountains lies the Néouvieille Nature Reserve which is now part of the Pyrénées Occidentales National Park created as recently as 1967, accessible by the road running south of Arreau through St Lary and up to the Lac d'Oredon.

The Gave de Pau is the next valley which penetrates deep into the heart of the Pyrenees, to Gavarnie and the celebrated Cirque de Gavarnie, famous as a botanical hunting ground. On the cliffs of the cirque grow two particularly interesting plants: **Pinguicula longifolia†* –a cliff-face is covered with this rare plant –and the strange *Dioscorea pyrenaica†*, the only member of the genus, which is a large tropical one, that occurs in Europe; it is endemic to the Pyrenees and probably an ancient relict. Two other Pyrenean endemics grow in great profusion: **Ramonda myconi†* and **Saxifraga longifolia†* which make a wonderful display on the rocks. Other interesting plants of this region are:

Arenaria purpurascens†	**Anthyllis montana*
Saponaria caespitosa†	**Geranium cinereum†*
Ranunculus thora	*Androsace cylindrica†*
*Aquilegia pyrenaica†	**Scutellaria alpina*
*Oxytropis pyrenaica	**Lonicera pyrenaica†*
Vicia pyrenaica	**Narcissus requienii*

On the rocks below the statue of the Virgin are **Sedum brevifolium*, *Potentilla alchimilloides*, **Astragalus sempervirens*, and *Asperula hirta†*.

On the Port de Gavarnie are *Ranunculus parnassifolius*; **Thymelaea dioica*, *Hippocrepis glauca*, and *Potentilla nivalis*.

A fine long walk can be taken up the Vallée d'Ossoue towards the Pic de Vignemale, 3298 m, and its glacier; and a mule track continues high over the Hourquette d'Ossoue, Lac de Gaube, and Pont d'Espagne to Cauterets where many good Pyrenean alpine plants can be found including *Ranunculus glacialis*† which is rare in the Pyrenees.

From the Gave de Pau the Route des Pyrénées continues westwards over the Col d'Aubisque (1710 m) where on limestone rocks below the summit are *Saxifraga aretioides*†, *Thymelaea tinctoria*†, and *Antirrhinum* species; also near Gourette on Penemedad grow *Petrocallis pyrenaica*, *Androsace pubescens*, and *Lithospermum gastonii*†.

Gabas, on the main road leading up to the Col de Portalet into Spain is another good centre; from here the Pic de Sagette and another part of the Pyrenean National Park can be reached to the east, while to the south lies the Pic du Midi d'Ossau a further important part of the park, both very fine areas for plant-hunting. The country around Gabas

is surrounded by well-wooded valleys in which grow . . . *Adonis pyrenaica*†, *Thalictrum aquilegifolium*, *Lilium martagon*, *Pyrola rotundifolia*, *Dianthus sylvestris* and *D. pyrenaicus* (*requienii*). Here too flourish *Allium victorialis*, *Geranium sylvaticum* and *G. phaeum*, *Eryngium bourgatii*† and *Saxifraga umbrosa*† . . . The woods round Gabas are rich in orchids, including *Orchis ustulata*, *Platanthera bifolia*, *Spiranthes aestivalis*, *Nigritella nigra* (*angustifolia*), and *Gymnadenia conopsea*.[17]

The Pic de Sagette can be reached by cable car, and from there a small train continues to the Lac d'Artouste in the magnificent country of the Park. On the Pic the following occur in addition to the more widespread plants:

Dianthus monspessulanus	*Potentilla rupestris*
Saxifraga hirsuta subsp.	*Vicia orobus*
paucicrenata	*Hypericum richeri*
S. nervosa†	*Primula integrifolia*†
S. aretioides†	*Veronica fruticulosa*
Rosa pendulina	*V. spicata*
R. pimpinellifolia	*Campanula linifolia*
Allium montanum	

From Gabas a road leads up to the reservoir of Bious-Artigues and the track above this leads into fine country round the base of the great isolated conical peak of the Pic du Midi d'Ossau, 2885 m. In the Nature Reserve where great herds of cattle still graze the alpine meadows intensively, there is a rich botanical feast, including *Gentiana burseri*†,

UPPER PYRENEES

1. *Gentiana lutea* 996 **2.** *Fritillaria pyrenaica* 1623 **3.** *Corydalis solida* 279
4. *Hyacinthus amethystinus* (1643) **5.** *Horminum pyrenaicum* 1151 **6.** *Bartsia alpina*
1236 **7.** *Oxytropis pyrenaica* 533a **8.** *Primula hirsuta* (945) **9.** *Dactylorhiza*
majalis 1898

10. *Ononis rotundifolia* 570   **11.** *Lathyrus montanus* 562   **12.** *Narcissus pallidiflorus*
1665j   **13.** *Ranunculus pyrenaeus* (248)   **14.** *Pulsatilla alpina* 220   **15.** *Erigeron
uniflorus* (1370)   **16.** *Geranium cinereum* 641a   **17.** *Androsace carnea* 953
18. *Gentiana acaulis* 992.

Swertia perennis, *Horminum pyrenaicum, *Iris xiphioides†, while higher up grow:

Arenaria purpurascens†	**Teucrium pyrenaicum†*
Lychnis alpina	*Stachys alopecuros*
Dianthus monspessulanus	*Veronica gouanii†*
Alchemilla saxatilis	*Pedicularis pyrenaica†*
Sempervivum arachnoideum	**Lonicera pyrenaica†*
Potentilla alchimilloides†	*Valeriana pyrenaica†*
*Oxytropis campestris	*V. montana*
*O. pyrenaica	*Leontopodium alpinum*
*Hypericum nummularium	**Tanacetum alpinum*
Helianthemum oelandicum	*Doronicum grandiflorum*
subsp. *alpestre*	**Dactylorhiza majalis*
Gentiana brachyphylla	

22. The Massif Central of France

The Massif Central offers some of the most varied and beautiful scenery in Europe, especially when the spring flowers are at their best in May and June. It contains the volcanic mountains of the Auvergne in the north-west, the granite mountains of Forez in the north-east, the Causses and limestone gorges of the south-west, and the Cévennes in the south-east. The flora has four elements: the Mediterranean element, which predominates in the south; the central European element which is characterized by silver fir and beech forests; the Atlantic element characterized by *Erica-Genista* heaths; and the Boreal element, consisting of arctic-alpine species.

The Mediterranean element is represented by the *Quercus ilex* forest or scrub, and if one approaches the Massif Central from the south one leaves the plains of Bas-Languedoc, where vineyards stretch to the horizon, and enters the Cévennes, a region dominated by a Mediterranean scrub covering the calcareous plateau and low ridges of the country. The Mediterranean vegetation may commonly reach altitudes of 500 m or more, while above it is a characteristic zone of sweet chestnut plantations at altitudes to 800 m.

A string of villages lies along this zone. Seen from a distance they make a white patch in the green cloak of the chestnuts, generally ensconced in a tributary valley, or *cale*, or else on the edge of one of the benches that break the slope of the valley-sides. The houses mount one above the other; the highest are often falling into ruins. The long winters and the hard mountain life do not suit the younger generation nowadays, since they have got to know the plain and its towns. But for hundreds of years men have lived here, contenting themselves with chestnut porridge and goats' cheese.[18]

The highest part of the Cévennes is a confused succession of denuded ridges and deep valleys, running from Mont Lozère, (1702 m) to Mont Aigoual (1567 m). For centuries huge flocks of sheep have been driven from Bas-Languedoc when their pastures in the *garrigues* dry out in early summer up into the high-level grazing grounds; the routes or *drailles* are traceable from a distance by the white ribbon they make on the mountain-side. Except for the highest summits, these mountains were once thickly forested, mainly with beech, but charcoal-burning for the glass-making industry, and excessive grazing by sheep have reduced the once extensive forests to a few small remnants. The last century has seen the replanting and extension of these forests on the slopes of Mont Aigoual. These higher regions of the Cévennes possess a typically central European assemblage of species.

The Causses are arid limestone tablelands bounded by precipitous ramparts and divided by deep gorges. The four largest are the Causse de Sauveterre, Causse Méjean, Causse Noir, and Causse du Larzac, and they are bounded by the Cévennes on the east, the valley of the Lot on the north, and the plateau of Rouergue or Ségala on the west.

The rocky, dry limestone soil is everywhere visible; the fields are strewn with great heaps of stones, to which the process of clearing the ground makes constant additions . . . The air quivers in the summer noons above the superheated soil and the step of the walker who ventures out upon the Causse raises clouds of grasshoppers. The native requires neither geography nor geology to tell him exactly how far the Causses extend in this region; the limits he sets to them are the limits of the Jurassic limestone. . . . What strikes the native most forcibly is the dryness of the soil, for he is forced to look for water in the deep valleys to be found here and there, and to establish his homestead and his cultivated fields in the irregular hollows of the ground. The geographer associates with this type of landscape . . . the enclosed hollows, which have the look of valleys that have ceased to function and are here called *sotches*; the funnel-shaped depressions, which in some places riddle the whole surface, and for which we have borrowed from the Karst districts the name *doline*; the yawning chasms in the bottom of the *dolines*, or even sometimes on the surface of the plateaux, which are here called *avens* . . . and the caves that penetrate the steep sides of the valleys, which can be reached by way of the *avens*.

The surface of the great Causses forms a kind of stony desert 700–1000 m high. Soil there is none, apart from the *sotches*, where the red clay resulting from the dissolution of the limestone collects. The melting snows of winter often form pools in them, and every effort is made to conserve the water as late into summer as possible. The land yields a few crops under the plough and stone walls endeavour to retain the soil on the sides. . . . Till you stand on the edge of them, you would never guess the existence of the great valleys that cut so deeply into the surface. There are only three: those of the Tarn, Jonte and the Dourbie, each

THE MASSIF CENTRAL OF FRANCE

1. *Campanula persicifolia* 1336 2. *Convolvulus cantabrica* 1036 3. *Scorzonera purpurea* 1529 4. *Campanula glomerata* 1333 5. *Saxifraga paniculata* 400 6. *Dryas octopetala* 441 7. *Pulsatilla vulgaris* 223 8. *Legousia speculum-veneris* 1347 9. *Ophrys insectifera* 1877

10. *Phyteuma spicatum* 1350 **11.** *Cytisus sessilifolius* 504 **12.** *Saponaria ocymoides* 180 **13.** *Ajuga genevensis* 1090 **14.** *Onosma arenaria* 1076 **15.** *Anthyllis montana* 621 **16.** *Orchis ustulata* 1888 **17.** *Astragalus purpureus* 528b **18.** *Adonis vernalis* 231

of them a gorge 500–700 m deep. At the bottom, the river winds between pebble banks of startling whiteness, or disappears into narrows between the vertical or overhanging walls of the massive limestone formations. . . .[19]

To reach the Causse country from the south the best road to take is the N 9 through Lodève and up the steep escarpment of the Pas de l'Escalette to the Causse du Larzac. The Causses have a rich flora, despite the excessive nibblings of innumerable sheep.

Plants of interest are:

*Anthyllis montana	*Ajuga genevensis
*Linum campanulatum	Chaenorhinum rubrifolium
*Daphne cneorum	Aster alpinus
Aphyllanthes monspeliensis	

A characteristic plant of the Causse is *Buxus sempervirens*, while round Roquefort *Amelanchier ovalis* and *Cytisus sessilifolius* are common, and the dead remains of the large stemless thistle *Carlina acanthifolia*, which had flowered the previous summer, remain dotted about everywhere. The following orchid species are often found:

*Orchis morio	*O. militaris
O. palustris	*Ophrys sphegodes
*O. ustulata	*O. insectifera
O. purpurea	O. apifera
*Aceras anthropophorum	

Roquefort is an intriguing place in which to spend a few days, sniffing the all-pervading smell of the famous cheese made from the milk of close on a million sheep that browse on the high-level pastures. The cheese is matured in limestone caves in the town itself, which is built against a cliff. From here also the famous Gorges du Tarn can be conveniently explored. Many of the lower Pyrenean plants grow around here, such as:

*Saponaria ocymoides	Erinus alpinus
*Vicia onobrychioides	*Campanula glomerata
*Astragalus purpureus	*C. persicifolia
*Convolvulus cantabrica	*Phyteuma spicatum

Leaving the Cévennes for the Auvergne one crosses the Monts d'Aubrac, a rolling basalt country with pastures stretching to the horizon, lying at no great height; the Signal de Mailhebiau—1471 m—is the highest point.

The volcanic area of the Auvergne—the Monts Dore, Dômes and Cantal—is very different again. The Monts Dore are the remains of a volcanic system that was originally made up of three cones close together, attaining 2500 m in height and a surface area six times greater than that of Vesuvius. The Monts Dômes are the youngest of the volcanoes of the Auvergne, only recently extinct, and preserving almost intact the form they had while they were still active. About 60 ancient volcanoes, extending over a distance of 30 km, rise from a plateau of crystalline rock 900–1000 m above sea level; the highest cone is the Puy-de-Dôme, 1465 m. The Plomb du Cantal represents the remnant of a huge volcano.

To the east of this volcanic region is the granite range of the Monts du Forez and its northward continuation the Montagne de la Madeleine. They are cultivated up to about 800–1000 m, above which there is a belt of beech and silver fir forest. At 1200–1300 m the forest gives way to the high level pastures known as the Hautes-Chaumes, by analogy with those of the Vosges.

Of particular interest are the sub-alpine and alpine plants which grow in the Massif Central. Though there is little terrain above 1500 m an altitude which approximates to the lower limit of sub-alpine species, in the Massif Central this zone tends to occur at lower altitudes, due in part to the influence of Atlantic weather and heavy summer rainfall. It is not for nothing that the mountains of the Massif Central have been christened 'the bald head of France'. The forest—usually beech, more rarely silver fir and beech, or exceptionally silver fir alone—have their natural upper limit at around 1500 m. However, this limit has often been lowered where the forest borders on pasture, by fellings and by fire; it is often difficult to separate genuine alpine pastures from pseudo-alpine pasture won at the expense of forest.

It is of interest to list the alpine species which occur in the Massif Central:

Juniperus communis	**S. oppositifolia*
Salix (two species)	*S. paniculata*
Polygonum viviparum	**S. stellaris*
Cerastium alpinum	**Dryas octopetala*
Sagina saginoides	*Geum montanum*
Minuartia recurva	*Potentilla aurea*
M. verna	*P. crantzii*
*Silene rupestris	*Alchemilla alpina*
*Pulsatilla alpina	*A. flabellata*
P. vernalis	*Trifolium alpinum*
Cardamine resedifolia	*T. badium*
Murbeckiella pinnatifida	*T. pallescens*
Sedum alpestre	*Galium asperum*
Saxifraga androsacea	*Epilobium anagallidifolium*
S. bryoides	*Astrantia minor*

Androsace carnea	Phyteuma hemisphaericum
Soldanella alpina	Jasione humilis
Myosotis alpestris	Aster alpinus
Veronica alpina	Erigeron alpinus
V. fruticans	Gnaphalium norvegicum
Bartsia alpina	G. supinum
Pedicularis foliosa	Senecio doronicum
P. verticillata	Leontodon pyrenaicus
Euphrasia alpina	Crepis conyzifolia
E. hirtella	Hieracium aurantiacum
E. minima	Nigritella nigra
Plantago alpina	Orchis globosa

This alpine flora reaches its maximum development in the volcanic mountains of Auvergne, on the Puy de Sancy, and the Plomb du Cantal (1858 m). The Puy de Sancy (1886 m), the highest mountain of the Massif Central, has a colourful display of flowers at the end of June above Super Besse, particularly of: *Anemone nemorosa*, *Pulsatilla alpina* (nearly all of the white subspecies), *Narcissus pseudonarcissus*, and *N. poeticus*, with bluebells, *Endymion non-scriptus* in the wood just above the village. Other notable plants include *Anemone narcissiflora* in wet places; a belt of *Cytisus purgans*; bogs with dwarf *Salix* species, *Trifolium alpinum*, *Androsace carnea*, *Arnica montana*, and *Polygonatum verticillatum*. The volcanic tops of Aubrac are much lower and have only ten alpine species. The Montagne de la Margeride, which connects Mont Lozère with the Auvergne, is a plateau with immense and monotonous pastures interrupted by forests of pine, silver fir, and beech. Despite its average altitude of 1400 m, its highest point, the Signal de Randon, is only 1554 m. Only seven alpine species are found there, probably because of the uniformity of the ancient granite rocks. In the Haut Vivarais there are eight alpine species and the following can be found on Mont Mezenc:

Silene rupestris	Euphrasia minima
Pulsatilla vernalis	Nigritella nigra
Alchemilla flabellata	Orchis globosa

Colonies of alpines, summit colonies, may be found on high-level rocky crests, but others, gorge colonies, may occur at comparatively low altitudes in humid gorges protected from Atlantic winds, and from intensive grazing. In consequence the highest summits are not always the richest in alpine species. On Mont Lozère, for example, the most interesting summit colony is on the rocks of Malpertus (1685 m), and not on the higher Signal de Finiels 10 km to the west, which is comparatively poor in alpine species. The most important colony on the Mont Aigoual is on the escarpments of the Pic de la Fajeole and not on the principal summit. A remarkable gorge colony is preserved at 600–900 m on the northern edge of the Causse Noir in the savage canyon of la Jonte, upstream from Peyreleau.

Arctic-alpine species are found in bogs where Sphagnum moss and sedges and dwarf willows are dominant, with associated plants such as:

Salix (four species)	**Geum rivale*
Caltha palustris ·	*Potentilla palustris*
Cardamine pratensis	*Viola palustris*
Saxifraga stellaris	*Menyanthes trifoliata*
Sanguisorba officinalis	*Pedicularis silvatica*
Vaccinium uliginosum	

all of which were counted in a bog on Monts Dore at 1450 m. Other interesting plants of this group are *Nuphar pumila* and *Saxifraga hieracifolia†*.

The Massif Central is a vast and very varied region which merits many exploratory visits; it is not possible in a short account of this nature to do more than touch on some of the interesting botanical features.

23. The Camargue

The flat low-lying delta lands of the Rhône river are known as the Camargue and in many ways they åre similar to the *marismas* of the Coto de Doñana. Two of Europe's greatest rivers, the Guadalquivir and the Rhône, produce two of the most wild and untameable areas in the south of Europe, and two of the most important wild-life areas. Botanically, however, they do not compare in interest with isolated mountains, islands, or steppes mainly because of the salt water, which only certain species can tolerate. Many of these plants—or helophytes—are widespread in similar situations in temperate Europe and Asia, wherever these conditions prevail, and such areas are not areas where the evolution of species is active or where endemic species prevail, although there are exceptions.

Both the Coto de Doñana and the Camargue are characterized by the annual flooding of large inland areas by salt water, producing the shallow lakes and salt marshes called *sansouires* in France, and separated from the sea by sandy bars and extensive sand dunes stretching across the seaward horizon. The Camargue is a wide windswept marshland becoming progressively less saline and more cultivated as one moves inland. It was a country of shepherds and herdsmen, where fighting bulls were reared and white horses roam, but since the last war its

157

whole character is rapidly being changed and the old ways of life are retreating. By the diversion of fresh water from the two arms of the Rhône, formerly saline areas have been rendered fertile, and now one third of the area is under rice cultivation and supports 2,000 rice farms which supply most of France's rice, which was previously imported from Indo-China. Consequently, not only have the *sansouires* been greatly reduced in area, but fresh water draining through the rice fields has also played its part in de-salifying much of the remaining area.

The drainage and metamorphosis of the remaining wetlands of Mediterranean Europe are very serious threats to the diminishing numbers of birds, and particularly to those species such as the pelicans, spoonbills, flamingoes and some terns, gulls and ducks, which obtain their food in shallow saline waters. These used to flourish long ago in vast numbers along the shallow shores of the ancient inland Sarmatic Sea. The Black Sea and the Caspian Sea are all that now remain of this once enormous sea.

Most of the Camargue is composed of silt and fine sand deposited by the two branches of the Rhône, and it is only to the east of the main channel that stony ground occurs and forms the quite different flat stony landscape of the Crau. These later deposits are from the much more tumultuous Durance river, which brings down enormous quantities of debris from the Alps; the Crau lies just outside our area.

The shore-line and coastal dunes stretch for about 100 km along the seaward side of the Camargue and sandy shores continue interruptedly westward along the coast of Languedoc. The dunes of the Camargue are the most important in the Mediterranean, but they do not compare with those of the Atlantic. They form an almost continuous belt immediately inland, in some places several kilometres wide—as at Beaudoc and L'Espiguette. Among the dunes are dune slacks, with freshwater pools and on the old consolidated dunes, woods of stone pine occur.

Plants of the shore-line of the Camargue and Languedoc are:

Polygonum maritimum	*Euphorbia paralias*
Salsola kali	*Eryngium maritimum*
Matthiola sinuata	*Echinophora spinosa*
Cakile maritima	*Crucianella maritima*
Medicago marina	*Calystegia soldanella*
Euphorbia peplis	*Otanthus maritimus*
Pancratium maritimum	

The fore-dunes are colonized by the Marram grass *Ammophila arenaria*, and the Sand Couch *Agropyron junceiforme*, but where they become stabilized and in the hollows and on the sheltered side *Crucianella maritima* becomes abundant, with such other species as *Silene italica*,

Teucrium polium, Scrophularia canina, Scabiosa atropurpurea, Artemisia campestris parasitized by *Orobanche arenaria*, and **Pancratium maritimum*.

These often form a dense sward which is soon colonized by garigue species and pines, including *Psorolea bituminosa, *Cistus salvifolius, Helianthemum hirtum*, and *Thymus vulgaris*; and by taller shrubs such as *Juniperus phoenicea, Tamarix* species, *Phillyrea* species, and *Rhamnus alaternus*. In places the woods of *Pinus pinea* have also *Dorycnium suffruticosum, *Daphne gnidium, Rosmarinus officinalis*, and *Asphodelus ramosus*. The dunes also contain a number of adventitive species which flourish in the sand, and where seed germination is easy. Examples are:

Silene conica	**Onosma arenarium*
Hypecoum procumbens	*Alkanna tinctoria*
Hymenolobus procumbens	*Linaria commutata*
Ononis natrix subsp.	*Bellardia trixago*
ramosissima	*Chondrilla juncea*
Melilotus species	*Crepis bulbosa*
Tribulus terrestris	*Romulea columnae*
*Coris monspeliensis	

The low-lying *sansouires* of the Camargue occupy about 500 sq. km and are dominated almost exclusively by glassworts, *Salicornia* species, which in summer are bright green, but turn brilliant reddish-brown in autumn. The shallow lagoons are often clogged with dense masses of the Tassel Pondweed, *Ruppia spiralis*. During winter they are flooded by salt water, and as summer advances they gradually dry out, leaving dry, parched, salty rims. It is in this area that artificial salt pans have been made, thus further restricting the natural areas. Where debris accumulates in these low-lying saline regions, the following succulent plants may occur commonly: *Suaeda maritima, Bassia hirsuta, Suaeda splendens* and *Salsola soda*; sometimes in addition, here and on the Languedoc *sansouires* are *Polygonum maritimum, Spergularia media, Trifolium squamosum, Lotus preslii*, the North American *Heliotropium curassavicum, Cressa cretica, Limonium vulgare, Inula crithmoides*, and *Aster tripolium*.

The upper *sansouires* have been much reduced in recent times, but there still remain two large untouched areas round the south-west and south-east of the saline Etang de Vaccares. Here the vegetation is more varied with low shrubby bushes ½ m high covering the ground. The most widespread succulent shrubby species are: *Arthrocnemum fruticosum, A. glaucum, A. perenne, Halimione portulacoides*, and *Suaeda vera*. On drier saline ground still grow *Inula crithmoides* and a sward of sea lavenders, such as *Limonium vulgare, L. oleifolium (virgatum)*, and *L. bellidifolium*.

CISTUS SCRUB

1. *Cistus ladanifer* 793   **2.** *Cistus salvifolius* 790   **3.** *Erica australis* 932   **4.** *Cistus monspeliensis* 791   **5.** *Lavandula stoechas* 1110   **6.** *Anemone palmata* 217   **7.** *Scilla monophyllos* 1636a   **8.** *Lithospermum diffusum* 1074   **9.** *Cytinus hypocistis* 78 (on *Halimium halimifolium*)

10. *Myrtus communis* 824   **11.** *Daphne gnidium* 758   **12.** *Calicotome villosa* 498
13. *Clematis cirrhosa* 226   **14.** *Phlomis purpurea* 1122b   **15.** *Serapias cordigera* 1903
16. *Cistus crispus* 789   **17.** *Arisarum vulgare* 1821   **18.** *Ornithogalum umbellatum*
1639

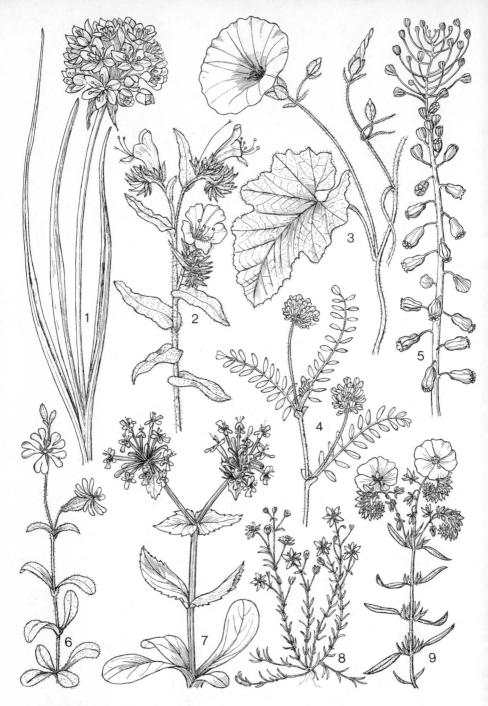

ROADSIDES

1. *Allium roseum* 1611   **2.** *Echium plantagineum* 1083   **3.** *Convolvulus althaeoides* 1039   **4.** *Astragalus echinatus* 527b   **5.** *Muscari comosum* 1645   **6.** *Silene colorata* (174)   **7.** *Fedia cornucopiae* 1312   **8.** *Spergularia purpurea* 156b   **9.** *Helianthemum lavandulifolium* 801a

10. *Lavatera cretica* 741   **11.** *Erodium laciniatum* 652a   **12.** *Cerinthe major* 1079
13. *Verbascum laciniatum* 1196b   **14.** *Ophrys speculum* 1875   **15.** *Tuberaria guttata*
798   **16.** *Tetragonolobus purpureus* 617   **17.** *Cladanthus arabicus* 1414c

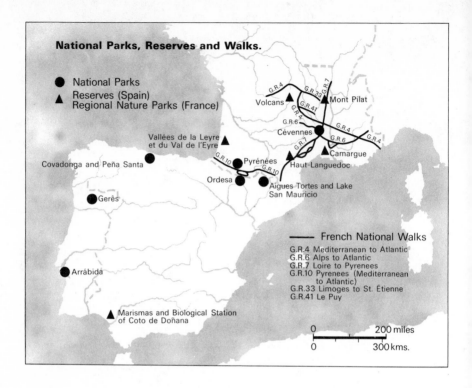

National Parks, Reserves and Walks.

● National Parks

▲ Reserves (Spain)
   Regional Nature Parks (France)

Volcans

Mont Pilat

G.R.4
G.R.33
G.R.7
G.R.41

G.R.6

Cévennes

G.R.4
G.R.6

G.R.4

Vallées de la Leyre
et du Val de l'Eyre ▲

Covadonga and Peña Santa ●

G.R.10
Pyrénées
G.R.10

G.R.7

Camargue ▲

Haut-Languedoc ▲

Ordesa ●

▲ Gerês

● ●
Aigues Tortes and Lake
San Mauricio

——— French National Walks

G.R.4 Mediterranean to Atlantic
G.R.6 Alps to Atlantic
G.R.7 Loire to Pyrenees
G.R.10 Pyrenees (Mediterranean
        to Atlantic)
G.R.33 Limoges to St. Étienne
G.R.41 Le Puy

● Arrábida

▲ Marismas and Biological Station
   of Coto de Doñana

0   200 miles

0   300 kms.

National Parks, Nature Reserves and National Walks

In western Europe the establishment and maintenance of nature reserves is in its infancy, and at the time of writing little serious attention has been given to the conservation of plants and animals and the biosystems in which they live. So far it has been little more than a paper agreement, and a matter of sign-posting some of the access routes, and as far as the visitor is concerned information about the natural history of the plants of the reserves listed below is either not available, or very difficult to obtain. The mountain reserves often seem to be little more than a preserve for game, and many of the traditional practices of forest clearing, grazing, and so on are continued as intensively as ever.

No doubt there will be a gradual realization of the importance of Nature Reserves both from the point of view of preserving the unique plant communities and their dependent animals, and as a necessity for humans for recreation, study, and education. Of the three countries concerned in our area, France is the most aware of the problems, and the movement there is rapidly gaining momentum.

PORTUGAL

Parc National do Peneda-Gerês (see p. 108).
Parc National de l'Arrábida

SPAIN

Parque Nacional de la Montaña de Covadonga o de Peña Santa (see p. 129).
Parque Nacional de Ordesa (see p. 137).
Reserve Marismas, Coto de Doñana (see p. 57).
Parque Nacional de Aigues Tortes y Lago de San Mauricio (see p. 142).

FRANCE

National Parks

Le parc national des Pyrénées occidentales (including the Nature Reserve of Néouvieille) (see p. 146).
Le parc national des Cévennes.

Regional Nature Parks

Le parc naturel régional de Camargue.
Le parc naturel régional des vallées de la Leyre et du Val de l'Eyre (Landes and Gironde).
Le parc naturel régional des Volcans (Auvergne).
Le parc naturel régional du Haut-Languedoc
Le parc naturel régional du Pilat (Loire)
Further details can be obtained from La Documentation Française, 29–31 Quai Voltaire, Paris VIIe.

The **National Walks** of France will also be of considerable interest to the naturalist. Information concerning these can be obtained from the Comité National des Sentiers de Grande Randonnée, 65 Avenue de la Grande-Armée, Paris XVIe.
The principal walks in our area are:

Sentier G.R. 4 Mediterranean to Atlantic
        G.R. 6 Alps to Atlantic
        G.R. 7 Loire to Pyrenees
        G.R. 10 Pyrenees (Mediterranean to Atlantic)
        G.R. 33 Limoges to St Etienne
        G.R. 41 Le Puy

REFERENCES TO CHAPTER 2

[1] Stocken, 'Plant hunting in Southern Spain', *Journ. Royal Horticultural Soc.*, LXXXIX (1), 1964, p. 14.

[2] Stocken, 'The Serranía de Ronda', *Quart. Bull. Alpine Garden Soc.*, 34(3), 1966, p. 258.

[3] Ripley, 'A Journey through Spain', *Quart. Bull. Alpine Garden Soc.*, 12(1), 1944, pp. 47–8.

[4] Stocken, 'The Serranía de Ronda' *Ibid.*

[5] Fernandez, 'Sierra Nevada' quoted in Way, *A Geography of Spain and Portugal* (London, 1962), p. 271.

[6] Stocken, 'The Spanish Sierra Nevada', *Quart. Bull. Alpine Garden Soc.*, 33(2), 1965, p. 162.

[7] Ripley, 'A Journey through Spain', pp. 43–4.

[8] Farrer, *The English Rock Garden* (Edinburgh, 1918), p. 235.

[9] Knocke, *Flora Balearica* (Montpellier, 1921–23). See Vol. III, p. 308.

[10] Houston, *The Western Mediterranean World*, (London, 1964), pp. 308–9.

[11] Dresser, 'Notes on the pre-alpine flora of the Picos de Europa, Spain', *Notes Royal Bot. Gdn. Edinburgh*, XXIII (1959) and XXIV (1962).

[12] Meadows, 'The Valley of Benasque', *Quart. Bull. Alpine Garden Soc.*, 38(3), p. 327.

[13] Taylor, 'Eastern Pyrenees', *Quart. Bull. Alpine Garden Soc.*, 33(1), 1965, p. 68.

[14] Taylor, *Ibid.*, p. 72.

[15] Losa España & Montserrat, 'Aportación al conocimiento de la flora de Andorra', *Primo Congreso Internacional del Pirineo del Instituto de Estudios Pirenaicos: Monograph Botanica*, 6, 1951.

[16] Albury, 'Plant hunting in Andorra and the Pyrenees', *Quart. Bull. Alpine Garden Soc.*, 24(1), 1956, p. 69.

[17] McClintock, 'Gabas and Gavarnie', *Quart. Bull. Alpine Garden Soc.*, 18(1), 1950, p. 24.

[18] de Martonne, *Geographical Regions of France* (London, 1933), p. 110.

[19] de Martonne, ibid., pp. 97–8, 100.

3. The Identification of Species

DICOTYLEDONES
PINACEAE | Pine Family

ABIES | 1. *A. alba* Miller SILVER FIR. Mountains. E. Spain, Pyrenees, Massif Central. 1a. *A. pinsapo* Boiss. SPANISH FIR. Like 1 but leaves stiff, pointed, 1–1½ cm, arranged evenly round the hairless twigs; buds resinous. Cones with bracts shorter than the scales. Mountains. S. Spain (Serranía de Ronda).

PINUS | Pine   The most commonly planted foreign species are: the three-leaved (10) *Pinus radiata* D. Don MONTEREY PINE, and the five-leaved (10) *P. strobus* L. WEYMOUTH PINE.

1. Cones large 8–22 cm long.
  2. Seeds large 1½–2 cm, wingless or nearly so; cones blunt before scales
    separate.   *pinea*
  2. Seeds small ½–1 cm, winged; cones more or less acute before scales separate.
    3. Twigs light grey; lvs. thin, flexible.   *halepensis*
    3. Twigs reddish-brown; lvs. stiff, spiny-tipped.   *pinaster*
1. Cones medium or small, less than 7 cm long.
  4. Lvs. usually more than 8 cm long, dark green, straight.   *nigra*
  4. Lvs. usually less than 8 cm long, not dark green.
    5. Lvs. glaucous; cones acute, pendulous.   *sylvestris*
    5. Lvs. bright green; cones blunt, spreading.   *uncinata*

4. *P. pinaster* Aiton MARITIME PINE. A dense, dark green, pyramidal tree with long, very stiff spiny-tipped leaves 10–25 cm. Cones 8–22 cm, clustered, scales often spiny-tipped. Coastal sands, dunes, in mountains inland. Widespread; often planted. **5. *P. pinea* L. STONE or UMBRELLA PINE. A dense, bright green, umbrella-shaped tree, bearing heavy and nearly globular, shining mahogany-brown cones 8–15 cm. Leaves rigid, 10–20 cm. Coastal areas, hills. S. Portugal, Spain, Med. France.

**6. *P. halepensis* Miller ALEPPO PINE. Usually a slender tree with silvery-grey branches and delicate pale green foliage. Leaves 6–15 cm, less than 1 mm broad. Cones conical, 5–12 cm. Coasts and hills. Med. region; introd. S. Portugal.

**7. *P. nigra* Arnold BLACK PINE. A very variable tree. Subsp. *salzmannii* (Dunal) Franco (*P. pyrenaica*) is found in the mountains of C. and E. Spain, Pyrenees and Cévennes. It is distinguished by its narrow pyramidal or cylindrical crown and its straight trunk with pale greyish-brown bark. Leaves rigid, not spiny-tipped, dark green, straight, 8–16 cm. Cones oval, 4–6 cm, shining light brown.

8. *P. sylvestris* L. SCOTS PINE. Readily distinguished by the bright ochre-red colour of its upper trunk and larger branches. Mainly in the hills. N. and E. Iberia (also S. de Guadarrama and S. Nevada), France.

9a. *P. uncinata* Mirbel. Like *P. mugo* MOUNTAIN PINE of Central Europe, but usually a tree to 25 m. Cones 5–7 cm, usually with prominent hooked or rounded tips to the scales. Mountains. C. Spain, Pyrenees. Page 171.

CUPRESSACEAE | Cypress Family

CUPRESSUS | **11. *C. sempervirens* L. FUNERAL CYPRESS. Often planted in gardens and cemeteries, particularly in its tall columnar form, 'the vegetable minarets of the Mediterranean landscape'. Twigs covered with small triangular scale-like leaves, recalling 14, but cones 2½–4 cm across, opening by flat-topped, angular scales. (11) *C. macrocarpa* Hartweg MONTEREY CYPRESS, and *C. lusitanica* Miller are also sometimes planted.

TETRACLINIS | 11a. *T. articulata* (Vahl) Masters BARBARY ARBOR-VITAE. An open, much-branched, pyramidal evergreen tree, with tiny scale-like leaves in 4 rows. Distinguished from other conifers by the small cones, 8–12 mm, consisting of a single whorl of 4 woody triangular scales. Spain (only near Cartagena).

JUNIPERUS | Juniper
1. Lvs. all needle-like; fr. axillary.
  2. Lvs. with 1 broad pale band above; fr. at length black.   *communis*
  2. Lvs. with 2 pale bands above; fr. at length reddish-brown or purple.   *oxycedrus*
1. Lvs. of adult plants scale-like; fr. terminal.
  3. Scale lvs. blunt, with narrow papery border; ripe fr. dark red.   *phoenicea*
  3. Scale lvs. acute, without papery border; ripe fr. dark purple or blackish.
    4. Spreading shrubs; mature fr. 4–6 mm.   *sabina*
    4. Pyramidal trees; mature fr. 7–8 mm.   *thurifera*

**12. *J. communis* L. JUNIPER. The erect pyramidal subsp. *communis* is widespread in heathlands, dry scrublands and in the hills of our region. The prostrate mat-forming subsp. *nana* Syme DWARF JUNIPER occurs in the highest mountains, often covering extensive areas. Widespread. **13. *J. oxycedrus* L. PRICKLY JUNIPER. Coastal hills. Portugal, Med. region.

**14. *J. phoenicea* L. PHOENICIAN JUNIPER. A dense, conical, dark green shrub or small tree. Fruit 8–14 mm, dark red when ripe. Sandy and rocky places, hills. S. Portugal, S., C. and E. Spain, S. France. 15. *J. sabina* L. SAVIN. A low spreading, strong-smelling shrub with older branches with 4–6 rows of tiny blunt, overlapping scale-like leaves and distinctive young branches with spiny-tipped, spreading scale-like leaves. Fruit bluish-black, bloomed. Mountains. S. and E. Spain, Pyrenees. Page 171. 16. *J. thurifera* L. SPANISH JUNIPER. Like 15, but a pyramidal tree with a rounded crown, to 20 m. Fruit dark purple when ripe. Mountains. S., C. and E. Spain. Page 171.

TAXACEAE | Yew Family

TAXUS | **17. *T. baccata* L. YEW. Rocky places in hills and mountains. Scattered throughout.

EPHEDRACEAE | Joint-pine Family

EPHEDRA | 3 species in the Med. region, including 1 species in Portugal.

SALICACEAE | Willow Family

SALIX | **Willow**   27 species are found in our area, including most of the common and widespread willows of Europe, while the dwarf alpine and northern species occur in the higher mountains of Spain and the Pyrenees. Species such as *S. babylonica*, *S. viminalis*, *S. purpurea* and *S. fragilis* are often planted in damp areas. For descriptions and drawings of all Spanish willows see *Saliceas de España*, C. Vicioso, Inst. Forestal, Madrid. Bol. 57, 1951.

POPULUS | **Poplar**   A number of species, hybrids and cultivated varieties are commonly planted on the banks of rivers and in low-lying areas.   **32. *P. tremula* L. ASPEN. The only native species, occurs scattered throughout.

MYRICACEAE | Bog Myrtle Family

MYRICA | **34. *M. gale* L. BOG MYRTLE. A very aromatic, deciduous shrub, rarely more than 1½ m high, growing in bogs and damp ground. Catkins ovoid, erect, and borne on reddish-brown shining twigs before the leaves expand. N. Portugal, N–W. Spain, W. France.   34a. *M. faya* Aiton A small evergreen tree or shrub, with foliage like *Arbutus unedo*, but darker green, with narrower leaves, and the twigs without rust-coloured scales. Leaves 4–11 cm, oblance-shaped, entire with inrolled margin, hairless, without conspicuous glands. Catkins borne amongst the expanded leaves, somewhat branched. Fruits fleshy. Sandy soils S. and C. Portugal; introd. elsewhere (Pinhal Leiria).

JUGLANDACEAE | Walnut Family

JUGLANS | **35. *J. regia* L. WALNUT. A native of S–E. Europe; widely planted in S–W. Europe.

BETULACEAE | Birch Family

BETULA | **36. *B. pendula* Roth COMMON SILVER BIRCH. Distinguished by its pendulous, hairless twigs with numerous resin-glands, its usually doubly toothed leaves, and its hairless nutlets. Mountains. N. Portugal, N. and C. Spain, Pyrenees, France.   37. *B. pubescens* Ehrh. BIRCH. Branches usually erect, often hairy and without resin-glands; leaves irregularly toothed; nutlets finely hairy at apex. Mountains. C. and N. Spain, Pyrenees, France; introd. Portugal.

ALNUS | **40. *A. glutinosa* (L.) Gaertner ALDER. Throughout, but less common in the south.

CORYLACEAE | Hazel Family

CARPINUS | **42. *C. betulus* L. HORNBEAM. Woods and thickets. Pyrenees, W. France.

CORYLUS | 44. *C. avellana* L. HAZEL, COB-NUT. Shady thickets in hills and mountains. N. Portugal, Spain, France.

FAGACEAE | Beech Family

FAGUS | 45. *F. sylvatica* L. BEECH. Forms forests in hills and mountain valleys. C. and N. Spain, Pyrenees, France.

CASTANEA | **46. *C. sativa* Miller SWEET or SPANISH CHESTNUT. Widely planted, in hills and mountains. N. Portugal, Spain, France.

QUERCUS | **Oak** Very important in our area, often dominating the landscape. Oaks may hybridize freely, often making the identification of Iberian species difficult See *Revision del Genero Quercus en España*, Vicioso, C. *Inst. Forestal Madrid*, Bol. No. 51, 1950.

1. Lvs. evergreen, leathery.
   2. Mature lvs. hairless beneath.   *coccifera*
   2. Mature lvs. hairy beneath.
      3. Mid-vein of lv. straight; bark not thick and corky.   *ilex, rotundifolia*
      3. Mid-vein of lv. wavy; bark thick and corky.   *suber*
1. Lvs. deciduous or semi-evergreen, seldom leathery.
   4. Fr. clusters long-stalked; scales of cup fused together except at their tips. (Widespread except in south).   **52. *Q. robur* L.
   4. Fr. clusters short-stalked or stalkless; scales of cup distinct, not fused together.
      5. Lvs. deciduous, falling in the autumn.
         6. Young twigs with curled woolly hairs; lv. stalks not grooved.
            7. Lvs. densely white-haired beneath.   *pyrenaica*
            7. Lvs. with woolly hairs, greyish beneath, becoming nearly hairless.   *pubescens*
         6. Young twigs hairless, or with straight silky hairs; lv. stalks grooved.
            8. Lvs. with *c.* 10 pairs of lateral veins; fr. stalks with adpressed hairs.   *mas*
            8. Lvs. with 6–8 pairs of lateral veins; fr. stalks hairless. (Widespread except in south).   53. *Q. petraea* (Mattuschka) Liebl.
      5. Lvs. semi-evergreen, remaining on branches over winter.

1. *Juniperus thurifera* 16
3. *Pinus uncinata* 9a
5. *Arceuthobium oxycedri* 73a

2. *Quercus faginea* 55a
4. *Quercus fruticosa* 55b
6. *Juniperus sabina* 15

9. Low shrubs usually below·½ m; lvs. nearly
   stalkless.   *fruticosa*
9. Shrubs above ½ m, or trees; lv. stalks more
   than ½ cm.   *faginea*; *canariensis*

**47. *Q. coccifera* L. KERMES or HOLLY OAK. A holly-like shrub, or small tree, forming dense thickets over many thousands of square miles. S. and C. Portugal, Med. region.   48. *Q. ilex* L. HOLM OAK, ILEX. Forms forests and extensive thickets. C. Iberia, Med. region.   (48) *Q. rotundifolia* Lam. Very like 48 and replacing it in the south-west, but leaves broadly egg-shaped or almost rounded, greyish-glaucous above and with fewer (5–8) pairs of lateral veins; stipules papery, almost hairless. Nuts sweet, not bitter. S. and C. Portugal, S. and C. Spain, Med. France.   49. *Q. suber* L. CORK OAK. Easily distinguished by its thick, usually deeply furrowed, grey corky bark. Leaves like 48, hairy beneath, entire, toothed or spiny. Grown in very extensive forests. S. and C. Portugal, S., C. and E. Spain. Med. France.

53a. *Q. mas* Thore A deciduous tree with twigs, buds and lower surface of leaves covered at first with silky hairs, later becoming hairless. Leaves large 8–18 cm, with 8–10 pairs of rather narrow, forwardly-directed lobes. Fruit stalk with silky hairs; scales of cup broadly egg-shaped, with warty swellings. N. Spain, S–W. France.   **54. *Q. pubescens* Willd. WHITE OAK. A small deciduous tree or shrub, with densely woolly-haired twigs and young leaves which later become finely hairy beneath. Leaves 4–12 cm. Fruit cups grey-haired, scales lance-shaped, closely adpressed. Dry hills, woods. S. Portugal. Med. region. Pl. 1.   **55. *Q. pyrenaica* Willd. (*Q. toza*) PYRENEAN OAK. Like 54 with woolly-haired twigs, but distinguished by its large leaves 8–16 cm, which are densely white-woolly beneath and deeply cut into 4–8 pairs of narrow acute lobes; young leaves characteristically pink in spring. Scales of cup narrow lance-shaped, blunt, loosely overlapping. Forming forests in the mountains. Scattered throughout Pl. 1.

55a. *Q. faginea* Lam. (*Q. lusitanica* auct.) LUSITANIAN OAK. A semi-evergreen shrub or tree with oval to long-elliptic leaves 4–10 cm, which have 5–12 pairs of more or less forwardly-directed shallow, triangular, tooth-like lobes, almost hairless and somewhat shining above, with dense felt-like hairs beneath. Fruit cups large 25 mm across, scales broadly lance-shaped or egg-shaped. Mountains. Portugal, Spain except in North. Page 171.   55b. *Q. fruticosa* Brot. (*Q. humilis*) A suckering shrub to 2m with semi-evergreen leaves. Leaves 3–5 cm, leathery, nearly hairless, with 4–6 pairs of forwardly-directed teeth towards the tip. Cup 8–12 mm across. Sandy places in lowlands and mountains. S. and C. Portugal, S–W. and C. Spain. Page 171.   55c. *Q. canariensis* Willd. A semi-evergreen tree with twigs and young leaves densely woolly-haired, but at length nearly hairless. Leaves large 6–18 cm, with 9–14 pairs of teeth, glaucous beneath; leaf stalk 8–30 mm. Upper scales of cup much smaller than lower. Lowlands and mountains. S. Portugal, S–W. and N–E. Spain.

ULMACEAE | Elm Family

ULMUS | **56. *U. glabra* Hudson WYCH ELM, and 57. *U. procera* Salisb. ENGLISH ELM are scattered throughout Spain and France.   58. *U. minor* Miller SMOOTH-LEAVED ELM. Widespread.

CELTIS | **60. *C. australis* L. NETTLE TREE. A slender graceful tree with a smooth greyish trunk, and oval-lance-shaped long-pointed, acutely toothed leaves. Flowers green. Fruits

9–12 mm, blackish, long-stalked, edible. Hedges, banks, sandy places. Native in the south and east of our area; widely planted.

MORACEAE | Mulberry Family

BROUSSONETIA | 61a. *B. papyrifera* (L.) Vent. PAPER MULBERRY. A native of E. Asia, sometimes planted in parks, gardens and roadsides in Iberia. A small tree with hairy young branches, and oval double-toothed leaves, which are rough above and grey-felty beneath. Plants one-sexed; male flowers in catkins; female flowers in dense globular woolly heads. Fruit globular, fleshy, *c*. 2 cm, orange-red.

MORUS | 61. *M. nigra* L. COMMON MULBERRY and **(61) *M. alba* L. WHITE MULBERRY, are widely grown in our area.

FICUS | 62. *F. carica* L. FIG. Widely cultivated in Portugal and the Med. region in a variety of forms; often becoming naturalized in rocky places.

CANNABACEAE | Hemp Family

HUMULUS | **63. *H. lupulus* L. HOP. Hedges, riversides, thickets. North of our area; often escaping from cultivation.

CANNABIS | **64. *C. sativa* L. HEMP. A slender erect leafy annual with leaves deeply cut into 3–9 narrow acute, toothed segments, and with branched clusters of green 'dock-like' flowers. Native of Asia; often cultivated in the south for its oil-producing seeds and stem fibres; and for drugs.

URTICACEAE | Nettle Family

URTICA | Nettle   5 species. The following are mostly widespread: 65. *U. dioica* L. NETTLE;   66. *U. urens* L. SMALL NETTLE;   **67. *U. pilulifera* L. ROMAN NETTLE;   **(66) *U. dubia* Forskål.

PARIETARIA | **Pellitory-of-the-wall**   (68) *P. diffusa* Mert. & Koch and 69. *P. lusitanica* L. are widespread. 68a. *P. mauritanica* Durieu An annual with the lowest leaf stalks usually longer than the leaf-blade. S. Iberian peninsula.

PROTEACEAE | Protea Family

HAKEA | Shrubs or small trees, natives of Australia and Tasmania; sometimes planted in Spain and Portugal for land reclamation.   69a. *H. sericea* Schrader (*H. acicularis*) A very spiny shrub with long, stiff, stout needle-like leaves up to 7 cm long. Flowers in axillary clusters; corolla white or pinkish, 4–5 mm across, lobes 4; stamens 4. Fruit very woody, bean-shaped, 2–3 cm across.   69b. *H. salicifolia* (Vent.) B.L. Burtt has oblong to narrow lance-shaped, flattened leaves 6–18 mm wide. Flowers white.

SANTALACEAE | Sandalwood Family

OSYRIS | **70. *O. alba* L. A small switch-like shrub to 1 m, with slender green branches and narrow leathery leaves, 15 mm by 2–5 mm. Flowers yellowish, with 3 spreading petals. Fruit fleshy, bright red, 5–7 mm. Dry rocky places. Portugal, Med. region.   70a. *O. quadripartita* Decne Like 70 but bracts subtending flowers papery, soon falling. Leaves larger ½–4 cm by ½–2 cm. Fruit larger, 7–10 mm. Hills, rocks, hedges. S. Iberian peninsula.

THESIUM | **Bastard Toadflax**   5 species. Inconspicuous, semi-parasitic plants of low slender habit, pale yellowish-green foliage, narrow leaves and tiny whitish or greenish flowers.

71a. *T. divaricatum* Mert. & Koch An erect, smooth perennial with a pyramidal leafy inflorescence of tiny white flowers. Flowers tubular, with 5 lobes. Leaves linear, 1–2 mm broad; bracts shorter than fruit. Stock woody. Dry places. Widespread in south.

LORANTHACEAE | Mistletoe Family

VISCUM | **73. *V. album* L. MISTLETOE. A pale green, branched shrub found growing parasitically on the branches of many different kinds of trees. Berries fleshy, usually white, but sometimes yellow. Widespread.   (73) *V. cruciatum* Boiss. Differs from 73 in having red berries and shortly stalked inflorescences and yellow-green foliage. Usually parasitic on olive trees but attacks many other species. S. Portugal, S–W. Spain.

ARCEUTHOBIUM | 73a. *A. oxycedri* (DC.) Bieb. A tiny, densely branched, yellowish-green shrublet 5–20 cm, growing parasitically on the branches of Juniper. Leaves triangular, scale-like, sheathing the stem; branches jointed. Flowers yellowish, one-sexed, the males with 3 petals. Fruit green, exploding when ripe to eject the sticky seeds. S., C. and E. Spain, S. France, Page 171.

ARISTOLOCHIACEAE | Birthwort Family

ASARUM | **74. *A. europaeum* L. ASARABACCA. Shady places in mountains, rare elsewhere. N–E. Spain, Pyrenees.

ARISTOLOCHIA | **Birthwort**

1. Fls. pale yellow, 2 or more in the axil of the lvs.
   (Scattered throughout).   **76 *A. clematitis* L.
1. Fls. not pale yellow, solitary in the axil of the lvs.
   2. Stems woody, usually climbing; lvs. evergreen,
      leathery.   *baetica*

2. Stems herbaceous, not climbing; lvs. not evergreen, soft.
3. Lvs. ·with hard-pointed teeth or swellings on margin and lower surface; tubers numerous, slender.   *pistolochia*
3. Lvs. without hard teeth; tuber solitary, globular or cylindrical.
    4. Lvs. stalkless, basal lobes of lvs. encircling stem.   *rotunda*
    4. Lvs. stalked, basal lobes of lvs. not encircling stem.
       5. Lvs. not more than 2 cm by 1 cm; fls. brownish-yellow with brown stripes and reddish lip. (Balearic Islands).   77b. *A. bianorii* Sennen & Pau
       5. Largest lvs. at least 2 cm by 2 cm; fls. brownish to yellowish-green, lip brownish-purple.   *longa*

77. *A. rotunda* L. ROUND-LEAVED BIRTHWORT. Distinguished by its almost stalkless leaves which have rounded basal lobes encircling the stem. Flowers with a yellowish corolla tube and a dark brown strap-shaped lip. Hedges, cultivated ground. Portugal, N. and N–W. Spain, S. France. (77) *A. pistolochia* L. Distinguished by its leaves which are glaucous and rough below, and with finely toothed margins. Flowers 2–5 cm, brownish with a dark purple upper lip. Cultivated ground, olive groves. Mostly south and centre of our area. Pl. 1. (77) *A. longa* L. Flowers brownish or yellowish-green, with brown-purple lip shorter than the tube which may be 3–6 cm long. Leaves triangular-oval with a heart-shaped base; leaf stalks about as long as flower stalks. Cultivated ground, waysides. Portugal, Spain, Med. France.

77a. *A. baetica* L. An evergreen climber to 2–3 m, with glaucous leaves, and with strongly curved dark brownish- or blackish-purple, long-stalked flowers. Leaves somewhat leathery, hairless, with basal lobes about one quarter as long as the blade. Dry uncultivated ground. Portugal, S. and S–E. Spain. Pl. 2. Page 62.

RAFFLESIACEAE | Rafflesia Family

CYTINUS | Quite unmistakable parasitic plants growing on the roots of *Cistus* and *Halimium*, and usually found under these bushes. **78. *C. hypocistis* (L.) L.** has dense rounded heads of yellow flowers contrasting with the orange or red bracts. Common. Portugal, Med. region. Page 160. (78) *C. ruber* (Fourr.) Komarov Grows on pink-flowered *Cistus* species; like 78 but flowers ivory-white or pale pink with contrasting crimson or carmine bracts. Much of our area except in the north. Pl. 1.

BALANOPHORACEAE | Balanophora Family

CYNOMORIUM | 78a. *C. coccineum* L. A remarkable parasitic plant, looking like a dark red or blackish-purple phallus arising direct from the soil. The erect fleshy stem 15–30 cm high, bears numerous scale-leaves at the base, and a terminal club-shaped cluster of many tiny flowers. Parasitic on the roots of salt-loving plants, particularly members of the *Chenopodaceae*. S. Portugal, S. and E. Spain. Pl. 1.

175

POLYGONACEAE | Dock Family

POLYGONUM | **Persicaria, Knotgrass, Bistort**   About 19 species occur in our area, many of which are widespread European species. Widespread weeds: 80. *P. aviculare* L. KNOTGRASS; 82. *P. persicaria* L. PERSICARIA, REDLEG; **(82) *P. lapathifolium* L. PALE PERSICARIA. Marsh plants: **83. *P. amphibium* L. AMPHIBIOUS BISTORT, and **81. *P. hydropiper* L. WATER-PEPPER. Mountain plants: **84. *P. bistorta* L. BISTORT, SNAKE-ROOT, and 86. *P. alpinum* All. ALPINE KNOTWEED.

**79. *P. maritimum* L. SEA KNOTGRASS. Frequent on the mobile sands of the Atlantic and Med. coasts.   79a. *P. equisetiforme* Sibth. & Sm. A beautiful slender plant with 'long wiry branches, set with myriads of tiny white flowers'. Leaves narrow, falling early, leaving slender erect or spreading stems with pale sheaths. Cultivated places, waste ground, hedges. Portugal, S. and E. Spain. Page 181.

FAGOPYRUM | **89. *F. esculentum* Moench BUCKWHEAT and (89) *F. tataricum* (L.) Gaertner GREEN BUCKWHEAT; sometimes planted and occasionally escaping.

RUMEX | **Dock**   About 25 species occur in our area; many are widespread European species.   95. *R. bucephalophorus* L. HORNED DOCK. A small annual of dry cultivated ground and olive groves, which may sometimes occur in such abundance as to give the ground a reddish flush. Distinguished by its fruit which has conspicuous straight or hooked teeth on the perianth. Widespread.

CHENOPODIACEAE | Goosefoot Family

A family well adapted to soils which are rich in salt, either by the sea or inland. They are usually shrubby, and their monotonous grey-green foliage may cover extensive areas around lagoons, estuaries, mud-flats, salt steppes and any ground periodically flooded with salt water. The GLASSWORTS: *Arthrocnemum* and *Salicornia* species have fleshy cylindrical stems which 'crackle underfoot'. The SEABLITES: 115. *Suaeda vera* J. F. Gmelin, and **(115) *S. maritima* (L.) Dumort. have fleshy, narrow, pointed leaves and grow with the Glassworts nearest the open saline water. The SALTWORTS: **116. *Salsola kali* L., **(116) *S. soda* L. and   116a. *S. vermiculata* L. are common on sands and salt-flushed mud; they are distinguished by their stiff, half-cylindrical, usually spiny-pointed leaves. Two shrubs with silvery-white, flattened leaves are: **112. *Halimione portulacoides* (L.) Aellen SEA PURSLANE, and 107. *Atriplex halimus* L., which grow on mud-flats. Common weeds of cultivation, waste places, and road-sides are the GOOSEFOOTS: *Chenopodium* (15 species), and ORACHES: *Atriplex* (8 species).

A number of rare salt-steppe plants, members of this family, have their only European station in the steppes of Iberia. See page 89.

AMARANTHACEAE | Cockscomb Family

AMARANTHUS | 10 species. Usually weeds of cultivation and waste places; mainly introduced from America.

PHYTOLACCACEAE | Pokeweed Family

PHYTOLACCA | ****119.** *P. americana* L. VIRGINIAN POKE. A tall leafy perennial 1–3 m with large oval-lance-shaped leaves and dense cylindrical spikes of greenish or pinkish flowers arising from the angles of the branches. Fruit reddish then purplish-black. Native of N. America. Grown for ornament and for its berries which are used in dyeing; often naturalized in the south.

AIZOACEAE | Mesembryanthemum Family

1. Petals absent; ovary superior.
  2. Fr. opening by 2 valves; lvs. usually alternate.   AIZOON
  2. Fr. opening by a circular lid; lvs. opposite.   SESUVIUM
1. Petals present; ovary inferior.
  3. Stigmas 8–20; fr. fleshy; seeds embedded in mucilage.   CARPOBROTUS
  3. Stigmas 5 or less; fr. woody; seeds not embedded in mucilage.
    4. At least some lvs. stalked.
      5. Lvs. heart-shaped at base; stigmas 4.   APTENIA
      5. Lvs. not heart-shaped at base; stigmas 5.   MESEMBRYANTHEMUM
    4. Lvs. all stalkless.
      6. Fls. pinkish-purple.   DISPHYMA
      6. Fls. white or yellow.
        7. Lvs. with waxy bloom; fls. bright yellow or magenta.   LAMPRANTHUS
        7. Lvs. green; fls. white or yellowish-white.   MESEMBRYANTHEMUM

AIZOON | 120b. *A. hispanicum* L. A small annual covered all over with tiny swellings, and with solitary, nearly stalkless, yellowish flowers in the forks of the branches. Sepals 5, pointed; petals absent; stamens about 20, in bundles. Leaves blunt, oblong-lance-shaped, stalkless. Sandy and salt-rich clays on the littoral, and on the steppes of Aragon. Med. and E. Spain.

SESUVIUM | 120c. *S. portulacastrum* L. A prostrate, branched perennial with somewhat fleshy, opposite, oblong-spathulate leaves, and with solitary, axillary, yellowish flowers. Sepals 5, papery; petals absent; anthers pink. Seeds black. A native of the tropics; naturalized on sands near Lisbon.

CARPOBROTUS | 120. *C. edulis* (L.) N.E.Br. HOTTENTOT FIG. A robust, creeping perennial often forming mats, with narrow fleshy leaves which are triangular in section, and with large brightly coloured yellow or lilac flowers 8–10 cm across. Stamens numerous, yellow. Rocks and sands near the sea, sometimes planted. Native of S. Africa; naturalized on the Atlantic and Med. coasts. ****(120)** *C. acinaciformis* (L.) L. Bolus Like 120 but flowers larger, up to 12 cm across, red-carmine; stamens purple. Distribution similar to 120. Pl. 2. 120a. *C. chilensis* (Molina) N.E. Br. Like 120 but smaller with purple flowers 2½–5 cm across. Native of W. America; naturalized on the littoral of Spain.

MESEMBRYANTHEMUM | 121. *M. nodiflorum* L. A procumbent annual with fleshy cylindrical leaves, and small white or yellowish flowers 1½ cm across, with numerous narrow petals shorter than the sepals. (121) *M. crystallinum* L. ICE PLANT. Differs from

121 in having flat, fleshy, oval leaves which are densely covered with shining crystalline swellings, like hoar-frost. Flowers 2–3 cm across; petals longer than the sepals. Both grow on sands and in salt marshes. Portugal, Med. region. Pl. 1.

APTENIA | 121a. *A. cordifolia* (L.fil.) N.E. Br. A spreading perennial, with oval-heart-shaped, flat, fleshy leaves, which are covered with fine swellings. Flowers purple, 1–1½ cm across. Native of S. Africa; naturalized in S. Europe.

DISPHYMA | 121b. *D. crassifolium* (L.) L. Bolus A perennial with woody-based, spreading and rooting branches, and pinkish-purple axillary, stalked flowers 1–1¼ cm across. Leaves 2½ cm, bluntly triangular, flat above, rounded or somewhat keeled beneath, dark green with transparent dots. Native of S. Africa; naturalized near Lisbon.

LAMPRANTHUS | 121c. *L. glaucus* (L.) N.E. Br. A shrubby, bright yellow-flowered plant with opposite, narrow cylindrical, slightly triangular-sectioned leaves, which are bloomed and covered with some crystalline swellings. Native of S. Africa; naturalized near Lisbon. 121d. *L. roseus* (Willd.) Schwantes has pale pink flowers about 3 cm across. It is commonly planted in gardens, and sometimes escapes in the south.

MOLLUGINACEAE | Mollugo Family

GLINUS | 121e. *G. lotoides* L. A woolly-haired, spreading annual with opposite or whorled, obovate, stalked leaves and yellowish flowers in dense axillary clusters. Sepals yellow; 'petals' white, numerous; stamens 12; stigmas 5. Fruit opening by 5 valves. Winter-flooded areas. Portugal, S., W. and C. Spain.

PORTULACACEAE | Purslane Family

PORTULACA | **122. *P. oleracea* L. PURSLANE. A creeping, fleshy-leaved annual with one or several yellow flowers. Widespread.

CARYOPHYLLACEAE | Pink Family

A large and quite important family in our area, comprising 33 genera and about 250 species. Many are small annuals and perennials with insignificant white flowers often difficult to distinguish from each other. The following genera are not included in the text: *Moehringia* (18 species); *Minuartia* (about 20 species); *Honkenya* (1 species); *Bufonia* (6 species, including 4 endemic species); *Stellaria* (7 species); *Holosteum* (1 species); *Moenchia* (1 species); *Myosoton* (1 species); *Sagina* (12 species); *Scleranthus* (3 species); *Corrigiola* (2 species); *Herniaria* (11 species); *Illecebrum* (1 species); *Polycarpon* (4 species); *Ortegia* (1 species); *Loeflingia* (3 species); *Spergula* (4 species); *Velezia* (1 species).

ARENARIA | **Sandwort**   Leaves opposite; stipules absent. Sepals 5, free; petals 5, not notched; stamens usually 10; styles usually 3. Fruit usually splitting by 6 teeth. 28 species, including 17 endemic species.

**126. *A. montana* L. A hairy, greyish-green, spreading or erect perennial, with relatively large white flowers 1½–2 cm, either solitary or several. Petals twice as long as sepals, which are 6–9 mm long and with or without papery margins. Leaves soft, oblong-lance-shaped to linear, 1–2 cm, one-veined. Heaths, bushy places. Throughout our area. Subsp. *intricata* (Dufour) Pau has narrower, linear leaves 1–2 mm broad; glandular flower stalks, and sepals usually without papery margins. S. Spain. Page 102.   126a. *A. grandiflora* L. Like 126 with flowers about half as large, and distinguished by its stiff narrower leaves ending in a long point, with prominent midrib and margin which are horny beneath. Sepals 2–5 mm, glandular-hairy, with prominent midvein; petals 1½–2½ times as long as sepals. Rocks, stony places in mountains. Portugal, S., C. and E. Spain, France.   126b. *A. pungens* Lag. A greyish-green perennial forming dense rounded prickly cushions of stiff, spine-tipped leaves, and with white flowers with petals a little shorter than the sepals. Flowers about 2 cm, usually solitary; sepals 6–13 mm, stiff, long-pointed, sticky. Screes of high mountains. S. Spain (S. Nevada). Pl. 3.

126c. *A. tetraquetra* L. A perennial, usually forming very dense flattened cushions, 1–2 cm high and 5–25 cm across, made up of tiny quadrangular rosettes. Flowers white, borne in or just above the cushion; petals a little longer than the sepals. Leaves 1–4 mm, oval-blunt, densely crowded in 4 ranks, recurved, stiff. Rocky places, scree in high mountains. S. and E. Spain, Pyrenees. Pl. 3.   126d. *A. aggregata* (L.) Loisel. Like 126c in habit, but leaves lance-shaped and fine-pointed, recurved and folded. High mountains. Iberian peninsula, S. France.   126e. *A. tomentosa* Willk. It forms silvery cushions 5–10 cm across, covered with short crisped white hairs. Leaves 1–2 mm, egg-shaped, blunt, recurved. Flowers 1–5, held above cushion. Limestone mountains, S–E. Spain.

126f. *A. purpurascens* DC. A small, neat perennial with spreading or ascending stems with dark glossy green, oval-acute leaves, and usually delicate rose-lilac flowers which appear darker at the centre. Flowers in rather dense clusters of 2–4; petals 1½ to 2 times as long as sepals, rarely white. Leaves ½–1 cm, strongly one-veined. Rock crevices. Endemic in the Cantabrian mountains and Pyrenees.

(128) *A. balearica* L. A delicate, very slender-stemmed, mat-forming perennial with tiny, broadly egg-shaped leaves 2–4 mm. Flowers white, solitary on very slender stems; petals twice as long as the bluntly egg-shaped, sparsely hairy sepals. Shady limestone rocks. Balearic Islands. Pl. 3.

126g. *A. cerastioides* Poiret A rather robust, pale green, glandular hairy annual with a lax inflorescence of 4 to many, relatively large white flowers about 1 cm, and spathulate to obovate leaves 1–2 cm. Petals slightly notched, twice as long as sepals, which are egg-shaped, and usually black-tipped; anthers blue. Damp places in fields and by paths. S–W. Spain.

CERASTIUM | Chickweed etc.   17 species, including 3 restricted to our area.   140a. *C. boissieri* Gren. A loosely tufted perennial with stiff, spreading or recurved, white- or grey-woolly leaves, and lax few-flowered clusters of white flowers. Petals to 1½ cm, cleft to one third; sepals with wide papery margins; inflorescence glandular hairy. Stony and rocky places in mountains. S. and E. Spain. Pl. 2.

**PARONYCHIA | 9 species.   **149. *P. argentea* Lam. Widespread and common in the lowlands.   149a. *P. aretioides* DC. It forms a shining, silvery, rounded cushion with the loosely overlapping silvery stipules covering the minute green leaves. Flower clusters very conspicuous; calyx lobes without awns and of nearly equal length. Limestone mountains of S. and E. Spain.   150. *P. capitata* (L.) Lam. Distinguished from 149a by the leaves and stipules, which are more or less equal in length, and flowers in dense rounded, terminal, silvery heads *c.* 1 cm across. Bracts 6–10 mm, very much longer than the flowers; calyx lobes orange, very unequal. Sandy and stony places, mainly in mountains. S., C. and E. Spain, Pyrenees, France. Page 181.

SPERGULARIA | **Spurrey** 14 species. Widespread coastal species are: 155. *S. media* (L.) C. Presl GREATER SEA SPURREY; (156) *S. marina* (L.) Griseb. LESSER SEA SPURREY. **157. *S. rubra* (L.) J. & C. Presl SAND SPURREY is a common sand-loving species found largely inland. 156a. *S. fimbriata* Boiss. A spreading perennial with robust stems, glandular-hairy in the upper part, and with fleshy leaves with long terminal bristles. Flowers cobalt-violet; petals equalling or longer than the sepals. Stipules silvery, very long, 6–10 mm and with bristly apex. Coastal sands. S. Portugal, S. Spain. **156. *S. rupicola* Le Jolis CLIFF SPURREY. A woody-based, glandular-hairy perennial like 156a, but differing in having shorter stipules, 2–6 mm, which are oval-triangular and not bristle-tipped. Petals pink, equalling the sepals. Leaves not bristle-tipped. Rocky coasts. Portugal, N–W. Spain, France. 156b. *S. purpurea* (Pers.) G. Don fil. A slender erect annual or biennial with linear leaves and silvery lance-shaped stipules half as long as the internodes. Flowers rose-purple; petals 3–5 mm, longer than sepals, which are green, glandular-hairy on the back, and with distinctive papery margins. Sandy waste places. Portugal, Spain. Page 162.

TELEPHIUM | Distinguished by the 3 separate styles, and the three-angled fruits, which split into 3 valves. 156c. *T. imperati* L. A hairless, glaucous, prostrate shrublet with herbaceous stems bearing numerous nearly stalkless, fleshy, oval leaves ½–1½ cm long, and dense terminal clusters of small white flowers. Petals little longer than sepals, which have conspicuous white papery margins. Sandy or gravelly places. S., C. and E. Spain, S. France. Page 181.

LYCHNIS | Catchfly.
1. Lvs. and stems covered with dense woolly hairs.
   (Locally naturalized)   158. *L. coronaria* (L.) Desr.
1. Lvs. and stems hairless or nearly so.
   2. Petals deeply four-lobed, or two-lobed and toothed.
      (Widespread)   **160. *L. flos-cuculi* L.
   2. Petals entire, notched or shallowy two-lobed.
      3. Inflorescence clustered; calyx less than 6 mm.
         (Pyrenees)   **162. *L. alpina* L.
      3. Inflorescence spreading, branched; calyx more than
         6 mm. (Scattered throughout)   **161. *L. viscaria* L.

AGROSTEMMA | **163. *A. githago* L. CORN COCKLE. Widespread and fairly frequent cornfield weed, rarer in the south.

PETROCOPTIS | Delicate, distinctive perennials, usually with glaucous leaves, and with lax clusters of white, reddish or purplish flowers. Calyx tubular-bell-shaped, five-lobed, ten-veined; petals 4, with petal-scales; stamens 10; styles 5. Fruit with 5 teeth; seeds with a tuft of hairs. 7 species, distinguished by small botanical differences, all restricted to the Pyrenees and Cantabrian mountains of N. Spain.

163a. *P. pyrenaica* (J. P. Bergeret) A. Braun A loosely tufted, fragile perennial, with a rosette of oval-lance-shaped stalked leaves, and slender flowering stems with a lax cluster of white or sometimes pale pink flowers. Flowers 1–2 cm across; petals notched; calyx whitish, 5–8 mm long. Shady rocks and walls. W. Pyrenees. Page 181. 163b. *P. glaucifolia*

1. *Petrocoptis pyrenaica* 163a
2. *Silene pseudovelutina* 166a
3. *Telephium imperati* 156c
4. *Polygonum equisetiforme* 79a
5. *Paronychia capitata* 150
6. *Silene ciliata* 170b

(Lag.) Boiss. (*P. lagascae*) Like 163a but stems without basal rosettes. Flowers purplish or white; calyx 7–9 mm: bracts 1–2 mm, entirely papery. Mountains. N. Spain. Pl. 3. Page 125.

SILENE | Campion, Catchfly A large and quite important genus, with about 60 species in our area. Small differences of seed, fruit, calyx and inflorescence often distinguish otherwise similar-looking plants.

Mountain perennials, often cushion or mat-forming.
**170. *S. acaulis* (L.) Jacq. MOSS CAMPION. Damp rocks, stony pastures in mountains. Pyrenees. 170a. *S. borderi* Jordan A dwarf mat-forming perennial with spathulate, ciliate leaves covered with minute swellings and pink flowers borne on sticky stems up to 12 cm. Petals deeply two-lobed, petal scales egg-shaped; calyx 9–10 mm, hairy, teeth blunt. Fruit ovoid. Alpine rocks. Pyrenees. 170b. *S. ciliata* Pourret A tufted plant with dense rosettes of bristly-haired leaves, and white or pink flowers, with petals usually greenish or reddish beneath. Flowers borne in stalked clusters of 1–3, on erect stems to 20 cm; calyx 11–20 mm, hairy, veins branching. Mountains. Portugal, C., N., and E. Spain, France. Page 181.

170c. *S. saxifraga* L. It forms a loose, glossy green mat from which arise numerous slender erect stems to 20 cm, bearing 1–2 dull whitish-green flowers, which are greenish or reddish beneath. Calyx club-shaped, hairless, usually green. Stems sticky above. Limestone grasslands and screes in mountains. C., N., and E. Spain, Pyrenees. 170d. *S. boryi* Boiss. A lax mat-forming, greyish-green perennial, with sticky stems, usually bearing large, solitary pink flowers with conspicuously coloured and veined calyx up to 3 cm long. Petals two-lobed, usually rolled up, red beneath, claw of petal longer than calyx which is glandular, red-veined. Carpophore longer than the ripe capsule. Rocks, sandy places, in mountains. N. Portugal, S. Spain. Page 77.

**171. *S. rupestris* L. ROCK CAMPION. Rocks and screes in mountains. S. and E. Spain, Pyrenees, Massif Central. Pl. 3.

Annuals, biennials and perennials of the lowlands and hills.
Widespread species are: 164. *S. alba* (Miller) E. H. L. Krause WHITE CAMPION; **165. *S. dioica* (L.) Clairv. RED CAMPION; **167. *S. nutans* L. NOTTINGHAM CATCHFLY; **168. *S. otites* (L.) Wibel SPANISH CATCHFLY; **169. *S. vulgaris* (Moench) Garcke BLADDER CAMPION; **174. *S. gallica* L. SMALL-FLOWERED CATCHFLY.

Annuals
174a. *S. littorea* Brot. A small, sticky, hairy annual of sandy beaches, with bright rose-coloured flowers. Flowers few; petals spreading, deeply notched; calyx 10–19 mm, not contracted at the mouth. Lower leaves oblong-obovate to narrow oblong, rather fleshy. Portugal, S. and W. Spain. Pl. 3. 174b. *S. psammitis* Sprengel A glandular-hairy, much-branched annual with pink or rarely white flowers, which open in the afternoon, with petals deeply two-lobed. Calyx 13–18 mm, contracted at the mouth, inflated in fruit. Lower leaves linear 1–3 cm, somewhat fleshy. Sandy and stony places. C. and N. Portugal, S. and C. Spain. 174c. *S. scabriflora* Brot. (*S. hirsuta* Lag.) A densely shaggy-haired annual with a lax cluster, or one-sided cluster, of short-stalked, bright pink flowers. Petals deeply lobed, limb of petal 8–10 mm; calyx 13–25 mm; carpophore 7–16 mm. Buds and upper stems shaggy-haired; lower leaves oblong-spathulate, to 4 cm long, upper leaves smaller and narrower. Sandy and heathy places. Portugal, Spain (common on Costa del Sol). 174d. *S. obtusifolia* Willd. Distinguished from 174c by the leaves which are, all except the uppermost, spathulate and rather thick, ciliate. Flowers pink or white; calyx 10–13 mm, finely hairy, veins usually reddish. Carpophore 3–5 mm. Rocky places. S–W. Portugal, S–W. Spain.

**(174) *S. colorata* Poiret A finely hairy annual with pink or white flowers 1–2 cm across in lax clusters. Limb of petal 5–9 mm, deeply two-lobed; calyx cylindrical, becoming club-shaped in fruit, calyx teeth egg-shaped, blunt, ciliate. Seeds with 2 undulate wings. Cultivations, sandy waste places, rocks. Portugal, Med. region. Page 162.

Perennials

166a. *S. pseudovelutina* Rothm. A rather robust, woolly-haired, woody-based perennial with rosettes of greyish-green leaves and dense, terminal, flat-topped clusters of numerous whitish flowers. Flowers about 2 cm across; petals deeply lobed, claw much longer than calyx, petal scales absent; calyx densely glandular hairy, 2–2½ cm. Lower leaves spathulate, upper linear-lance-shaped. Shady limestone rocks. S. Spain. Page 181. 167a. *S. viridiflora* L. Distinguished by its drooping greenish-white flowers 2 cm across, with 3 violet styles. Claw of petals longer than calyx, petal-lobes linear. A softly hairy, robust perennial; stems sticky above; lower leaves oblong-spathulate, long-stalked. Shady bushy places in hills. Spain, France. 165a. *S. foetida* Sprengel A very glandular-sticky perennial with rosy pink flowers recalling a stunted form of *S. dioica*. Flowers few, terminal, often in threes, about 1½ cm across; petals usually reflexed, shallowly notched, petal-scales white, conspicuous; calyx glandular-hairy, somewhat inflated, teeth long-pointed. Stem leaves oval-lance-shaped, stalkless. Mountains. N. Portugal, N–W. Spain. Page 102. 169a. *S. legionensis* Lag. A tufted perennial with rosettes of narrow leaves and slender inflorescences with few, almost stalkless flowers. Petals white or pink and often reddish or greenish beneath; calyx slender 13–23 mm long, finely hairy. Carpophore longer than the capsule. Sandy places, roadsides, rocks. N. Portugal, Spain.

CUCUBALUS | **177. *C. baccifer* L. BERRY CATCHFLY. Shady places, field verges, river banks. Widespread.

GYPSOPHILA | 5 species, including 3 endemic to our area. **178. *G. repens* L. CREEPING GYPSOPHILA. Mountains. N. Spain, Pyrenees, Massif Central.

SAPONARIA | **Soapwort** **180. *S. ocymoides* L. ROCK SOAPWORT. A lax, mat-forming perennial with attractive, bright pink flowers and conspicuous, usually purplish, cylindrical calyx. Flower stalks and calyx glandular-hairy. Sandy places, rocks in mountains. Spain, France. Page 153. (180) *S. bellidifolia* Sm. A tufted plant distinguished by its pale yellow flowers in a dense head, surrounded by a pair of bracts. Stem 20–40 cm, unbranched, usually with one pair of linear stem leaves; basal leaves spathulate. Alpine rocky meadows. Pyrenees.

180a. *S. caespitosa* DC. A tufted perennial with few or several bright pink flowers borne on a short stem. Petals oboval, with scales; calyx reddish, densely hairy, with acute teeth. Leaves linear, about 4 mm wide. Alpine rocks and meadows. Pyrenees. **181. *S. officinalis* L. SOAPWORT. By cultivations, hedges, streamsides. Widespread. 181a. *S. glutinosa* Bieb. An annual or biennial to ½ m, readily distinguished by its very sticky, hairy stems and leaves. Flowers purple, in branched pyramidal clusters; petal limb ½ cm, bilobed, reflexed; calyx narrowly tubular, 2–2½ cm. Bushy places in mountains. E. Spain.

VACCARIA | 182. *V. pyramidata* Medicus COW BASIL. Frequent in cornfields, hedges, waste place. Widespread.

PETRORHAGIA | 4 species. 183. *P. saxifraga* (L.) Link TUNIC FLOWER. Dry areas. N. Portugal, N–E. and E. Spain, S. France. 184. *P. prolifera* (L.) P. W. Ball & Heywood PROLIFEROUS PINK. Dry sandy places. Widespread.

DIANTHUS | Pink About 29 species, including 13 species which are restricted to our area in Europe. Many are very similar in general appearance; the dimensions and details of the calyx and epicalyx scales are important in distinguishing species.

Flowers in densely clustered heads
**185. *D. barbatus* L. SWEET WILLIAM. Meadows and woods. Pyrenees, Massif Central. Pl. 4.
**186 *D. armeria* L. DEPTFORD PINK. Roadsides, shady places. Widespread, except south.
**187. *D. carthusianorum* L. CARTHUSIAN PINK. Pyrenees, France.

Flowers solitary, few or in lax clusters
**190. *D. monspessulanus* L. FRINGED PINK. Flowers sweet-scented, pink or whitish, rather large, with petals deeply cut into a fringe of narrow lobes, the uncut part of the blade oval. Epicalyx scales 4, egg-shaped, long-pointed; calyx 18–25 mm. Woods, dry pastures in mountains. N. Portugal, N. and N–E. Spain, Pyrenees, France. 191. *D. seguieri* Vill. Flowers 1 or several, magenta with a darker centre often whitish-spotted at base; petal-limb 7–17 mm, sharply toothed, hairy above. Calyx 14–20 mm; epicalyx scales 2–6, egg-shaped and narrowed to a fine point one third to three-quarters as long as calyx. Mountains, field margins, rocks, shady places. C. and N. Spain, Pyrenees, Massif Central. Pl. 4. 191a. *D. furcatus* Balbis (*D. requienii*) Like 191 but flowers pink or whitish, unspotted but dark at base, petal-limb ½–1 cm, shallowly cut or nearly entire, usually hairless; calyx 10–17 mm. Stony and rocky places in mountains. N–W. Spain, Pyrenees. Page 124.

**194. *D. sylvestris* Wulfen WOOD PINK. A tufted plant with pale pink flowers often with entire, hairless petals, and glaucous calyx, but flowers very variable. Epicalyx scales 2–5, one quarter as long as the calyx, which is 12–20 mm. S. and E. Spain, France. 194a. *D. lusitanus* Brot. A slender perennial forming large, dense glaucous tufts, often with brittle woody basal stems, usually with solitary pink spotted flowers. Petals with claw longer than the calyx, and blade oval, strongly toothed, with some hairs at the base. Epicalyx scales egg-shaped, long-pointed; calyx 20–23 mm, tapering markedly from about the middle. Rocks in the hills and lower mountains. Portugal, W., S–E., and C. Spain. Pl. 4. Page 133. 194b. *D. subacaulis* Vill. A low, densely tufted, woody-based perennial with rigid, glaucous leaves and small solitary, pale pink flowers. Petals 3–5 mm, hairless, rounded and scarcely toothed; epicalyx scales 4, about one third the length of the calyx, which is 6–10 mm. Rocky places in mountains. Portugal, Spain, France. Pl. 4. Page 132.

194c. *D. hispanicus* Asso A shrubby perennial with a stout woody stock and solitary pink flowers with widely spaced petals with small oval blades 3–5 mm long. Epicalyx scales 4, obovate and tapering to a rigid point; calyx 13–20 mm tapering above the middle but becoming cylindrical or even widening upwards in fruit. Dry grassy places, stony arid ground. S., C., and E. Spain. Page 138.

NYMPHAEACEAE | Water-Lily Family

**NYMPHAEA | **196. *N. alba* L. WHITE WATER-LILY. Throughout our area except in the south. 196a. *N. candida* C. Presl Like 196 but petals fewer 15–18; stigma strongly concave with 6–14 rays. N. Spain.

**NUPHAR | **197. *N. lutea* (L.) Sibth. & Sm. YELLOW WATER-LILY. Throughout our area.

RANUNCULACEAE | Buttercup Family

HELLEBORUS | **199. *H. foetidus* L. STINKING HELLEBORE. Lowlands and hills, mountains in the south. Widespread. (199) *H. lividus* Aiton subsp. *lividus* is endemic to the Balearic Islands. It has pale pink-flushed, open flowers, and leaves with only 3 broad white-veined leaflets, with entire or spiny margins. Limestone cliffs and stony places. Pl. 7. 200. *H. viridis* L. GREEN HELLEBORE. Meadows, thickets, and woods in mountains. C., N., and E. Spain, Pyrenees, France.

CALLIANTHEMUM | Perianth segments in 2 whorls, the inner petal-like, with nectaries at the base. Leaves twice-pinnate. Carpels numerous, one-seeded. 200a. *C. coriandrifolium* Reichenb. A delicate perennial of the high Pyrenees with parsley-like leaves and small solitary white flowers 1½–3½ cm across with orange centres, reddish outside. Stems 5–15 cm; rootstock fibrous. Rocky ground, pastures watered by melting snow.

NIGELLA | Love-in-a-Mist.
1. Fls. surrounded by a leafy involucre similar to the lvs. (Widespread)   **203. *N. damascena* L.
1. Fls. stalked, without a leafy involucre directly below.
  2. Carpels three-veined, united for half their length. (Portugal, Med. region)   (203) *N. arvensis* L.
  2. Carpels one-veined (at least at the apex), united for two-thirds their length or more.
    3. Fls. medium-sized 2–3½ cm, pale blue; fr. smooth or sparsely glandular. (C. and N. Spain, S. France)   203a. *N. gallica* Jordan
    3. Fls. large 4–5 cm, bright blue; fr. densely glandular. (S. Portugal, S. and S–E. Spain)   203b. *N. hispanica* L. Page 188.

TROLLIUS | **204. *T. europaeus* L. GLOBE FLOWER. Damp mountain meadows in mountains. C., N., and E. Spain, France.

ACTAEA | **206. *A. spicata* L. BANEBERRY. Damp woods in mountains. C. and E. Spain, Pyrenees, France.

CALTHA | 207. *C. palustris* L. MARSH MARIGOLD. Common in marshes and by streams, and in mountains further south. Throughout our area.

ACONITUM | Monkshood
1. Hood about three times as high as wide, conical-cylindrical; nectary-spurs more or less spirally curved.
  2. Lv. segments divided to the middle into 3 lobes; terminal inflorescence small and few-flowered. (Pyrenees, France)   **208. *A. vulparia* Reichenb.
  2. Lv. segments divided to beyond the middle into several lobes; terminal inflorescence large and many-flowered. (Spain, France)   208a. *A. lamarckii* Reichenb.

185

1. Hood never more than twice as high as wide,
   rounded or hemispherical; nectary-spurs
   straight.
3. Fls. almost always yellow. (N. and N–E.
   Spain, Pyrenees)   209. *A. anthora* L.
3. Fls. blue or purple.
   4. Hood usually higher than wide; seeds
      winged on one angle and with transverse
      wings on the side.   *paniculatum*
   4. Hood usually wider than high; seeds
      winged on 3 angles but smooth or rough
      and unwinged on the sides.   *napellus*; *compactum*; *nevadense*

210a. *A. paniculatum* Lam. Distinguished by the glandular-hairy, flexuous-branched inflorescence, with linear bracts, and its usually blue flowers with a broad hood, scarcely higher than wide. Carpels 4–5. Mountains. N. Spain, Pyrenees.

**210. *A. napellus* L. COMMON MONKSHOOD. Inflorescence simple or branched. Leaves divided to the base into narrow segments which are further divided to more than halfway into linear lobes. Carpels usually 3. Pyrenees, France. 210b. *A. nevadense* Gáyer An endemic species of the mountains of S. Spain (S. Nevada). Distinguished from 210 by its hairy leaves which have the segments divided almost to the midrib into narrow linear lobes. Flowers blue; inflorescence densely covered with glandular hairs. 210c. *A. compactum* Reichenb. Like 210 but leaves crowded below the simple inflorescence. Leaf segments cut into narrow linear lobes. Pyrenees.

DELPHINIUM | Delphinium   9 species.

Perennials; inner lateral perianth segments with hairs on upper surface.
211a. *D. montanum* DC. Roots not tuberous. Fruits densely hairy; seeds not covered with scales. Pyrenees. 211b. *D. pentagynum* Lam. Roots with tubers. Carpels 5, more or less spreading; seeds covered with long scales. S. Portugal, S. Spain. Page. 188. 211c. *D. nevadense* G. Kunze Roots with tubers. Carpel 3, hairless; seeds with scales. S. Spain (S. Nevada).

Annuals or biennials; inner lateral perianth segments hairless on upper surface.
212. *D. staphisagria* L. STAVESACRE, LICEBANE. Distinguished by its very short spur and its perianth segments which are similar-coloured. Carpels rather swollen. Med. region. Small details of the shape of the innermost perianth segments, or honey-leaves, distinguish the 3 most widespread species in our area: 211d. *D. obcordatum* DC. Limb of lateral honey-leaves as long as or a little shorter than the claw, distinctly exserted, not heart-shaped at base. Spain. 211e. *D. gracile* DC. Limb of lateral honey-leaves half to three-quarters as long as the claw and 1½ to two times as long as wide, and heart-shaped at base. Portugal, Spain. 211f. *D. verdunense* Balbis Limb of lateral honey-leaves shorter than claw, and more or less rounded and heart-shaped at base. Scattered throughout. Pl. 7.

CONSOLIDA | Larkspur   Like *Delphinium* but inner perianth segments only 2, and fused at the base into a single long spur which is included in the outer spur. Carpel solitary.

1. Fr. hairless. (Widespread)   **(213) *C. regalis* S. F. Gray
1. Fr. hairy.
   2. Bracteoles reaching to or beyond
      the base of the fl.; spur not more
      than 12 mm. (S. and E. Iberia)   (213) *C. orientalis* (Gay) Schrödinger. Page 116

2. Bracteoles at least of lower fls. not
reaching the base of the fl.; spur 12
mm or more. (Portugal, Med.
region) **213. *C. ambigua* (L.) P. W. Ball & Heywood

ANEMONE | Anemone

1. Stem lvs. shortly stalked and more or less similar to
basal lvs.
  2. Fls. yellow. (Spain, France) (214) *A. ranunculoides* L.
  2. Fls. white, pinkish, or rarely blue.
    3. Fr. very woolly; plants without a creeping rhizome. *pavoniana*
    3. Fr. not woolly; plant with a creeping rhizome.
      4. Stem lvs. once-divided into 3 oval-acute, toothed
leaflets. *trifolia*
      4. Stem lvs. two to three times divided into lobed
leaflets. (Spain, France) **214. *A. nemorosa* L.
1. Stem lvs. stalkless, and dissimilar to basal lvs.
  5. Stem lvs. bract-like, entire or lobed.
    6. Petals 12–19, narrowly elliptic. (S. France) **(216) *A. hortensis* L.
    6. Petals 7–12, oval. (S–W. France) **(216) *A. pavonina* Lam.
  5. Stem lvs. deeply divided into narrow segments.
    7. Fls. in an umbel of 2–8, white. (N–E. Spain,
Pyrenees) **218. *A. narcissiflora* L.
    7. Fls. solitary
      8. Fls. red, purple or white; basal lvs. deeply divided.
(Introd. Med. region) **216. *A. coronaria* L.
      8. Fls. yellow; basal lvs. shallowly lobed. *palmata*

214a. *A. trifolia* L. Like *A. nemorosa* with white or pinkish flowers, but stem leaves divided into 3 leaflets, and with buds in their axils, basal leaves usually absent. Subsp. *albida* (Mariz) Tutin with white anthers, elliptic petals, nodding fruits, is the Iberian subsp. Fields, damp meadows. N. Portugal, N. and E. Spain. Page 188. 216a. *A. pavoniana* Boiss. A small perennial usually with solitary white flowers with 7–8 blunt petals and deeply divided stem leaves. Fruiting head globular, very woolly-haired. Mountain pastures. Endemic. N. and C. Spain. Pl. 5.

**217. *A. palmata* L. PALMATE ANEMONE. Readily distinguished by its yellow flowers with 10–15 narrow, blunt-tipped petals often flushed with red outside. Basal leaves rounded, shallowly lobed, stem leaves deeply cut into 3–5 narrow lobes. Open ground, heaths and pastures. S. Portugal, S. and C. Spain. Pl. 6, Page 160.

HEPATICA | **219. *H. nobilis* Miller HEPATICA. Woods and banks in mountains. Spain, France. Page 116.

PULSATILLA | Like *Anemone*, but fruit with long feathery styles.

1. Stem lvs. shortly stalked resembling the basal lvs. but smaller. *alpina*; *alba*
1. Stem lvs. stalkless, divided into narrow segments and not like basal lvs.
  2. Basal lvs. evergreen, once-cut; fls. usually white. *vernalis*
  2. Basal lvs. withering in autumn, 2–4 times cut; fls. usually pale or dark
purple. *vulgaris*; *rubra*

**220. *P. alpina* (L.) Delarbre ALPINE ANEMONE. Flowers white tinged with violet, 4–6 cm across. Subsp. *apiifolia* (Scop.) Nyman has yellow flowers and is less common in our area.

1. *Delphinium pentagynum* 211b
2. *Ceratocephalus falcatus* 251a
3. *Anemone trifolia* 214a
4. *Nigella hispanica* 203b
5. *Thalictrum macrocarpum* 255a

Alpine pastures, rocks, pine woods. C., N. and E. Spain, Pyrenees, Massif Central. Page 149. 220a. *P. alba* Reichenb. Like 220, but white flowers smaller, 2½–4½ cm across, and leaves hairless or almost so above, with the terminal segments of the mature leaves divided to the midvein. C. France.

**221. *P. vernalis* (L.) Miller SPRING ANEMONE. Flowers white with golden stamens within, and violet with golden silky hairs outside. High mountains. S. and C. Spain, Pyrenees, and Massif Central. Page 153.

**223. *P. vulgaris* Miller PASQUE FLOWER. Flowers erect, dark or pale purple. Grassy places. France. Page 152. 223a. *P. rubra* (Lam.) Delarbre. Flowers nodding, dark violet, dark reddish-brown to blackish-red; perianth segments two to three times as long as stamens. Stem leaves with about 20 lobes. Limestone hills. Endemic to C. and E. Spain, S. and C. France. Pl. 8.

CLEMATIS | Clematis

1. Plant erect, not climbing; stem not woody.   *recta*
1. Plant climbing or scrambling; stem usually woody.
   2. Fls. bluish, purple or violet.   *viticella*; *campaniflora*
   2. Fls. white, yellowish or rarely reddish.
     3. Fls. solitary, 4–7 cm across, with a leafy involucre.   *cirrhosa*
     3. Fls. in branched clusters, *c.* 2 cm across, without an involucre.
       4. Lvs. once-pinnate. (Widespread)   **224. *C. vitalba* L.
       4. Lvs. twice-pinnate. (S. Portugal, Med. region)   **225. *C. flammula* L.

226. *C. cirrhosa* L. VIRGINS' BOWER. A climbing or scrambling plant distinguished by its large, solitary, hairy, cream-coloured, rarely reddish nodding flowers, with a conspicuous green cup-shaped involucre. Leaves clustered, shiny, very variable, with entire, three-lobed or divided leaves often occurring on the same plant. Rocks, bushy places. Portugal, S. Spain. Pl. 5, Page 161.

226a. *C. viticella* L. A deciduous, partly woody climber with solitary fragrant, bluish or purple, long-stalked, nodding flowers, with spreading petals. Petals 1½–3½ cm; styles hairless. Leaves pinnate, with three-lobed leaflets. Native of S–E. Europe; often grown for ornament and locally naturalized in the south. 226b. *C. campaniflora* Brot. Like 226a but flowers pale violet, broadly bell-shaped and petals smaller, 12–20 mm. Styles hairy below. Stems very slender; leaf segments smaller. Hedges and uncultivated ground. Endemic Portugal, S. and W. Spain.

228. *C. recta* L. Flowers white, in a terminal branched cluster; petals hairless except on the margin. Stems hollow; leaves with 5–7 oval leaflets. Shady hills, hedges, dry fields. Spain, S. France.

ADONIS | 6 species, including 4 similar-looking annual species. Widespread annual species are: **230. *A. annua* L. PHEASANT'S EYE; (230) *A. microcarpa* DC.

**231. *A. vernalis* L. YELLOW ADONIS. Flowers solitary with many spreading, narrow, shining yellow petals, and hairy sepals half their length. Basal leaves reduced to scales; stem leaves cut into linear segments. Beak of fruit curved. Rocky pastures, calcareous hills. Spain, France. 231a. *A. pyrenaica* DC. Like 231, but the sepals hairless. The basal leaves all stalked. Beak of fruit crooked. Chestnut forests, rocks and screes. Pyrenees. Pl. 4.

189

RANUNCULUS | About 56 species of Buttercups and Crowfoots occur in our area. They can be divided into: (a) White-flowered species of ponds, ditches, lakes and damp mud, often with both submerged dissected leaves as well as rounded floating leaves. (11 species) (b) White-flowered terrestial plants of the mountains, or with flowers occasionally pink or purplish. (9 species) (c) Shining yellow-flowered plants, ranging from lowland marshes and meadows to the highest mountains. (About 36 species).

The white-, pink- or purplish-flowered montane species.

1. Basal lvs. arrow-shaped.	*acetosellifolius*
1. Basal lvs. not arrow-shaped.	
2. Some or all basal lvs. entire or toothed.	
3. Stem lvs. lance- to egg-shaped, clasping the stem.	
4. Sepals hairless; fr. strongly veined.	*amplexicaulis*
4. Sepals hairy; fr. smooth. (N–E. Spain, Pyrenees)	**248. *R. parnassifolius* L.
3. Stem lvs. linear to lance-shaped, not clasping the stem.	*pyrenaeus*
2. All lvs. more or less deeply lobed.	
5. Sepals densely reddish-haired. (S. Spain, Pyrenees)	**(247) *R. glacialis* L.
5. Sepals hairless or nearly so.	
6. Stem lvs. simple or rarely three-lobed, differing markedly from basal lvs. (N. Spain, Pyrenees)	247. *R. alpestris* L.
6. Stem lvs. lobed or toothed, similar to basal lvs.	
7. Plant rarely as much as 20 cm; fls. solitary or few. (N. Spain)	247a. *R. seguieri* Vill.
7. Plant usually much more than 20 cm; fls. several to many in a branched inflorescence.	
8. Middle segment of basal lvs. free to the base. (C. and N. Spain; Pyrenees, Massif Central)	**246. *R. aconitifolius* L.
8. Middle segment of basal lvs. not free to the base. (Pyrenees, France)	(246) *R. platanifolius* L.

248a. *R. amplexicaulis* L. A small plant with a rosette of flat, glossy, oval leaves and narrow upper leaves and short stems bearing 1 to several white flowers up to 3 cm across. Flower stalks and sepals hairless. Damp alpine meadows by snow. N., N–E., and E. Spain. Page 124. **(248)** *R. pyrenaeus* L. PYRENEAN BUTTERCUP. Distinguished by its linear to lance-shaped, stalkless leaves and white flowers, *c.* 2 cm across, borne on hairy flower stalks. Sepals hairless, whitish. Alpine screes. S. and N–E. Spain, Pyrenees. Page 149. **248b.** *R. acetosellifolius* Boiss. Like (248) but distinguished by its fleshy, arrow-shaped, sorrel-like leaves with 1 or more pairs of backward-pointing lobes at the base of the blade. Flowers several or many, 1½–2½ cm across, on short, usually unbranched stems; sepals hairless, green, soon falling. By melting snow. S. Spain (S. Nevada). Pl. 5, Page 77.

Yellow-flowered species
(a) Leaf blades about as long as broad, shallowly lobed to less than half their width.
Widespread species are: **232. *R. muricatus* L. SPINY-FRUITED BUTTERCUP; **(232) *R. parviflorus* L. SMALL-FLOWERED BUTTERCUP; **233. *R. ficaria* L. LESSER CELANDINE.

234a. *R. bullatus* L. Flowers yellow, scented, 1–2, rather large 2½ cm across, borne on leafless stems. Leaves distinctive, all basal, broadly egg-shaped, margin with rounded teeth, bristly beneath and more or less embossed (bullate). Dry roadsides, olive groves, rocky places (forming sheets of colour on the Costa del Sol in the autumn). Portugal, N–W. Spain, Med. region. Page 62.

239a. *R. rupestris* Guss. A rather robust, downy perennial up to $\frac{1}{2}$ m with few, large, showy yellow flowers *c.* 4 cm across. Leaves rounded or kidney-shaped, shallowly three-lobed with lobes further shallowly lobed and rounded-toothed, all densely covered with ad-pressed hairs on both sides; basal leaves more deeply lobed. Beak of carpels recurved and hooked. Rock fissures in mountains. S. Portugal, S. Spain.

(*b*) *Leaf blades about as long as broad, deeply lobed to more than half their width or to the base*
Widespread species are: 235. *R. arvensis* L. CORN BUTTERCUP;   236. *R. scleratus* L.
CELERY-LEAVED BUTTERCUP;   237. *R. bulbosus* L. BULBOUS BUTTERCUP;   **(237) *R. sardous*
Crantz HAIRY BUTTERCUP;   238. *R. repens* L. CREEPING BUTTERCUP;   **(238) *R. polyanthe-mos* L.;   239. *R. acris* L. MEADOW BUTTERCUP.
239b. *R. gregarius* Brot. A slender hairy perennial with cylindrical tubers and 1–4 yellow flowers about 2$\frac{1}{2}$ cm across. Basal leaves kidney-shaped or pentagonal, three- to five-lobed, lobes further lobed or toothed. Fruit head about 1 cm across; carpels usually with few bristly hairs, beak short, hooked. Shady places, rocks in mountains. Portugal, Spain.
239c. *R. demissus* DC. A hairless perennial with solitary flowers borne on slender spreading stems. Sepals hairless. Basal leaves rounded three- to five-lobed, with narrow blunt, shortly stalked lobes; stem leaves stalkless. Fruits 2$\frac{1}{2}$ mm, beak very short, hooked. Stony mountains. S. Spain (S. Nevada). Pl. 5.

(*c*) *Leaf blades several times longer than broad, usually entire*
Fairly widespread in damp places or watersides are: 243. *R. flammula* L. LESSER SPEAR-WORT;   **(243). *R. ophioglossifolius* Vill. SNAKESTONGUE CROWFOOT.   **244. *R. lingua* L. GREAT SPEARWORT.

245. *R. gramineus* L. GRASS-LEAVED BUTTERCUP. An erect, usually hairless perennial with glaucous, grass-like basal leaves and a few narrower stalkless stem leaves. Flowers yellow, 2–3 cm across; petals 5. Dry grasslands, rocky pastures in hills and mountains. South of our area. Pl. 7, Page 116.   245a. *R. abnormis* Cutanda & Willk. A low-growing perennial of the alpine regions with distinctive grass-like leaves which are linear-lance-shaped and hooded at the apex, and mostly basal. Flowers yellow, 2–2$\frac{1}{2}$ cm across, solitary or 2–3; petals 8–10. Carpels with short, curved beaks. Damp alpine meadows. S. Portugal, W. and C. Spain. Pl. 5, Page 111.   245b. *R. bupleuroides* Brot. An endemic species of N–W. Portugal like (248) *R. pyrenaeus* L. but with pale yellow flowers. Leaves egg-shaped to lance-shaped, with slender leaf-stalks. Carpels transversely wrinkled. Dry hills.

CERATOCEPHALUS | Like *Ranunculus* but carpels with an empty cell on either side of the seed, and with a long-pointed, up-curved beak.   251a. *C. falcatus* (L.) Pers. BEAKWORT. A small downy annual 2–10 cm with solitary yellow flowers 1–1$\frac{1}{2}$ cm across, and a distinctive cylindrical fruiting head of numerous carpels with long sickle-shaped beaks. Leaves deeply cut into many narrow oblong segments, all basal. Fields, waste places. Med. region, C. Spain. Page 188.

AQUILEGIA | **Columbine**   About 12 species; they are often difficult to distinguish from one another.

Flowers one-coloured
**253. *A. vulgaris* L. COLUMBINE. Spur strongly hooked, 15–22 mm long. Woods, mountain pastures, shady rocks. Scattered throughout.   (254) *A. pyrenaica* DC. PYRENEAN COLUM-BINE. Spur straight or slightly curved, 10–16 mm. Pasture and rocky places. Pyrenees. Pl.

6.   253a. *A. nevadensis* Boiss. & Reuter Readily distinguished by its very glandular leaves. Perianth segments with a green apex. Fruit viscid. Shady banks in mountains. S. Spain (S. Nevada etc.).

Flowers bi-coloured

253b. *A. dichroa* Freyn Outer perianth segments blue, inner blue with a white apex. Spur about 12 mm, hooked. Carpels glandular-hairy. Shady damp places. Portugal, N–W. Spain.

THALICTRUM | Meadow Rue   10 species.   **256. *T. minus* L. LESSER MEADOW RUE. Sunny hills, rocky bushy places, pastures. Scattered throughout.   **255. *T. aquilegifolium* L. GREAT MEADOW RUE. Flowers numerous; filaments of stamens stout, brightly coloured, lilac or cream. In mountains, bushy places. C., N., and E. Spain, Pyrenees, Massif Central.   255a. *T. macrocarpum* Gren. A rare endemic of damp limestone rocks of the Pyrenees. It has few, long-stalked, large yellow flowers and is distinguished by its large fruits with long curved styles. Page 188.

255b. *T. tuberosum* L. Readily distinguished from all other species by its rather large showy, cream-coloured flowers about 3 cm across. A hairless, erect perennial to 40 cm; leaves two to three times pinnate with rounded, toothed leaflets. Fruits spindle-shaped, numerous, stalkless. N. and E. Spain, Pyrenees. S–W. France. Pl. 5, Page 116.

**257. *T. flavum* L. COMMON MEADOW RUE. A tall, often slender plant up to 1½ m, with dense masses of fluffy yellow flowers and compound leaves. Leaflets broadly oval or wedge-shaped with 3–4 deep teeth towards the apex. A very variable species. Damp places, streamsides. Scattered throughout. Subsp. *glaucum* (Desf.) Batt. is a tall striking, glaucous plant with dense terminal clusters of yellow flowers borne on slender, sparsely leafy stems. Leaflets with prominent veins beneath. Spain, Portugal. Pl. 4.

PAEONIACEAE | Peony Family

PAEONIA |

1. Most leaflets divided, the lower lvs. with 17–30 narrowly elliptic to lance-shaped segments.

  2. Lvs. glaucous, hairless beneath; filaments of stamens yellow.   *broteroi*

  2. Lvs. green, brownish and downy beneath; filaments of stamens red.   *officinalis*

1. Most leaflets undivided, the lower lvs. with 9–16 elliptic, egg-shaped or circular segments.

  3. Carpels 5–8, *c.* 6 cm, hairless, purple. (Balearic Islands)   259b. *P. cambessedesii* (Willk.) Willk.

  3. Carpels less than 5, less than 6 cm, usually downy, never purple, but sometimes reddish.

    4. Carpels usually 2, *c.* 4½ cm, hairless, narrowed to the stigma. (S. Spain)   259a. *P. coriacea* Boiss.

    4. Carpels usually more than 2, 2½–4 cm, usually downy, with a blunt or rounded apex and stalkless stigma.

5. Lvs. hairless beneath; leaflets 2–4 cm
  wide, narrowly elliptic.   *broteroi*
5. Lvs. usually downy beneath; leaflets 5–10
  cm wide; elliptic. (C. France)   **259. *P. mascula* (L.) Miller

**258. *P. officinalis* L. PEONY. Flowers red with red filaments, 7–13 cm across. Leaves hairless above and finely hairy beneath, leaf-stalk deeply channelled on the upper surface. Carpels 2–3. Subsp. *humilis* (Retz) Cullen & Heywood occurs in our area. Distinguished by its leaflets which are only cut to one third their length, and its hairless fruit and finely hairy stems and flower stalks. Meadows and woods in mountains. Portugal, S. and C. Spain, France. Pl. 6. (258) *P. broteroi* Boiss. & Reuter Distinguished from 258 by the usually glaucous, hairless, narrower leaflets, and more or less rounded leaf-stalk. Flowers 8–10 cm across, red; filaments yellow. Carpels 2–4, densely woolly-haired. Woods, bushy places in mountains. Portugal, S. and W. Spain. Pl. 7.

BERBERIDACEAE | Barberry Family

BERBERIS | **261. *B. vulgaris* L. BARBERRY. Hedges, rocky and bushy places. C. and E. Spain, France; probably introduced elsewhere. 261a. *B. hispanica* Boiss. & Reuter Like 261 but leaves smaller 8–18 mm, usually without marginal bristles, and usually shorter than the spines. Flower clusters short, to 3 cm, with 6–16 flowers. Berries dark red or black. Scrub-covered mountain slopes. S. and E. Spain. Pl. 8. Page 117.

LAURACEAE | Laurel Family

LAURUS | 263. *L. nobilis* L. LAUREL. Often occurring in shady places, river margins, cork oak woods and on the littoral of our area.

PAPAVERACEAE | Poppy Family

Sub-family **PAPAVEROIDEAE.** Flowers regular, petals large, stamens numerous. Fruit top-shaped or cylindrical.

PAPAVER | **Poppy** 9 species. Widespread, red-flowered annual species of cornfields and disturbed ground are: **265. *P. rhoeas* L. CORN POPPY; (265) *P. dubium* L. LONG-HEADED POPPY; (266) *P. hybridum* L. PRICKLY ROUND-HEADED POPPY.

266a. *P. rupifragum* Boiss. & Reuter Like (265) but a perennial with non-flowering shoots present at flowering and smaller pale brick-red flowers 2–2½ cm across. Fruit oblong-club-shaped, hairless, 2½ cm. Limestone crags. Endemic in S. Spain (S. de Ronda).

**264. *P. somniferum* L. OPIUM POPPY. Distinguished by its large white, lilac or purple flowers and its broad lobed or toothed, clasping leaves. Subsp. *setigerum* (DC.) Corb. with bristly leaves is probably the native plant of the Western Mediterranean region. Widespread.

**267. *P. rhaeticum* Leresche ALPINE POPPY. A mountain perennial of the E. Pyrenees, usually with golden-yellow, cup-shaped flowers 4–5 cm across; stamens longer than the ovary. Leaves soft to the touch, greyish.   267a. *P. suaveolens* Lapeyr. Like 267 but flowers smaller, 2½–3½ cm across, ranging from orange to pink, with petals spreading in a Maltese Cross, not overlapping at base. Stamens shorter than or as long as the ovary. Leaves usually bristly, green. Screes and rocks in mountains. S. Spain (S. Nevada), Spanish Pyrenees.

MECONOPSIS | **269. *M. cambrica* (L.) Vig. WELSH POPPY. Mountain woods, shady places. N. Spain, Pyrenees, Massif Central.

ARGEMONE | **270. *A. mexicana* L. PRICKLY POPPY. A glaucous annual with spiny leaves, and pale yellow or orange flowers 5–6 cm across. Leaves coarsely lobed. Fruit spiny. Native of America; naturalized in waste places in the south.

ROEMERIA | **271. *R. hybrida* (L.) DC. VIOLET HORNED-POPPY. Readily distinguished by its deep violet flowers with petals that soon fall, and by its long slender fruits. A poppy-like annual of cornfields and waste ground. Portugal, Med. region.

GLAUCIUM | **272. *G. flavum* Crantz YELLOW HORNED-POPPY. Sands and gravels, mainly on the littoral. Widely scattered.   **273. *G. corniculatum* (L.) J. H. Rudolph RED HORNED-POPPY. Like 272 but flowers orange or red, often with a dark basal blotch to the petals. Fruit hairy, to 20 cm. Cornfields, fallow, steppes. Widespread except in north.

CHELIDONIUM | 274. *C. majus* L. GREATER CELANDINE. Walls, banks, hedges, waysides. Widespread.

HYPECOUM | Flowers small, yellow; petals 4, at least 2 of which are distinctly three-lobed. Fruit slender, often jointed.

1. Stems ribbed: outer petals almost as wide as long,
   more or less three-lobed.
   2. Lateral lobes of outer petals much smaller than the
      middle lobe. (Portugal, Med. region)   276. *H. procumbens* L.
   2. Lateral lobes of outer petals as large as or larger
      than the middle lobe. (Throughout)   **(276) *H. imberbe* Sibth. & Sm.
1. Stem smooth; outer petals almost twice as long as
   wide, entire. (C. and S-E. Spain, France)   276a. *H. pendulum* L.

Sub-family **FUMARIOIDEAE**   Flowers small, clustered, irregular, often shortly spurred; stamens few; fruits oval or rounded.

1. Fr. 2 to many-seeded, splitting.
   2. Fr. splitting into 2 one-seeded units; placenta not persistent.   *SARCOCAPNOS*
   2. Fr. splitting into 2 valves; placenta persistent.   *CORYDALIS*
1. Fr. one-seeded, not splitting.
   3. Upper petal not spurred; stigma three-lobed.   *PLATYCAPNOS*
   3. Upper petal spurred; stigma two-lobed.
      4. Anns.; stem lvs. numerous; fls. in spike-like clusters.   *FUMARIA*
      4. Perenns.; stem lvs. few; fls. in flat-topped clusters.   *RUPICAPNOS*

CORYDALIS | **277. *C. claviculata* (L.) DC. CLIMBING CORYDALIS. A delicate, climbing annual, with leaf tendrils, and clusters of few white or cream-coloured flowers. Shady places. North of our area. **278. *C. lutea* (L.) DC. YELLOW CORYDALIS. A yellow-flowered, non-climbing perennial with flowers 12–20 mm long, in clusters of 6–16. Leaves twice-pinnate, without tendrils. Rocks, walls; introd. N. and E. Spain, France.

**279. *C. solida* (L.) Swartz A small hairless perennial with solitary erect stems bearing a terminal cluster of rather few purple, or cream-coloured flowers, and a pair of dissected fern-like leaves. Lowest floral bracts deeply lobed, the upper entire. Shady places. Pyrenees, France. Page 148. (279) *C. bulbosa* (L.) DC. Like 279 but flowers larger, 2–3 cm long, and bracts subtending flowers always entire. Mountain woods. Scattered throughout, uncommon in south.

SARCOCAPNOS | Delicate tufted perennials growing from cracks and crevices in shaded cliffs and walls; restricted to the mountains of Spain and the Pyrenees in Europe. Flowers in lax clusters, white, yellow or pale pink, symmetrical in one plane only, the upper petal spurred, or sac-like. Leaves one to three times dissected, glaucous, fern-like. Fruit oval, flattened, usually two-seeded.

1. Fls. spurred.
  2. Fls. usually 8–10 mm (white to yellow-
    ish, purple-tipped); lvs. 2–3 times cut
    into threes. (S–C. and E. Spain, E.
    Pyrenees)   279a. *S. enneaphylla* (L.) DC. Pl. 8, Page 117
  2. Fls. 1½–2 cm, (usually pinkish); lvs.
    simple, ternate or sometimes twice
    cut into threes. (S. and S–E. Spain)   279b. *S. crassifolia* (Desf.) DC. Pl. 8, Page 84
1. Fls. not spurred.
  3. Lvs. simple, fleshy. (S. Spain)   279c. *S. integrifolia* (Boiss.) Cuatrec.
  3. Lvs. cut into 3 segments, thin. (S. and
    S–E. Spain)   279d. *S. baetica* (Boiss. & Reuter) Nyman

FUMARIA | **Fumitory** Small botanical differences distinguish the 20 or so species of our area, 6 of which are restricted to the Iberian peninsula.

**280. *F. capreolata* L. RAMPING FUMITORY. A climbing annual without tendrils and with long-stalked axillary clusters of creamy-white flowers with reddish-purple tips, which often turn reddish with age. Flowers 1–1½ cm. Leaves compound, segments oval-oblong, glaucous. Cultivated and disturbed ground, walls, hedges. Widespread.

PLATYCAPNOS | 282. *P. spicata* (L.) Bernh. SPIKED FUMITORY. A delicate glaucous annual, distinguished by its very dense rounded clover-like heads of numerous tiny pink flowers which are usually tipped with dark purple. Leaves finely cut, segments linear. Fields, disturbed ground, tracksides. Widespread, except in north.

RUPICAPNOS | 282a. *R. africana* (Lam.) Pomel A robust, fleshy but fragile glaucous perennial growing in limestone rocks in S–W. Spain, with white or pinkish flowers, in lax, long-stalked clusters. Petals 12–16 mm, spur about 4 mm. Leaves mainly basal, long-stalked, fleshy, once or twice cut.

CAPPARIDACEAE | Caper Family

CAPPARIS | ****283.** *C. spinosa* L. CAPER. Easily recognized by its large white flowers 5-7 cm with numerous projecting violet stamens, its rounded and sometimes fleshy leaves, and its usually spiny, trailing stems. Fruit a large fleshy berry, with pink flesh and dark purple seeds. Rocks, waste places, on the littoral. Portugal, Med. region. (283) *C. ovata* Desf. Very like 283 but leaves oblong to elliptic, or egg-shaped, with a fine projecting point, the continuation of the midrib. Flowers 4–5 cm across, distinctly irregular. Med. region.

CLEOME | Differs from *Capparis* in having small flowers in clusters; 6 stamens; a slender two-valved fruit.

283a. *C. violacea* L. A sticky glandular-hairy annual to 60 cm, with clusters of small, irregular, purplish-brown flowers. Corolla of 2 erect, violet and yellow-spotted petals, 4–6 mm long and 2 lateral, shorter and broader petals; stamens and stigma down-curved. Lower leaves ternate, upper simple. Fruit pendulous, 5–10 cm. Dry hillsides, field verges. Portugal, S. and W. Spain. Page 207.

CRUCIFERAE | Cress Family

An important family in south-western Europe, with 75 genera and about 280 species in our area. Many species have relatively inconspicuous white or yellow flowers, and are distinguished by small botanical characters of the ripe fruit. In general, only conspicuous pink-, violet-, or purple-flowered species have been included in the following account.

ERYSIMUM | **Treacle Mustard**   8 species.   295a. *E. linifolium* (Pers.) Gay An attractive perennial with numerous narrow leaves and a dense terminal cluster of violet or dark purple flowers. Inflorescence variable, branched or simple, long or short. Petals to 2 cm; sepals purplish; anthers greenish. Leaves often in a basal rosette, green or grey, threadlike to linear. Rocks, stony places in mountains. Portugal, Spain. Pl. 11, Page 85.

295b. *E. myriophyllum* Lange A yellow-flowered, tufted perennial, with narrow, white or greyish leaves, the lower ½–2 mm wide. Petals 12–18 mm; sepals 6–10 mm. Fruit closely pressed to stem. Mountains of S. Spain (Valencia and Andalusia). Pl. 11.   295c. *E. grandiflorum* Desf. A variable tufted perennial with grey or whitish linear leaves and yellow petals 12–18 mm long; sepals 6–10 mm. Calcareous rock and screes. S., C., and E. Spain, S. France.

HESPERIS | ****296.** *H. matronalis* L. DAME'S VIOLET. A robust, leafy perennial or biennial with a conspicuous terminal cluster of very sweet-scented, pale purple, lilac or white flowers. Leaves lance-shaped, toothed, all stalked. Grassy places, hedges. C., N., and E. Spain, France; naturalized elsewhere.   (296) *H. laciniata* All. CUT-LEAVED DAME'S VIOLET. Like 296 but flowers reddish-violet, or rarely yellow. Upper leaves stalkless, lower leaves deeply lobed. Rocky places in mountains. Widespread. Page 69

MALCOLMIA | Distinguished by its narrow elongated fruit, with three-veined valves and two-lobed stigma with lobes erect and pressed together. Flowers pink or violet.

1. Sepals not or slightly pouched at the base.
  2. Lower fr. stalks 2–7 mm; fr. 12–35 mm by 1 mm, constricted between the seeds and with ad-pressed hairs. (Portugal, Med. region)   297a. *M. ramosissima* (Desf.) Thell.
  2. Lower fr. stalks to 2 mm; fr. larger, not con-stricted, hairless or with spreading hairs. (Med. region)   297b. *M. africana* (L.) R.Br.
1. Sepals, at least the lower 2, strongly pouched at the base.
  3. Basal lvs. more or less stalkless; fr. 3–6½ cm, not constricted between seeds. (Portugal, Med. region)   297. *M. littorea* (L.) R.Br. Pl. 9
  3. Basal lvs. shortly stalked; fr. 2–4 cm, constricted between seeds. (Portugal, Med. region)   297c. *M. lacera* (L.) DC. Pl. 8. Page 53.

**298. *M. maritima* (L.) R.Br. VIRGINIA STOCK. Naturalized in sands on the littoral.

MATTHIOLA | **Stock** Distinguished by its two-lobed stigmas, the lobes erect and thickened with knobs or horns.

1. Fr. with 3 conspicuous apical horns, all at least 2 mm. (Med. region)   (301) *M. tricuspidata* (L.) R.Br.
1. Fr. without horns, or with only 2 horns more than 2 mm.
  2. Fr. flattened in section; lower fr. stalks usually 1 cm or more.
    3. Lower lvs. deeply lobed; fr. with conspicuous glands. (Coasts)   **301. *M. sinuata* (L.) R.Br.
    3. Lower lvs. entire; fr. without conspicuous glands. (Atlantic coasts)   **300. *M. incana* (L.) R.Br.
  2. Fr. rounded in section; lower fr. stalks usually less than ½ cm.
    4. Perenns. with basal rosettes of lvs.; horns blunt, usually less than 2 mm. (Portugal, Spain, Med. France)   **(301) *M. fruticulosa* (L.) Maire Pl. 11, Page 117
    4. Anns. or bienns.; without rosettes of lvs.; horns usually more than 2 mm.
      5. Fls. small; petals 6–12 mm.   *parviflora*
      5. Fls. large; petals 1½–2½ cm.   *lunata*

301a. *M. parviflora* (Schousboe) R.Br. Distinguished by its small flowers, which are purple to brownish-purple, and its elongated fruit with 2 horns up to ½ cm long, which are at first curved downwards like rams' horns, and later become horizontal. Stony and sandy places, abandoned cultivations. S. Portugal, S. Spain.   301b. *M. lunata* DC. Like 301a but flowers much larger, purple, and fruits with 2 short upcurved horns and conspicuously constricted between the seeds when dry. Sandy and stony places. S. Spain. Pl. 8.

CARDAMINE | **Bitter-Cress, Coral-Wort** 13 species.   309. *C. heptaphylla* (Vill.) O. E. Schulz SEVEN-LEAVED CORAL-WORT. Woods in mountains. E. Spain, Pyrenees, Massif

Central.   **(309) *C. pentaphyllos* (L.) Crantz FIVE-LEAVED CORAL-WORT. Woods in mountains. E. Spain, Pyrenees, France.

**310. *C. pratensis* L. LADY'S SMOCK, CUCKOO FLOWER. Damp meadows, woods. Throughout except in south.   310a. *C. raphanifolia* Pourret Like 310 but usually more robust and more showy, with reddish-violet or rarely white petals up to 12 mm. Leaflets all more or less similar, oval to kidney-shaped, but terminal leaflet at least twice as large as lateral leaflets. Fruit without a thickened border. Damp places and streamsides in mountains. N. Spain, Pyrenees, Massif Central. Pl. 10.   **(310) *C. amara* L. LARGE BITTER-CRESS. Damp and shady places, marshes. Scattered throughout, except in south.

ARABIS |*c.* 16 species   **(319). *A. verna* (L.) R.Br. A tiny annual with a rosette of oval leaves and simple erect, usually unbranched stems, bearing violet flowers with a yellowish base. Petals 5–8 mm. Stem leaves clasping. Stony places, rock fissures, in lowlands and mountains. Med. region. Page 69.

LUNARIA | **Honesty**   321. *L. rediviva* L. PERENNIAL HONESTY. A robust perennial with pale purple or violet flowers and distinctive large, wafer-thin, usually elliptic fruits. Petals 12–20 mm. Leaves all stalked, finely toothed. Shady thickets in mountains. Portugal, E. Spain, France.   **322. *L. annua* L. HONESTY. Like 321 but a biennial with the upper leaves stalkless and coarsely toothed, and flowers usually a bright reddish-purple. Fruit usually almost circular in outline. Often grown for ornament. Naturalized throughout.

ALYSSUM | A difficult genus with about 12 species in our area.   325. *A. montanum* L. A montane perennial preferring limestone, with spreading or erect branches bearing small narrow greyish leaves and numerous tiny bright yellow flowers in a lax cluster. Petals notched, 3–6 mm, twice as long as sepals. Leaves densely covered with star-shaped hairs with 6–24 rays. Fruit rounded, inflated. Cliffs, rocks. Spain, France.· Page 201.   325g. *A. serpyllifolium* Desf. A spreading or erect, variable perennial with numerous non-flowering rosettes and short stems. Flowers yellow; petals 2–3 mm, entire; sepals 1–2 mm. Leaves grey or white beneath, grey or grey-green above, pleated on the non-flowering stems. Fruit 3–4 mm, densely white hair, hairs with 12–16 rays. Stony rocky places in limestone mountains. Scattered throughout except in north.

PTILOTRICHUM | Like *Alyssum* but flowers white to purple, never yellow. Hairs star-shaped or scale-like. A small genus found largely in the Iberian peninsula in Europe.

1. Fls. purple; stems not more than 5 cm long.   *purpureum*
1. Fls. white, or rarely pink; stems 10–30 cm long.
  2. Branches becoming spiny.   *spinosum*
  2. Branches not spiny.
    3. Stems slender, brittle; basal lvs. spathulate.   *longicaule*
    3. Stems tough woody; basal lvs. narrower.
      4. Fr. hairy. (E. Pyrenees)   325e. *P. pyrenaicum* (Lapeyr.) Boiss.
      4. Fr. hairless.
        5. Lvs. broadly spathulate. (S. de Cazorla)   325f. *P. reverchonii* Degen & Hervier
        5. Lvs. oblong or linear.   *lapeyrousianum*

325a. *P. purpureum* (Lag. & Rodr.) Boiss. A low tufted perennial growing in alpine screes with dense clusters of tiny rosy-purple flowers. 'One of the loveliest crucifers anywhere!' (Farrer). Leaves about ½ cm, silvery-grey, with a dense felt of scale-like and star-shaped hairs. Fruit ellipsoid, with star-shaped hairs. High mountains. S. and S–E. Spain. Pl. 9. 325b. *P. spinosum* (L.) Boiss. Readily distinguished by its spiny older branches which form a dense stiff domed shrublet to about ½ m. Flowers white or pinkish, in dense terminal clusters. Leaves about 1 cm, densely covered with silvery scale-like hairs, the lower broadly spathulate, the upper narrower. Fruit hairless, style nearly as long. Rocks and screes in mountains. S. and E. Spain, S. France. Pl. 9, Page 77. 325c. *P. longicaule* (Boiss.) Boiss. A woody-based shrublet, with rosettes of silvery leaves and slender herbaceous stems to 60 cm, bearing a terminal cluster of tiny white flowers. Rosette leaves obovate-spathulate; stem leaves few, linear-lance-shaped. Fruit rounded, about ½ cm, hairless. Limestone rocks. S. Spain. 325d. *P. lapeyrousianum* (Jordan) Jordan A shrublet to 30 cm, with silvery-haired, oblong, blunt leaves and a lax elongated cluster of white flowers, with conspicuous petals twice as long as the sepals. Fruit hairless, spreading or deflexed, in a lax elongated cluster. Rock fissures in limestone mountains. N–C., and E. Spain, Pyrenees. Page 201.

LOBULARIA | **329. *L. maritima* (L.) Desv. SWEET ALISON. Dry rocky and sandy areas on the littoral (often flowering throughout the winter). S. Portugal, Med. region.
329a. *L. libyca* (Viv.) Webb & Berth. Like 329 but an annual with fruit with flattened valves and containing 4–5 seeds (not 1) in each cell. Leaves blunt. Maritime sands. Spain.

DRABA | A difficult genus with about 12 species occurring in our area. **331. *D. aizoides* L. YELLOW WHITLOW-GRASS. A small cushion-forming plant, distinguished by its stiff rosettes of narrow, bristly-haired leaves, and its rather dense clusters of golden-yellow flowers on short hairless stems. Petals 4–6 mm, equalling stamens. Fruit usually hairless. Limestone rocks and stony places. Pyrenees, Massif Central. 331a. *D. hispanica* Boiss. Like 331 but flowering stems hairy, and petals longer, 5–9 mm, and longer than the stamens. Fruit with dense rough hairs. Alpine rock fissures. S. and E. Spain.

332. *D. tomentosa* Clairv. DOWNY WHITLOW-GRASS. Distinguished by its soft rosettes of white-woolly oval leaves, covered with star-shaped hairs, forming a loose cushion, and its white flowers. Petals 3–5 mm. Flower stalks and fruit densely hairy. Rock fissures on acidic mountains. Pyrenees. 332a. *D. dedeana* Boiss. & Reuter An endemic species of rock fissures in calcareous mountains of N. and E. Spain (including Picos de Europa). It has a very woody stock bearing dense tufts of rosettes with rigid linear, keeled, bristly leaves, and white flowers. Petals notched, 4–6 mm, much longer than stamens; flowering stem hairy. Fruit bristly-haired. Page 201.

PETROCALLIS | Like *Draba* but leaves deeply lobed; hairs all unbranched; fruit with 1 or 2 seeds in each cell.

335a. *P. pyrenaica* (L.) R.Br. A low-growing perennial forming tight rounded cushions of tiny greyish rosettes, each bearing nearly stalkless clusters of very sweet-scented, pale lilac and pink flowers. Petals 4–5 mm. Leaves with 3–4 lobes stiff, bristly-haired. Alpine rocks and screes. Pyrenees. Page 201.

IONOPSIDIUM | 3 species. Distinguished by its fruit, which is a laterally compressed silicula with keeled valves. Seeds 2–6 in each cell, covered with transparent glandular protuberances.

341a. *I. acaule* (Desf.) Reichenb. A tiny hairless, rosette-forming annual bearing several solitary, long-stalked, lilac-purple flowers with yellow centres. Flowers about ½ cm across. Leaves long-stalked, with a rounded or three-lobed blade. Sandy places, open ground, pine woods near the coast. Portugal; naturalized elsewhere. 341b. *I. prolongoi* (Boiss.) Batt. Differs from 341a in having unequal petals and inflorescence in a spike-like cluster, becoming lax and more or less flat-topped. Sepals with a narrow white margin. S. Spain.

THLASPI | Pennycress   8 species.   345a. *T. nevadense* Boiss. & Reuter. A tufted rosette perennial bearing dense terminal clusters of white or very pale pink flowers on a short leafy stem. Rosette leaves ½–1½ cm, elliptic, stalked, stem leaves oblong, clasping with rounded lobes. Petals 6–7 mm, twice as long as sepals. Alpine screes and rock fissures. Spain (S. Nevada, S. de Guadarrama).

**AETHIONEMA | **346. *A. saxatile* (L.) R.Br. BURNT CANDYTUFT. A lowly hairless annual or perennial with glaucous leathery leaves and a dense terminal cluster of white, pale violet or pink flowers. Petals 2–5 mm, longer than sepals. Fruit flattened and broadly winged, notched. Stony and rocky places in mountains. S. and C. Spain, S. France.

IBERIS | Candytuft   Fruit compressed, winged and often notched; seed solitary in each cell. 16 species occur in our area, including 6 restricted to the Iberian peninsula; they are distinguished by small botanical differences.

Fruit in elongated clusters.

347. *I. sempervirens* L. EVERGREEN CANDYTUFT. An evergreen shrublet with thick, blunt, oblong-spathulate leaves 2–5 mm wide and a flat-topped cluster of white flowers which are borne on lateral stems. Fruit 6–7 mm, broadly winged. Alpine rocks. S. and E. Spain, Pyrenees.   (347) *I. saxatilis* L. Like 347 but with narrower linear leaves 1–2 mm wide. Flower clusters terminal. Calcareous rocks in hills and mountains. S. Spain, S. France. Subsp. *cinerea* (Poiret) P. W. Ball & Heywood is a greyish-green velvety shrublet, growing on gypsum hills of S. and C. Spain.   348d. *I. intermedia* (see below).

Fruit in flat-topped clusters.

(a) *Lowland perennials, usually 10–40 cm tall.*

347a. *I. linifolia* Leofl. A robust woody-based perennial with erect herbaceous stems to 30 cm, bearing terminal flat-topped clusters of deep purple, pink, or white flowers. Leaves rather fleshy, the lower oblong-wedge-shaped, with few teeth, the upper stem leaves linear entire. Sandy places near the sea, and lime-rich soils. Portugal, S–W. and C. Spain. Pl. 10.

347b. *I. procumbens* Lange A bushy shrublet with a stout woody base and bearing lilac flowers in a flat-topped cluster; innermost flowers sometimes mauve, or all flowers rarely white. Flower clusters 2–2½ cm across, contracted in fruit; outer petals 8–10 mm. Differing from 347a in the broadly spathulate stem leaves, which are entire or with 1–2 pairs of teeth near the apex. Sea-shores, cliffs and hills near sea. Portugal, N–W. Spain. Pl. 9, Page 44.

347c. *I. gibraltarica* L. A much more robust and showy perennial with reddish-purple to white flowers in a broad flat-topped cluster 4–5 cm across; outer petals 1½–2 cm. Shady cliffs on the Rock of Gibraltar. Page 201.   348d. *I. intermedia* (see below).

(b) *Montane perennials (usually above 800 m); up to 10 cm tall.*

347d. *I. pruitii* Tineo. A very variable perennial or annual, usually with small lilac but sometimes with white flowers. Leaves rather fleshy, entire or with a few teeth, blunt, the lower obovate-spathulate, the upper narrower. Fruit 6–8 mm, rectangular-elliptical, with wide, erect, acute lobes. Rock crevices in mountains. Med. region (common in S. de Cazorla). Page 201.

(c) *Annuals (without non-flowering rosettes).*

348a. *I. crenata* Lam. A rough-haired annual, distinguished by its lower linear-spathulate leaves which have conspicuous regular comb-like teeth, contrasting with the upper linear entire leaves. Flowers white, in dense convex clusters up to 3½ cm across; outer petals up to 1 cm. Fruit with rough swellings; lobes broad, triangular, acute. Dry soils. S. and C. Spain. Page 69.   348b. *I. pinnata* L. A cornfield weed of the Med. region with leaves deeply cut into 1–3 pairs of linear segments. Flowers white or lilac, fragrant. Fruit 5–6 mm, hair-

1. *Ptilotrichum lapeyrousianum* 325d
2. *Petrocallis pyrenaica* 335a
3. *Draba dedeana* 332a
4. *Vella pseudocytisus* 365b
5. *Alyssum montanum* 325
6. *Iberis gibraltarica* 347c
7. *Iberis pruitii* 347d

less, lobes blunt. 348c. *I. sampaiana* Franco & P. Silva A procumbent, rough annual endemic to the coasts of S–W. Portugal. Flowers white, outer petals wide-spreading. Stems spreading; leaves roughly hairy, spathulate, regularly cut into 3–4 pairs of oblong, blunt segments. Fruit 5 mm by 4 mm; lobes large, acute. 348d. *I. intermedia* Guersent Flowers a beautiful pale pink, in flat-topped clusters about 4 cm across, sometimes elongated into cylindrical heads. Upper leaves linear entire, up to 2½ cm long, stalkless. An erect hairless annual to 80 cm, or a woody-based perennial with numerous flowering shoots. Fruit 5–9 mm. Bushy and rocky places, calcareous slopes. N–E. Spain, S–C. France.

TEESDALIOPSIS | Like *Iberis* but style inconspicuous. Flowers irregular, the outer petals large, white, radiating, the inner petals small. Fruit scarcely winged at apex, notched; seeds single, pendant.

348e. *T. conferta* (Lag.) Rothm. Like a white-flowered *Iberis*, up to 15 cm tall, but flowering stem without leaves, which are confined to a crowded basal rosette. Mountains. N. Portugal, N–W. Spain. Page 125.

BISCUTELLA | A very difficult genus with about 32 species occurring in our area, including 21 which are restricted to it.

349a. *B. vincentina* (Samp.) Guinea An endemic woody-based perennial of S–W. Portugal (Cape St Vincent) with woolly-haired rosettes of oval-oblong wavy-margined leaves and erect almost leafless stems bearing a dense cluster of yellow flowers. Petals 6 mm. Fruit very conspicuous, 1½ cm or more across, with 2 large disk-shaped valves. Pl. 10, Page 44. 349b. *B. frutescens* Cosson Similar to 349a but whole plant covered with dense white interwoven hairs. Inflorescence much-branched; petals yellow, 4 mm. Fruit about 7 mm across, valves with small swellings. Limestone rocks. S–W. Spain (especially Serranía de Ronda), Balearic Islands. 349c. *B. intermedia* Gouan Basal leaves with a few weak lobes, usually rough-haired, 1–1½ cm broad. Petals yellow, 4½ mm. Fruit 5–6 mm by 9–10 mm. Endemic in mountains of C. and N. Spain, Pyrenees. 349d. *B. valentina* (L.) Heywood A woody-based perennial with simple or slightly branched stems 30–50 cm. Petals yellow, 5–6 mm; sepals 2–3 mm. Leaves linear, deeply toothed, bristly-haired. Fruit 3–6 mm by 6–11 mm. Dry places. S., C. and E. Spain (Serranía de Cuenca, etc.).

CARDARIA | **353. *C. draba* (L.) Desv. HOARY PEPPERWORT. Waysides and disturbed ground, often in masses and demanding attention by the roadside. Probably throughout.

MORICANDIA | **356. *M. arvensis* (L.) DC. VIOLET CABBAGE. An annual or woody-based perennial, easily distinguished by its showy clusters of rather large violet flowers, and its broad, glaucous, hairless leaves, the upper clasping the stem. Flowers about 2½ cm across. Fruit four-angled. Fields, banks. Med. region. 356a. *M. moricandioides* (Boiss.) Heywood (*M. ramburei*) Similar to 356 but a more robust plant with larger inflorescences bearing more numerous (20–40) flowers, and the outer pair of sepals strongly hooded. Fruit more or less rounded in section. Rock fissures, sandy places, dry hills, and mountains. S., C., and E. Spain. Page 68.

EUZOMODENDRON | Filaments of inner stamens united in pairs. Fruit beaked.

356b. *E. bourgaeanum* Cosson A small much-branched prickly shrub to ½ m, with pinnately lobed leaves and clusters of rather showy pale yellow flowers with brown veins. Petals 12–16 mm. Leaves bristly-haired, with 2–3 pairs of narrow, blunt, fleshy lobes. Fruit oblong 2–4 cm, beak to 1 cm. Calcareous soils. S. Spain (Almeria province). Page 90.

DIPLOTAXIS | Wall Rocket   8 species in our area.   358. *D. erucoides* (L.) DC. WHITE WALL ROCKET. Usually an annual with white flowers with violet veins, but flowers often at length becoming violet. Basal leaves in a lax rosette, entire, toothed or pinnately lobed, the upper leaves clasping. Fruit linear. Common in fields. Portugal, Med. region.
358a. *D. catholica* (L.) DC. An annual with yellow flowers and no basal rosette. Petals usually 7–8 mm. Leaves hairless or rough-haired on margin, pinnately cut or twice-cut into narrow, more or less equal, toothed lobes. Plant hairless above. Dry hills, vineyards, fields, walls, on the littoral. Portugal, Spain.   358b. *D. siifolia* G. Kunze Like 358a, but leaves hairy with short conical hairs on both sides, irregularly cut, the terminal lobe larger than the lateral lobes. Cultivated ground, vineyards, walls, waste places. S. Portugal, S. Spain.

BRASSICA | Cabbage   10 species in our area, including the following which are often cultivated: 359. *B. oleracea* L. WILD CABBAGE;   360. *B. napus* L. RAPE, SWEDE;   (360) *B. rapa* L. TURNIP;   361. *B. nigra* (L.) Koch BLACK MUSTARD.

Key to remaining species.
1. Stem lvs. absent or few.   *balearica, repanda*
1. Stem lvs. many.
  2. Beak of fr. nearly as long as basal part; fls. at length
    nodding.   *barrelieri*
  2. Beak much shorter than basal part; fls. remaining erect.
    3. Petals 5–7 mm, small; beak of fr. 1–2½ cm. (Med. region)   361d. *B. tournefortii* Gouan
    3. Petals 6–12 mm, conspicuous; beak of fr. 3–10 mm.
      4. Lower lvs. pinnately cut, segments saw-toothed.   *barrelieri*
      4. Lower lvs. pinnately cut, segments not saw-toothed.
        5. Fr. stalkless. (Med. region)   361e. *B. fruticulosa* Cyr.
        5. Fr. stalked. (introd. weed)   361f. *B. juncea* (L.) Czern.

361c. *B. barrelieri* (L.) Janka Usually an annual up to ½ m with lax clusters of white or pale yellow flowers, and with lowest leaves in a rosette. Petals 7–12 mm. Basal leaves cut into 7–10 pairs of narrow acute lobes, rough-hairy, stem leaves few, entire, clasping, hairless. Beak of fruit long or short. Sandy uncultivated fields, rocks in mountains, walls. Portugal, Spain.   361a. *B. balearica* Pers. A hairless perennial with a thick woody stock and a rosette of rounded-lobed fleshy leaves recalling those of the Common Oak in outline. Flowers bright yellow; petals 1–1½ cm. Endemic on limestone cliffs. Majorca. Pl. 10.   361b. *B. repanda* (Willd.) DC. A very variable perennial up to ½ m, of rocks and stony places, usually with leafless stems and yellow flowers with petals varying from 7–30 mm. Plants densely or loosely tufted, with or without a woody stock, spiny or hairless; basal leaves entire or lobed. Fruit variable, 1–8 cm. Spain, France.

**ERUCA | **363. *E. vesicaria* (L.) Cav. A common annual weed of cornfields and waste ground. Portugal, C. and E. Spain, Med. region.

**VELLA | ** Shrubs with entire leaves, distinguished by their fruits which have a sterile flattened beak and a lower two-valved section.

365a. *V. spinosa* Boiss. A low, very dense, spiny 'hedgehog' plant forming rounded bushes, and bearing conspicuous white or yellowish flowers with violet veins. Flowers about 1 cm across, in clusters of 3–5; petals narrowed to a long claw longer than the sepals. Leaves stiff, all with bristly margins, the upper linear, the lower broader; branches ending in rigid spiny points. Dry rocky slopes on high limestone mountains. S. and S–E. Spain. Pl. 11, Page 77.   365b. *V. pseudocytisus* L. A spineless shrub 30–100 cm, with somewhat leathery, obovate leaves, and elongated clusters of 10–35 yellow flowers. Petal-limb rounded, indistinctly veined. Leaves with rough, bristly hairs on both sides. Gypsum hills. S. and C. Spain. Page 201.

SUCCOWIA | 365c. *S. balearica* (L.) Medicus An annual with small dull yellow flowers, readily distinguished by its paired globular fruits beset with stiff spines and topped with a long curved beak. Petals 7–10 mm. Leaves two to three times cut into deeply lobed, lance-shaped segments, hairless. Shady rocks, stony ground on the littoral. Med. Spain.

CAKILE | ***366. *C. maritima* Scop. SEA ROCKET. Coasts of our area.

CRAMBE | ***368. *C. maritima* L. SEAKALE. A robust, glaucous, cabbage-like plant of the Atlantic coast of N. Spain and France. Flowers white, borne in conspicuous spreading clusters. Fruit globular.   368a. *C. hispanica* L. A slender, densely rough-haired annual to 1 m with spreading branches and tiny white flowers. Petals 3–4 mm. Lower leaves pinnately lobed, with a large broad terminal lobe, upper leaves toothed or wavy-margined. Damp places, hedges, rocks on the littoral. S. Portugal, Med. region.   368b. *C. filiformis* Jacq. Like 368a but lower segment of fruit longer than upper segment. A slender, spindly perennial to 1 m. Walls, stony places, rock crevices. In mountains. S. Spain.

RESEDACEAE | Mignonette Family

RESEDA | **Mignonette**   17 species, 11 are restricted to our area in Europe. Widespread are: 371. *R. luteola* L. DYER'S ROCKET, WELD;   ***372. *R. lutea* L. WILD MIGNONETTE.

Leaves all entire or with a few minute teeth at base.
372a. *R. lanceolata* Lag. An erect annual or biennial to 1 m or more, with oblance-shaped to spathulate leaves, usually entire but sometimes with 1–2 pairs of lateral lobes. Sepals 6–8; petals yellow, the 2 upper three-lobed; flower stalks ½ cm or more. Fruit 16–28 mm, three to four times as long as wide. Arid fields, rocks, hills. S. and E. Spain. Page 207.
372b. *R. glauca* L. Distinguished by its narrow, entire glaucous leaves 1½–6 cm long. Flowers white in numerous slender spikes; flower stalks short; petals 6, 3–5 mm. Fruit wider than long, four-toothed. Rock and screes in mountains. N. Spain, Pyrenees.   372c. *R. gredensis* (Cutanda & Willk.) Müller Arg. C. Portugal, W–C. Spain (S. de Gredos) and distinguished from 372b by smaller leaves 3–12 mm long, which are in clusters on the stems. Mountain pastures and stony places.

***374. *R. phyteuma* L. RAMPION MIGNONETTE. Like 372a. with usually entire linear-spathulate leaves, but fruits distinctive, drooping three-lobed, up to twice as long as wide, with 6 conspicuous persisting and enlarged sepals. Flowers whitish; upper petals with numerous lobes. Rocks, rough stony places, hills. Mainly in the S. and C. of our area.   374b. *R. virgata* Boiss. & Reuter Usually a very glaucous perennial with linear leaves and with small whitish flowers. Petals 3 mm, the upper ones three-lobed for one-third their length, the lobes toothed. Leaves 1–4 cm by ½–1 mm, with 2–5 pairs of small whitish teeth near the base. Fruit 3–4 mm. Waste sandy fields, roadsides. N. Portugal, C. Spain.

Some leaves ternate or pinnately lobed.
374a. *R. media* Lag. Like 374 but leaves dark green, many with 1–4 or more pairs of lobes and a small terminal lobe. Sepals not enlarging in fruit to more than 6 mm. Heaths, fields, hills. Portugal, Spain. Pl. 12.

**373. *R. alba* L. UPRIGHT MIGNONETTE. A robust annual, biennial or perennial, distinguished by its long slender spikes of whitish flowers and its deeply pinnately cut leaves with 5–15 pairs of narrow undulate lobes. Sepals and petals usually 5; petal limb three to five times as long as claw. Stamens 10–12. Sandy places by the sea, disturbed ground. Portugal, Med. region.   373a. *R. suffruticosa* Loefl. The finest of all the European Mignonettes, growing up to 2 m, with closely massed white flowers in a long wand-like spike, and with contrasting orange anthers. Leaves usually cut into about 25 pairs of oblong-acute lobes which are further lobed. Sepals and petals 6; limb of petal rectangular or triangular up to twice as long as the claw. Fruit 11 mm, ovoid, covered with swellings. Limestone and gypsum hills. N. Portugal, S., S–C. and E. Spain. Pl. 12.

SESAMOIDES | Like *Reseda* but with 4–7 more or less free, spreading carpels.

375. *S. pygmaea* (Scheele) O. Kuntze (*Astrocarpus sesamoides*) A slender-stemmed shrubby plant, much-branched from the base and bearing long, often interrupted spikes of tiny white flowers. Calyx divided to at least halfway into narrow blunt lobes; petals 5–6, much cut; stamens 7–9. Leaves narrow, entire, the basal leaves often in a rosette. Fruit of 4–6 carpels, spreading in a star. Alpine meadows, rocky places. N–W. Spain, Pyrenees, Massif Central. Pl. 9.   375a. *S. canescens* (L.) O. Kuntze (*Astrocarpus purpurascens*) Like 375 but calyx divided to less than half way into broad triangular, acute lobes; stamens usually 12–14. Fruit 4–5, styles lateral. A shrub, herbaceous annual, or biennial. Dry sandy, uncultivated places, roadsides, pine woods. Scattered throughout.

DROSERACEAE | Sundew Family

DROSERA | Sundew   376. *D. rotundifolia* L. COMMON SUNDEW. Bogs and wet peaty places mainly in the mountains. Widespread except in south.   **377. *D. anglica* Hudson GREAT SUNDEW. Leaf-blades linear-oblong, about 3 cm, leaf stalk long, and flowering stem much longer than the leaves. Rare. N. and E. Spain, Pyrenees, Massif Central.   **378. *D. intermedia* Hayne LONG-LEAVED SUNDEW. Leaf-blades like 377 but smaller, about 7 mm. Flowering stems scarcely longer than the leaves. Damp peaty, ground. Portugal, N–W. Spain, France.

DROSOPHYLLUM | **(378) *D. lusitanicum* (L.) Link YELLOW SUNDEW. One of the strangest and most interesting of European insectivorous plants, with very long slender leaves, covered with red-tipped glandular hairs, and coiled like octopus tentacles. Flower bright yellow, about 2½ cm across, several borne on a stem 15–30 cm tall. 'Sits perched on a dense mat of its own dried leaves, usually on top of a bush of heather, etc., in the undergrowth of cork woods.' Mainly in coastal regions. S. and C. Portugal, S–W. Spain. Pl. 13, Page 52.

CRASSULACEAE | Stonecrop Family

UMBILICUS | Pennywort   **379. *U. rupestris* (Salisb.) Dandy WALL PENNYWORT. Frequent on walls, rocks, cliffs. Widespread.   (379) *U. horizontalis* (Guss.) DC. Like 379 with greenish-white or straw-coloured flowers but inflorescence not more than half the length of the stem; corolla horizontal, about 7 mm by 3 mm, lobes triangular lance-shaped, acute.

Stem leaves numerous, many linear. Walls, rocks. S., W. and C. Spain.   379a. *U. heylandianus* Webb & Berth. Distinguished by its long, dense, spike-like cluster of bright yellow horizontal or drooping flowers which are 10–12 mm long. Corolla tube distinctly five-angled, narrowed at the throat, lobes pointed, about one third as long as tube; stamens 5. Sandy places in cork oak woods. Portugal, W., C., and N. Spain.

PISTORINIA | Flowers in a dense, somewhat flat-topped inflorescence; corolla with a long narrow tube and 5 spreading lobes; stamens 10, exserted.

379b. *P. hispanica* (L.) DC. A small annual 5–15 cm, with fleshy glandular leaves, often turning reddish, and bright rosy-purple starry flowers. Corolla with a slender tube and 5 narrow, pointed spreading lobes each with a darker purplish blotch towards the tip; stamens much longer than the tube. Leaves ½–1 cm oblong. Sandy and stony places in mountains. Portugal, Spain. Pl. 13.   379c. *P. breviflora* Boiss. Like 379b. but a more robust plant with yellow flowers, tinged with reddish-brown on the tips of the lobes, and corolla tube broadening towards the throat. Leaves to 2 cm, more or less pointed. Lowlands. S. Spain.

MUCIZONIA | Like *Pistorinia* but corolla bell-shaped, divided to about the middle into 5 lobes; stamens 10, included.

379d. *M. hispida* (Lam.) A. Berger A greyish, glaucous, usually glandular-hairy, much-branched fleshy annual to 15 cm, with a lax cluster of yellowish-green to pinkish flowers. Flowers ½–1 cm across; corolla lobes spreading, rounded. Leaves fleshy, oblong, 1–2 cm, streaked with red. Walls, rocks in the lowlands and mountains. Portugal, S. Spain. Pl. 13. 379e. *M. sedoides* (DC.) D.A. Webb An unbranched, hairless, reddish annual 2–6 cm, usually growing in dense tufts formed by numerous tiny plants, with clusters of small pinkish-purple flowers nestling amongst the leaves. Flowers 6–7 mm; petals pointed, erect, paler at base. Leaves globular, 2–4 mm, numerous and over-lapping. Rock crevices in high mountains. Portugal, Spain, Pyrenees. Pl. 13.

SEMPERVIVUM | **Houseleek**   7 species in our area; hybrids are common, but only in the presence of both parents. Usually plants of rocky places in the mountains.

1. Mature lvs. downy on both surfaces.
   2. Apical hairs of lvs. very long and flexuous,
      interwoven to form a spider's web over the
      rosette. (Pyrenees)   **381. *S. arachnoideum* L.
   2. Apical hairs short, not like a spider's web.
      3. Plant with resinous scent; marginal
         bristles of rosette lvs. scarcely longer than
         the other hairs; petals 12–20 mm. (Pyrenees)   **382. *S. montanum* L.
      3. Plant without distinctive scent; marginal
         bristles of rosette lvs. at least twice as long
         as the other hairs; petals 7–11 mm.
         4. Rosette lvs. sparsely downy.
            5. Stem usually more than 20 cm, bearing
               at least 40 fls. (Pyrenees; naturalized
               elsewhere)   383. *S. tectorum* L. Pl. 12.
            5. Stem 8–16 cm, bearing 12–30 fls. (Picos
               de Europa)   383a. *S. cantabricum* J. A. Huber Pl. 9.
         4. Rosette lvs. densely downy. (N–W. Spain
            (Peña de Espiguete))   383b. *S. giuseppii* Wale

1. *Reseda lanceolata* 372a
2. *Saxifraga erioblasta* 411a
3. *S. haenseleri* 404b
4. *Sedum rubens* (2 drawings) 392a
5. *Cleome violacea* 283a
6. *Saxifraga latepetiolata* 411g

1. Mature lvs. hairless apart from marginal
   bristles (or rarely faintly downy).
   6. Stem *c*. 12 cm, bearing less than 30 fls.
      7. Inner lvs. of rosette pressed together,
         forming a cone; anthers yellow. (E.
         Pyrenees (Sierra del Cadi))   383c. *S. andreanum* Wale
      7. Inner lvs. of rosette erect; anthers reddish-
         brown. (S. Spain (Sierra Nevada))   383d. *S. nevadense* Wale
   6. Stem usually more than 20 cm, bearing more
      than 30 fls. (Pyrenees; naturalized else-
      where)   383. *S. tectorum* L. Pl. 12

AEONIUM | **386. *A. arboreum* (L.) Webb & Berth. A branched succulent plant with
thick brown stems, terminal flattened rosettes of overlapping, broadly strap-shaped leaves
and a terminal compact rounded cluster of numerous bright yellow flowers. Native of
Morocco; naturalized on the rocky coasts of Portugal, Med. and N–W. Spain.

SEDUM | **Stonecrop**   30 species. Most species grow in dry stony and rocky places
ranging from the lowlands to the high mountains.

Leaves fleshy, cylindrical, (not flattened)
(a) Flowers yellow, cream or greenish-white
(i) Carpels erect; petals usually 6–9; inflorescence erect, branched only near the top

1. Fls. few, in lax one-sided, flat-topped
   clusters.
   2. Lvs. of non-flowering shoots clasping
      stem; petals 6–8 mm.   (388) *S. tenuifolium* (Sibth. & Sm.) Strobl
   2. Lvs. not clasping the stem; petals 8–10
      mm. (C. and N. Portugal)   388b. *S. pruinatum* Brot.
1. Fls. numerous, in crowded flat-topped
   clusters.
   3. Living lvs. on non-flowering shoots con-
      fined to a terminal tassel-like cluster;
      dead lvs. persistent on lower part of
      shoot. (Throughout)   388a. *S. forsteranum* Sm.
   3. Not as above.
      4. Sepals 5–7 mm, glandular-hairy; in-
         florescence flat-topped in bud and fr.
         (Spain, France)   387. *S. ochroleucum* Chaix
      4. Sepals 2–5 mm, hairless; inflorescence
         rounded in bud, concave in fr.
         5. Inflorescence erect in bud; sepals egg-
            shaped; lvs. *c*. 4 mm wide. (Portugal,
            N. Spain, Med. region)   (387) *S. sediforme* (Jacq.) Pau Pl. 13
         5. Inflorescence drooping in bud; sepals
            lance-shaped; lvs. 2 mm wide. (Spain,
            France)   **388. *S. reflexum* L.

(*ii*) *Carpels spreading; petals 5; inflorescence short, spreading*
**389. *S. acre* L. STONECROP. Widespread. (389) *S. alpestre* Vill. ALPINE STONECROP. Pyrenees.
390. *S. annuum* L. ANNUAL STONECROP. Spain, France.

(*b*) *Flowers white, pink, red, or violet*
(*i*) *Perennials with non-flowering shoots at flowering*
**391. *S. album* L. WHITE STONECROP. Widespread. (391) *S. anglicum* Hudson ENGLISH STONECROP. Scattered throughout. (391) *S. dasyphyllum* L. THICK-LEAVED STONECROP. Spain, France. **394. *S. villosum* L. HAIRY STONECROP. Spain, France. 391a. *S. gypsicola* Boiss. & Reuter Very like 391 but leaves greyish, densely and minutely downy, and in 5 closely overlapping rows on the sterile shoots. Gypsum hills. C. and S. Spain. 394a. *S. hirsutum* All. Glandular hairy like 394 but buds drooping; carpels white, and plants of dry rocks not wet places. Flowers white or pink. Throughout. Pl. 12. 391b. *S. brevifolium* DC. Like (391) but globular leaves quite hairless, usually closely overlapping in 4 rows on non-flowering shoots. Stony ground and rocks in mountains. Throughout. Pl. 13, Page 133.

(*ii*) *Annuals or biennials without non-flowering shoots at flowering*
(*x*) *Stamens twice as many as petals*
392. *S. atratum* L. DARK STONECROP. Mountains. N. Spain, Pyrenees, France. 396a. *S. lagascae* Pau A glandular-hairy, bushy annual with numerous drooping bell-shaped, rather long-stalked, pale pink flowers with yellowish bases. Petals about 7 mm. Leaves broadly linear, rounded, 6–9 mm. Portugal, S. and C. Spain. 396b. *S. pedicellatum* Boiss. & Reuter A hairless annual with an erect stem, bearing rather few white flowers, in a lax flat-topped cluster. Petals 3 mm, with a central pink line. Leaves spreading, oval-cylindrical, glaucous, 2–5 mm. Rocks and stony places in mountains. N. Portugal, C. Spain.

(*xx*) *Stamens equal in number to petals.*
392a. *S. rubens* L. RED STONECROP A glandular-hairy, usually reddish, somewhat glaucous annual with flat-topped clusters of white or pink flowers. Petals 5 mm, sharply pointed; stamens usually 5, rarely 10. Leaves 1–2 cm, cylindrical-linear. Carpels glandular-hairy, long-pointed, spreading. Sandy fields. Scattered throughout. Page 207. 392b. *S. caespitosum* (Cav.) DC. A tiny, hairless, usually reddish annual with small stalkless white flowers tinged with pink. Petals 3 mm, pointed; stamens 4 or 5. Leaves ovoid 3–6 mm. Carpels hairless, spreading. Sandy places. Throughout. 392c. *S. andegavense* (DC.) Desv. Like 392b but flowers stalked, white or pinkish; carpels erect, somewhat wrinkled. Flowers white and pinkish. Leaves globular or ovoid, overlapping, shortly spurred. Dry sandy places and rocks. Scattered throughout. 392d. *S. nevadense* Cosson Distinguished from 392b. by its linear-oblong leaves which are not spurred. Flowers white or pale pink, stalked; petals 3–4 mm, acute, fused at base. Carpels erect. Mountains. S., N., and E. Spain.

Leaves flattened, fleshy
**395. *S. telephium* L. ORPINE, LIVELONG. Scattered throughout; sometimes cultivated. 395a. *S. anacampseros* L. Distinguished by its rather small, flat, egg-shaped entire glaucous leaves and short ascending stems each bearing a dense rounded cluster of reddish-violet flowers. Petals dull deep red within, glaucous-lilac outside, 4–5 mm; stamens 10. Siliceous rocks. Pyrenees. 396. *S. cepaea* L. Shady and bushy places. Spain, France. (396) *S. stellatum* L. STARRY STONECROP. Stony ground. Med. region, N. Spain.

RHODIOLA | *397. *R. rosea* L. ROSEROOT, MIDSUMMER-MEN. Rocks and screes. Pyrenees.

SAXIFRAGACEAE | Saxifrage Family

SAXIFRAGA | Saxifrage Approximately 58 species are found in our area, of which 32 species are endemic to the Iberian peninsula and the Pyrenees. Many endemic species, though often distinctive in the field are restricted to limited regions, and are distinguished by small botanical characters. For full treatment, keys and diagnoses see *Flora Europaea*, Vol. I, pages 364–380.

Leaves with lime-glands on margin
(a) Flowers purple or pink
**398. *S. oppositifolia* L. PURPLE SAXIFRAGE. Mountains. Spain, France. Page 76.   398a. *S. retusa* Gouan Like 398 but more compact. Flowers purplish-red, with stamens longer than petals; anthers orange (not bluish); sepals hairless or glandular-hairy (without cilia). Shady alpine rocks, exposed ridges. Pyrenees.

410a. *S. media* Gouan 'It forms columnar tufts of small oval grey leathery leaves, recurving and overlapping, with a broad rim of limey cartilage. The stems are 7½–10 cm high, red, and densely clad in glandular red-tipped hairs, that are also thick on all parts of the few-flowered flower-spray. The blossoms are small and pink, hardly emerging from the great crimson-velvet furry bells of the calyx, which in itself is as attractive as any flower.' (Farrer) Mountain rocks, preferring limestone. E. Pyrenees. Pl. 15.

(b) Flowers white
**410. *S. caesia* L. BLUISH-GREY SAXIFRAGE. Rock, screes. Pyrenees.   **400. *S. paniculata* Miller LIVELONG SAXIFRAGE. Distinguished by its medium-sized, greenish-grey, lime-encrusted rounded rosettes 1–5 cm across, and its erect, sparsely branched inflorescence, bearing white flowers. Branches occurring only from the upper third of stem, wide-spreading, glandular-hairy, bearing 1–3 flowers. A very variable plant. Common on rocks and screes in mountains. N. Spain, Pyrenees, Massif Central. Page 152.

**(399) *S. longifolia* Lapeyr. PYRENEAN SAXIFRAGE. 'Crown-royal they call this in Spain, where it abounds on the limestones all through the Pyrenees from the vine-level upwards. It is one of the grandest of the race; the huge silver star-fish rosette splayed and hard against the cliffs is superb enough picture in itself, even without those dominating regal fox-brush spires of white standing stiffly straight out from the face of the rock . . . then dies making no offsets.' (Farrer) Leaves linear, 3–8 mm broad, lime-encrusted. Mountain rocks. E. Spain, Pyrenees. Pl. 14.   399. *S. cotyledon* L. PYRAMIDAL SAXIFRAGE. Like (399) but leaves strap-shaped, 9–15 mm broad, finely toothed, forming large rosettes. Flowers white, in a large pyramid; petals 6–10 mm. Damp rocks. N. Spain, C. Pyrenees.   399a. *S. callosa* Sm. LIMESTONE SAXIFRAGE Rosettes glaucous, lime-encrusted, often dark red near the base, with linear to oblance-shaped leaves 1½–3 cm by 4–6 mm. Flowers white, numerous in an elongated cluster; petals 6–9 mm, often crimson-spotted; inflorescence glandular, flower stalks slender, often reddish. E. Spain (S. de Monserrat).

Leaves without lime-glands
(a) Flowers bright yellow
410b. *S. aretioides* Lapeyr. An unmistakable plant forming hard, compact, greyish-green cushions of numerous columnar rosettes, and with short leafy stems carrying 3–4 bright yellow flowers. Petals notched, wavy-margined; calyx densely glandular. Leaves stiff with a broad white border. Limestone cliffs. N. Spain, Pyrenees. Pl. 15, Page 125.   **402 *S. aizoides* L. YELLOW MOUNTAIN SAXIFRAGE. Mountain springs. Pyrenees.

(b) Flowers white or pink
(i) Bulbils present in the axils of the basal leaves, at or below ground level
404. *S. granulata* L. MEADOW SAXIFRAGE. Grasslands, rocky places. Widespread. Page 68.
404a. *S. corsica* (Duby) Gren. & Godron Like 404 but basal leaves more or less three-lobed, with rounded teeth. Flowers white, in a lax inflorescence of spreading branches, the lowest arising from the middle of the flowering stem, or below. Shady rocks, walls and screes. E. Spain.   404b. *S. haenseleri* Boiss. & Reuter A white-flowered species with a very lax inflorescence of wide-spreading branches and few, long-stalked flowers to about 1 cm across. Leaves forming basal tufts, with bulbils underground, blades deeply three- to five-lobed, glandular-hairy. Spain (Sierras of Granada and Murcia). Page 207.

(ii) Bulbils absent, or bulbils in axils of stem leaves or bracts
(x) Ovary superior (arising above the other parts of the flower)
**403. *S. rotundifolia* L. ROUND-LEAVED SAXIFRAGE. Shady places in mountains. Spain, France.   **405. *S. umbrosa* L. WOOD SAXIFRAGE. Easily distinguished by its large flat rosette of somewhat leathery leaves with rounded, deeply toothed racquet-shaped blades which have conspicuous papery margins. Flowers white, red-spotted, in a tall, lax inflorescence. Leaf stalk broad and flat, densely ciliate, scarcely as long as blade. Damp shady rocks, mossy places. W. and C. Pyrenees.   405a. *S. spathularis* Brot. ST PATRICK'S CABBAGE. Like 405 but distinguished by its leaves which have flattened leaf stalks which are very sparsely hairy, and at least as long as the blade. Shady rock fissures, mossy places. N. Portugal, N–W. Spain. Pl. 14.   **(405) *S. hirsuta* L. KIDNEY SAXIFRAGE. Like 405 but distinguished by its leaves which have narrow rounded, hairy leaf stalks, and leaf blades which are hairy on both sides. Damp shady rocks, mossy places. N. Spain, Pyrenees.

406. *S. cuneifolia* L. Shady and mossy places, in mountains. N. Spain, Pyrenees.   **407. *S. stellaris* L. STARRY SAXIFRAGE. Mountain streamlets, grassy places. Locally scattered. 407a. *S. clusii* Gouan A sticky glandular-hairy plant with numerous white flowers, in a very fragile, much-branched spreading, pyramidal inflorescence. Petals unequal, each with 2 yellow spots. Basal leaves large, 6–15 cm, obovate, coarsely toothed, long-stalked in a lax rosette. Mountain streams, damp shady rocks. N. Portugal, N. Spain, Pyrenees, Massif Central.   408. *S. aspera* L. ROUGH SAXIFRAGE. Damp rocks. C. and E. Pyrenees. (408) *S. bryoides* L. MOSS SAXIFRAGE. Damp rocks in mountains. Pyrenees, C. France.

(xx) Ovary half-inferior (sepals, petals and stamens arising level with the middle of the ovary)
(y) Leaves mostly entire
411a. *S. erioblasta* Boiss. A small white-flowered species distinguished by its cushion-like clumps of numerous, globular, densely white-haired summer-resting buds, recalling miniature cobweb-houseleeks. Flowers 3–5, about 8 mm across. Leaves 3–6 mm, entire or shortly three-lobed, glandular-hairy, lobes blunt. Limestone rocks and screes in mountains. S–E Spain. Page 207.   411b. *S. conifera* Cosson & Durieu A silvery-green, matted plant, distinguished by its numerous silvery, spindle-shaped summer-resting buds formed of long-pointed leaves entwined in cobweb-like hairs. Flowers white, tiny about $\frac{1}{4}$ cm across; flowering stem often deep crimson, glandular-hairy. Limestone rocks. N. Spain. Pl. 14, Page 124.   **409. *S. androsacea* L. Mountain screes, pastures. Pyrenees, Massif Central.

(yy) Leaves toothed or lobed; perennials
**411. *S. moschata* Wulfen MUSKY SAXIFRAGE. Mountain screes. N. Spain, Pyrenees, France.   411c. *S. globulifera* Desf. (incl. *S. granatensis*) Very like 411a but summer-resting buds obconical and less hairy, and plant forming a loose, not a dense cushion. Flowering stem 7–12 cm, bearing 3–7 white flowers. Leaves semi-cicular in outline, deeply three- to seven-lobed, with acute lobes. Mountain rocks. S. Spain (common S. de Ronda, S. Blanca). Pl. 14.   411d. *S. rigoi* Porta Like 411b but larger in all its parts, with striking

white, bell-shaped flowers, with petals 12–15 mm. Summer-resting buds numerous, oblong-conical with long white bristly hairs. Damp places in mountain rocks. S–E. Spain. 411e. *S. biternata* Boiss. A lax tufted perennial with an almost woody base, bearing numerous bulbil-like buds in the axils of the lower leaves. Flowers few, white, rather large, narrowly bell-shaped; petals 1½–2 cm. Leaves ferny, twice-cut into lobed or deeply toothed segments, which are covered with velvety glandular hairs. Dry limestone cliffs. S. Spain (El Torcal de Antequera). Page 61.

411i. *S. boissieri* Engler A loosely tufted, cushion-forming plant with glandular-hairy leaves deeply cut into threes and further lobed or cut, and reddish winter-dormant resting buds among the lower leaves. Flowers rather numerous in a leafy cluster, white with a yellow centre; petals 4–7 mm, hairless. Damp, shady, limestone rocks in mountains. S–W. Spain (S. de Ronda). 411f. *S. continentalis* (Engler & Irmescher) D. A. Webb Distinguished by its stalked summer-resting buds borne in the axils of many leaves, and which are covered by silvery papery-scales. Flowers white, borne on slender flowering stems 10–30 cm; buds nodding; petals 7–10 mm. Leaves entire or three- to seven-lobed, lobes ovate-lance-shaped, sometimes arched. Damp rocks in mountains. N. Portugal, N. Spain, S. France. 411h. *S. nevadensis* Boiss. A lax or dense cushion-forming plant densely covered with long glandular hairs. Flowers white, sometimes red-veined; petals about 4 mm by 2 mm. Leaves usually three-lobed. S. Spain (S. Nevada, above 2600 m).

S. pentadactylis group Distinguished from others by the complete absence of hairs on the normal plant. Flowers white, rather few, in a somewhat flat-topped and fairly compact cluster. Leaves leathery and persistent when dead, deeply lobed, covered like the calyx and flower stalks with stalkless glands. Usually forming loose rounded cushions from a somewhat woody base. The following, often distinctive, species are included in this group: 404c. *S. pentadactylis* Lapeyr. Mountains of C. and N–C. Spain, E. Pyrenees. Pl. 14, Page 110; 404d. *S. corbariensis* Timb.-Lagr. Mountains of N–E. and E. Spain (S. de Cazorla, S. de Cuenca), S. France (Corbières); 404e. *S. camposii* Boiss. & Reuter Mountains of S. and S–E. Spain; 404f. *S. trifurcata* Schrader Mountains of N. Spain (Asturias to Navarra); 404g. *S. canaliculata* Engler Mountains of N. Spain.

404h. *S. cuneata* Willd. Distinguished by its shining, leathery, somewhat sticky, stalked leaves in a basal rosette, with broadly wedge-shaped blades 1½–3 cm, deeply cut to one third into 3–5 broadly triangular lobes, hairless except on the margins. Flowers large, white, in a lax branched cluster; petals 6–7 mm. Shady rocks. N. Spain, W. Pyrenees. Pl. 14, Page 133. 404i. *S. geranioides* L. 'One of the most delightful among the mossy saxifrages, strongly sweet-scented, and with snow-white flowers, narrow-petalled and never expanded widely, closely set in rather tight clusters at the top of 15–20 cm stems.' (Farrer) Sepals distinctive, linear-acute. Leaves variously cut into grooved segments, leaf stalks long-sheathing at the base. Rocks and screes. N–E. Spain, E. Pyrenees.

(404) *S. aquatica* Lapeyr. PYRENEAN WATER SAXIFRAGE. Unmistakable with its large, soft deep cushion-like growths by the side of streams and in marshy ground in the Pyrenees. Inflorescence robust 30–60 cm, leafy, branched above and bearing numerous large white flowers, with petals 6–9 mm. Leaf-blades shining, glandular-hairy, semi-circular in outline and deeply divided into numerous triangular-pointed segments. Pl. 15.

(yyy) *Leaves toothed or lobed; annuals or biennials*
**(411) *S. tridactylites* L. RUE-LEAVED SAXIFRAGE. Walls, rocks, sandy places. Widespread. 411g. *S. latepetiolata* Willk. A glandular-hairy biennial forming large, domed rosettes of rounded, deeply lobed leaves, which produce in the second year a branched pyramidal leafy

cluster of white flowers 15–30 cm high. Flowers somewhat bell-shaped; petals 8–9 mm. Stems and leaf stalks reddish. Local on limestone rocks. E. Spain (S. de Cuenca, S. de Chiva, etc.). Page 207.

CHRYSOSPLENIUM | **(412) *C. oppositifolium* L. OPPOSITE-LEAVED GOLDEN SAXIFRAGE. Leaves all opposite. Damp places in the lowlands and mountains. Widespread except in Med. region. 412. *C. alternifolium* L. ALTERNATE-LEAVED GOLDEN SAXIFRAGE. Like (412) but leaves of flowering stem alternate. Pyrenees, France.

PARNASSIACEAE | Grass of Parnassus Family

PARNASSIA | **413. *P. palustris* L. GRASS OF PARNASSUS. Marshes, damp grassy places, in mountains. C., N., and E. Spain, France.

GROSSULARIACEAE | Gooseberry Family

RIBES | **414. *R. rubrum* L. RED CURRANT. Native in montane regions of N–E Spain. Cultivated and sometimes naturalized elsewhere. 415. *R. nigrum* L. BLACK CURRANT. Cultivated in gardens and sometimes naturalized in our area. **416. *R. uva-crispa* L. GOOSEBERRY. Native in our region. Sometimes cultivated and naturalized. (415) *R. petraeum* Wulfen ROCK RED CURRANT. A deciduous shrub 1–3 m, distinguished by its broad leaves 7–15 cm wide, and its dark red-purple, drooping, acid fruit. Flower clusters at length drooping; flowers pinkish, hermaphrodite; bracts 1–2 mm. Woods, streamsides. Pyrenees, Massif Central. **(414) *R. alpinum* L. MOUNTAIN CURRANT. Like (415) but leaves smaller 2–6 cm, longer than broad, usually hairless and deeply three-lobed. Flower clusters ascending, either male or female; flowers greenish; bracts 4–10 mm. Fruit scarlet, hairless, insipid. Open woods, rocky places in mountains. N. Spain, Pyrenees, Massif Central.

PITTOSPORACEAE | Pittosporum Family

PITTOSPORUM | 417. *P. tobira* (Thunb.) Aiton fil. A robust evergreen shrub with oval-oblong blunt, leathery leaves, and flat-topped clusters 5–8 cm across, of very sweet-scented, creamy-white flowers with blunt petals. Native of the Far East; often grown for ornament in the west and south. 417a. *P. undulatum* Vent. A small tree with shiny green, wavy-margined, thin evergreen leaves, and fragrant clusters of white or creamy flowers with lance-shaped acute petals. Native of Australia; grown for ornament in the west and south; sometimes naturalized.

PLATANACEAE | Plane Tree Family

PLATANUS | **418. *P. orientalis* L. ORIENTAL PLANE. Native of Eastern Europe; widely grown for ornament and shade. (418) *P. hybrida* Brot. LONDON PLANE. Distinguished from 418 by the leaves which are lobed to the middle at most; female heads usually 2. More widely grown, particularly in the north.

213

ROSACEAE | Rose Family

SPIRAEA | ****(419)** *S. hypericifolia* L. A small shrub with hairless, obovate leaves 1–2½ cm, and short-stalked clusters of white flowers, forming long terminal leafy inflorescences. Bushy places in limestone mountains. N. Portugal, Spain, France.

ARUNCUS | **420. *A. dioicus* (Walter) Fernald GOAT'S-BEARD. A tall leafy perennial to 2 m, distinguished by its conspicuous branched pyramidal inflorescence of very numerous tiny white flowers, and its large fern-like leaves. Shady valleys. Pyrenees.

FILIPENDULA | **421. *F. ulmaria* (L.) Maxim. MEADOW-SWEET. Marshes and riversides. Throughout, except in south.   **422. *F. vulgaris* Moench DROPWORT. Grasslands, mainly in hills and mountains. Widespread.

ALCHEMILLA | 423. *A. vulgaris* L. agg. LADY'S-MANTLE. A hairless perennial with lax or dense clusters of tiny greenish-yellow flowers. Leaves rounded, palmately five- to eleven-lobed to half the width of the blade, green above and below. Shady places, meadows, in the mountains. Scattered throughout.   **424. *A. alpina* L. ALPINE LADY'S-MANTLE. Leaves deeply cut to the base into 5–7 lobes, which are green above and silvery-hairy beneath and on the margin. Mountain pastures and rocks. Scattered throughout.

RUBUS | 428. *R. idaeus* L. RASPBERRY. Bushy places in mountains, Spain, France; cultivated elsewhere.   429. *R. fruticosus* L. agg. BLACKBERRY. Widespread.   427. *R. saxatilis* L. ROCK BRAMBLE. Rocky places in mountains. Pyrenees, Massif Central.   **(429) *R. caesius* L. DEWBERRY. Bushy places. Throughout.

ROSA | About 29 species are found in our area, including 7 in the *canina* group, 4 in the *tomentosa* group, and 5 in the *rubiginosa* group.   (430) *R. sempervirens* L. It has flat-topped clusters of white flowers, and evergreen, shining, long-pointed leaflets. Hedges, waysides, bushy places. S. Portugal, S., N., and E. Spain, France.   **434. *R. pimpinellifolia* L. BURNET ROSE. Distinguished by its solitary creamy-white flowers, small rounded leaflets, and very bristly stems. Dry hills and rocks. N., S–E. and E. Spain, France.   **435. *R. pendulina* L. ALPINE ROSE. Easily recognized by its beautiful bright carmine flowers and its usually spineless branches. Rocks, shady places in mountains. Pyrenees, France.

AGRIMONIA | 436. *A. eupatoria* L. AGRIMONY. Bushy places. Widespread.

SANGUISORBA | Burnet

1. All fls. bisexual; stamens 4.   *officinalis*
1. Upper fls. female; stamens numerous.
  2. Plant sticky, glandular. (W. Iberian peninsula)   438a. *S. hybrida* (L.) Nordborg
  2. Plant not sticky, without glandular hairs.
    3. Rhizome clothed with the sheaths of old lvs.   *ancistroides*
    3. Rhizomes not clothed with the sheaths of old lvs.
      (Widespread)   439. *S. minor* Scop.

**438. *S. officinalis* L. GREAT BURNET. Distinguished by its large dark red oval heads of flowers borne on long leafless stems; and its pinnate leaves which are glaucous beneath. Meadows in mountains. C. and N. Spain, Pyrenees, France.   438b. *S. ancistroides* (Desf.) Cesati A low, rather delicate, cushion-like plant with a thick woody stock and basal leaves with 3–10 pairs of leaflets each ½–1 cm. Flowering stems slender, usually leafless; flower heads often solitary, about 1 cm across. Limestone cliffs. S. Iberian peninsula.

DRYAS | **441.** *D. octopetala* L. MOUNTAIN AVENS. Its solitary, usually eight-petalled white flowers, its attractive plume-like fruits, and its delicately scalloped leaves make this creeping mountain shrub unmistakable. N. Spain, Pyrenees, Massif Central. Page 152.

GEUM | Avens
1. Style persisting in its entirety, not hooked.
  2. Upper part of style covered with stiff, deflexed bristles, lower
    part hairless.   *heterocarpum*
  2. Whole style covered with long, ascending, soft hairs.   *montanum*
1. Upper part of style soon falling off and leaving behind a
  hooked lower part.
    3. Receptacle raised on a distinct stalk ½–1 cm.
      4. Petals narrowed to a stalk (clawed), pink to cream.   *rivale*
      4. Petals not or scarcely clawed, yellow.   *sylvaticum*
    3. Receptacle not stalked.
      5. Petals usually not more than 8 mm.
        6. Stipules not more than 1 cm. (N–E. Spain)   442b. *G. hispidum* Fries
        6. Stipules more than 1 cm. (Widespread)   442. *G. urbanum* L.
      5. Petals usually more than 8 mm.   *pyrenaicum*

443. *G. rivale* L. WATER AVENS. Marshes and damp pastures in mountains. S., C., and E. Spain, Pyrenees, France. Page 125.

444. *G. montanum* L. ALPINE AVENS. Flowers bright yellow, 3 cm across, solitary. Terminal leaf lobe much larger than unequal lateral lobes; stolons absent. Pastures and stony places in mountains. N. and N–E. Spain, Pyrenees, Massif Central.   444a. *G. pyrenaicum* Miller. Very like 444 but style jointed and upper part soon falling. Flowers bright yellow; petals rounded, 10–14 mm. Basal leaves with 4–6 pairs of small unequal leaflets and a very large rounded heart-shaped and lobed terminal leaflet to about 10 cm across. Rocky pastures. Pyrenees, Corbières.

444b. *G. heterocarpum* Boiss. An erect, branched, hairy perennial to ½ m, with a cluster of pale yellowish, somewhat bell-shaped flowers, each about 1 cm across, with petals about as long as the calyx. Carpels spreading, 5–15. Shady places in mountains. S. and E. Spain. Page 76.   442a. *G. sylvaticum* Pourret Like 442, *G. urbanum* but stem leaves small, simple (not lobed as in 442), and yellow flowers erect, larger, to about 2 cm, with spreading, rounded petals. Carpels 15–30; stalk of receptacle ½ cm. Damp shady places, woods, hills. Widespread. Pl. 15, Page 117.

POTENTILLA | **Cinquefoil**   A difficult genus with about 30 species in our area; hybridization is common. Distinguished from *Ranunculus*, with which it is sometimes confused, by the presence of an additional row of lobes, or epicalyx, which alternates with the true calyx.
Widespread lowland and montane species are: 445. *P. sterilis* (L.) Garcke BARREN STRAWBERRY; 449. *P. erecta* (L.) Räuschel COMMON TORMENTIL; 450. *P. reptans* L. CREEPING CINQUEFOIL; 456. *P. anserina* L. SILVERWEED.

Flowers white or reddish

(a) *Leaves trifoliate*

445a. *P. montana* Brot. A spreading, lax, tufted hairy perennial with pale green, trifoliate leaves and rather larger white flowers 1½–2 cm across, with notched petals. Flowers 1–4, borne on slender stems; petals longer than sepals. Leaflets narrowly obovate, silvery-haired beneath; creeping stems hairy, brown to blackish. Sunny rocks, field verges, lowlands and montane. N. Iberian peninsula, France. Pl. 15.

(b) *Leaves with 5–7 leaflets*

(447) *P. caulescens* L. SHRUBBY WHITE CINQUEFOIL. A very variable plant 5–30 cm with a woody stock, leaves with 5–7 oblong leaflets, and a lax terminal cluster usually of many white flowers. Petals 6–10 mm, spreading, little longer than sepals, not or indistinctly notched; epicalyx as long as or slightly longer than sepals. Rock crevices, predominantly on limestone mountains. S. and E. Spain, Massif Central.   447a. *P. petrophila* Boiss. Like (447) but a densely tufted, silvery-haired perennial with short flowering stems to 10 cm, bearing a dense cluster of white flowers, with petals 5–7 mm. Epicalyx segments distinctly shorter than sepals. Mountains of S. Spain.   447b. *P. nivalis* Lapeyr. Like some dwarf cushion forms of (447) but distinguished by its somewhat bell-shaped flowers which have white ascending (not spreading) petals, and petals shorter than the narrower epicalyx and as long as the calyx lobes. Leaflets silvery, usually 7. Rocks in the mountains. N. and E. Spain, Pyrenees.

447c. *P. alchimilloides* Lapeyr. Distinguished by its silky-silvery stems, inflorescence, and undersides and margin of leaflets. Flowering stems 10–30 cm; flowers usually numerous, white; petals spreading, notched, 8–10 mm, longer than the sepals and linear-lance-shaped epicalyx. Leaflets 5–7, narrow elliptical, with 3 small teeth at apex, hairless above. Mountain rocks and screes. Pyrenees.

**446. *P. palustris* (L.) Scop. MARSH CINQUEFOIL. Flowers reddish-purple; petals shorter than calyx. Marshes and bogs. Spain (rare), Pyrenees, France.

**448. *P. rupestris* L. ROCK CINQUEFOIL. A robust erect perennial to 60 cm, distinguished by its rather large white flowers 1–2½ cm across, borne in a branched cluster. Petals longer than sepals; epicalyx shorter than sepals. Leaves pinnate with 5–7 oval, toothed or double-toothed leaflets. Rock fissures in lowlands and mountains. N. Portugal, Spain, France.

Flowers yellow

(a) *Leaves, at least some, pinnate*

**457. *P. fruticosa* L. SHRUBBY CINQUEFOIL. The only European species with woody branches, sometimes creeping, but often forming an erect bush to 1 m. Flowers golden-yellow, 2–2½ cm. Leaves pinnate with 5 entire elliptic leaflets, woolly-haired beneath. Alpine pastures and damp rocks. Pyrenees.

456a. *P. pensylvanica* L. Distinguished from all other species by its pinnate leaves with 5–18 oblong leaflets, which are deeply toothed or lobed, and grey-woolly beneath. Flowers in a dense cluster; petals golden-yellow, notched, at least 1½ times as long as sepals. A greyish softly-hairy perennial to 60 cm, with stout stock. Grassy places and mountains. S. and C. Spain.

(b) *Leaves trifoliate*

453c. *P. grandiflora* L. A robust erect plant to 40 cm, with trifoliate basal leaves with broadly oval, sharply toothed leaflets, and lax branched, terminal clusters of large yellow flowers, each to 3 cm across. Petals 1–1½ cm, about twice as long as sepals. Rocky pastures in mountains. C. and E. Pyrenees.

(c) *Leaves mostly digitate (with 5–7 leaflets arising from the apex of the leaf stalk)*
452. *P. argentea* group Distinguished by its digitate leaves with 5 toothed leaflets, which are green or grey above and densely white-woolly beneath. Flowers rather small, 1–1½ cm across, in spreading clusters, on woolly-haired stems; petals as long as the woolly-haired sepals. Grassy, sandy and rocky places in mountains. C. and E. Spain. 453. *P. recta* L. SULPHUR CINQUEFOIL. A robust perennial to 70 cm, distinguished by its digitate leaves with 5–7 large oblong, deeply toothed or lobed leaflets, each 1½–10 cm long. Flowers numerous, in a lax cluster; petals as long as or longer than sepals. C. Spain, France. 453a. *P. hirta* L. Like 453 but stem and leaves with long spreading white hairs and without glandular hairs; leaflets linear to oblong, with 3–7 blunt teeth or lobes at the apex. Flowers large, about 2½ cm; petals longer than sepals. Sunny rocks. S., C., and E. Spain, Med. France. 453b. *P. pyrenaica* DC. A robust, erect perennial recalling 453 but basal leaves with 5 oblong to narrowly obovate leaflets with acute teeth. Flowers 2–2½ cm across in a lax branched cluster; petals 1–1½ cm. 1½ to twice as long as sepals. Rocky pastures in mountains. C. and N. Spain, Pyrenees.

FRAGARIA | 459. *F. vesca* L. WILD STRAWBERRY. Woods, hedges, and scrub. Widespread. (459) *F. moschata* Duchesne HAUTBOIS STRAWBERRY. Like 459 but creeping runners absent or few, and flowering stem longer than the leaves. Flowers white, about 2 cm across. Shady, grassy hills. N. Spain, France. (459) *F. viridis* Duchesne Differs from 459 in having short, very slender runners. Flowers creamy-white; sepals adpressed after flowering. Fruit about 1 cm; receptacle without carpels near its base. Woods and banks. E. Spain, France.

MESPILUS | 460. *M. germanica* L. MEDLAR. Sometimes cultivated in our area and naturalized in hedges and bushy places.

CRATAEGUS | **Hawthorn** 461. *C. monogyna* Jacq. HAWTHORN. Hedges, thickets. Widespread. (461) *C. laevigata* (Poiret) DC. Less common, hedges and thickets. Pyrenees. 462a. *C. laciniata* Ucria Distinguished by its young twigs, leaves and flower stalks which are densely woolly-haired, and its leaves which are deeply cut into 3–7 narrow, oblong-acute, toothed lobes. Flowers 1½–2 cm across in dense flat-topped clusters; styles 3–5. Fruit erect, red to orange-red. Mountains. S. Spain (S. Nevada, S. de Velez Blanco). **462. *C. azarolus* L. AZAROLE. Like 462a with downy twigs and young leaves, but leaves usually with 3 bluntish entire lobes; styles 1–2. Cultivated and sometimes naturalized.

CYDONIA | **463. *C. oblonga* Miller QUINCE. Naturalized in bushy places, hedges; widely cultivated in S. and C. Iberian peninsula.

PYRUS | **Pear** The leaves described below are the middle leaves of the short shoots.

1. Fr. with calyx which soon falls.   *cordata*
1. Fr. with persistent calyx.
 2. Fr. large, 6–17 cm, fleshy, sweet-tasting. (Widely cultivated) 464. *P. communis* L.
 2. Fr. small, not more than 5½ cm, hard, not sweet.
  3. Lvs. not more than 1½ times as long as wide. (Widespread) (464) *P. pyraster* Burgsd.
  3. Lvs. more than 1½ times as long as wide.
   4. Mature lvs. hairless or with small swellings beneath.   *bourgaeana,*
                                                                    amygdaliformis
   4. Mature lvs. downy or hairy beneath. (C. France)   464d. *P. nivalis* Jacq.

464a. *P. cordata* Desv. A thorny shrub or small tree with purplish twigs. Leaves 2½–5½ cm, oval-lance-shaped, densely woolly when young, like the young flower clusters. Fruit globular, 8–15 mm, red and densely covered with lenticels. Woods and hedges. C. and N. Portugal, N. Spain.   464b. *P. amygdaliformis* Vill. A shrub or small tree which is sometimes spiny and distinguished by its dull grey twigs which are woolly when young. Leaves narrowly lance-shaped to obovate, usually entire. Fruit 1½–3 cm, usually globular, tawny, borne on a stout stalk about as long as the fruit. Rocky places, banks, waysides. Med. region.   464c. *P. bourgaeana* Decne A tree with spiny lower branches, grey stoutish twigs and heart-shaped leaves with rounded teeth. Fruit 17–25 mm, dull yellow with brown spots, to almost brown when mature. By seasonal streams. Portugal, W. Spain.

MALUS | (465) *M. sylvestris* Miller CRAB APPLE. Shady places in mountains. Widespread. 465. *M. domestica* Borkh. APPLE. Widely cultivated in our area, except in south.

SORBUS | The leaves of short shoots are referred to in the key below.

1. Lvs. pinnate.
  2. Bark shedding: styles 5; ripe fr. greenish or
    brownish. (S. and E. Spain, France)   **(466) *S. domestica* L.
  2. Bark smooth; styles 3–4; ripe fr. scarlet.
    (Widespread)   **466. *S. aucuparia* L.
1. Lvs. entire or lobed, sometimes deeply so.
  3. Lvs. green at maturity, similar on both surfaces.
    4. Trees; lvs. lobed; fr. brown. (Local)   469. *S. torminalis* (L.) Crantz
    4. Shrubs; lvs. not lobed; fr. scarlet. (Pyrenees,
      France)   **468. *S. chamaemespilus* (L.) Crantz
  3. Lvs. white or grey woolly-haired beneath at
    maturity.
    5. Lvs. not or only very shallowly lobed. (Wide-
      spread)   **467. *S. aria* (L.) Crantz
    5. Lvs. distinctly lobed.   *mougeotii, latifolia*

467a. *S. latifolia* (Lam.) Pers. A tree distinguished by its conspicuously lobed leaves, which are grey-green with woolly hairs beneath and hairless above; lobes triangular, once or twice toothed, the teeth terminating the main veins straight. Flowers few, about 2 cm across. Fruit nearly globular, 12–15 mm, yellowish-brown with large lenticels. C. and E. Portugal, Spain, France.   467b. *S. mougeotii* Soyer-Willemet & Godron A shrub of the Pyrenees with leaves shallowly lobed to about a quarter the width to the midvein, and with whitish-grey woolly hairs on the undersides. Flowers about 1 cm across. Fruit about 1 cm, slightly longer than wide, red with few small lenticels. Rocky slopes.

ERIOBOTRYA | **470. *E. japonica* (Thunb.) Lindley LOQUAT. An unmistakable small tree with its large, dark green, strongly veined leaves, rusty-haired beneath, and with orange edible plum-like fruits. Native of China: widely cultivated in the Med. region, and S. and C. Portugal.

AMELANCHIER | **471. *A. ovalis* Medicus SNOWY MESPILUS. Easily recognized when in flower by its narrow, widely separated, white petals, and its 'snowy' young foliage covered with dense white hairs which later disappear. Flowers in clusters; petals 1–1½ cm. Fruit bluish-black, sweet. Open woods, rocks, in mountains. Southern part of Iberian peninsula, France. Page 102.

COTONEASTER | **473. *C. integerrimus* Medicus A branched shrub, distinguished by its neat oval leaves 2–5 cm, which are densely grey-woolly beneath and green and hairless above, and its short axillary clusters of 2–4 small drooping pinkish flowers. Calyx and flower stalks hairless. Fruit shining red. Open rocky places in mountains. C. and E. Spain, Pyrenees, Massif Central. **474. *C. nebrodensis* (Guss.) C. Koch Like 473 but calyx and inflorescence densely hairy. Flower clusters with 3–12 flowers. Leaves 3–6 cm, at first white, hairy, but later hairless above, densely white or grey-woolly beneath. Rocky slopes. E. Spain, Pyrenees, Massif Central. 474a. *C. granatensis* Boiss. An endemic shrub or small tree of S. Spain (S. Nevada), distinguished by its lax, many-flowered inflorescence of white flowers with spreading rounded petals. Calyx somewhat hairy. Leaves elliptic or orbicular 2–5 cm, hairless above, paler and with scattered hairs beneath. Fruit pear-shaped, red, 6–9 mm. Rocks.

PYRACANTHA | 475. *P. coccinea* M. J. Roemer FIRE THORN. Bushy places, hedges. Med. region; introduced Portugal.

PRUNUS | The following fruit trees are widely cultivated and often occur naturalized in our area: 479. *P. dulcis* (Miller) D. A. Webb ALMOND; 478. *P. persica* (L.) Batsch PEACH; (478) *P. armeniaca* L. APRICOT; 477. *P. domestica* L. PLUM, BULLACE, DAMSON, GREENGAGE; **480. *P. avium* (L.) L. CHERRY. The remaining *Prunus* species can be identified as follows:

1. Fls. in elongated spike-like clusters.
   2. Lvs. evergreen; fr. ovoid-conical.   *lusitanica, laurocerasus*
   2. Lvs. deciduous; fr. globular.   *padus*
1. Fls. solitary, or in rounded clusters or umbels.
   3. Fr. stalk longer than ripe fr.
      4. Fls. in a stalkless umbel, with papery bud-scales
         at base. (Naturalized)   481. *P. cerasus* L.
      4. Fls. in a short more or less elongate cluster,
         without papery scales at base.   *mahaleb*
   3. Fr. stalk usually shorter than ripe fr.
      5. Petals pink.   *prostrata*
      5. Petals white.
      6. Bark blackish; lvs. dull; fls. mostly solitary,
         appearing before the lvs. Fr. 1–1½ cm. (Wide-
         spread except in south)   **476. *P. spinosa* L.
      6. Bark silvery-grey; lvs. glossy above; fls. mostly
         in clusters of 2–3, appearing with lvs. Fr. 6–8
         mm. (S. Nevada, S. de Gador)   476a. *P. ramburii* Boiss.

476b. *P. prostrata* Labill. A low twisted, tough, woody spreading shrub with bright pink to pale pink solitary, stalkless flowers appearing before the leaves. Leaves oblong to obovate, 9–12 mm, finely toothed, grey-woolly beneath. Fruit red, about 8 mm. Limestone mountains. S. Spain. Pl. 15, Page 76.

**484. *P. mahaleb* L. ST LUCIE'S CHERRY. Usually a deciduous shrub with short erect clusters of 3–10 white flowers in the axil of each leaf. Fruit small, about ½ cm, red, then black. Woods and dry rocky places in hills. Scattered throughout except in the south. **482. *P. padus* L. BIRD-CHERRY. A handsome small tree with long pendulous or inclined, spike-like clusters of many small white flowers. Leaves deciduous, dull, long-pointed, toothed. Fruit black, astringent. Shady thickets in mountains. N. Portugal, C., N. and E. Spain.

**(483) *P. lusitanica* L. PORTUGAL LAUREL. An evergreen shrub or tree with dark green shining leathery leaves, and erect spike-like clusters of numerous white flowers, which are considerably longer than the leaves. Twigs and leaf stalks dark red. Fruit blackish-purple. Mountains. Iberian peninsula, S–W. France; sometimes grown for ornament.   **483.
P. laurocerasus L. CHERRY-LAUREL. Distinguished from (483) by its pale green twigs and leaf stalks, and by its flower spikes which are shorter or little longer than the leaves. Native of Eastern Europe; sometimes grown for ornament and locally naturalized.

LEGUMINOSAE | Pea Family

A very important family in S–W. Europe with many species occurring in Europe only in the Iberian peninsula. In large genera such as *Cytisus, Genista, Astragalus, Ononis, Anthyllis* etc. it has only been possible to describe the most attractive and widespread species. In other large genera with small flowers, such as *Trifolium, Vicia, Melilotus, Medicago*, only a few species have been described. There are 55 genera of the *Leguminosae* in our area.

CERCIS | **485. *C. siliquastrum* L. JUDAS TREE. Native of S–E. Europe, often grown for ornament in S. Portugal and Med. region.

CERATONIA | **486. *C. siliquae* L. CAROB. A robust but low-growing evergreen tree with dark green leaves with 4–10 pairs of oval leathery leaflets, and numerous broad, strap-shaped pods arising from the old branches. Flower clusters green. Native of the Med. region; widely cultivated and naturalized on the littoral of S. and C. Portugal, and Spain.

GLEDITSIA | 487. *G. triacanthos* L. HONEY LOCUST. Widely grown in parks and gardens in S–W. Europe; sometimes naturalized.

ACACIA | **Wattle, Mimosa**   Most of the species brought into cultivation into Western Europe have originated from Australia. Because of their drought-resistant character, they are largely planted for ornament and shade in the south. 488, 489, 490 are valuable for stabilizing dunes and eroded slopes; 492, 490 are sometimes grown for timber, and 490a, 492b for tanning. The S. African species 491a is a useful hedge and boundary plant in arid regions.

1. Lvs. all twice-pinnate.
  2. Lvs. deciduous, with 2–8 pairs of primary divisions; spines present.
    3. Leaflets 3–5 mm: spines 2½ cm on older branches.   491.   *A. farnesiana* (L.) Willd.
    3. Leaflets 6–10 mm; spines very stout, up to 10 cm on older branches.   491a. *A. karoo* Hayne Pl. 16
  2. Lvs. evergreen, with 8–20 pairs of primary divisions; spines absent.
    4. Twigs and young lvs. whitish-tomentose; leaflets 3–4 mm; fr. 10–12 mm wide, not or scarcely constricted between seeds.   490.   *A. dealbata* Link
    4. Twigs and young lvs. yellowish-villous; leaflets 2 mm; fr. 5–7 mm wide, distinctly constricted between seeds.   490a. *A. mearnsii* De Willd.

1. Adult lvs. reduced to flattened blades, occasionally mixed with some twice-pinnate lvs.
5. Fls. in axillary spikes; fr. cylindrical.   **488.   A. *longifolia* (Andrews) Willd.
5. Fls. in rounded heads, usually arranged in
   elongate clusters; fr. compressed.
   6. Leaf-blades with 2–6 longitudinal veins; fr.
      twisted, or curled in ring.
      7. Procumbent shrubs; leaf-blades 3–8 cm; fls.
         bright yellow.   492a. A. *cyclops* G. Don fil.
      7. Trees; leaf-blades 6–13 cm. fls. creamy-
         white.   492.   A. *melanoxylon* R.Br.
   6. Leaf-blades with a single longitudinal vein;
      fr. almost straight.
      8. Leaf-blades strongly sickle-shaped; 10–20
         fl. heads in each cluster.   492b. A. *pycnantha* Bentham
      8. Leaf-blades not or scarcely sickle-shaped;
         2–10 fl. heads in each cluster.
         9. Fl. heads 10–15 mm in diameter; fr. distinct-
            ly constricted between seeds.   489.   A. *cyanophylla* Lindley
         9. Fl. heads 4–6 mm in diameter; fr. not or
            scarcely constricted between seeds.   489a. A. *retinodes* Schlecht. Pl. 16

SOPHORA | 494. *S. japonica* L. PAGODA-TREE. Grown for ornament in gardens and parks;
sometimes naturalized.

ANAGYRIS | 495. *A. foetida* L. BEAN TREFOIL. A fetid deciduous shrub with trifoliate
leaves, green twigs and clusters of yellowish laburnum-like flowers borne on last year's
branches. Standard dark-blotched at base, about half as long as wings. Hedges, bushy
places. S. Portugal, Med. region. Page 44.

GORSES, BROOMS and their allies with yellow or white flowers, including: *CALICO-
TOME, CYTISUS, CHAMAECYTISUS, CHRONANTHUS, TELINE, GENISTA,
CHAMAESPARTIUM, ECHINOSPARTUM, LYGOS, SPARTIUM, ULEX, STAURA-
CANTHUS, ADENOCARPUS, LOTONONIS, ARGYROLOBIUM*. About 73 species
belonging to these genera occur in our area. They are usually switch-like shrubs, often
spiny and with small leaves which soon fall, and are often almost undistinguishable from
each other when not flowering or fruiting. Many species are common and locally dominant
in the Iberian peninsula.

GROUP A. *Plants spiny. Lvs. of adult plants all reduced to spines (except Ulex densus* Webb)
1. Lvs. and branches mostly alternate; fr. scarcely longer than
   calyx.   *ULEX*
1. Lvs. and branches mostly opposite; fr. conspicuously longer
   than calyx.   *STAURACANTHUS*

GROUP B. *Plants spiny. Lvs. not reduced to spines, though often falling early; stems spiny.*
1. Calyx with 5 short teeth, the upper portion breaking away at
   flowering time to leave a cup-like remnant.   *CALICOTOME*
1. Calyx two-lipped, the upper part not breaking away.
   2. Calyx not more than 7 mm long.   *GENISTA*
   2. Calyx 7 mm or more long.

498 *Calicotome villosa*

502 *Cytisus purgans* T.S. Stem

504 *Cytisus sessilifolius*

507 *Chamaecytisus supinus*

507a *Chronanthus biflorus*

507b *Teline monspessulana*

511e *Genista pumila*

513a *Chamaespartium tridentatum*

513d *Echinospartum lusitanicum*

514 *Lygos monosperma*

515 *Spartium junceum*

(517) *Ulex parviflorus*

517a *Stauracanthus boivinii*

518b *Adenocarpus decorticans*

518c *Lotononis lupinifolia*

522 *Argyrolobium zanonii*

3. Lvs. and branches mostly alternate; calyx not inflated.   *GENISTA*
3. Lvs. and branches mostly opposite; calyx somewhat inflated.   *ECHINOSPARTUM*

GROUP C. *Plants not spiny.*
1. Young stems broadly winged.   *CHAMAESPARTIUM*
1. Young stems not broadly winged.
  2. Fr. with prominent glandular swellings.   *ADENOCARPUS*
  2. Fr. without glandular swellings (sometimes with glandular hairs).
    3. Lvs. simple or one-bladed, sometimes very small.
      4. Calyx split to the base.   *SPARTIUM*
      4. Calyx not split to the base.
        5. Upper lip of calyx with short teeth.   *CYTISUS*
        5. Upper lip of calyx two-lobed or deeply toothed.
          6. Fr. not inflated.   *GENISTA*
          6. Fr. inflated.   *LYGOS*
    3. At least some lvs. with 3 leaflets, or 5 leaflets.
      7. Lvs. with 5 leaflets.   *LOTONONIS*
      7. At least some lvs. with 3 leaflets.
        8. Calyx-tube distinctly shorter than lips.   *ARGYROLOBIUM*
        8. Calyx-tube as long as or longer than lips.
          9. Upper lip of calyx deeply two-lobed.
            10. Fl. stalks 5–10 mm; fr. hairless.   *CYTISUS*
            10. Fl. stalks 1–3 mm; fr. hairy.
              11. Fls. in umbel-like heads.   *GENISTA*
              11. Fls. axillary, or in axillary clusters.
                12. Standard distinctly shorter than keel.   *GENISTA*
                12. Standard longer than keel.   *TELINE*
          9. Upper lip of calyx with 2 short teeth.
            13. Calyx tubular.   *CHAMAECYTISUS*
            13. Calyx bell-shaped.
              14. Fls. axillary, arranged in leafy elongated clusters.   *CYTISUS*
              14. Fls. in leafless, terminal heads, or in leafless, elongated clusters.
                15. Fls. usually in heads of 2–4 fls.   *CHRONANTHUS*
                15. Fls. in elongated, many-flowered clusters.   *CYTISUS*

CALICOTOME | Spiny Broom   Distinguished by its calyx in which the terminal part breaks off.   (498) *C. spinosa* (L.) Link A very spiny, much-branched shrub with stout lateral spines, small trifoliate leaves, and usually solitary, yellow, gorse-like flowers. Leaflets silvery-haired beneath. Fruit nearly hairless. Bushy places, scrub. Med. region. Pl. 17. **498. *C. villosa* (Poiret) Link Very like (498) but fruit densely covered with silvery hairs. Lower surface of the leaflets, young twigs and calyx silvery-haired. Flowers in clusters of 2–15. Bushy places, scrub. S. Portugal, Med. region. Page 161, 222.

CYTISUS | Broom   Not easy to distinguish in the field, particularly from some species of *Chamaecytisus* and *Genista*. Shrubs without spines. Leaves with 1 or 3 leaflets, alternate, sometimes crowded on old branches. Flowers axillary, forming leafy terminal racemes. Calyx two-lipped, the upper lip with 2 teeth, the lower lip with 3 teeth, the teeth usually much shorter than the lips. Corolla yellow to white; stamens with all filaments fused in a tube; stigma club-shaped or turned towards the axis (*introrse*). Fruit linear or oblong,

223

splitting. Seeds usually numerous, usually with an external swelling. The 'true' brooms (section *Sarothamnus*) possess styles that are coiled after pollination and comprise a group of 8 closely related species restricted largely to the south-west. 15 species in our area, including 9 species restricted to it.

1. Fls. in leafless elongate clusters.	*sessilifolius*
1. Fls. in leafy elongate clusters.	
2. Twigs five- to ten-angled, the angles wing-like and more or less T-shaped in transverse section.	
3. Fls. white.	*multiflorus*
3. Fls. yellow.	*purgans*
2. Twigs rounded or angled but angles not T-shaped in transverse section.	
4. Style coiled in a ring after the fls. are fully open.	
5. Lvs. with 1 leaflet, except sometimes on new growth.	
6. Branches incurved to form a rounded bush; calyx hairless; fr. hairless. (S–E. Spain)	505f. *C. reverchonii* (Degen & Hervier) Bean
6. Branches erect, not incurved; calyx silvery-haired; fr. hairy. (N. Spain).	505c. *C. commutatus* (Willk.) Briq.
5. Lvs. mostly trifoliate, except on new growth.	
7. All lvs. stalked; keel straight on upper side, not beaked; style hairless.	
8. Branches and twigs more or less rounded, slightly lined; fr. sparsely hairy or nearly hairless. (South of our area)	505d. *C. malacitanus* Boiss.
8. Branches and twigs seven- to eight-angled; fr. densely woolly- or shaggy-haired. (S. Portugal, S–W. Spain)	505e. *C. baeticus* (Webb) Steudel
7. All lvs., or at least the single leaflets, stalkless; keel strongly curved on upper side, more or less beaked; style with spreading hairs below.	
9. Fr. more or less inflated, the valves densely hairy; twigs usually eight- to ten-angled.	*striatus*
9. Fr. strongly compressed, the valves hairless or with adpressed hairs; twigs usually five-angled.	
10. Lvs. all stalkless, those of young twigs with 1 leaflet.	*grandiflorus*
10. Single leaflet stalkless; trifoliate lvs. stalked.	
11. Twigs five- to eight-angled, the angles somewhat rounded; fr. densely hairy. (N. Spain, S–W. France)	505g. *C. cantabricus* (Willk.) Reichenb. fil.
11. Twigs almost always five-angled, the angles flattened and ridge-like, sometimes slightly winged laterally; fr. hairless.	*scoparius*
4. Style straight or curved after the fls. are fully open.	
12. Lvs. mostly with 1 leaflet.	

13. Leaflets 1–1½ cm; fls. greenish-yellow. (N. Spain)   505c. *C. commutatus* (Willk.) Briq.

13. Leaflets 2–3 cm; standard cream, wing and keel yellow.   *ingramii*

12. Lvs. mostly with 3 leaflets.

14. Lvs. stalkless or very shortly stalked; bracteoles egg-shaped. (S–W. Spain)   505i. *C. tribracteolatus* Webb

14. Lvs. distinctly stalked; bracteoles narrowly elliptical or linear, often falling early.

15. Leaflets obovate, sparsely hairy beneath; fr. 12–15 mm, hairless. (E. Spain)   505h. *C. patens* L.

15. Leaflets oblong or elliptical, villous beneath; fr. 2–4½ cm, hairy or villous, becoming hairless.   *villosus*

**504. *C. sessilifolius* L. Easily recognized by its stalkless trio of broadly rhombidal leaflets ranged along the flowering stems, and its terminal leafless clusters of few, clear yellow flowers. Standard rounded, about 1 cm. A quite hairless, leafy, branched shrub to 2 m; lower leaves stalked. Scrub and mountain woods. E. Spain, S. France. Page 153, 222.

501. *C. villosus* Pourret An erect, very leafy, hairy shrub to 2 m with yellow flowers arising singly, or in groups of 2–3, and forming a leafy terminal spike-like cluster. Flowers rather large 1½–2 cm, veined with red at the base; flower stalks hairy. Leaflets elliptic 1½–3 cm, the central leaflet longer, all with long silky hairs beneath. Woods and scrub, preferring acid soils. S., C.,and E. Spain, Med. France.

502. *C. purgans* (L.) Boiss. A much-branched shrub 30–100 cm; distinguished by its very numerous stiff, erect, densely packed branches, which are conspicuously ribbed and usually leafless. Flowers deep yellow, smelling of vanilla, small about 1 cm, 1 or 2 at the ends of the stiff branches. Ribs of branches 5–10, obscurely T-shaped in section. Fruit black when ripe, with adpressed silky hairs. Widespread and often occurring in extensive masses throughout our area. Page. 110, 222.   503. *C. multiflorus* (L'Hér.) Sweet WHITE SPANISH BROOM. Like 502 with numerous erect, ribbed branches but taller to 3 m, and with white flowers about 1 cm long. Leaves to 1 cm, silvery when young, with 1 or 3 leaflets; young twigs and calyx silvery-haired. C. and N. Portugal, N–W. and C. Spain. Pl. 16.   503a. *C. ingramii* Blakelock The only bi-coloured species in our region, with a cream-coloured standard and yellow wings and keel. Open treeless country between Corunna and Oviedo, N–W. Spain. Page 103.

**505. *C. scoparius* (L.) Link BROOM. Widespread and locally abundant throughout our area.   505a. *C. grandiflorus* DC. (*Sarothamnus virgatus*). S. and C. Portugal, S. Spain. 505b. *C. striatus* (Hill) Rothm. (*Sarothamnus patens*). Portugal, W. and C. Spain. Page 227. The most important distinguishing characters are:

Species	Leaves	Calyx	Pod
scoparius	Trifoliate leaves stalked.	Hairless.	Compressed, hairless except on the margins.
grandiflorus	All leaves stalkless.	Hairless.	Compressed, densely white hairy.
striatus	Trifoliate leaves stalked.	With adpressed hairs.	Inflated, densely white-hairy.

225

All three species are erect shrubs to 3 m, forming thickets in scrub, hedges, and woods, and are spectacular in late spring and early summer with masses of showy golden-yellow flowers, each up to 2½ cm. On the granite mountains of W. Spain and N. Portugal, they may be found growing together and are easily confused. In this area *C. striatus* is usually easily recognized by its much narrower leaflets, but elsewhere in its range the leaflets are variable in shape.

CHAMAECYTISUS | Not easily distinguished from *Cytisus* and *Genista* but in our species the flowers are in terminal heads and are subtended by leafy involucres. Calyx tubular, two-lipped, the upper lip with 2 teeth, the lower with 3 teeth, the teeth much smaller than the lips (upper lip deeply divided in *Genista*).   **507. *C. supinus* (L.) Link CLUSTERED BROOM. A woolly-haired shrub with trifoliate leaves with elliptic, hairy leaflets and a dense terminal cluster of 2–8 yellow flowers blotched with brown. Calyx and fruit with spreading, woolly hairs; fruit becoming black. Thickets, banks. S–E. Spain, Pyrenees, Massif Central. Page 222.

CHRONANTHUS | Distinguished by its flattened fruit enclosed by the persistent corolla and containing 1–3 seeds. Calyx bell-shaped, two-lipped, the upper lip with 2 teeth not divided to the base, lower with 3 teeth.   507a. *C. biflorus* (Desf.) Frodin & Heywood An erect, densely branched shrub to ½ m, with angled stems, trifoliate leaves with narrow hairy leaflets, and terminal clusters usually of 2–4 conspicuous yellow flowers. Standard heart-shaped, hairless; keel nearly as long, with a rounded beak. Leaflets 4–9 mm; flowering stems and leaves with spreading hairs. Fruit 1–1½ cm, valves translucent, hairless. Woods and scrub in lowlands and hills. S. and E. Spain (common in Andalusia). Page 222, 227.

TELINE | Distinguished by its narrow, flattened fruits which do not split, containing 2–6 seeds, and each seed with a swelling. Upper lip of calyx deeply two-lipped, lower finely three-toothed.   507b. *T. monspessulana* (L.) C. Koch A much-branched, erect, leafy shrub 1–3 m, with numerous trifoliate leaves with obovate leaflets, and axillary cluster of rather small yellow flowers. Recalling 501 *Cytisus villosus* but differing in having pale silvery calyx and darker yellow flowers. Standard 10–12 mm, hairless like the wings; keel sparsely silvery-haired. Young branches hairy. Fruit about 2 cm, white-woolly. Woods, thickets, river margins. S. Portugal and Med. region. Pl. 18. Page 222.   507c. *T. linifolia* (L.) Webb & Berth. Readily distinguished from 507b by its stalkless leaves with long, narrow leaflets to 2cm, which are conspicuously silky-haired beneath and contrasting darker silky-haired above, and its silvery hairy stems and calyx. Flowers yellow, about 2 cm, in leafy clusters; standard silky-haired. Woods and scrub. Med. region. Page 227.

GENISTA | Distinguished by the calyx which is two-lipped with the upper lip deeply two-lobed and the lower lip three-toothed. Flowers yellow. Members of the genus are not always readily identified in the field because they show a considerable range of form. About 33 species occur in our area, 22 of which occur only in the Iberian peninsula in Europe. See *Genisteas Españolas* I and II, Vicioso, C. Inst. Forestal Madrid, Bol. No. 67, 72 1953–55.

Spineless shrubs with simple leaves
508. *G. pilosa* L. HAIRY GREENWEED. A small, much-branched shrub with spreading often rooting stems with slender ribbed branches, simple narrowly oval leaves and small yellow axillary flowers about 1 cm. Calyx, standard and keel with dense silvery hairs. Fruit 1½–2½

1. *Cytisus striatus* 505b
3. *Echinspartum horridum* 513b
5. *Teline linifolia* 507c

2. *Lygos sphaerocarpa* 514a
4. *Chronanthus biflorus* 507a

cm, covered with dense adpressed hairs. Heaths, forests, rocky places, often in mountains. Widespread in Spain and France.   508a. *G. cinerea* (Vill.) DC. Similar to 508 but flowers mostly in pairs and borne directly on the main branches; bracts clustered. Standard hairless, or with a median ridge of hairs. Hillsides in woods and scrub. Widespread.   508b. *G. florida* L. A tall, graceful, silvery-leaved shrub to 3 m, distinguished by its relatively large oblance-shaped, shortly stalked leaves ½–1½ cm by 2–5 mm, and yellowish branches. Flowers yellow, about 1½ cm, borne singly in the axil of each bract and forming long, lax axillary clusters. Standard hairless; lowermost bracts leaf-like, upper bracts linear, not leaf-like. Bushy places and woods in lowlands and mountains. N. Portugal, Spain. Pl. 17.

508c. *G. spartioides* Spach A very sweetly honey-scented, tufted switch-like shrub to 1 m, with lateral clusters of 3–4 flowers subtended by small papery bracts. Standard 7–9 mm, rhombic, hairless or sparsely silvery-haired; calyx silvery-haired. Leaves very small, linear, 3–8 mm, widely spaced on non-flowering stems, absent on flowering stems. Fruit ovoid-acute to sickle-shaped, with 1–2 seeds. Dry hills, preferring limestone. S. Spain.

508d. *G. umbellata* (L'Hér.) Poiret Easily recognized by its numerous, densely tufted, slender erect, almost leafless stems which bear tight, globular terminal heads of pale yellow flowers with a conspicuous silvery-haired calyx. Flowers 4–16; standard oval, 8–12 mm, densely silvery-haired. Leaves ½–1½ cm, simple, thus distinguishing it from 513d. Dry stony hillsides. S. and S–E. Spain.   508e. *G. obstusiramea* Spach A spreading or erect, much-branched spineless shrub, with simple leaves, and solitary or paired flowers borne near the apex of each branch. Standard 12–14 mm, densely silky-haired; calyx 5–6 mm. Leaves 2–8 mm, stalkless, hairless above, finely hairy beneath. Mountain heaths. C. Portugal, N–W. Spain.   508f. *G. micrantha* Ortega A spreading spineless shrub with simple leaves and flowers in terminal spike-like clusters. Flowers 5–7 mm; standard hairless; keel with sparse silky hairs; calyx 2–5 mm, hairless. Bracteoles borne just below calyx, about 2 mm. Leaves about 12 mm, hairless. Heaths, thickets. N. Portugal, N. and E. Spain.

Spiny shrubs
**(511) *G. hispanica* L. SPANISH GORSE. A low, much-branched, gorse-like shrub, easily distinguished by its two kinds of branches, and its dense terminal heads of small yellow flowers. Present year's flowering branches softly hairy, spineless and with narrow-elliptic hairy leaves, contrasting with last year's stiff, dark green branches with numerous lateral, branched spines. Flowers about 8 mm; standard hairless, as long as keel; bracts minute. Open stony hills and scrub in lowlands and mountains. N. and E. Spain, France. Page 133.   **511. *G. germanica* L. Like (511) but flowers in lax clusters; standard half to two-thirds as long as keel. An erect shrub. S–W. France. Pl. 17.

511a. *G. hirsuta* Vahl A low, domed, spiny, gorse-like shrub with dense elongated and tapering clusters of yellow flowers, with conspicuous woolly-haired calyx. Flowers 12–15 mm; standard hairy or silky-haired; calyx with lower lip longer; bracts leaf-like, borne just below bracteoles which are 3–5 mm and just below calyx. Branches with long straight spines 1½–5 cm; leaves 6–15 mm, narrow-lance-shaped, hairless above, with long spreading hairs beneath and on margin. Scrub and open woods in lowlands and mountains. S. Portugal, S. and W. Spain. Pl. 17, Page 45.   511f. *G. tournefortii* Spach Like 511a but spines much-branched. Flowers smaller 8–15 mm; standard notched, silky-haired; upper lip of calyx as long as lower; bracts borne at base of flower stalk. Heaths, pine woods, hedges. Portugal, S. and C. Spain.

511b. *G. scorpius* (L.) DC. An erect, very spiny, much-branched gorse-like shrub with small orange-yellow flowers borne mostly on the spines. Flowers 7–12 mm, calyx nearly hairless, with triangular teeth shorter than the calyx tube. Leaves inconspicuous, 3–11 mm; spines stout, spreading. Fruit 1½–4 cm, constricted between seeds, hairless. Sunny, bushy hills, in lowlands and mountains. Spain, France. Pl. 17.

511c. *G. triacanthos* Brot. Like 511b, but differing in having trifoliate, hairless leaves and stiff, spreading, often three-pronged lateral spines. Flowers orange-yellow, with keel much longer than wings or standard; calyx hairless, teeth as long as or longer than calyx tube. Fruit oval, about 6 mm. Heaths and bushy hills, in lowlands and mountains. Western part of Iberian peninsula.

511d. *G. hystrix* Lange A shrub with spiny branches either spreading out to 30 cm or erect to 1½ m, with 2 or more flowers in the axil of each bract, in lax spike-like clusters 4 cm or more long. Flowers about 1 cm; flower stalks 2–5 mm with bracteoles 1 mm borne at the middle. Leaves simple 3–5 mm, with adpressed hairs beneath. N. Portugal, N. Spain   511e. *G. pumila* (Hervier) Vierh. A much-branched spiny shrub with stout rigid branches, simple leaves, and short clusters of flowers borne in the axil of each bract. Flowers 8–12 mm, standard silvery-haired; flower stalks 3–4 mm. Mountains. S., S–E., and E–C. Spain. Pl. 17. Page 222.

512. *G. anglica* L. NEEDLE FURZE. An almost hairless, small spiny shrub with simple axillary spines on young and old branches, and short clusters of small yellow flowers borne in the axils of leafy bracts. Flowers 5–8 mm; standard hairless; calyx hairless. Leaves simple, elliptic, ½–1 cm. Fruit inflated, hairless, 1½–2 cm. Bushy places in the hills. Throughout our area.

CHAMAESPARTIUM | Distinguished by the young stems which are conspicuously winged and flattened; leaves simple or absent.   **513. *C. sagittale* (L.) P. Gibbs WINGED BROOM. Unmistakable with its broadly winged and flattened green stems and branches, and its dense terminal cluster of yellow flowers. Flowers about 1 cm; calyx silvery-haired. A low spreading and mat-forming dwarf shrub. Woods, scrub, rocky places mainly in the mountains. S., N., and E. Spain, Pyrenees, France. Pl. 18, Page 138.   513a. *C. tridentatum* (L.) P. Gibbs A much-branched, thorny shrub to 70 cm with winged stems and clusters of orange-yellow flowers. Wings of stem tough, undulate and contracted at each node to form 3 lobes or teeth; leaves absent. Flowers and fruit about 1 cm; keel silvery-haired. Heaths and scrub on acid soils. W. Iberian peninsula (characteristic of the matorral of Andalusia). Pl. 18, Page 102, 222.

ECHINOSPARTUM | Distinguished by its inflated, bell-shaped, two-lipped calyx with all teeth longer than the calyx-tube. Fruit ovoid, pointed, splitting.   513b. *E. horridum* (Vahl) Rothm. A neat, tightly branched extremely spiny, greenish-white, hedgehog-like shrub to 40 cm, with pairs of terminal yellow flowers with woolly-haired calyx. Standard 12–16 mm, hairless or silvery-haired. Leaflets narrow lance-shaped, silvery-haired; young branches silvery-haired, older branches becoming hairless and forming very sharp, often curved spines. Fruit 9–14 mm. Exposed mountain slopes and rocks on limestone. Pyrenees, Massif Central. Page 227.   513c. *E. boissieri* (Spach) Rothm. Like 513b. but flowering branches terminated by a small spine; standard densely silvery-haired. Limestone mountains. S. and S–E. Spain.   513d. *E. lusitanicum* (L.) Rothm. Like 513b, but a bush up to 2 m in height and flowers in a dense terminal cluster of 3–9, with conspicuous, very woolly-haired calyx. Young branches covered with woolly hairs, later hairless and ribbed. Fruit 1½–2 cm. Acidic mountains. W. Iberian peninsula. Pl. 18, Page 111, 222.

LYGOS | Retama   Distinguished by its more or less globular, usually one-seeded fruits which do not as a rule split. Tall, slender, spineless shrubs.   **514. *L. monosperma* (L.) Heywood WHITE BROOM. A beautiful slender, nearly leafless shrub up to 3 m bearing masses of small white, sweet-scented flowers with crimson calyx, in lax pendulous clusters. Flowers about 1 cm; standard hairy; calyx hairless, soon falling. Branches silvery when young, pendent. Fruit obovoid 12–16 mm, wrinkled. Atlantic coasts, S. Portugal, S–W. Spain, Page 44, 222.   514a. *L. sphaerocarpa* (L.) Heywood Distinguished from 514 by its smaller yellow flowers borne in erect lateral clusters on ascending, hairless, finely ribbed branches that have a silvery sheen. Flowers tiny, 5–8 mm; calyx 3 mm, persistent, hairless or hairy. Fruit ovoid 7–9 mm, smooth. Dry hills, mainly on sandy soils. E. Portugal, S. and C. Spain. Pl. 18, Page 227.

SPARTIUM | Distinguished by the calyx which is deeply cleft on the upper side.   **515. *S. junceum* L. SPANISH BROOM. Unmistakable with its erect cylindrical, glaucous, rush-like, leafless stems 1–3 m, and large showy, sweet-scented, rich-yellow flowers. Flowers 2–2½ cm, in lax terminal spikes. Leaves narrow, soon falling. Fruit flattened, 5–8 cm by 7 mm. Widespread in bushy places, hedges. S. and C. Portugal, Med, region. Page 222.

ERINACEA | Distinguished by its blue-violet flowers and inflated calyx.   **(515) *E. anthyllis* Link HEDGEHOG BROOM. A tough, extremely spiny, tightly branched hummock-forming shrub bearing blue-violet flowers near the ends of the spine-tipped branches. Calyx 'baggy, all shaggy with silver fluff'; flowers 'fading through a series of azure tints so that it seems to be producing flowers of two different kinds and colours'. Exposed stony limestone mountains. S. and E. Spain, E. Pyrenees. Pl. 18, Page 76.

ULEX | Gorse, Furze   Distinguished by the yellow, deeply two-lipped calyx, with 2 small yellowish bracts at the base. Fruit scarcely longer than calyx. 7 rather similar-looking species occur, including 3 endemic in our area. See *Revision del Genero 'Ulex' en España*, Vicioso, C. Inst. Forestal, Madrid, Bol. No. 80, 1962. Widespread species are: 516. *U. europaeus* L. GORSE;   **517. *U. minor* Roth DWARF FURZE;   (517) *U. parviflorus* Pourret Pl. 19 Page 222.   (517) *U. gallii* Planchon Portugal, N–W. Spain.

STAURACANTHUS | Very like *Ulex*, but spiny leaves or *phyllodes*, (not stems) often more or less opposite. Calyx less deeply divided into 2 lips, not to the base in bud, at least. Fruit conspicuously longer than the calyx and not enclosed by it.

Bracts subtending spines, spine-tipped; standard much longer than calyx; fr. 8–12 mm. (S. Portugal, S.W Spain)   517a. *S. boivinii* (Webb) Samp. Page 222

Bracts subtending spines scale-like, not spine-tipped; standard about equalling the calyx; fr. 1½–2½ cm. (S. and W. Portugal, S–W. Spain)   517b. *S. genistoides* (Brot.) Samp. Page 232

ADENOCARPUS | Distinguished by the fruits which are covered with conspicuous glandular swellings.

1. Leaflets 1–1½ mm wide, markedly inrolled, appearing linear.   *decorticans*
1. Leaflets usually more than 3 mm wide, not or only slightly inrolled.
　2. Fl. stalk 7–15 mm; standard 15–23 mm.   *hispanicus*
　2. Fl. stalk not more than 5 mm; standard 10–16 mm.

3. Bracteoles 2–4 mm, ovate or lance-shaped; calyx 8–11 mm; the lower
teeth one third to half the total length of the lip.   *telonensis*
3. Bracteoles 1 mm or less, linear; calyx 5–8 mm, the lower teeth not more
than one third the total length of the lip.   *complicatus*

**518. *A. complicatus* (L.) Gay An erect, straggly, spineless shrub to 4 m, with small tri-
foliate leaves, terminal elongate clusters of deep yellow flowers and characteristic fruits
covered with raised glandular swellings. Standard 1–1½ cm, silvery-haired; calyx with or
without glandular swellings. Leaflets ½–2½ cm. Twigs silvery-haired or nearly hairless.
Bushy places, hills, hedges. Throughout. Page 102. (518) *A. telonensis* (Loisel.) DC. Like
518 but flowers in a terminal umbel-like cluster of 2–7. Leaflets smaller 3–8 mm, densely
clustered on stem. Bushy places. S. Portugal, Med. region. 518a. *A. hispanicus* (Lam.) DC.
Readily distinguished by its larger, broader leaflets 1½–3 cm by 3–8 mm, which are either
silvery-haired on both sides or sparsely hairy above. Flowers larger, in conspicuous
terminal clusters; standard 15–23 mm, calyx densely hairy, with or without glandular
swellings. Bushy places in mountains. S. Portugal, S–W. and C. Spain. Page 232.

518b. *A. decorticans* Boiss. A very handsome shrub or small tree found growing in the
Sierras of S. Spain, with numerous clusters of silvery leaves with linear leaflets, topped
with clusters of brilliant yellow flowers with woolly calyx. Standard about 1½ cm, silvery-
haired. Leaflets 1–1½ cm, inrolled; bark peeling in long strips. Fruit covered with very
conspicuous raised swellings. Thickets, dry slopes. Pl. 19, Page. 77, 222.

LOTONONIS | Distinguished by the leaves which have 5 leaflets. Calyx weakly two-lipped,
with the upper lip deeply four-toothed and the lower lip with 1 tooth. 518c. *L. lupinifolia*
(Boiss.) Bentham A low, silvery or greyish, cushion-like shrub without spines, with
leaflets in fives, and axillary clusters of 2–4 yellow flowers. Corolla about 12 mm, silvery-
haired; calyx 8–10 mm. Fruit up to twice as long as calyx. Dry arid hills, rocks. S. Spain.
Page 222.

LUPINUS | **Lupin** Distinguished by its leaves which have 5–11 leaflets arising from the
apex of a stout leaf-stalk (*palmate*).

1. Upper lip of calyx with 2 shallow teeth; seeds
   8–14 mm. Fls. white, blue-tipped. (Cultivated)   **521. *L. albus* L.
1. Upper lip of calyx deeply divided into two; seed
   not more than 1 cm. Fls. not white.
  2. Fls. yellow, cream, pale pink or lilac; fls.
     regularly arranged in whorls.
    3. Fls. bright yellow; leaflets sparsely silky
       above. (Widespread, also cultivated)   **519. *L. luteus* L. Pl. 19
    3. Fls. at first cream, becoming pale pink or
       lilac; leaflets hairless except on margin.   519a. *L. hispanicus* Boiss. & Reuter
       (Portugal, W. Spain)   Pl. 19, Page 110
  2. Fls. blue or bluish; fls. alternately arranged or
     irregularly whorled.
    4. Fls. 15–17 mm; fr. 13–20 mm wide.   *varius*
    4. Fls. 10–14 mm; fr. not more than 12 mm wide.
      5. Leaflets 2–5 mm wide, linear; lower lip of
         calyx 6–7 mm.   *angustifolius*
      5. Leaflets 5–15 mm wide, obovate; lower lip of
         calyx 10–12 mm.   *micranthus*

1. *Astragalus incanus*
   subsp. *nummularioides* 526a
4. *Astragalus granatensis* 528c

2. *Astragalus alopecuroides* 527a
3. *Stauracanthus genistoides* 517b
5. *Adenocarpus hispanicus* 518a

**520. *L. angustifolius* L. NARROW-LEAVED LUPIN. Distinguished by its narrow leaflets and rather dark blue flowers which are arranged alternately. Sandy fields and uncultivated ground. Throughout.   (520) *L. micranthus* Guss. HAIRY LUPIN. Flowers blue with the standard white in the centre, and the tip of the keel blackish-violet. Plant covered in brown spreading hairs. Grassy places on acid soils. S. and C. Portugal, Med. region.   520a. *L. varius* L. Like (520) but blue flowers larger, irregularly whorled; standard blue with a white and yellow, or pale purple blotch. Plants silky- or shaggy-haired. Sandy acid cultivated soils. Portugal, Med. region.

ARGYROLOBIUM | Distinguished by its calyx which is deeply two-lipped, the upper lip longer than the calyx tube and deeply two-lobed, the lower lip three-toothed.   **522. *A. zanonii* (Turra) P. W. Ball A silvery-grey shrublet with solitary or clusters of 2–3 terminal golden-yellow flowers, and narrow elliptical leaflets which are densely silvery-haired beneath. Flowers 9–12 mm. Fruit 1½–3½ cm, silvery-haired. Dry places, preferring limestone. Widespread. Page 222.

ROBINIA | **523. *R. pseudacacia* L. FALSE ACACIA. Native of N. America; widely planted as a wayside tree for ornament and shade.

GALEGA | **524. *G. officinalis* L. GOAT'S RUE. Fields, ditches, and damp places. Scattered throughout.

COLUTEA | **525. *C. arborescens* L. BLADDER SENNA. A hairless shrub to 3 m, distinguished by its inflated bladder-like fruits which become brittle and parchment-like. Flowers yellow 1½–2 cm, in pendulous clusters of 3–8. Bushy places, dry stony ground. Med. region.   525a. *C. atlantica* Browicz Like 525 but young twigs woolly-haired. Flowers usually in clusters of 1–3; ovary densely silvery-haired. Sunny bushy places, rocky ground. S. and C. Spain.

ASTRAGALUS | **Milk-Vetch**   An important but difficult genus with 42 species in our area, including 19 which are restricted to it.

Spiny plants

(528) *A. massiliensis* (Miller) Lam. An extremely spiny, stiff, low rounded shrublet with tiny silvery leaflets and few rather large white flowers in a lax rounded cluster. Flowers 13–17 mm; calyx with appressed hairs. Leaflets 6–12 pairs, 4–6 mm, soon falling and leaving a stout spine. Hairs on leaves and stems branched. Fruit about 1 cm, hairy, exposed. Sunny rocky places. S. Portugal, Med. region. Pl. 19, Page 44.   528a. *A. sempervirens* Lam. MOUNTAIN TRAGACANTHA. Distinguished from (528) by the somewhat inflated calyx not more than 1 cm wide with teeth as long as the calyx-tube; bracts longer than the flower stalks, and fruit included in the calyx. Flowers white flushed with lilac, rarely yellow, in clusters of 4–8; standard 1–2 cm. A lax, woody, spiny, cushion-forming shrublet. Stony pastures in mountains. Spain, Pyrenees. Page 132.   528b. *A. clusii* Boiss. Similar to 528a but calyx strongly inflated in fruit up to 1½ cm wide, and densely covered with white hairs. Flowers whitish, in clusters of 2–3; standard 1½–2 cm. Arid places in lowlands and mountains. S. and E. Spain.   528c. *A. granatensis* Lam. A very distinctive, silvery-leaved, spiny tussock-forming plant with dense clusters of whitish-yellow flowers with very woolly-haired calyx, set amongst the leaves. Flowers stalkless; standard 12–15 mm, white with pinkish veins; calyx completely hidden by dense hairs. Arid hills, stony and rocky places in mountains. S. and C. Spain. Page 232.

Plants without spines

(527) *A. lusitanicus* Lam. IBERIAN MILK-VETCH. A conspicuous leafy plant to ½ m or more, with axillary, stalked clusters of large white flowers and longer leaves with numerous oval leaflets. Flowers 2–3 cm; calyx reddish conspicuous 1–1½ cm. Fruit 6 cm or more, inflated. Cultivated and uncultivated ground, pine woods. Portugal, S–W. Spain. Pl. 20. 527a. *A. alopecuroides* L. A very striking herbaceous plant to ½ m or more, usually with 3 globular heads of large pale yellow flowers in the axils of the uppermost leaves. Flowers 22–27 mm; calyx with long dense, shaggy hairs. Leaves with 12–20 pairs of oblong leaflets, hairless above, often densely hairy beneath; stem densely hairy. Sandy hills, sunny fields on limestone in lowlands and mountains. S–C. and E. Spain, S–W. France. Page 232.

528d. *A. glaux* L. A robust perennial 5–30 cm, with globular clusters of numerous purplish flowers borne on flowering stems half as long as to a little longer than the leaves. Flowers 10–12 mm; calyx about ½ cm, teeth as long as tube. Leaves with 12–15 pairs of linear to oblong leaflets which are hairless or nearly so above, and hairy beneath. Fruit 5–8 mm with white hairs; beak short, hooked. Dry pastures. Scattered throughout except in north. Pl. 20. 528e. *A. purpureus* Lam. A slender ascending perennial with globular clusters of many purplish, or rarely white, flowers on flowering stalks as long or twice as long as the leaves. Flowers about 18 mm; standard deeply notched; calyx 8–10 mm, the teeth two-thirds as long as the tube. Leaves with 7–15 pairs of oblong, notched leaflets. Fruit 1–1½ cm, inflated white-hairy. Dry sandy or stony places. E. Spain, W. France. Page 153.

530a. *A. penduliflorus* Lam. MOUNTAIN LENTIL. A rather robust herbaceous plant to ½ m, with long-stalked clusters of bright yellow flowers longer than the subtending leaves. Flowers rather numerous, in axillary clusters; standard about 1 cm. Leaves with 7–15 pairs of oblong, sparsely hairy, bright green leaflets; stipules to 1 cm. Fruit 2–3 cm, ovoid, strongly inflated beneath (recalling *Colutea*), densely blackish-hairy at first, later hairless. Meadows, screes. Pyrenees. **530. *A. alpinus* L. ALPINE MILK-VETCH. Differs from 530a in having whitish flowers with bluish-violet keel. Fruit to 1½ cm, scarcely inflated. Pyrenees.

**526. *A. monspessulanus* L. MONTPELLIER MILK-VETCH. Readily distinguished by its rosy-purple flowers, in dense ovoid clusters, borne on flowering stems which arise directly from the rootstock. Inflorescence longer than the leaves; standard about 2 cm, upturned, longer than the wings. Leaves with 10–15 pairs of oblong leaflets. Fruit 2½–4½ cm, slightly curved, sparsely hairy. Arid hills and rocky ground in the mountains. S., C., and E. Spain, S. France. Pl. 19. 526a. *A. incanus* L. Differs from 526 in having 10 or less pairs of whitish, silky-haired leaflets, and whitish to purple flowers. Fruit 1–2½ cm, whitish with adpressed hairs, often purple-spotted. Arid places on limestone. S., C., and E. Spain, S–W. France. Page 232.

Plants with distinctive fruits

527b. *A. echinatus* Murray Flowers purplish in long-stalked rounded heads about 1½ cm across. Fruits distinctive, triangular flattened with a curved beak and covered with raised bristles and contorted scales. Portugal, Med. region. Page 162. 527c. *A. epiglottis* L. An annual with axillary clusters of pale yellowish flowers, and rounded clusters of triangular heart-shaped, flattened fruits. Portugal, Med. region. 527d. *A. stella* Gouan An annual with globular heads of yellowish flowers, and star-shaped clusters of short straight cylindrical-acute fruits. Spain. (527) *A. hamosus* L. An annual with yellowish flowers and a cluster of spreading, upcurved, laterally compressed sickle-shaped fruits. Southern half of our area.

OXYTROPIS | Distinguished from *Astragalus* by the keel of the flowers which has a tooth at the apex. Fruit oblong to ovoid. 6 similar-looking species. **533. *O. campestris* (L.) DC. MEADOW BEAKED MILK-VETCH. Rocks and pastures. E. Spain, Pyrenees. Pl. 21. 534. *O. halleri* Koch PURPLE BEAKED MILK-VETCH. Pastures. E. Spain, Pyrenees. 533a. *O. pyrenaica* Godron & Gren. PYRENEAN BEAKED MILK-VETCH. Flowers purplish to bluish-violet in a globular cluster of 8–20 arising directly from the stock on a stout flowering stalk 3–20 cm. Stalk of carpel half as long as or equalling the calyx-tube. Leaflets elliptic to lance-shaped, 12–20 pairs. Stony, grassy places in mountains. Pyrenees. Page 148.

BISERRULA | 533b. *B. pelecinus* L. A small annual with clusters of bluish or pale yellow flowers with a blue tip, which can be mistaken for no other species on account of its unique fruits, which look like a two-edged saw. Leaves with 7–15 pairs of oblong, notched leaflets. Sandy, arid places. S. Portugal, Med. region.

GLYCYRRHIZA | **536. *G. glabra* L. LIQUORICE. A robust perennial ½–1 m, with pinnate leaves and lax spike-like axillary clusters of bluish or violet flowers. Leaflets 9–17, large elliptic, often sticky beneath. Fruit to 3 cm, hairless or glandular bristly. Sandy and stony places, cultivated ground. Throughout; sometimes cultivated. 536a. *G. foetida* Desf. A smaller erect plant ¼–½ m, smelling of bitumen, with yellowish flowers with a pale yellow standard. Fruit about 1½ cm long, densely bristly and glandular. Sandy ground. S. Spain.

PSORALEA | **537. *P. bituminosa* L. PITCH TREFOIL. Readily distinguished by its compact clover-like heads of blue-violet flowers, its trifoliate leaves, and its strong penetrating smell of tar when crushed. Flowers 1–1½ cm. Leaflets of upper leaves lance-shaped. Waysides, sandy and stony ground. Portugal, Med. region. 537a. *P. americana* L. Distinguished from 537 by its smaller white flowers in elongated clusters. Flowers 8 mm; keel with a violet tip. Leaflets broadly egg-shaped, toothed; whole plant covered with glandular swellings. S–W. part of Iberian peninsula. Page 241.

PHASEOLUS | 538. *P. vulgaris* L. KIDNEY BEAN. Widely cultivated in our area. 539. *P. coccineus* L. SCARLET RUNNER. Often cultivated in our area.

CICER | **541. *C. arietinum* L. CHICK-PEA. Distinguished by its small, solitary, axillary whitish flowers borne on jointed flower stalks, and its inflated fruits. Often cultivated for its edible seeds known as *garbanzo*.

VICIA | Vetch About 32 species in our area, the majority of which are widespread, particularly in the south. 5 species are restricted to our area.

542a. *V. vicioides* (Desf.) Coutinho A rather delicate, densely hairy annual to ½ m, with small elongate axillary clusters of pale lavender flowers fading to blue. Flowers 4–8 mm, in clusters of 5–20, and shorter than the subtending leaves. Leaves with 6–9 pairs of oblong leaflets, tendril branched. Fruit yellowish, usually hairless. Dry stony places, limestone fissures. S. Portugal, S. Spain.

543a. *V. argentea* Lapeyr. SILVERY VETCH. A rare endemic of stony bushy montane pastures of the Pyrenees. An erect, silvery-haired perennial 10–40 cm, with few-flowered axillary clusters of white flowers with violet veins. Flowers 18–25 mm. Leaflets 7–9 pairs, linear, without a tendril. Fruit brown, woolly-haired. **(544) *V. onobrychioides* L. FALSE SAINFOIN. A handsome plant recalling *V. cracca* L., but with larger violet or deep blue flowers in loose one-sided clusters, borne on a long flowering stem longer than the subtending leaves. Flowers 17–24 mm; keel paler. Fruit reddish-brown, hairless. Fields, hedges, rocky places in mountains. South of our area. Page 84. (547) *V. lutea* L. YELLOW VETCH Distinguished by its solitary, or up to 3, axillary yellow flowers often reddish-tinged. Standard hairless on the back. Leaves with tendrils. Fruit usually densely hairy. Grassy places. Widespread. Pl. 21.

**545. *V. cracca* L. TUFTED VETCH. Grassy and bushy places, hedges. Widespread.

**(545) *V. villosa* Roth Very variable species. Waysides, fields, hedges. Widespread. Pl. 21.

549a. *V. pyrenaica* Pourret PYRENEAN VETCH. A low-growing, spreading perennial with violet-purple flowers, endemic in alpine pastures and screes in the Pyrenees. Flowers rather large, solitary 1½–2½ cm, arising stalkless from the leaf axils. Leaves with 3–6 pairs of leaflets and usually an unbranched tendril. Fruit 2½–5 cm, black, hairless.

LENS | Lentil **552. *L. culinaris* Medicus LENTIL. Distinguished by its few tiny, whitish flowers about ½ cm, borne in the axils of the leaves, and its flattened, almost rectangular fruits. Leaflets 3–8 pairs, entire, tendril usually present; stipules entire. Widely cultivated in the south for its edible seeds. (552) *L. nigricans* (Bieb.) Godron Like 552 but stipules toothed or half hastate. Grassy hills, sandy places. South of our area.

LATHYRUS | Pea, Vetch About 32 species, most of which are widespread in the south; 2 are restricted to our area.
Flowers yellow, or cream-coloured
(i) Stems winged
**554. *L. ochrus* (L.) DC. WINGED VETCHLING. Cornfields, dry places. Portugal, Med. region. (555) *L. annuus* L. ANNUAL YELLOW VETCHLING. Fields, waysides, waste ground. Portugal, Med. region.

(ii) Stems not winged
**553. *L. aphaca* L. YELLOW VETCHLING. Fields, dry places. Widespread. 555. *L. pratensis* L. MEADOW VETCHLING. Grassy places, waysides. Widespread. 565a. *L. laevigatus* (Waldst. & Kit.) Gren. A striking robust perennial with long-stalked clusters of pale yellow flowers which turn to an ochre colour as they mature. Flowers 2–20, 1½–2½ cm, pendulous and out-turned. Leaflets large 3–10 cm, two- or six-paired, pinnately veined, without tendrils. Fruit hairless. Montane pastures and woods. N. Spain, Pyrenees. Pl. 20, Page 138. 565b. *L. pannonicus* (Jacq.) Garcke Like 565a with pale cream-coloured flowers, but smaller about 1½ cm, in clusters of 3–8. Leaflets oblong-lance-shaped or elliptical, 1½–3 cm, in 1–4 pairs, parallel-veined, without tendrils. Tubers fleshy, spindle-shaped. Montane meadows. C. and E. Spain, Pyrenees.

Flowers red, purple, blue or two-coloured (rarely yellowish)
(a) Stems not winged
(i) Some leaves with tendrils
1. Fl. clusters two- to many-flowered; calyx teeth more or less distinctly unequal.
  2. Leaflets 1 pair. (Spain, France)   **559. *L. tuberosus* L.
  2. Leaflets 2–5 pairs.
    3. Leaflets linear to lance-shaped, distinctly parallel-veined, lateral veins extending nearly to the apex. (Locally scattered)   (562) *L. palustris* L.
    3. Leaflets narrow elliptical to rounded-elliptical, pinnately veined. (Locally scattered)   **565. *L. japonicus* Willd.
1. Fls. solitary (rarely 2); calyx teeth usually equal.

4. Fl. stalks 2–5 mm. (Med. region)   557d. *L. inconspicuus* L.
4. Fl. stalks 5–70 mm.
  5. Fr. 7–11 mm wide, downy when immature. (Wide-
     spread)   (557) *L. setifolius* L.
  5. Fr. 3–7 mm wide, hairless.
    6. Fl. stalks ½–2 cm; fr. 4–7 mm wide, with prominent
       longitudinal veins.   *sphaericus*
    6. Fl. stalks 2–7 cm; fr. 3–4 mm wide, with indistinct net-
       work of veins.   *angulatus*

557a. *L. sphaericus* Retz. A slender annual with very narrow, pointed leaflets and small solitary, short-stalked orange-red flowers, usually 6–13 mm. Flower stalks ending in an awn; calyx teeth unequal. Leaflets 2–6 cm; stipules linear. Fruit brown, with prominent longitudinal veins. Vineyards, fields, among crops. South of our area.   557b. *L. angulatus* L. Like 557a, but flowers purple or pale blue, on longer stalks, 2–7 cm. Cornfields, fields, uncultivated places. Throughout.

(ii) All leaves without tendrils
1. All lvs. without leaflets. (Locally scattered)   **556. *L. nissolia* L.
1. Lvs. with 2 or more leaflets.
  2. Fls. solitary, with stalks 2–10 mm; calyx teeth more or
     less equal.
    3. Hairy plant; leaflets of upper lvs. 7–20 mm; fr. 5–7 mm
       wide, hairless.   *saxatilis*
    3. Hairless plant; leaflets of upper lvs. 2½–4 cm; fr. 2–5
       mm wide, densely hairy when young. (Med. region)   557d. *L. inconspicuus* L.
  2. Fls. in clusters, rarely solitary and then with stalks
     more than 1 cm, and calyx teeth unequal.
    4. Leaflets pinnate-veined or very feebly parallel-veined,
       the lateral veins much weaker than the midvein.
      5. Leaflets fine-pointed; stipules 1½–2 cm. (Spain,
         France)   **564. *L. vernus* (L.) Bernh.
      5. Leaflets blunt or nearly acute; stipules 4–10 mm.
         (Locally scattered)   **563. *L. niger* (L.) Bernh.
    4. Leaflets parallel-veined, the lateral veins reaching
       nearly to the apex of the leaflets.
      6. Keel more or less winged at apex; style swollen at
         apex.   *filiformis*
      6. Keel acute, not winged at apex; style not swollen at
         apex. (Pyrenees)   565d. *L. bauhinii* Genty

557c. *L. saxatilis* (Vent.) Vis. Flowers solitary, pale blue or yellowish, 6–9 mm. Leaflets 1–3 pairs, those of the lower leaves obcordate, those of upper leaves linear. Sandy, rocky places. Med. region.   565c. *L. filiformis* (Lam.) Gay Flowers bright reddish-purple, or flowers blue, or mauve, wings and keel paler, 14–22 mm, in clusters of 4–10. A hairless perennial with an angled stem and 2–4 pairs of linear-lance-shaped, pointed leaflets 3–6 cm. Montane meadows. N. and N–E. Spain, Pyrenees. Page 124.

(b) Stems winged
(i) Lower leaves without leaflets
**561. *L. clymenum* L. Flowers crimson with wings violet or lilac. South of our area.   561a. *L. articulatus* L. Flowers crimson with wings white or pink. South of our area.

(ii) All leaves with 2 or more leaflets

1. At least some lvs. with 2 or more pairs of leaflets; fl. clusters two- to many-flowered.
  2. Lvs. without a tendril.   *montanus*
  2. Lvs. with a tendril.
    3. Leaf-axis (*rachis*) at least 4 mm wide, broadly winged. *heterophyllus*
    3. Leaf-axis (*rachis*) not more than 3 mm wide.
      4. Lowest tooth of calyx about as long as tube. (Locally (562) *L. palustris* L.
        scattered)   *cirrhosus*
      4. Lowest tooth of calyx distinctly shorter than tube.
1. All lvs. with only 1 pair of leaflets; or rarely some with 2 pairs and then fls. solitary.
  5. Fl. clusters five- to many-flowered (rarely 3).
    6. Stipules less than half as wide as stem. (Widespread) 560. *L. sylvestris* L.
    6. Stipules at least half as wide as stem.   *latifolius*
  5. Fl. clusters one- to three-flowered (rarely 4).
    7. Fls. 2 cm or more.
      8. Fl. stalks not more than 6 cm; fr. with 2 wings on back.
        (Widespread)   **(557) *L. sativus* L.
      8. Fl. stalks more than 7 cm; fr. not winged.
        9. Leaflets 1–4 mm wide, linear-lance-shaped; calyx teeth
          longer than tube.   *tremolsianus*
        9. Leaflets 4–18 mm wide, egg-shaped to lance-shaped; calyx
          teeth shorter than tube.   *tingitanus*
    7. Fls. less than 2 cm.
     10. Calyx teeth not or only slightly longer than tube.
      11. Sparsely downy plant; fls. crimson with blue wings.
        (Widespread)   **558. *L. hirsutus* L.
      11. Hairless plant; fls. orange-red. (Widespread) (557) *L. setifolius* L.
     10. Calyx teeth 1½–3 times as long as tube.
      12. Fls. red; fr. with 2 wings on the upper and lower margin. *amphicarpos*
      12. Fls. white, pink or purple.
        13. Fr. with 2 keels on upper margin; fl. stalks 1–3 cm long.
          (Widespread)   557. *L. cicera* L.
        13. Fr. with 2 wings on upper margin; fl. stalks 3–6 cm.
          (Widespread)   **(557) *L. sativus* L.

**(560) *L. latifolius* L. EVERLASTING PEA. Easily recognized by its large, bright reddish-carmine flowers 2–3 cm long, and its large, oval to oblong paired leaflets, and its broadly winged stem and leaf stalks. Flowers in clusters of 5–18. Stipules leaf-like 3–6 cm. Hedges, vineyards, fields and uncultivated places. Scattered throughout. 560a. *L. heterophyllus* L. Like (560) but upper leaves with 2–3 pairs of leaflets, and flowers pink, smaller 12–22 mm. Vineyards, fields and hedges. Scattered throughout. 560b. *L. tingitanus* L. Like (560) but an annual with long-stalked clusters of 1–3 bright rosy-purple flowers. Flowers 2–3 cm. Stipules 12–25 mm. Hedges and bushy places. S. and E. part of Iberian peninsula. Pl. 20. 560c. *L. tremolsianus* Pau Flowers 2–3 cm, pink with blue wings, in clusters of 1–3. Flowers 2–3 cm. Leaflets linear-lance-shaped; stipules 2–3 cm, linear-lance-shaped. Hedges and bushy places. Endemic to S–E. Spain.

562. *L. montanus* Bernh. BITTER VETCH. A low-growing, non-climbing perennial recognized by its rather small flowers about 1½ cm which are at first bright crimson, becoming greenish-blue with age. Leaves usually with 2–4 pairs of leaflets, without tendrils. Fruit red-brown.

Mountain pastures, bushy places. N. Portugal, C., N., and E. Spain, France. Page 149. 562a. *L. cirrhosus* Ser. A climbing perennial with clusters of 4–10 pink flowers, each 12–15 mm, and leaves with much-branched tendrils. Leaflets 2–3 pairs, narrow-elliptical. Fruit pale brown, with 3 keels on the upper margin. Mountain woods and banks. Pyrenees, Cévennes.

Annuals

557e. *L. amphicarpos* L. An annual readily distinguished by its broad brown fruits with 2 conspicuous wings on the upper and lower margin. Flowers red, solitary; calyx teeth equal, about 1½ times as long as the tube. Leaves with a pair of oval to linear leaflets, with or without a tendril. Dry and stony places. S–W. Iberian peninsula.

ONONIS | **Restharrow**   41 species occur in our area; 17 are restricted to it; more than half are annuals with inconspicuous flowers.

Perennials
(a) Flowers pink
(i) Fruit two to four times as long as calyx
**570. *O. rotundifolia* L. ROUND-LEAVED RESTHARROW. A glandular-hairy, dwarf shrub to ½ m with axillary clusters of very attractive large pink flowers. Flowers 1½–2 cm, in long-stalked clusters of 2–3, rarely whitish. Leaves trifoliate with almost circular, coarsely toothed leaflets, usually 2½ cm. Fruit 2–3 cm. Rocky places and woods in hills. S. and E. Spain, S–W. France. Pl. 21, Page 149.   570a. *O. tridentata* L. Distinguished from 570 by its woolly-haired stems and linear to obovate leaflets, usually less than 1½ cm. Flowers pink, 1–2 cm, borne on short primary branches to 1 cm, and forming a lax cluster. Dry arid fields and hills. S., C., and E. Spain.   **(569) *O. fruticosa* L. SHRUBBY RESTHARROW. Differs from 570 in having terminal clusters of rather numerous, showy pink flowers borne on long primary branches 10–30 cm. A handsome dwarf shrub with hairless, somewhat leathery trifoliate leaves, with stalkless oblance-shaped, finely toothed leaflets. Fruit to 2 cm. Dry rocky places, principally on limestone. Spain, Pyrenees, Cévennes. Pl. 21. 568a. *O. cristata* Miller (*O. cenisia*) A spreading, rhizomatous, montane and alpine perennial with jointed flowerstalks longer than the leaves. Flowers pink, in clusters of 1–6, each 1–1½ cm. Leaflets ½–1 cm, oblong to oblance-shaped, strongly toothed, stalkless, somewhat leathery. Rocks and pastures. S. and E. Spain, Pyrenees.

(ii) Fruit less than twice as long as calyx
568. *O. spinosa* L. SPINY RESTHARROW. Dry grassy places, waysides. Widespread.   569. *O. repens* L. RESTHARROW. Dry places, pastures, uncultivated ground. Widespread.   569a. *O. pinnata* Brot. A handsome, greyish, sticky-hairy, much-branched erect dwarf shrub with pinnate lower leaves with 5–9 leaflets. Flowers large 15–23 mm, pink, one to each node, in a compact cluster which later elongates. Leaflets egg-shaped, 1–1½ cm. Scrub, river margins. S. Portugal, S. and W. Spain.

(b) Flowers yellow
(i) Flowers mostly in branched, axillary clusters; or long-stalked and solitary
**571. *O. natrix* L. LARGE YELLOW RESTHARROW. An attractive small shrub with lax leafy clusters of conspicuous, often large yellow flowers, variously veined with red or violet. Flowers and leaflets variable in size; leaves glandular-sticky, egg-shaped to linear. Fruit 1–2½ cm. Dry stony places, on the littoral. Widespread.   571a. *O. crispa* L. Like 571 with yellow flowers, often with red veins, but leaflets 7–9 mm, rounded, with undulate margins. Flowers 1½–2 cm, in dense clusters. Fruit 1½–2 cm. Sandy and rocky places. S–W. and E. Spain.   571c. *O. aragonensis* Asso A low, twisted, grey-stemmed shrub, like 571b but flowers in long, lax, terminal branched clusters, and leaflets smaller 4–10 mm and leathery. Flowers

yellow, 12–18 mm. Fruit 7–8 mm. Rocky, stony and bushy places, largely on limestone hills and mountains. S. and E. Spain, Pyrenees. Page 241.   571e. *O. reuteri* Boiss. Like 571c. but flowers smaller, not more than 1 cm; leaflets 2–4 mm. Mountains. S–W. Spain.

571f. *O. striata* Gouan A somewhat woody, rhizomatous, ascending perennial with yellow flowers in few-flowered spike-like clusters. Flowers 10–13 mm, longer than calyx; flower stalks short. Leaves trifoliate, leaflets 3–6 mm, oblance-shaped to rounded, the veins very prominent when dry. Fruit 6–7 mm. Rocks in mountains. Spain, Pyrenees, France. Pl. 21.

(ii) Flowers in spike-like clusters, usually solitary; flower stalks short, shorter than the leaves.

571b. *O. speciosa* Lag. The most spectacular yellow-flowered species forming a sticky, twiggy bush to 1 m, with dense, elongated, terminal clusters of golden-yellow flowers. Flowers 1½–2 cm; calyx very hairy; bracts papery, soon falling. Leaves trifoliate, leaflets elliptic to rounded, 1½–2½ cm, toothed; stems densely glandular-hairy. Stony hills and scrub. S. and S–E. Spain. Page 91.   571d. *O. minutissima* L. A dwarf shrub with spreading, often rooting stems, and dense clusters of small yellow flowers not longer than the whitish calyx. Flowers 8–10 mm. Leaflets 3–6 mm, oblong to obovate, soon falling. Fruit shorter than calyx. Arid stony places, preferring limestone. Med. region.   572c. *O. saxicola* Boiss. & Reuter A slender spreading perennial to 20 cm with yellow flowers like 572. *O. pusilla*, but flowers larger about 12 mm and longer than calyx. Flowers in lax leafy spikes. Leaves trifoliate, leaflets rounded 5–13 mm. Shady north-facing rocks. S–W. Spain (Serranía de Ronda).

Annuals

(a) Flowers yellow

567a. *O. viscosa* L. A very variable, densely softly hairy and glandular annual with solitary, long-stalked axillary yellow flowers forming a long leafy terminal cluster. Flowers often red-veined, longer or shorter than calyx; flower stalk ending in a short awn, shorter than the flower. Leaves mostly with 1 elliptic to obovate leaflet 1–2 cm. Dry banks and fields. S. Portugal, Med. region.   567b. *O. variegata* L. Distinguished from 567a by its short-stalked flowers in a lax spike-like cluster which is leafy only at the base; larger flowers, twice as long as calyx; and fruit longer than calyx. Leaves mostly with 1 leaflet ½–1 cm. Maritime sands. Portugal, Med. region.

(b) Flowers pink.

567c. *O. subspicata* Lag. A delicate and attractive, sticky, pink-flowered, erect annual. Flowers solitary, short-stalked, in a dense leafy cluster which later elongates; petals 8–14 mm, longer than calyx. Leaves with 3 toothed leaflets. Fruit about 7 mm. Maritime sands. Portugal, W. Spain.

MELILOTUS | Melilot   10 similar-looking, widespread species occur in our area, 9 with yellow flowers.

TRIGONELLA | Fenugreek   5 species.   578. *T. monspeliaca* L. STAR-FRUITED FENU-GREEK. It has stalkless clusters of 4–14 small yellow flowers which are shorter than the subtending leaves, and a star-shaped cluster of fruits. Dry arid places. Portugal, Med. region.   **580. *T. foenum-graecum* L. FENUGREEK. A leafy annual with solitary or paired yellowish-white flowers, tinged with violet, and very long erect or spreading, somewhat curved fruits 6–11 cm. Naturalized in field verges and uncultivated ground; sometimes grown for fodder.

1. *Anthyllis barba-jovis* 620
3. *Anthyllis tejedensis* 621a
5. *Psoralea americana* 537a

2. *Ononis aragonensis* 571c
4. *Anthyllis cytisoides* (619)

MEDICAGO | Medick About 27 very similar-looking species are to be found in our area. The fruits are very distinctive and show small but constant differences in shape, degree of coiling, and spininess. ****589. *M. marina* L. SEA MEDICK.** An easily recognized littoral plant, usually growing prostrate over the sand, with silvery or almost white woolly foliage and stems. Flowers pale yellow, in short-stalked clusters; fruit woolly-haired, coiled, with or without spines. Portugal, Med. region. ****583. *M. sativa* L. LUCERNE, ALFALFA.** Frequently cultivated for fodder in dryer regions and often naturalized.

TRIFOLIUM | Clover, Trefoil About 56 species are found in our area, only one of which is restricted to it. Many species are widespread, many more are common in the Med. region, while others are found in the high mountain ranges.

**DORYCNIUM | Distinguished by its clusters of whitish or pinkish flowers, each with a dark red or blackish keel. Leaves with 5 leaflets; usually shrubby plants.

1. Fls. 1–2 cm. (South, Med. region)   **609. *D. hirsutum* (L.) Ser.
1. Fls. 3–7 mm.
  2. At least the lower and stem lvs. with *rachis* of leaflets
      at least ½ cm long. (Throughout)   **610. *D. rectum* (L.) Ser.
  2. Lvs. with *rachis* of leaflets very short or absent.
      (Throughout)   611. *D. pentaphyllum* Scop.

LOTUS | Birdsfoot-Trefoil About 20 rather similar-looking species occur in our area.

Calyx tubular-bell-shaped, with 5 more or less equal teeth
613. *L. corniculatus* L. BIRDSFOOT-TREFOIL. Cultivated ground, grassy, stony places. Widespread. **614. *L. uliginosus* Schkuhr LARGE BIRDSFOOT-TREFOIL. Damp places and marshes. Widespread. 614a. *L. edulis* L. Readily distinguished by its inflated fruit which is grooved on the back. A spreading, sparsely hairy annual, with 1–2 yellow flowers, borne on a stalk longer than the subtending leaves. Petals 1–1¼ cm, curved. Sandy, stony and rocky places. S. Portugal, Med. littoral.

Calyx two-lipped, teeth unequal

613a. *L. glareosus* Boiss. & Reuter A variable but often a small spreading, densely hairy or silvery-haired perennial of stony mountains. Flowers like 613, 8–10 mm, usually reddish, in heads of 1–6; calyx two-lipped, the teeth unequal, the lateral and upper teeth curved. Leaves and calyx covered with silvery adpressed or spreading hairs. C. Portugal, S. Spain. Pl. 22.

**615. *L. creticus* L. SOUTHERN BIRDSFOOT-TREFOIL. A densely silvery-haired, spreading perennial with heads of 2–6 yellow flowers. Calyx distinctive, two-lipped, the upper 2 teeth curved upwards, and the lateral 2 teeth acute, about as long as the upper 2 and shorter than the lower tooth. Flowers 12–18 mm, wings much longer than the keel which has a long, straight, purple beak. Maritime sands. Portugal, Med. region. Pl. 22. 615a. *L. cytisoides* L. Like 615 but 2 lateral teeth of calyx blunt, much shorter than the upper 2 teeth. Flowers 8–14 mm, standard notched, wings slightly longer than the keel which has a short, curved, purple-tipped beak. Rocky, stony and sandy places. Med. region.

615b. *L. tetraphyllus* L. A tiny endemic of the Balearic Islands, readily distinguished by its 4 (not 5) wedge-shaped to inversely heart-shaped leaflets. Flowers solitary, yellow with red or purple streaking. Limestone crevices.

612. *L. ornithopodioides* L. A hairy annual with long-stalked clusters of yellow flowers, each ½–1 cm. Distinctive in fruit with 2–5 long, curved fruits hanging below 3 conspicuous rounded leafy bracts. Leaflets broadly oval-rhomboid. Grassy, sandy and stony places. S. Portugal, Med. region.

TETRAGONOLOBUS | Distinguished from *Lotus* by its angled or winged fruit. Leaves trifoliate. ****616.** *T. maritimus* (L.) Roth WINGED PEA. A low creeping perennial, easily distinguished by its large solitary, long-stalked, pale yellow flowers 2½–3 cm, and its large four-winged fruits. Fruit 3–6 cm by 3–5 mm. Fields, damp grassy places, sands on the littoral, brackish soils. Spain, France. Pl. 22. ****617.** *T. purpureus* Moench ASPARAGUS PEA. Unmistakable with its large, short-stalked solitary or paired crimson flowers 15–22 mm, with a blackish keel, and its large four-winged fruit. Calyx teeth as long or up to twice as long as calyx tube. Cultivated places, vineyards. S., C., and E. Spain, S. France; introd. Portugal. Page 163. 617a. *T. requienii* (Sanguinetti) Sanguinetti Distinguished from 617 by its smaller bright red flowers 13–15 mm, and its fruits with 2 wings on the upper side only. S. Portugal, Med. Spain.

ANTHYLLIS | Fruit often not splitting and usually encircled by the persistent calyx. 17 species in our area.

Shrubs with woody branches
(619) *A. cytisoides* L. A dense shrub, usually to 60 cm, with erect, little-branched, white-felted stems, greyish or whitish leaves, and long interrupted spikes of pale yellow flowers. Calyx 3½–7 mm, shaggy-haired; bracts egg-shaped. Lower leaves with 1 leaflet, the upper trifoliate with a much larger central leaflet. Sands, rocky places, sunny hills. S. and E. Spain, S. France. Page 241. 619a. *A. terniflora* (Lag.) Pau A less robust, finely silky-haired shrub like (619) but with all leaves with only 1 oblong to narrowly elliptical leaflet. Calyx 3–4½ mm. Arid hills. S. and S–E. Spain. 620. *A. barba-jovis* L. JUPITER'S BEARD. A very handsome, dense, silvery shrub to 1 m, with pinnate leaves and bright yellow flowers in compact globular heads, 1–2 cm across, borne on silvery stems. Flowers usually in heads of 10 or more. Leaflets numerous, silvery-white. Rocks by the sea. E. Spain, S. France. Page 241.

620a. *A. henoniana* Batt. A scraggy shrub, distinguished from 620 by the much smaller leaves which have 3–5 unequal leaflets, and old leaf-stalks persisting in spine-like projections. Flower heads yellow, globular, five- to eight-flowered. Arid places in the hills. S. and E. Spain.

Perennials with herbaceous branches
(a) Calyx not inflated
**621. *A. montana* L. MOUNTAIN KIDNEY-VETCH. A low, spreading mat-forming plant with globular heads of pink or purple flowers borne on short, nearly leafless stems. Standard much longer than other petals; calyx tubular, with awl-shaped, feathery teeth. Leaves with numerous leaflets, the terminal often more rounded. Rock fissures. S. Spain, Pyrenees. Page 153.

621a. *A. tejedensis* Boiss. A very handsome greyish or brownish, densely hairy perennial with spreading stems forming dense tufts, each stem bearing 1–3 globular clusters about 3 cm across, of yellow or orange flowers, sometimes flushed with brownish-violet. Calyx densely hairy. Leaflets 9–15, egg-shaped with short dense more or less adpressed soft hairs. Fissures of limestone rocks, in mountains. S. Spain. Page 241. 621b. *A. polycephala* Desf. A very handsome, woody-based, softly hairy, erect perennial to 60 cm, with stems bearing 2–4 dense rounded heads of pale or orange-yellow flowers, arranged in a spike-like inflorescence, the lower heads stalked. Petals slightly longer than the densely woolly-haired calyx, which is 6–9 mm, with teeth as long as the tube. Leaflets 13–15, narrowly elliptic, shaggy-haired. Rocky crevices on limestone in the lowlands and mountains. S. Spain.

(b) Calyx inflated

**622. *A. vulneraria* L. KIDNEY VETCH. An extremely variable annual, biennial or perennial with globular heads of yellow, red, orange, purple, whitish or multi-coloured flowers sub-tended by 2 deeply lobed bracts closely surrounding the base of the dense flower heads. Its conspicuous swollen calyx, which is constricted at the apex into an oblique mouth with 5 unequal teeth, is distinctive. Fruit usually one-seeded. A number of distinctive and often attractive subspecies are restricted to the Iberian peninsula and the Pyrenees, but there are intermediate forms which make their identification difficult. Lowlands and mountains. Widespread. Pl. 22.

Annuals

**623. *A. tetraphylla* L. BLADDER VETCH. A low spreading annual, readily distinguished by its very swollen calyx, which enlarges and becomes almost globular in fruit. Flowers pale yellow, tipped with red, in dense stalkless axillary clusters. Leaflets 3–5, the terminal leaflet much larger. Fruit usually two-seeded. Cultivated ground, grassy places. S. Portugal, Med. region.   623a. *A. lotoides* L. Distinguished from 623 by its rather slender tubular calyx, and its upper leaves which are deeply and unequally cut into 5–7 lobes. Flower heads dense, stalked, with 4–8 yellow or orange flowers. Fruit straight, with 6–10 seeds. Un-cultivated ground, on the littoral. Portugal, Spain.   623b. *A. cornicina* L. An erect, hairy annual similar to 623a but distinctive in fruit. Calyx swollen, woolly-haired, completely enveloping the fruit, which is curved almost into a complete circle. Flower heads globular, compact, stalked; flowers small, orange-yellow. Dry undisturbed ground. Portugal, S. and C. Spain.   623c. *A. hamosa* Desf. Distinguished from 623a by the sickle-shaped, upcurved fruits with a long awn-shaped beak which projects beyond the hairy, tubular calyx. Leaflets up to 11. Sandy fields, bushy places, waysides, pine woods on the littoral. Portugal, S. Spain.

ORNITHOPUS | Fruit elongate, splitting into one-seeded portions, usually constricted between the seeds, with a network of veins.

Flower heads without bracts

623d. *O. pinnatus* (Miller) Druce A small annual, distinguished by its fruits which are slender, cylindrical and curved into three-quarters of a circle, so that the fruiting heads look like uncoiling springs. Flowers 6–8 mm, yellow, in long-stalked clusters of 1–5. Leaves with 3–7 pairs of linear to oblance-shaped leaflets and a terminal leaflet. Cultivated and grassy places, fields, sandy ground. Portugal, Med. region. Page 247.

Flower heads with leafy bracts

(a) Flowers yellow

623e. *O. compressus* L. Like 623d but fruits curved, flattened, and ending in a long sickle-shaped beak, which is jointed and wrinkled when dry. Leaflets elliptic, 7–18 pairs, hairy. Dry sandy places, cultivated ground. Portugal, Med. region.

(b) Flowers white or pink

623f. *O. sativus* Brot. Flowers pink or white. Fruits 'knobbly', conspicuously constricted between the seeds. Subsp. *sativus* with straight fruits. Native, but often cultivated for fodder in the northern part of the Iberian peninsula. Subsp. *isthmocarpus* (Cosson) Dostál with strongly curved fruits, occurs in the south-west.   623g. *O. perpusillus* L. BIRDSFOOT. Sandy ground, tracksides. Widespread, except in the south.

CORONILLA | Distinguished by its long slender, jointed fruits which are not constricted between the seeds, and compact 'umbels' of flowers.

1. Small anns.
  2. Upper lvs. with 1 or 3 very unequal leaflets.   *scorpioides*
  2. Upper lvs. with 3–9 more or less equal leaflets. (South)   628a. *C. repanda* (Poiret) Guss.
1. Shrubs or robust perenns.
  3. Stalk of standard (claw) 2–3 times as long as calyx.   *emerus*
  3. Stalk of standard (claw) equalling or slightly longer
    than calyx.
    4. Fls. yellow.
      5. Leaflets with a narrow papery margin.   *minima*
      5. Leaflets without a papery margin.
        6. Stems not rush-like; lvs. persisting.   *valentina*
        6. Stems rush-like; lvs. soon falling.   *juncea*
    4. Fls. white, pink or purple.   *varia*

**624. *C. emerus* L. SCORPION SENNA. A robust leafy shrub to 1 m or more, with long-stalked globular clusters of yellow flowers, and long, slender, pendulous, jointed fruits 5–11 cm. Leaflets 5–9, obovate, 1–2 cm. Thickets, rocky places in the hills. E. Spain, Pyrenees, France. 625. *C. valentina* L. SHRUBBY SCORPION-VETCH. Like 624 but leaflets shallowly notched at apex and claw of petals little longer than calyx. Fruit shorter, to 5 cm, angled. Subsp. *glauca* (L.) Batt. occurs in our area. Scrub. S. Portugal, Med. region. Pl. 23. 625a. *C. minima* L. A small bushy shrub to 30 cm, with long-stalked, globular clusters of up to 10, or sometimes more, small yellow flowers, each 5–8 mm. Leaflets 2–4 pairs, often very small but up to 1½ cm, mostly obovate. Fruit 1–3½ cm, conspicuously jointed. Dry sands and stony places, rocks in lowlands and hills. S. and C. of our area.

626. *C. juncea* L. RUSH-LIKE SCORPION-VETCH. A slender, branched, often nearly leafless shrub, distinguished by its rush-like, compressible, usually glaucous stems. Flowers yellow, 6–12 mm in long-stalked, globular clusters about 2 cm across. Leaflets narrow, fleshy, linear, 2–3 pairs. Dry open habitats. Portugal, Med. region. Page 247.

628. *C. scorpioides* (L.) Koch ANNUAL SCORPION-VETCH. A common weed of cultivation with rounded glaucous leaflets, long-stalked clusters of tiny yellow flowers and long curved pendulous jointed fruits. Flowers 4–8 mm. Widespread.

**627. *C. varia* L. CROWN VETCH. A robust, spreading, leafy herbaceous perennial, distinguished by its globular, long-stalked clusters of white, pink or often multi-coloured flowers. Leaflets large, oblong, to 2 cm. Fruit four-angled, to 6 cm. Grassy and bushy places; sometimes cultivated. N. and E. Spain, France.

HIPPOCREPIS | **Horseshoe Vetch**   About 10 very similar-looking species occur in our area. The distinctive fruits often look like a row of horseshoes placed edge to edge. 631a. *H. glauca* Ten. A woody-based perennial to 40 cm with heads of 2–8 yellow flowers on stems two to three times as long as the leaves. Flowers 6–12 mm, claw of standard about as long as calyx. Leaflets 4–7 pairs, densely white-woolly beneath. Fruit 3–4 cm, without broad flattened regions between the seed protruberances. Dry limestone rocks and screes. Med. region. 631b. *H. balearica* Jacq. A woody-based perennial to 50 cm with long-stalked conspicuous, sweet-scented, globular heads of numerous flowers, each flower 1–1½ cm. Claw of standard about twice as long as calyx. Leaflets 5–10 pairs, linear or oblong. Fruit 1½–4½ cm. Limestone rocks. Balearic Islands. Pl. 22.

SCORPIURUS | Readily distinguished by their fruits which look very like swollen, coiled caterpillars. Leaves parallel-veined, entire.

1. Fr. smooth or with swellings or spines on the outer ridges;
   fls. usually 2–5 in a head. (South)   **(633) *S. muricatus* L.
1. Fr. with club-shaped swellings on the outer ridges; fls.
   usually 1. (South)   633. *S. vermiculatus* L.

HEDYSARUM | Distinguished by its flattened fruits which are strongly constricted between the seeds with segments which look like a series of disks placed edge to edge.

1. Anns.
   2. Leaflets usually 1–3 pairs, terminal leaflet 1½–4½
      cm by 1–4 cm.   *flexuosum*
   2. Leaflets usually 4–8 pairs, terminal leaflet ½–1½
      cm by 2–5 mm.
      3. Fls. 8–11 mm, 1½–2 times as long as calyx.   *spinosissimum*
      3. Fls. 1½–2 cm, 2½–5 times as long as calyx.   *glomeratum*
1. Perenns.
   4. Leaflets hairless or nearly so beneath; fr. hairless,
      or sparsely downy, without spines. (Pyrenees)   **635. *H. hedysaroides* (L.) Schinz
                                                             & Thell.
   4. Leaflets downy or silky beneath; fr. with spines or
      bristles.
      5. Leaflets 3–5 pairs, large; fr. hairless.   *coronarium*
      5. Leaflets 6–16 pairs, small; fr. downy.   *humile*

634. *H. coronarium* L. ITALIAN SAINFOIN, FRENCH HONEYSUCKLE. A very striking perennial with cylindrical clusters of large bright carmine flowers and leaves with 3–5 pairs of large elliptical or rounded leaflets. Fruit with 2–4 spiny segments. Cultivated ground, rich soils. S–W. Spain; cultivated for fodder and naturalized elsewhere in the south. Pl. 22, Page 52. (634) *H. spinosissimum* L. A low, spreading annual with few-flowered clusters of pinkish-purple to whitish flowers. Leaflets 4–8 pairs, often oblong. Fruit with 2–4 spiny and finely hairy segments. Limestone hills, dry stony places. S. Portugal, Med. region.   **(634) *H. glomeratum* F. G. Dietrich Like (634) *H. spinosissimum* but flowers more showy, pinkish-purple; leaflets broader, sometimes obovate. Fruit distinctive, with netted segments covered with fine grey hairs and pale hooked spines. Dry places. S. Portugal, Med. region. page 247.

634a. *H. flexuosum* L. A rather robust annual with only 1–3 pairs of large, oblong or obovate leaflets, the terminal larger. Flowers pink or purple, numerous in a cylindrical cluster; petals 8–12 mm. Fruit with 2–8 spiny segments. Sandy places near the sea. S–W. Portugal, S–W. Spain.   634b. *H. humile* L. A showy, erect annual with cylindrical clusters of rather numerous, conspicuous purple or pink flowers borne on long stems, much longer than the leaves. Flowers 8–12 mm, about 3 times as long as the calyx. Leaflets small, 6–16 pairs. Fruit strongly netted, with few spines. Sunny hills, arid ground and waste places. Med. region. Page 117.

ONOBRYCHIS | Distinguished by its rounded, non-splitting, somewhat flattened fruits, which have very conspicuous toothed margins and are netted and spiny on the flanks.

1. Anns.; fls. small, petals little longer than
   the calyx; inflorescence usually one- to
   six-flowered.

1. *Geranium malviflorum* 643a
2. *Hedysarum spinosissimum* (634)
3. *Ornithopus pinnatus* 623d
4. *Onobrychis peduncularis* 636c
5. *Coronilla juncea* 626

2. Fls. 7–8 mm. (Med. region)
2. Fls. 10–14 mm. (Med. region)
1. Perenns.; fls. conspicuous; inflorescence
   with at least 10 fls.
3. Standard at least $1\frac{1}{5}$ times as long as
   keel.
3. Standard shorter or slightly longer than
   keel.
  4. Standard at least 5 mm shorter than
    keel. (S–E. Spain)
  4. Standard not more than 2 mm shorter
    than keel.
    5. Wings of fls. 7–8 mm, distinctly longer
      than calyx.
    5. Wings of fls. not more than 6 mm,
      shorter than calyx.
      6. Fr. with spines 3–6 mm long.
      6. Fr. with spines less than 3 mm long.
        7. Calyx longer than keel. (N. Spain)
        7. Calyx shorter than the keel. (Locally
          naturalized)

637.   *O. caput-galli* (L.) Lam.
637a. *O. aequidentata* (Sibth. & Sm.) D'Urv.

supina

636e. *O. stenorhiza* DC.

saxatilis

peduncularis

636f. *O. reuteri* Leresche
636. *O. viciifolia* Scop.
    argentea

636a. *O. saxatilis* (L.) Lam. ROCK SAINFOIN. A slender, tufted, white-downy perennial with distinctive linear leaflets and salmon-coloured or pale pink flowers in a dense elongated spike. Flowers 9–14 mm, wings much shorter than standard and keel. Leaflets 8–15 pairs. Fruit without marginal spines, but with sunken hairy cavities on the flanks. Rocks, stony places, mainly on limestone. Med. region. Pl. 21.   636b. *O. supina* (Chaix) DC. Like 636a. but fruit with 3–4 teeth up to 2 mm long on the margin and short teeth on the flanks. Flowers pink or white with pink veins, smaller 7–10 mm, the standard distinctly longer than the keel. Leaflets oblong or elliptic. Dry stony places. E. Spain, Pyrenees, S. France.

636c. *O. peduncularis* (Cav.) DC. A handsome plant with spikes of large white to purplish, often purple-veined flowers and large, conspicuous, extremely spiny fruits up to $1\frac{1}{2}$ cm. Flowers 1–1½ cm. Lower leaves with 4–14 pairs of linear to elliptic leaflets, hairless above. Fruit with marginal spines up to 6 mm and shorter spines on the flanks. Dry hills and grassland. S. Portugal, S. and C. Spain. Page 247.   636d. *O. argentea* Boiss. Like 636c. but fruit smaller to 8 mm, with short marginal spines 1 mm or more, flanks netted, densely hairy, without spines. Flowers conspicuous, pink, often with darker veins. Leaflets either densely silvery-haired beneath, subsp. *argentea*; or hairy beneath but not silvery, subsp. *hispanica* (Širj.) P. W. Ball. Dry limestone mountains. S. and E. Spain, Pyrenees.

OXALIDACEAE | Wood-Sorrel Family

OXALIS | 2 native species, and 7 introduced species from America and S. Africa.

1. Fls. yellow.
2. Stalked fl. heads arising directly from the rootstalk;
   bulbils present at base.
2. Aerial stems present; fls. axillary; bulbils absent.

    pes-caprae

3. Stems rooting at the nodes; lvs. alternate. *corniculata*

3. Stems not rooting at the nodes; lvs. more or less opposite. (Naturalized)   (640) *O. europaea* Jordan

1. Fls. white, pink or purple.

  4. Stem rhizomatous.

    5. Fls. many in a flat-topped cluster; fls. pink. (Naturalized)   639a. *O. articulata* Savigny

    5. Fls. solitary; fls. usually white. (Mostly in north)   **638.   *O. acetosella* L.

  4. Stem not rhizomatous, erect or absent; bulbils present.

    6. Fls. solitary; petals 2½–3½ cm, pink. (Naturalized)   639b. *O. purpurea* L.

    6. Fls. in a flat-topped cluster; petals 1½–2 cm, pink.

      7. Leaflets widest at or below middle, hairy, dotted near margin beneath. (Naturalized)   639c. *O. corymbosa* DC.

      7. Leaflets widest near apex, nearly hairless, not dotted. (Naturalized)   639d. *O. latifolia* Kunth

**639. *O. pes-caprae* L. BERMUDA BUTTERCUP. A serious weed of cultivated ground, vineyards and olive groves. Readily distinguished by its umbel of large, bright yellow flowers and pale clover-like, long-stalked trifoliate leaves. The production of numerous bulbils accounts for its aggressiveness and persistence. Native of S. Africa; introd. to Portugal, Med. region. Page 63.   **640. *O. corniculata* L. PROCUMBENT YELLOW SORREL. A delicate spreading perennial with a cluster of 1–7 small yellow flowers, with petals 4–7 mm. Fruit 1–2½ cm, hoary. Disturbed ground, walls. Widespread but rarer in the south.

GERANIACEAE | Geranium Family

GERANIUM | Cranesbill   21 species are found in our area, including 10 small-flowered annual species, most of which are frequent and fairly widespread, including: 648. *G. dissectum* L. CUT-LEAVED CRANESBILL;   (648) *G. columbinum* L. LONG-STALKED CRANESBILL; 649. *G. rotundifolium* L. ROUND-LEAVED CRANESBILL;   (649) *G. molle* L. DOVE'S-FOOT CRANESBILL;   (649) *G. pusillum* L. SMALL-FLOWERED CRANESBILL;   650. *G. robertianum* L. HERB ROBERT;   (650) *G. purpureum* Vill.;   **651. *G. lucidum* L. SHINING CRANESBILL.

Key to perennial species

1. Petals with a distinct stalk (claw) at least ⅓ as long as limb.   *cataractarum*

1. Petals with a very short stalk (claw), or none.

  2. Petals entire, or slightly notched, or pointed.

    3. Petals spreading horizontally, or deflexed. (Pyrenees, France)   **646. *G. phaeum* L.

    3. Petals curving upwards, giving a more or less cup-shaped fl.

      4. Inflorescence compact, usually with more than 10 fls.

        5. Fl. stalk deflexed as fr. matures; sepals 11–15 mm. (N. Spain, France)   644. *G. pratense* L.

        5. Fl. stalk erect as fr. matures; sepals 6–12 mm. (C. and E. Spain, France)   **645. *G. sylvaticum* L. Pl. 23

      4. Inflorescence spreading, usually with less than 10 fls.

6. Rhizome long, slender, horizontal.	*endressii*
6. Rhizome short. (Pyrenees)	646a. *G. palustre* L.
2. Petals two- or three-lobed, or distinctly notched.	
7. Stock small and inconspicuous; petals usually less than 1 cm.	
8. Frs. smooth, hairy. (Widespread)	**642. *G. pyrenaicum* Burm. fil.
8. Frs. rough, hairless. (Widespread)	(649) *G. molle* L.
7. Plant with conspicuous tuber, rhizome or woody stock; petals usually more than 1 cm.	
9. Fls. solitary. (Widespread except in south)	**641. *G. sanguineum* L.
9. Fls. paired.	
10. Lvs. not more than 4 cm wide, usually all basal.	*cinereum*
10. Larger lvs. at least 5 cm wide, some of them borne on the stem.	
11. Lvs. divided for 70 per cent of their radius. (Pyrenees, France)	**(642) *G. nodosum* L.
11. Lvs. divided to the base.	*malviflorum*

643a. *G. malviflorum* Boiss. & Reuter A handsome erect leafy perennial with striking clusters of large rosy-purple flowers 3–4 cm across, recalling *Malva sylvestris*, and with distinctively lobed leaves. Petals 17–22 mm; calyx 8–10 mm. Basal leaves 7 cm wide, with 5 deeply cut lobes, each further divided into 3–4 linear-oblong pointed segments on each side. Rocky hillsides. S. Spain. Page 247.   645a. *G. endressii* Gay An endemic plant of wet places in the Pyrenees, with few large pink and strongly veined flowers, borne on long stalks. Petals about 1½ cm, oval, entire or slightly notched; sepals 9–10 mm. Leaves divided to 80 per cent of radius into 5 broad, nearly touching lobes, which are further irregularly cut and toothed.

650a. *G. cataractarum* Cosson An attractive perennial with bright pinkish-purple flowers with petals about 1½ cm, and twice as long as the sepals. Fruit hairless, wrinkled, separating into one-seeded units without a stylar beak. Damp, shady limestone rocks in mountains. S–E. Spain.   641a. *G. cinereum* Cav. A low-growing perennial with large lilac, or whitish and strongly veined, or deep reddish-purple flowers, arising paired on leafless stems directly from a very stout vertical stock. Petals about 1½ cm, shallowly notched, with a very short claw. Leaves all basal, circular, 2–3 cm across, divided to 80 per cent of radius into 5–7 lobes, which are further three-lobed, hairy and often silvery-grey beneath, but variable. Rocky and grassy places in mountains. Pyrenees. Pl. 24, Page 149.

ERODIUM | Storksbill Distinguished from *Geranium* by its fruits which have spirally twisted beaks derived from the style. 22 similar-looking species are found in our area. Important distinguishing features are the two depressions, or pits, on the one-seeded units, with or without furrows and ridges. Widespread are: 653. *E. cicutarium* (L.) L'Hér. COMMON STORKSBILL; Pl. 24, Page 61.   654. *E. moschatum* (L.) L'Hér. MUSK STORKSBILL.

(a) Leaves simple, toothed or shallowly lobed
**652. *E. malacoides* (L.) L'Hér. SOFT STORKSBILL. Distinguished by the pits on the fruits which are glandular and have a wide deep furrow on the lower margin lying between 2 ridges; beak of fruit 1½–3½ cm. Petals 5–9 mm, purplish. Waysides, dry fields, sandy places. Portugal, Med. region.   652a. *E. laciniatum* (Cav.) Willd. Distinguished by its shallow non-glandular pits in the fruits, without a furrow below; beak of fruit 3½–9 cm. Petals 7–10 mm, purplish. Maritime sands, dry fields. S. Portugal, Med. region. Page 163.

652b. *E. chium* (L.) Willd. Pits of fruits covered with glands but without a furrow below; beak of fruit 3–4 cm. Petals 5–9 mm, purplish. Sandy places, cultivated ground. Portugal, Med. region. 652c. *E. reichardii* (Murray) DC. A tiny mat-forming endemic of the Balearic Islands, with rounded crenate leaves ½–1 cm, and tiny solitary pink or white flowers with purple veins, borne on slender stems arising directly from the stock. Damp rocks. 652d. *E. guttatum* (Desf.) Willd. Flowers deep violet with a black centre; sepals 11–13 mm. Apical pits of fruit with a few small glands and with a shallow furrow at the base; beak of fruit long 6½–10 cm. Sandy or rocky places. S. Spain (Málaga province). Page 60.

652e. *E. boissieri* Cosson A stemless, densely white-haired perennial with umbels of 1–5 lilac flowers with purple veins. Petals 12–15 mm. Fruits 7–9 mm, apical pits without glands and without a furrow; beak 5–7½ cm. Limestone screes. S. Spain (S. Nevada). 653a. *E. botrys* (Cav.) Bertol. Flowers large, violet, with petals 1–1½ cm. Fruits 8–11 mm, with deep non-glandular pits with 2 furrows at the base and long beaks 5–11 cm. Leaves deeply lobed, lobes further toothed or cut. Dry places. Portugal, Med. region.

(*b*) *Leaves divided to the midvein; smaller leaflets alternating with the main leaflets absent*
E. acaule group, includes a number of closely related, stemless perennials with stalked umbels of 3–10 violet, purple or pink flowers, often with dark basal blotches to the petals. Calyx with a short point or none; bracts papery. Fruit with glandular or non-glandular pits, with or without a furrow at the base. 653b. *E. acaule* (L.) Becherer & Thell. Flowers lilac, without black blotches; petals equal; hairs on sepals adpressed, not glandular. Dry places. S. Portugal, Med. region. 653c. *E. carvifolium* Boiss. & Reuter An attractive species of this group, particularly characteristic of Soria (Spain), with large crimson flowers with the upper 2 petals larger and with a blackish basal patch. Roadsides, pine woods, mountain pastures. W–C. and N–C. Spain. Page 132. 653f. *E. daucoides* Boiss. A stemless perennial with a stalked umbel of pale pink or purplish flowers with the upper 2 petals with a dark basal patch. Distinguished from 653e by the pinnate leaves, without smaller alternating leaflets. Leaves glandular-hairy, or shaggy-haired. Fruits with a shallow pit with conspicuous glandular hairs and without a furrow at the base; beak 2½–4½ cm. Limestone rocks in mountains. Spain. Pl. 24, Page. 84.

(*c*) *Leaves divided to the midvein; smaller leaflets alternating with the main leaflets present*
653d. *E. petraeum* (Gouan) Willd. ROCK STORKSBILL. A very variable perennial, recognized by its stemless habit, very finely cut and usually densely hairy leaves, and usually strongly veined petals. Beak of fruit 18–33 mm. Subsp. *glandulosum* (Cav.) (*E. macradenum*) Bonnier has strong-smelling, densely glandular hairy leaves and large attractive, violet to purple flowers with the upper 2 petals larger, with dark blotches at the base. Subsp. *crispum* (Lapeyr.) Rouy (*E. cheilanthifolium*) has densely white-woolly usually non-glandular leaves, and white, pale pink or lilac flowers with red or purple veins, and with the 2 upper petals with black blotches at the base. Other subspecies have more or less equal petals without dark basal blotches and leaves variously cut, sparsely or densely hairy. Rocky places, mainly in mountains. Spain, S–W. France. Pl. 24, Page 139. 653e. *E. rupestre* (Cav.) Guittonneau A stemless perennial, with umbels of 1–3 pale pink flowers with darker veins, but distinguished from 653d by its leaves which are densely silvery-white above, and contrastingly green, almost hairless beneath. Beak of fruit short, 12–15 mm. Dry limestone rocks. N–E Spain. (654) *E. ciconium* (L.) L'Hér. Like 653a, with very long-beaked fruits, but petals much smaller about 8 mm, bluish or lilac, and leaves pinnate with smaller leaflets alternating with larger leaflets. Apical pits of fruits glandular and without a furrow at the base; beak long, 6–10 cm. Dry sandy, places, disturbed ground. Med. region.

ZYGOPHYLLACEAE | Caltrop Family

1. Lvs. alternate, divided to the base into linear-pointed lobes.   *PEGANUM*
1. Lvs. opposite with 2 or more distinct leaflets.
  2. Stipules spiny.   *FAGONIA*
  2. Stipules not spiny.
    3. Fls. yellow; fr. spiny.   *TRIBULUS*
    3. Fls. white; fr. not spiny.   *ZYGOPHYLLUM*

PEGANUM | 655a. *P. harmala* L. A somewhat fleshy-leaved, glaucous, erect and much-branched perennial with solitary, terminal, greenish-white flowers 1–2 cm across. Petals 5; sepals 5, linear, persisting; stamens 12–15. Leaves deeply and irregularly cut. Fruit a globular capsule, to 1 cm. Salt-rich semi-desert areas, steppes. S–E. and C. Spain. Page 254.

FAGONIA | 655b. *F. cretica* L. A prostrate, much-branched perennial with attractive solitary magenta flowers 1–1½ cm across, trifoliate leaves and spiny stipules. Sepals and petals 5; stamens 10. Leaflets linear ½–1½ cm, spine-tipped, the leaf resembling the imprint of a bird's foot; stems angled; stipules spiny. Fruit of 5 sharply angled carpels, angles ciliate. Arid regions. S–E. Spain. Pl. 23, Page 91.

ZYGOPHYLLUM | 655c. *Z. fabago* L. SYRIAN BEAN-CAPER A hairless, erect, somewhat glaucous perennial with leaves of 2 rather fleshy, rounded to elliptic, asymmetrical leaflets. Flowers usually paired, axillary, cream with orange spots at the base; filaments of stamens orange-yellow, much longer than the petals. Fruit 2–3½ cm, oblong-cylindrical, angled, pendulous. Dry places. Native of S–E. Europe; naturalized in Med. region. Page 254.   655d. *Z. album* L. fil. A small greyish shrub with leaves like 655c. but with cobweb-like hairs. Flowers solitary, white. Fruit ½–1 cm, erect or spreading downwards. Saline ground. N–E. Spain.

TRIBULUS | **655. *T. terrestris* L. MALTESE CROSS. A hairy, creeping annual with neat compound leaves with many elliptic leaflets, and small solitary yellowish axillary flowers about ½ cm across. Fruit very distinctive, five-starred, each lobe bearing 2 long and 2 short hard spines. Waste places, waysides, cultivated ground. S. and C. of our area.

LINACEAE | Flax Family

LINUM | **Flax**   15 species in our area.

Flowers yellow
(a) *Perennials*
(658) *L. campanulatum* L. The only species with large yellow flowers in our area. A woody-based perennial with angled stems, spathulate lower leaves and clusters of 3–5 somewhat tubular flowers with petals 2½–3½ cm. Sepals much longer than fruit. Limestone hills. E. Spain, S. France. Pl. 25.   658a. *L. maritimum* L. A tall perennial with yellow flowers, and with the lower leaves opposite, three-veined, the upper alternate, one-veined. Petals 8–15 mm; sepals egg-shaped, white; stigmas club-shaped. Salt-rich soils. S. Portugal, Med. region.

(b) Annuals

658b. *L. tenue* Desf. A slender erect, rather tall annual to 70 cm, with a lax branched inflorescence of many medium-sized yellow flowers. Petals 8–18 mm; sepals 3–5 mm, longer than fruit. Leaves linear to lance-shaped, 1–4 mm wide margin smooth, entire. Dry places, hills. S. and C. Portugal, S. Spain. Page 85.   657. *L. strictum* L. UPRIGHT YELLOW FLAX. Distinguished from 658b by its sepals which are glandular-hairy and minutely toothed on the margin. Petals 6–12 mm, yellow; stigma rounded. Leaves 1½–3 mm wide, margin minutely toothed, very rough and often inrolled. Dry hills, sandy and rocky places, vineyards. Portugal, Med. region.   657a. *L. setaceum* Brot. Like 657 but leaves bristle-like, ½ mm wide, with inrolled margin and clustered on the stem. Petals large 1–1½ cm, more than twice as long as the calyx. Sandy places, dry limestone hills and rocks. S. and C. Portugal, S. Spain.

Flowers white or pink

**(664) *L. suffruticosum* L. WHITE FLAX. A very attractive, often shrubby perennial with large white flowers with yellow, pink or violet centres. Petals 1–3 cm, 3 to 4 times as long as the sepals which have a glandular hairy, minutely toothed margin. Styles erect. Leaves to 1 mm wide, very variable, linear or bristle-like, rough, with inrolled margin. Banks, hills, rocks in dry places. S., C., and E. Spain, France. Pl. 25, Page 69.   663. *L. tenuifolium* L. Like (664) but leaves mostly flat ½–1 mm wide, and petals pink or almost white, 2 to 2½ times as long as the sepals. Styles spreading. Dry hills. C. and S–E. Spain, France.

662. *L. viscosum* L. STICKY FLAX. Easily recognized by its sticky-hairy leaves, 3–8 mm wide, its pink flowers about 3 cm across, and its glandular-hairy sepals. Shady and grassy places, by rivers, thickets. C., N., and E. Spain, Pyrenees. Pl. 24.   664. *L. catharticum* L. PURGING FLAX. Grassy places, heaths, dunes. Widespread.

Flowers blue

1. Bracts with papery margin; sepals 10–14 mm; petals 2½–4
   cm   *narbonense*
1. Bracts without papery margin; sepals 3½–9 mm; petals 1–2½
   cm.
   2. Stigmas rounded. (Widespread)   **659. *L. perenne* group
   2. Stigmas linear or club-shaped.
      3. Usually bienns. or perenns.; stems several; fr. 4–6 mm.
         (Widespread)   660. *L. bienne* Miller
      3. Anns.; stem solitary; fr. 6–9 mm. (Widespread)   (660) *L. usitatissimum* L.

661. *L. narbonense* L. BEAUTIFUL FLAX. A spectacular, somewhat glaucous perennial with large bright azure-blue flowers with petals 2½–3 times as long as the sepals. Sepals 10–14 mm, minutely toothed and with a papery margin; stigmas linear. Dry hills. N–E. Portugal, N. Spain, Med. region. Pl. 24, Page 84.

EUPHORBIACEAE | Spurge Family

SECURINEGA | 665b. *S. tinctoria* (L.) Rothm. A small spiny shrub with two-ranked, egg-shaped, blunt or notched leaves, 8–15 mm. Plants one-sexed; male flowers in clusters, with calyx and stamens 5–6, petals absent; female flowers solitary or 2–3. Fruit numerous, on slender stalks, three-lobed, about 3 mm. Sandy river banks. E. Portugal, S., W., and C. Spain. Pl. 26.

1. *Zygophyllum fabago* 655c
2. *Rhamnus lycioides* 720a
3. *Peganum harmala* 655a
4. *Ziziphus lotus* 725a

CHROZOPHORA | **665. *C. tinctoria* (L.) A. Juss. TURN-SOLE. Fields, cultivated ground. Southern part of our area.   665a. *C. obliqua* (Vahl) Sprengel Like 665 but a whitish perennial with cut-off or shallowly heart-shaped leaves (wedge-shaped in 665), and stamens 4–5, (9–11 in 665). Flowers yellowish, insignificant, one-sexed. S. and C. Spain.

MERCURIALIS | **Mercury**   **666. *M. perennis* L. DOGS MERCURY. Woods in the mountains. N. and E. of our area.

(666) *M. tomentosa* L. HAIRY MERCURY. A densely white-woolly perennial, with soft, almost stalkless, entire or shallowly toothed elliptic leaves. Fruit densely woolly-haired. Hedges, waysides, dry pastures. S. and C. of our area.   666a. *M. elliptica* Lam. A hairless perennial with elliptical to egg-shaped, toothed leaves. Fruit hairless. Hedges, waysides, undisturbed ground. S. and C. Portugal, S. and W. Spain.   667. *M. annua* L. ANNUAL MERCURY. Open ground, hedges. Widespread.

RICINUS | **668. *R. communis* L. CASTOR OIL PLANT. Usually a robust shrub or small tree-like plant in the south, less commonly an annual. Easily distinguished by its very large palmately five- to seven-lobed leaves and dense erect clusters of reddish flowers. Fruit large, covered with long conical spines; seeds large, smooth, mottled, with a swelling. A native of the tropics, widely naturalized in the south. Pl. 25.

EUPHORBIA | **Spurge**   About 60 species occur in our area; many are similar-looking and are not easily distinguished. Distinctive species are: **669. *E. dendroides* L. TREE SPURGE. A robust shrub forming dense rounded bushes up to 2 cm, with thick oblong-lance-shaped, blunt leaves. Glands triangular or rounded. Fruit smooth, or nearly so. Rocky slopes, usually near the sea. E. Spain.   **678. *E. characias* L. LARGE MEDITERRANEAN SPURGE. A stout, hairy perennial up to 1 m, with a terminal cylindrical head of numerous 'flowers', with dark reddish-brown glands with short rounded horns. Fruit densely woolly. Hills, rocky places, field verges. S. and C. Portugal, Med. region.

RUTACEAE | Rue Family

RUTA | **Rue**

1. Lv. segments linear *c.* 1 mm wide; petals not toothed or fringed.   *montana*
1. Lv. segments not linear; petals toothed or fringed.
   2. Petals fringed with long hairs.
      3. Bracts not or scarcely wider than branches which they subtend; plant glandular-hairy above.   *angustifolia*
      3. Lower bracts much wider than the branches which they subtend; plant hairless throughout. (South)   687. *R. chalepensis* L.
   2. Petals finely toothed, without long hairs. (Spain, France)   **686. *R. graveolens* L.

(686) *R. montana* (L.) L. A shrubby plant with feathery greyish leaves and erect, branched, sparsely leafy stems bearing dense, flat-topped, glandular heads of yellow flowers. Petals oblong, undulate; sepals long-pointed. Fruit about 3 mm. Dry places, arid rocky ground. Portugal, Med. region.   686a. *R. angustifolia* Pers. Distinguished by its small bracts and its strongly fringed petals with teeth often as long as the width of the petals. Dry places. S. Portugal, N. Spain, Med. region.

HAPLOPHYLLUM | Distinguished from *Ruta* by the simple, undivided or three-lobed leaves. Sepals and petals 5; stamens 10. Fruit five-lobed.   687a. *H. linifolium* (L.) G. Don fil. An erect, woody-based perennial with linear-lance-shaped, entire, stalkless leaves 1–3½ cm, and very dense flat-topped clusters of yellow flowers. Flowers 1½–2 cm across; petals blunt, gland-dotted; sepals hairless to white-felted. Fruit hairless or densely hairy, strongly warted. Arid hills and pastures. S., C. and E. Spain.

DICTAMNUS | **688. *D. albus* L. BURNING BUSH. An unmistakable, strongly aromatic, bushy perennial with a large lax terminal spike of pink or white flowers, and with flowering stems, bracts and sepals densely covered with conspicuous dark glands. Petals streaked and dotted with pink, unequal, 4 erect or spreading and 1 deflexed. Leaves pinnate, with 3–6 leathery leaflets. Bushy places. N–E. and C. Spain, France. Pl. 26, Page 110.

CITRUS | The following are commonly cultivated for their fruit and aromatic oils in S. Portugal and the Med. region: 689. *C. aurantium* L. SEVILLE ORANGE;   **690. *C. sinensis* (L.) Osbeck SWEET ORANGE;   691. *C. deliciosa* Ten. TANGERINE;   692. *C. medica* L. CITRON; **693. *C. limon* (L.) Burm. fil. LEMON;   693a. *C. limetta* Risso SWEET LIME;   693b. *C. bergamia* Risso & Poiteau;   693c. *C. paradisi* Macfadyen GRAPEFRUIT.

CNEORACEAE | Cneorum Family

CNEORUM | Fruit of 3 somewhat fleshy nuts attached to a central column. Sepals and petals 3–4.   694a. *C. tricoccon* L. An attractive, dense evergreen shrub up to 1 m, with small yellow flowers in axillary clusters and red fruits with 3 nuts which at length turn black. Flowers in axils of leaves; petals about 5 mm. Leaves 1–3 cm, oblong-blunt, stalkless. Rocky slopes, shady rocks. Med. Spain. Pl. 23.

SIMAROUBACEAE | Quassia Family

AILANTHUS | **694. *A. altissima* (Miller) Swingle TREE OF HEAVEN. Leaves ½–1 m, with 13–41 large oval long-pointed toothed leaflets. Fruit conspicuous, reddish-brown, three-winged. A native tree of China, often grown as an ornamental and for soil conservation, and often suckering. Pl. 25.

MELIACEAE | Mahogany Family

MELIA | 695. *M. azedarach* L. PERSIAN LILAC, INDIAN BEAD TREE. Flowers lilac. Leaves twice-pinnate. Fruit yellow, pea-sized. A native of China; commonly planted roadside tree in the south.

POLYGALACEAE | Milkwort Family

POLYGALA | Milkwort   About 17 similar-looking species are found in our area, 6 of which are restricted to the Iberian Peninsula in Europe. Wings refer to the 2 larger outer coloured sepals.   696b. *P. microphylla* L. A scrambling shrub with bare, broom-like branches, and elongated terminal and axillary clusters of large bright blue flowers. Petals and wings bright blue, up to 1 cm; keel without a crest. Leaves small, linear-lance-shaped, soon falling. 'Perhaps the finest flower of Northern Portugal and the neighbouring part of Spain, which hangs from the shady banks and is literally covered with blue flowers in myriads'. (Giuseppi) Pine and cork oak woods, scrub. C. and N. Portugal, W. Spain. Pl. 27, Page 103.

696a. *P. vayredae* Costa Like 696. *P. chamaebuxus* of the Alps, but with narrower linear to lance-shaped leaves and pinkish-purple flowers with a bright yellow keel. A prostrate shrublet 2½–15 cm, with very dark green leaves. Spain (E. Pyrenees).

697a. *P. rupestris* Pourret A small bushy woody-based plant with many finely-hairy stems, linear to oblong pointed leaves with inrolled margins. Flowers solitary or in few-flowered clusters of greenish-white tipped with purple. Wings greenish with papery margin; corolla about 5 mm, white, purple-tipped. Fruit winged, nearly twice as broad as wings. Rock crevices in lowlands and mountains. S. and S–E. Spain, S. France. Page 90.   699a. *P. boissieri* Cosson Recognized by its showy red-purple, or whitish, flowers up to 1½ cm long, in a terminal, one-sided cluster of 5–20 flowers. Wings about 1 cm, shorter than the straight corolla tube; keel exposed; bracts 1½–2 mm. A woody-based perennial with stems to 40 cm. Bushy and grassy places in mountains. S. and S–E. Spain. Page 77.   701. *P. calcarea* F. W. Schultz CHALK MILKWORT. An attractive, spreading perennial with lax rosettes of obovate leaves and conspicuous clusters of usually bright blue, or rarely white, flowers. Wings 5 mm, with 3–5 veins, shorter than the corolla and longer and broader than the fruit. Rocks, sandy places, (conspicuous roadside plant in N–E. Spain). S., N–E., and E. Spain, France. Pl. 27.

701a. *P. alpina* (Poiret) Steudel ALPINE MILKWORT;   698a. *P. alpestris* Reichenb. MOUNTAIN MILKWORT, are both alpine meadow plants of the Pyrenees.   **697. *P. monspeliaca* L. MONTPELLIER MILKWORT is a frequent annual in the Med. region.

CORIARIACEAE | Coriaria Family

CORIARIA | **702. *C. myrtifolia* L. MEDITERRANEAN CORIARIA. An erect and suckering shrub forming thickets, with quadrangular twigs bearing opposite, oval-lance-shaped leaves. Flowers in short lateral clusters, greenish with large red anthers and conspicuous projecting styles. Fruit reddish-purple, becoming shining black, five-lobed. Thickets, dry hills. Med. region: introd. Portugal.

ANACARDIACEAE | Cashew Family

RHUS | 707. *R. coriaria* L. SUMACH. A shrub or small tree to 3 m, distinguished by its softly hairy, pinnate leaves with a winged rachis and with 7–21 coarsely toothed, oblong to egg-shaped leaflets. Flowers whitish, in dense, erect, terminal elongated clusters. Fruit hairy, brownish-purple. Rocky places, waysides. Portugal, Med. region. Pl. 25.   (707) *R. typhina* L. STAGHORN SUMACH A N. American shrub, often grown in gardens in Spain.

PISTACIA | **703. *P. lentiscus* L. MASTIC TREE, LENTISC. A dense dark green, evergreen shrub with leaves with 8–12 shining elliptic leaflets with a winged rachis and without a terminal leaflet. Flowers in dense axillary clusters; fruit red, then black. One of the most abundant shrubs of the *matorral*. Portugal, C. Spain, Med. region.   **704. *P. terebinthus*

L. TURPENTINE TREE, TEREBINTH. Distinguished from 703 by its deciduous leaves with a terminal leaflet, and without a winged rachis; leaflets 5–11, reddish when young. Flowers reddish-purple in branched clusters; fruit red, then brown. Scrub and dry open woods. Portugal, C. Spain, Med. region. Pl. 25.

SCHINUS | 705. *S. molle* L. CALIFORNIAN PEPPER-TREE, PERUVIAN MASTIC-TREE. Planted for ornament in Portugal and the Med. region.   705a. *S. terebinthifolia* Raddi Recalling 703 with compound evergreen leaves, but distinguished from it by possessing a terminal leaflet, and usually a small tree. Distinguished from 705 by its evergreen leaves and its spreading, not pendulous, branches. Fruit smaller, 4–5 mm, bright red. Planted for ornament in the south-west.

ACERACEAE | Maple Family

ACER | Maple   The following are mostly widespread, but rarer in the south: **708. *A. pseudoplatanus* L. SYCAMORE;   **710. *A. platanoides* L. NORWAY MAPLE;   711. *A. campestre* L. COMMON MAPLE.

712. *A. monspessulanum* L. MONTPELLIER MAPLE. A small tree, easily recognized by its three-lobed leaves with the lobes untoothed and diverging almost at right angles to each other. Fruit with parallel or converging wings. Shady places in hills and mountains. Portugal, Med. region.   **709. *A. opalus* Miller Distinguished by its usually five-lobed leaves with lobes triangular-egg-shaped, toothed, and its fruits with wings diverging at an acute angle. Flowers yellowish, in flat-topped, nearly stalkless clusters, opening before the leaves. Shady places in hills and mountains. S. and E. Spain, Pyrenees, France.   709a. *A. granatense* Boiss. Like 709 but 3 main lobes of leaves more or less parallel-sided and 2 smaller subsidiary lobes present. Undersides of leaves, leaf stalks and young branches usually densely hairy. Montane regions S. Spain, Majorca. Pl. 26.

(712) *A. negundo* L. BOX-ELDER. A compound-leaved native tree of America; widely planted and sometimes naturalized as a wayside shade tree in the centre and south.

SAPINDACEAE | Soapberry Family

CARDIOSPERMUM | 712a. *C. halicacabum* L. HEART-SEED, HEART-PEA, BALLOON-VINE. A tendril-climbing annual, with twice-ternate leaves and small, white flowers in long-stalked, axillary clusters. Fruit dry, papery, three-lobed; seeds black with a white heart-shaped scar. Native of the tropics; grown for ornament and sometimes self-seeding in S. Spain.

KOELREUTERIA | 712b. *K. paniculata* Laxm. A picturesque tree, with pinnate leaves, erect broad clusters of many yellow flowers, and large inflated papery, three-lobed fruits. Native of China; often planted by roadsides for ornament in S. Spain.

HIPPOCASTANACEAE | Horse-Chestnut Family

AESCULUS | **713. *A. hippocastanum* L. HORSE-CHESTNUT. Widely grown for ornament and shade except in the Med. region.

BALSAMINACEAE | Balsam Family

IMPATIENS | Balsam   **714. *I. noli-tangere* L. TOUCH-ME-NOT. Shady woods, by streams. Pyrenees, France.   **716. *I. glandulifera* Royle POLICEMAN'S HELMET. A very robust, fetid annual with terminal clusters of large pink or white flowers. Native of the Himalaya; naturalized by streams. Pyrenees, France.

AQUIFOLIACEAE | Holly Family

ILEX | **717. *I. aquifolium* L. HOLLY. Mountains of N. Portugal, C., N., and E. Spain, France; often planted for ornament elsewhere.

CELASTRACEAE | Spindle-Tree Family

EUONYMUS | **718. *E. europaeus* L. SPINDLE-TREE. Throughout our area except in the south.

MAYTENUS | 718a. *M. senegalensis* (Lam.) Exell (*Catha europaea*) A much-branched, very spiny, grey-stemmed, evergreen shrub with somewhat glaucous leaves, and minute white flowers. Leaves narrow-egg-shaped to obovate, gradually narrowed to the base, 1–3 cm. Fruit shiny red, globular, about ½ cm. Rocky places. S. Spain (between Málaga and Almeria).

BUXACEAE | Box Family

BUXUS | 719. *B. sempervirens* L. BOX. Dry hills. Widespread, often abundant in the north. 719a. *B. balearica* Lam. Like 719, but distinguished by its larger leaves 2½–4 cm, broader inflorescence about 1 cm, and styles which are nearly as long as the fruit. S. Spain (near Nerja), Balearic Islands.

RHAMNACEAE | Buckthorn Family

RHAMNUS | Buckthorn   7 species.

**720. *R. alaternus* L. MEDITERRANEAN BUCKTHORN. An evergreen, spineless shrub, varying in habit, usually erect, but sometimes prostrate in the mountains, with lance-shaped to egg-shaped, leathery shining leaves. Flower clusters yellowish, small, axillary. Fruit globular, at first red then black. Common in *matorral*. Portugal, Spain, Med. France.   *R. myrtifolius* Willk. is a procumbent, much-branched shrub with small oblong-lance-shaped leaves, solitary flowers and smaller fruit. It is found in rock fissures in the mountains of S. Spain. It is probably only a form of 720. Pl. 28.   720a. *R. lycioides* L. A spiny shrub, with linear or oblong, evergreen or deciduous leaves, and obovoid, laterally compressed, yellowish or black fruits when ripe. Flowers usually with 4 sepals and stamens. Dry stony places, hedges. Portugal, Med. region. Page 254.

720b. *R. ludovici-salvatoris* Chodat A distinctive and handsome shrub found only in the Balearic Islands and E. Spain (near Valencia), with tough, rounded, leathery leaves 1–2¼ cm, with stiff spiny margins, shiny green above and silvery beneath. Flowers yellowish, in dense clusters. Fruit red. Limestone cliffs. Pl. 27.   **723. *R. catharticus* L. BUCKTHORN. In hills and mountains. Portugal, C., N., and E. Spain, France.

PALIURUS | 726. *P. spina-christi* Miller CHRIST'S THORN. An extremely spiny deciduous shrub with slender, often intricately intertwined branches, two-ranked egg-shaped leaves and distinctive fruits 'like miniature umbrellas'. Flowers tiny, yellow, in axillary clusters. Fruit flattened disk-like, ribbed and undulate on the margin. Sandy arid hills, hedges. S. and E. Spain, S. France.

ZIZIPHUS | 725. *Z. jujuba* Miller COMMON JUJUBE. A spiny shrub or small tree with green twigs, two-ranked leaves and conspicuous dark reddish or black ovoid-oblong edible fruits, 1½–3 cm. Flowers yellowish, few, in axillary clusters. Leaves oblong-blunt, with glandular-toothed margin. Native of Asia; often cultivated in the south.   725a. *Z. lotus* (L.) Lam. Distinguished from 725 by its grey, hairless, zig-zag twigs, very shallowly glandular-crenate oval leaves about 1½ cm, and globular deep yellow, pea-sized fruits. Dry sunny hills. S–E. Spain; occasionally cultivated and naturalized elsewhere. Page 254.

FRANGULA | 724. *F. alnus* Miller ALDER BUCKTHORN. Damp shady places. Scattered throughout, except in the Med. region.

VITACEAE | Vine Family

VITIS | 727. *V. vinifera* L. COMMON VINE. Cultivated throughout our area in a variety of forms, often grafted on to stock from American species.   Subsp. *sylvestris* (C. C. Gmelin) Hegi has small bluish-black acid fruits, and usually 3 seeds. Probably native of S–E. Europe; naturalized in our area.   Subsp. *vinifera* has larger fruits 6–22 mm, which are sweet and vary in colour from green, yellow, red or blackish-purple, with 2 or no seeds. It is the hybrid cultivated form used for winemaking, etc.

TILIACEAE | Lime Tree Family

TILIA | **Lime, Linden**   The following 3 species occur in the mountains of Spain and France: 729. *T. platyphyllos* Scop. LARGE-LEAVED LIME;   **731. *T. cordata* Miller SMALL-LEAVED LIME;   732. *T. × vulgaris* Hayne COMMON LIME.

MALVACEAE | Mallow Family

MALOPE | Fruits in a globular head; epicalyx-lobes wider than the sepals.   733. *M. malacoides* L. An erect or ascending, rough, hairy perennial with oval, irregularly lobed, long-stalked leaves and large rose-coloured flowers veined with purple (recalling 742). Petals 2–4 cm. Fruits strawberry-like but uncoloured, surrounded by broad-lobed epicalyx and longer, narrower calyx. Waste places, maritime sands. S–W. Spain, France.   733a. *M. trifida* Cav. Distinguished from 733 by the large leaves which are mostly broader than long, and larger flowers with deep reddish-purple petals 3½–6 cm. A nearly hairless annual, with a stout stem to 1½ m. Fields; sometimes grown for ornament. Rare. S–C. Portugal, S–W. Spain.

SIDA | Woody perennials; epicalyx absent.   733b. *S. rhombifolia* L. A low shrub with broadly egg-shaped to oblong, toothed leaves, and dull yellow, axillary flowers with petals 10–12 mm long. Native of Africa, America, and S–E. Asia; grown as a medicinal plant; sometimes naturalized in Portugal, S–W. Spain.

MALVELLA | Like *Malva* but fruit with individual nutlets inflated; petals not notched or two-lobed.   733c. *M. sherardiana* (L.) Jaub. & Spach A low, spreading, woolly-haired, woody-based perennial with small rounded leaves 1½–3½ cm. Flowers, small, deep pink with petals about 1 cm. Fruit about 1 cm across, woolly-haired. Cultivations, waste places. C. Spain. Page 263.

MALVA | **Mallow**   Epicalyx-lobes 2–3, in addition to the 5 calyx segments.

1. Sepals linear to narrowly triangular, more than 3 times as
   long as wide.
   2. Stamen-tube hairless. (S. and E. Spain)   736. *M. cretica* Cav.
   2. Stamen-tube downy.   *stipulacea*
1. Sepals egg-shaped or triangular, not more than 3 times as
   long as wide.
   3. Epicalyx-lobes egg-shaped to wedge-shaped, not more than
      3 times as long as wide.
      4. Ripe frs. smooth or faintly ribbed; lower fls. solitary in lv.
         axils.   *alcea*
      4. Ripe frs. distinctly net-veined; lower fls. 2 or more in each
         lv. axil.
         5. Petals 12–30 mm, 3–4 times as long as sepals; lower surface
            of sepals with numerous small, star-shaped hairs.
            (Widespread)   737. *M. sylvestris* L.
         5. Petals 10–12 mm, not more than 2½ times as long as sepals;
            lower surface of sepals with few star-shaped hairs or
            none. (Widespread)   **738. *M. nicaeensis* All.
   3. Epicalyx-lobes linear to narrowly egg-shaped, at least 3
      times as long as wide.
      6. Lower fls. solitary in lv. axils, or all fls. in a terminal
         cluster.
         7. Petals about equalling sepals.   *aegyptia*
         7. Petals at least twice as long as sepals.
            8. Middle stem lvs. lobed to at most one tenth of the radius;
               stamen-tube hairless.   *hispanica*
            8. Middle stem lvs. lobed to at least one fifth of the radius;
               stamen-tube hairy.
               9. Epicalyx-lobes 2, linear; frs. with flat back and sharp
                  angles.   *stipulacea*
               9. Epicalyx-lobes 3, narrowly oblong; frs. with rounded
                  back and rounded angles.
                  10. Fl. stalks with star-shaped hairs; frs. more or less
                      hairless.   *tournefortiana*
                  10. Fl. stalks without star-shaped hairs; frs. with
                      numerous white hairs.   *moschata*
      6. Lower fls. in groups in each lv. axil.
         11. Back of ripe frs. smooth, or only faintly ridged.

261

12. Petals at least twice as long as sepals; calyx not en-
larging in fr. (Widespread)   (738) *M. neglecta* Wallr.
12. Petals less than twice as long as sepals; calyx much
enlarging in fr. (France)   739. *M. verticillata* L.
11. Back of ripe frs. distinctly net-veined.
13. Petals at least 12 mm, usually bright purple or pink.
(Widespread)   737. *M. sylvestris* L.
13. Petals less than 12 mm, pale pink or lilac.
14. Calyx much enlarging in fr.; fr. stalks usually less
than 1 cm; angles of frs. winged. (Widespread)   738a. *M. parviflora* L.
14. Calyx only slightly enlarging in fr.; fr. stalks usually
more than 1 cm; angles of frs. not winged. (Wide-
spread)   **738. *M. nicaeensis* All.

736a. *M. hispanica* L. A hairy, branched annual with large, very pale pink flowers 5–6 cm
across, and rather small semi-circular, toothed leaves. Petals about 2½ cm; epicalyx-lobes
2, linear; sepals 6–9 mm, broadly egg-shaped. Roadsides, waste ground. S. and C. Iberian
peninsula. Pl. 26, Page 61.   736b. *M. aegyptia* L. A bristly annual with smaller lilac
flowers, mostly in terminal clusters, with petals about equalling the sepals. Epicalyx-lobes
usually 2, linear; sepals 7–11 mm, broadly triangular-egg-shaped. Leaves rounded, deeply
cut into narrow segments. Fruits usually hairless. Arid hills and salt-rich steppe areas. C.
and E. Spain.   736c. *M. stipulacea* Cav. Like 736b but petals twice as long as sepals, and
sepals twice as long as wide. Fruits densely hairy. Uncultivated fields. S., C., and E. Spain.

**735. *M. alcea* L. A very variable perennial to 1 m or more, with bright pink flowers with
notched petals, varying in size from 2½–3½ cm. Upper leaves usually deeply five-lobed with
lobes further toothed or lobed, lower leaves usually less deeply lobed. Flower stalks with
star-shaped hairs only; epicalyx-lobes broadly egg-shaped, densely hairy. Hedges, woods
and fields. Spain, France.

734. *M. moschata* L. MUSK MALLOW. Woods, hedges, thickets. Most of our area except
Portugal. Page 133.   734a. *M. tournefortiana* L. Like 734 but a more slender perennial with
upper leaves often deeply cut into narrow linear segments, and epicalyx-lobes usually
linear. Petals pale pink, 1½–2½ cm. Fallow fields, rocky and bushy places. Scattered throughout. Page 110.

LAVATERA | Very like *Malva* but usually distinguished by its epicalyx-lobes which are
fused together at the base, but some species, notably 741, have lobes almost free when in
full flower.

1. Fls. in clusters.
2. Simple, glandular hairs present in addition to other hairs;
stipules broad, sometimes clasping the stem.   *triloba*
2. Simple, glandular hairs absent; stipules narrow.
3. Epicalyx-lobes longer than sepals, at least in fr.; stems
woody below. (Widespread)   **740. *L. arborea*
3. Epicalyx-lobes shorter than sepals; stems all herbaceous.
4. Sepals much enlarging in fr.; frs. ridged, with flat back and
sharp, finely toothed angles.   *mauritanica*
4. Sepals only slightly enlarging in fr.; frs. smooth or slightly
ridged, with rounded back and rounded angles. (Wide-
spread)   741. *L. cretica* L.
                                                                     Page 163.

1. *Hypericum ericoides* 763b
2. *Malvella sherardiana* 733c
3. *Hypericum tomentosum* 764a
4. *Lavatera triloba* 740a
5. *Thymelaea lanuginosa* 754a

1. Fls solitary (rarely in pairs) in lv. axils.
  5. Anns.; stems not densely woolly-haired.
    6. Central axis of fr. expanded above to a disc that covers and
      conceals the ripe fr. segments; stem rough-hairy with
      simple or few-rayed, deflexed hairs.   *trimestris*
    6. Central axis of fr. not expanded above fr. segments; stem
      sparsely and minutely covered with star-shaped hairs. (S.
      France, Balearic Islands)   742a. *L. punctata* All.
  5. Perenns.; stems densely woolly-haired, at least when young.
    7. Lvs. about twice as long as wide, not lobed.   *oblongifolia*
    7. Lower lvs. nearly as long as wide; upper lvs. usually lobed.
      8. All lvs. nearly circular, scarcely lobed; frs. with concave
        back and very sharp angles.   *maritima*
      8. At least the middle lvs. distinctly three- to five-lobed; frs.
        with rounded back and angles.   *olbia*

740a. *L. triloba* L. A musk-scented, white or greyish shrubby perennial to 1 m or more, with clusters of purple, conspicuously veined flowers, sometimes flushed with yellow, or pure yellow. Epicalyx-lobes 8–15 mm, lance-shaped to broadly egg-shaped, shorter than the egg-shaped sepals which enlarge in fruit to about 2 cm. Leaves rounded or heart-shaped, slightly three-lobed. Damp sandy, often salt-rich places, ditches, streamsides. S. and C. Iberian peninsula. Page 263   740b. *L. mauritanica* Durieu An annual with smallish purple flowers with petals only 8–15 mm, and epicalyx-lobes nearly free and shorter than the sepals which greatly enlarge and cover the ripe fruit. Montane rocks. S. and C. Portugal, ?E. Spain.

**742. *L. trimestris* L. A tall annual with large showy, bright pink flowers 5–7 cm across and large baggy epicalyx. Flowers solitary; petals 2–5 cm; epicalyx-lobes shorter than sepals, enlarging in fruit, united for most of their length. Sandy areas, mainly by the sea. Portugal, Med. region.

743. *L. maritima* Gouan Readily distinguished by the pink or bluish-pink flowers, which have distinctive green, oval gaps where the sepals show between the bases of the petals. A whitish or greyish shrubby perennial, with small rounded shallowly five-lobed leaves, and with younger parts and leaves densely covered with star-shaped hairs. Petals 1½–3 cm, often with crimson-purple bases; epicalyx-lobes 3–8 mm, shorter than sepals and nearly free. Fruits hairless. Dry rocky places, often near the sea. Med. region. Page 90.

743a. *L. oblongifolia* Boiss. An endemic, grey-leaved, woody-based perennial to 1½ m of S. Spain (Granada and Almeria provinces), found in dry *Cistus*-scrub on limestone. Flowers pink, with a purplish base, in a spike-like cluster; petals 1½–2½ cm; epicalyx-lobes 6–8 mm, about half as long as the lance-shaped, erect sepals. Stems with dense tufts of yellowish, star-shaped hairs; leaves narrowly egg-shaped to oblong, toothed.

(743) *L. olbia* L. A robust handsome shrub to 2 m, with large purple-violet, short-stalked flowers in the axils of the upper leaves, forming an elongated spike-like inflorescence. Petals 1½–2 cm; epicalyx-lobes 7–13 mm, egg-shaped-pointed; sepals similar. Hedges, river banks, damp places. Portugal, Med. region.

ALTHAEA | Distinguished from *Malva* and *Lavatera* by the epicalyx-lobes which are 6 or more, and united at the base.

1. Anns.; both simple and star-shaped hairs present; anthers yellow.

2. Stipules entire; petals scarcely longer than sepals. (Widespread)   745. *A. hirsuta* L.
2. Stipules deeply divided into narrow lobes; petals twice as long as
   sepals.   *longiflora*
1. Perenns.; only star-shaped hairs present; anthers purplish-red.
3. Sepals erect in fr.; fr. segments hairless.   *cannabina*
3. Sepals curved over in fr.; fr. segments with star-shaped hairs.   *officinalis*

745a. *A. longiflora* Boiss. & Reuter An annual with conspicuous pinkish-lilac flowers and with the lower leaves rounded and toothed, the upper deeply three-lobed. Fruits keeled on the back. Hills and salt-rich areas. S. Spain.

**746. *A. cannabina* L. A tall perennial, readily distinguished by its deeply cut leaves which have 3–7 linear- to lance-shaped toothed or lobed segments. Flowers pink, solitary or in long-stalked axillary clusters; petals 1½–3 cm. Damp shady places, by rivers and ditches. S. and C. Iberian peninsula, S. France.   **747. *A. officinalis* L. MARSH MALLOW. A tall, densely grey-hairy perennial to 2 m, with upper leaves triangular-egg-shaped, toothed, entire or shallowly three- to five-lobed, and usually very pale lilac-pink flowers. Petals 1½–2 cm, rarely deeper pink. Damp places. Widespread.

ALCEA | (748) *A. rosea* L. HOLLYHOCK. Often grown for ornament; frequently self-seeding in our area.

ABUTILON | **749. *A. theophrasti* Medicus A weed of cultivation in E. Spain.

MODIOLA | Epicalyx segments 3. Fruits not detaching from fruiting head, but splitting and releasing seeds.   749a. *M. caroliniana* (L.) G. Don fil. A hairy, spreading annual with orange-scarlet axillary flowers, with petals 3–5 mm, slightly longer than the sepals. Epicalyx lobes 3. Leaves rounded, toothed to deeply three- to seven-lobed. Fruit black, with about 20 segments, each segment with 2 stout spines. Native of tropical America; naturalized in grassy places in N–W. Portugal and N. Spain.

GOSSYPIUM | **750. *G. herbaceum* L. LEVANT COTTON. Cultivated in the hotter regions of Spain; sometimes self-seeding on disturbed ground.

HIBISCUS | 751. *H. syriacus* L. SYRIAN KETMIA. Often planted for ornament in the south. 751a. *H. palustris* L. A tall herbaceous perennial to 1 m or more, with very large pink flowers, about 7 cm long (rarely white with a red base), and usually with 11 linear epicalyx lobes about half as long as the sepals which are fused for half their length. Leaves rounded to egg-shaped, long-pointed, toothed, densely and softly hairy beneath. Marshes and river banks. Portugal, S–W. France.   **752. *H. trionum* L. BLADDER KETMIA. An occasional weed of cultivation, in E. Spain.

THYMELAEACEAE | Daphne Family

THYMELAEA | 17 species occur in our area, 10 of which are restricted to S–W. Europe. All but 3 are dwarf evergreen shrubs with very tiny greenish or yellowish flowers and they are not easily distinguished from each other.   **753. *T. hirsuta* (L.) Endl. A distinctive shrub to 1 m with white woolly branches, numerous small fleshy, rather scale-like over-lapping, dark green leaves which are cottony above. Flowers small, yellowish, in clusters, densely hairy outside. Leaves 3–8 mm, usually blunt. Dry places, maritime sands. S–E. Portugal, Med. region.

**754. *T. tartonraira* (L.) All. A low branched, silvery-greyish shrublet with numerous oblong, silky leaves 1–2 cm, and small yellowish flowers in clusters amongst the upper leaves. Tube of flowers 5–6 mm, usually silky-haired. Dry stony places. Med. region. 754a. *T. lanuginosa* (Lam.) Ceballos & C. Vicioso A dwarf, silvery-greyish, woolly-leaved shrub like 754, with yellowish-green, strongly scented flowers with deep orange stamens. Distinguished by the tube of flowers which is 6½–8 mm, woolly-haired, and by its smaller leaves ½ cm or less. Maritime sands, rocks inland. S. Spain (frequent in the *Cistus*-scrub of Andalusia). Page 263.

745b. *T. dioica* (Gouan) All. A low, bushy, grey-stemmed shrub to ½ m, with numerous linear to oblance-shaped hairless leaves, and tiny hairless yellow flowers with narrow-triangular spreading lobes. Leaves 3–12 mm; young shoots hairless, older branches covered with raised leaf scars. Montane and alpine rocks. S–E. Spain, Pyrenees, Corbières. Pl. 27. 754c. *T. calycina* (Lapeyr.) Meissner Distinguished from 754b by the young shoots which are hairy, by the absence of prominent raised leaf scars, by the longer, inrolled leaves and the broadly egg-shaped lobes to the flowers. Alpine rocks. N. and N–E. Spain, Pyrenees. 754d. *T. tinctoria* (Pourret) Endl. Like 754b with rough, grey branches with prominent raised leaf scars and hairless yellow flowers, but distinguished from it by the young hairy shoots and usually hairy leaves. Flowers solitary; lobes broadly egg-shaped. Fruit hairless. Dry rocks in the hills. N–E. Spain, Pyrenees.

DAPHNE |

1. All fls. terminal, solitary, or in dense, more or less stalkless heads or clusters.
  2. Lvs. deciduous, not leathery. (Pyrenees, Cévennes)   ****757.** *D. alpina* L.
  2. Lvs. evergreen, more or less leathery.
    3. Sepals narrowly triangular, long-pointed; inflorescence without bracts.   *oleoides*
    3. Sepals egg-shaped or broadly triangular, obtuse or acute; fls. subtended by papery or leaf-like bracts.
      4. Fls. purple, tinged with yellow; lvs. fringed with hairs. (Minorca)   756a. *D. rodriguezii* Texidor
      4. Fls. pink; lvs. entire, not fringed with hairs.   *cneorum*
1. Fls. wholly or partly in axillary clusters, or in terminal branched clusters.
  5. Fls. greenish-yellow, hairless.   *laureola*
  5. Fls. white, cream or pink, often hairy.
    6. Lvs. deciduous, 8–25 mm wide.   *mezereum*
    6. Lvs. evergreen, leathery, 3–10 mm wide.   *gnidium*

(757) *D. oleoides* Schreber A small shrublet with leathery evergreen leaves and sweet-smelling white or cream-coloured flowers in terminal heads of 3–6. Fruit red. Mountains. S. Spain. ****756.** *D. cneorum* L. GARLAND FLOWER. Distinguished from 757 by its sweet-scented pink flowers in an almost stalkless head of 6–13. A low, evergreen shrublet with stalkless, hairless leaves 1–2 cm long. Dry stony places in the mountains. N. Spain, Pyrenees, France. Pl. 27.

760. *D. laureola* L. SPURGE LAUREL. Shady rocks, woods, bushy places in the hills and mountains. Spain, France. Subsp. *philippi* (Gren.) Rouy is a dwarf spreading shrub with yellowish flowers of N. Spain and Pyrenees. Pl. 26. ****759. *D. mezereum* L. MEZEREON. Woods and shrubby places in mountains. C. and N. Spain, Pyrenees, France. 758. *D.*

gnidium L. MEDITERRANEAN MEZEREON. An erect, evergreen shrub with linear pointed leaves, clusters of small cream-coloured flowers and red fruits, often present at the same time as the flowers. Widespread in the *matorral*. Portugal, Med. region. Pl. 27, Page 161.

ELAEAGNACEAE | Oleaster Family

HIPPOPHAE | **761. *H. rhamnoides* L. SEA BUCKTHORN. A spiny, much-branched, silvery-leaved shrub with striking orange fruits and tiny greenish flowers covered with rust-coloured scales. Leaves linear with silvery or rust-coloured scales; twigs with silvery scales. Sands, dunes, river gravels in the hills. S. and E. Spain, France.

ELAEAGNUS | **762. *E. angustifolia* L. OLEASTER. Native of Asia; often planted for ornament in the S. and S–E. and sometimes naturalized in the south.

GUTTIFERAE | St John's Wort Family

HYPERICUM | **St John's Wort** About 24 species are found in our area. The following are fairly widespread, particularly in the cooler northern and north-eastern regions, and in the mountains:

**764. *H. montanum* L. MOUNTAIN ST JOHN'S WORT; **765. *H. hirsutum* L. HAIRY ST JOHN'S WORT; 767. *H. tetrapterum* Fries SQUARE-STEMMED ST JOHN'S WORT; (767) *H. maculatum* Crantz IMPERFORATE ST JOHN'S WORT; **768. *H. perforatum* L. COMMON ST JOHN'S WORT; 769. *H. pulchrum* L. SLENDER ST JOHN'S WORT; 770. *H. humifusum* L. TRAILING ST JOHN'S WORT.
Other interesting and distinctive species are:

Shrubs
**763. *H. androsaemum* L. TUTSAN. Damp shady places, woods, riversides. Widespread except in the south. 763a. *H. balearicum* L. A very distinctive shrub restricted to the Balearic Islands, with small, oval, leathery leaves which are conspicuously crimped and covered with prominent resin glands. Flowers yellow, solitary, 1½–4 cm across; sepals rounded, spreading in fruit. A dense, much-branched shrub 15–120 cm. Dry woods and rocky places. Pl. 28.

763b. *H. ericoides* L. A dwarf heather-like shrub with tiny narrow leaves in whorls of 4, densely covered with minute swellings. Flowers yellow, about 1 cm across, in branched clusters; sepals with or without marginal glands. Sunny, limestone rocks. S., W., and E. Spain. Page 263.

Herbaceous perennials (sometimes woody-based)
(a) Hairless plants
770a. *H. linarifolium* Vahl A nearly hairless, usually erect perennial with linear to narrowly oblong, inrolled leaves, usually without translucent dots. Flowers yellow, in conspicuous rather compact clusters with orange-red buds. Petals 2–4 times as long as sepals; sepals with stalked black marginal glands and numerous black dots and streaks. Rocky places, heaths. Portugal, C., W., and N. Spain, France. 769a. *H. nummularium* L. A small erect or spreading perennial, creeping at the base, with neat pairs of rounded leaves which are green above and contrasting glaucous beneath. Flowers yellow, to 2 cm across, solitary

or in short flat-topped clusters; petals sometimes red-veined; sepals with black marginal glands. Leaves ½-1½ cm. Rock crevices, stony slopes. N. Spain, Pyrenees. Pl. 28.

**772. *H. richeri* Vill. ALPINE ST JOHN'S WORT. A more robust and erect perennial than 769a with larger clasping, oval leaves 1-5½ cm, and large yellow flowers with petals 1½-2½ cm, and covered with numerous black dots. Sepals acute, with black marginal glands, and numerous black streaks and dots. Leaves with black marginal dots. Montane meadows and woods. N. Spain, Pyrenees.

(b) Greyish woolly-leaved plants

764a. *H. tomentosum* L. A small, nearly prostrate, woody-based perennial with densely woolly-haired stems and leaves, and with small yellow flowers with petals 6-11 mm, with marginal black glands. Sepals 3-6 mm, woolly-haired, pointed, with glandular margin and usually with an apical gland. Leaves silvery-grey, ½-2½ cm, oblong to egg-shaped. Waysides, damp places. S. and C. Portugal, S. and C. Spain. Page 263.   764b. *H. pubescens* Boiss. Like 764a but flowers larger with petals 9-15 mm; sepals ½-1 cm, lance-shaped, long-pointed, with stalkless marginal glands but without an apical gland. Damp places. S. Portugal, S. Spain.   764c. *H. caprifolium* Boiss. A rather robust perennial 20-100 cm, with pairs of grey-haired leaves 2-5 cm, which are fused by their bases round the stem (perfoliate). Inflorescence usually dense; petals about 1 cm, with black marginal glands; sepals long-pointed, with numerous black dots over the surface and long-stalked glands on the margin. Damp shady places. S. and E. Spain. Pl. 28.

VIOLACEAE | Violet Family

VIOLA | **Violet, Pansy**   About 34 species are found in our area. The following are mostly widespread:

**773. *V. palustris* L. BOG VIOLET;   **774. *V. odorata* L. SWEET VIOLET;   775. *V. hirta* L. HAIRY VIOLET;   776. *V. canina* L. HEATH DOG VIOLET;   **777. *V. riviniana* Reichenb. COMMON DOG VIOLET;   (777) *V. reichenbachiana* Boreau PALE DOG VIOLET;   **783. *V. tricolor* L. WILD PANSY;   784. *V. arvensis* Murray FIELD PANSY.
Other interesting and distinctive species, often with limited distributions are:

Annuals

784a. *V. demetria* Boiss. A small spreading annual with very small, bright yellow flowers to 1 cm, and with a stout violet spur 2-3 mm, (upper petals sometimes violet). Lower leaves rounded, shallowly toothed, long-stalked, the upper ½-1½ cm, oval, coarsely toothed, and with deeply pinnately-lobed stipules nearly as long as the leaf. Grassy slopes, limestone rocks. W-C. Portugal, S-W. Spain. Page 68.

Herbaceous perennials

**785. *V. lutea* Hudson MOUNTAIN PANSY. An attractive montane perennial with large yellow, violet, or variegated flowers 1½-3 cm across, with the lower petal densely veined, and with a short slender spur 3-6 mm, not more than half as long as the petals. A creeping plant with slender rhizomes and slender erect stems; stipules palmately- or pinnately-lobed, with 3-5 lobes, the middle lobe larger. Mountain grasslands, rocky places. Pyrenees, Massif Central.   785a. *V. cornuta* L. A widespread endemic species of the Pyrenees distinguished by its sweet-smelling, violet or purple flowers which have long tapering spurs longer than the petals. Petals about 1 cm, narrow, not overlapping. Upper leaves oval-heart-shaped, toothed; stipules large, oval-triangular in outline, palmately-lobed, as long as or longer than the leaf stalk. Mountain pastures. Pyrenees. Pl. 28. *V. montcaunica* Pau an allied form with flowers half the size and with shorter spur, and deeply palmately lobed leaves C. and N. Spain (abundant in the mountains of Soria).

785c. *V. bubanii* Timb.-Lagr. Flowers blue-purple, large 2–3 cm across and spur slender, straight or slightly curved, about 1 cm. Leaves and stipules with spreading hairs; leaves small elliptic, with 2–3 teeth on margin; stipules deeply divided into linear-oblong segments, the lowest lobe directed downwards. Mountains. N. Spain, Pyrenees.

785b. *V. crassiuscula* Bory (*V. nevadensis*) A tiny cushion-forming perennial of the high screes of the Sierra Nevada, growing above 2500 m, with violet, pink to pale yellowish flowers 1–1½ cm across, and with a very short, stout spur. Lower petal golden at the base. Leaves rounded; stipules similar but smaller and entire. Pl. 29.

Woody-based perennials

781. *V. arborescens* L. SHRUBBY VIOLET. A distinctive shrubby plant, often growing up through more robust shrubs, with small whitish or pale violet flowers 1–1½ cm across, with a blunt curved spur. Leaves mostly linear-lance-shaped, entire; stipules narrow, lobed or deeply cut, one third as long as the leaves. Scrub in rocky places. S–W. Portugal, W. Med. region. Pl. 28, Page 52.   781a. *V. cazorlensis* Gand. A strikingly beautiful plant with intensely coloured pinkish-purple flowers with very long slender spurs 2–3 cm. A woody-based shrublet with slender herbaceous stems, narrow lance-shaped leaves and similar stipules. Limestone crevices in mountains. S–E. Spain (S. de Cazorla). Pl. 29, Page 85.

CISTACEAE | Rockrose Family

A very important family in S–W. Europe. Identification is sometimes difficult, particularly with *Helianthemum* species which are very variable; hybridization among *Cistus* species, and between *Cistus* and *Halimium* species occurs not infrequently. Two bulletins of the Ministerio de Agricultura, Madrid: *Jarales y Jaras* (*Cistografia Hispanica*) No. 49 (1949), and *Cistaceas Españolas* No. 71 (1954), describe and illustrate all species.

CISTUS | Fruit with 5, 6 to 10 cells; flowers white to reddish-purple; sepals all similar in size. Plants of dry open woodland, scrub and sunny rocky slopes.

1. Sepals 5.
   2. Style very slender, as long as the stamens.
      3. Petals white; fls. in one-sided clusters. (S. France) 791a. *C. varius* Pourret
      3. Petals purplish-pink or red; fls. not in one-sided clusters.
         4. Lvs. stalkless, with 3 more or less parallel veins.
            5. Lvs. not undulate; fl. stalks 5–20 mm. *albidus*
            5. Lvs. undulate; fl. stalks 1–5 mm. *crispus*
         4. Lvs. stalked, pinnately-veined.
            6. Lvs. usually 2–5 cm, variable in shape, the veins impressed above; sepals ovate-lance-shaped, long-pointed. (Balearic Islands) **787. *C. incanus* L.
            6. Lvs. ½–2 cm, elliptical, the veins not or scarcely impressed above; sepals ovate, acute. (S–E. Spain, (Cartagena)) 787a. *C. heterophyllus* Desf.
   2. Style absent or nearly so, thus stigma stalkless.

7. Lvs. stalkless or nearly so.
  8. Lvs. linear-lance-shaped or linear; outer sepals
    broadly wedge-shaped at base.   *monspeliensis*
  8. Lvs. oblong; outer sepals heart-shaped at base.   *psilosepalus*
7. Lvs. distinctly stalked.
  9. Lvs. 1–4 cm, rough above, rounded or wedge-shaped
    at base.   *salvifolius*
  9. Lvs. 4–10 cm, smooth above, heart-shaped at base.   *populifolius*
1. Sepals 3.
  10. Lvs. at least 6 mm wide; fls. at least 5 cm across.
    11. Inflorescence usually four- to eight-flowered; ovary
      five-celled.   *laurifolius*
    11. Fls. solitary; ovary six- to ten-celled.
      12. Lvs. linear-lance-shaped; ovary ten-celled.   *ladanifer*
      12. Lvs. oblance-shaped to spathulate; ovary six to
        ten-celled.*   *palhinhae*
  10. Lvs. not more than 4 mm wide, linear; fls. not more
    than 3 cm across.
    13. Fl. stalks and calyx clothed with long white hairs.   *clusii*
    13. Fl. stalks and calyx more or less hairless.   *libanotis*

**788. *C. albidus* L. GREY-LEAVED CISTUS. One of the prettiest of the genus with whitish-grey velvety leaves and large crumpled rose or magenta flowers 4–6 cm across. A compact bush to 1 m; leaves three-veined. Iberian peninsula except in the north, S. France. Pl. 30. 789. *C. crispus* L. Flowers purplish-red, crumpled, 3–4 cm across. Leaves greyish-green, undulate, three-veined. A rounded bush to ½ m. S. and C. Portugal, Med. region. Pl. 30, Page 161.

**790. *C. salvifolius* L. SAGE-LEAVED CISTUS. A sage-like shrub to 1 m, with rough, wrinkled, stalked leaves and long-stalked, white flowers often with orange centres, 3–5 cm across. Outer 2 sepals conspicuously larger and investing the inner 3. Widespread. Page 160. (790) *C. psilosepalus* Sweet (*C. hirsutus*) Easily confused with 790, having white flowers and large outer sepals, but the leaves are stalkless and the sepals have long spreading bristly hairs. Portugal, W. Spain. Page 102. **791. *C. monspeliensis* L. NARROW-LEAVED CISTUS. Distinguished by its linear to linear-lance-shaped, very sticky, aromatic, stalkless leaves, dark shining green above and grey-haired beneath. Flowers white, small, 2–3 cm across, in rather one-sided clusters. Portugal, Med. region. Page 160.

792. *C. populifolius* L. POPLAR-LEAVED CISTUS. A rather tall, woody shrub to 2 m, easily recognized by its broad green oval-heart-shaped leaves (which have an undulate margin in the southern part of its range). Flowers white, 4–6 cm across, in long-stalked clusters; outer sepals large, conspicuous, often coloured, with few or many long white hairs. Iberian peninsula except north, S. France. Pl. 31.

**793. *C. ladanifer* L. GUM CISTUS. A very sticky, fragrant, rather spindly shrub up to 3 m with large solitary, white flowers 7–10 cm, with or without a large purple blotch at the base of each petal. Leaves linear-lance-shaped, hairless above, densely white-woolly beneath. Often an important and dominant shrub in the southern part of our area. S. Iberian peninsula, S. France. Pl. 30, Page 160.

---

*Based on actual counts at Cape St Vincent; all previous authors state categorically that the ovary is six-celled. B.E.S.

793a. *C. palhinhae* Ingram A low, extremely sticky-leaved shrub like 793 with large, solitary, pure white or purple-blotched flowers. S–W. Portugal (Algarve). Pl. 31.

793b. *C. clusii* Dunal A much-branched, slender shrub with linear leaves with inrolled margins, dark green above and white-woolly beneath, and terminal umbel-like clusters of small white flowers. Inflorescence stalk, flower stalks and calyx densely covered with long white hairs; flowers 2–3 cm across. Easily mistaken for 795. *H. umbellatum* but fruit five-celled. The most drought-resistant and lime-loving *Cistus* species. S., C., and E. Spain. Pl. 30.   793c. *C. libanotis* L. (*C. bourgaeanus*) Very like 793b, but leaves larger and broader 2–3½ cm by 2–3 mm, and inflorescence stalks, flower stalks and calyx hairless but sticky. Sandy soils under pine trees, near the coast. S–W. Portugal, S–W. Spain. Page 272.

**794. *C. laurifolius* L. LAUREL-LEAVED CISTUS A dense, dark green shrub to 1½ m, with oval to lance-shaped, stalked leaves, and a terminal stalked umbel-like cluster (thus differing from 793) of large white flowers 5–6 cm across. N. and C. Portugal, S., C., and E. Spain, S. France. Pl. 31, Page 85.

HALIMIUM | A handsome genus of small *Cistus*-like shrubs distinguished by their three-celled capsules and yellow or white flowers. Restricted almost entirely in Europe to the Iberian peninsula. A difficult genus with some hybridization.

1. Lvs. egg-shaped to lance-shaped, more than 4 mm wide, flat, or with the margins only slightly inrolled.
   2. Lvs. of flowering shoots nearly hairless, stalk-less; those of sterile shoots smaller, stalked, and woolly-haired; inflorescence lax, long-stalked.   *ocymoides*
   2. Lvs. all alike.
      3. Sepals 5, covered with flat-topped (peltate) scales   *halimifolium*
      3. Sepals 3, hairy or woolly-haired.
         4. Flowering branches and fl. stalks densely woolly-haired and with spreading, purplish, sticky hairs.   *atriplicifolium*
         4. Flowering branches and fl. stalks silky-haired or shaggy-haired, without purplish hairs.
            5. Inflorescence shaggy-haired, with short hairs; sepals without purple bristles; petals unspotted.   *alyssoides*
            5. Inflorescence silvery, with long hairs; sepals often with long purple bristles; petals sometimes with a brown spot at base.   *lasianthum*
1. Lvs. linear, less than 4 mm wide, the margins strongly inrolled.
   6. Fls. yellow; sepals hairless.   *commutatum*
   6. Fls. white; sepals shaggy-haired.
      7. Stems 15–25 cm; branches tortuous; lvs. crowded at ends of branches.   *umbellatum*
      7. Stems 25–60 cm; branches straight; lvs. distributed along the branches.

1. *Tuberaria lignosa* 799
2. *Cistus libanotis* 793c
3. *Fumana scoparia* 805a
4. *Halimium atriplicifolium* 797a
5. *Helianthemum squamatum* 801b

8. Branches with dense, whitish hairs; fl. stalks
   unequal.   *viscosum*
8. Branches sparsely covered with star-shaped
   hairs; fl. stalks equal. (S–W. Portugal)   795b. *H. verticillatum* (Brot.) Sennen

**795. *H. umbellatum* (L.) Spach A low shrub with narrow heather-like leaves, white-felted twigs and terminal, sometimes whorled, clusters of rather small white flowers, recalling 793b. Petals 1–1½ cm; sepals woolly-haired. Bushy places, pine woods. N. Portugal, N. Spain, S–W. and C. France. 795a. *H. viscosum* (Willk.) P. Silva A taller, glandular-hairy shrub from 25–60 cm, with more or less straight erect branches, and leaves 1–2 cm evenly distributed along the branches. Flowers white, borne in 3–5 whorls, with 5–6 in each whorl. Heaths, oak and pine woods. E. Portugal, S–C. and E. Spain. Pl. 29. **796. *H. commutatum* Pau A low heath-like shrub with long-stalked, pale yellow flowers in terminal clusters of 1–3. Flowers 1–1½ cm across; sepals hairless. Leaves 1–3½ cm, recalling those of Rosemary, dark green above, inrolled and white-woolly beneath. Mainly coastal sands. S. and W. Portugal, S–W. Spain. Pl. 29, Page 53.

797. *H. halimifolium* (L.) Willk. An erect, much-branched shrub to 1 m, with white or grey-felted leaves, and much-branched clusters of medium-sized yellow, often black-spotted flowers. Flowers 2–3 cm across; sepals 5, densely covered with flat (peltate) scales. Leaves narrowly elliptic to spathulate, 1–4 cm. Sandy hills, on the littoral. S. Iberian peninsula.

797a. *H. atriplicifolium* (Lam.) Spach Probably the finest member of the genus with large golden-yellow flowers 4–5 cm across, borne on long sticky, hairy, branched inflorescences. Sepals 3. A tall shrub 1–1½ m; lower leaves stalked, broadly egg-shaped, 2–5 cm, upper leaves stalkless, three-veined, all silvery-grey. Open pine forests, *matorral*. S. and C. Spain. Page 272. 797b. *H. alyssoides* (Lam.) C. Koch A compact, tufted, greyish-green shrub 10–100 cm, with short terminal clusters of relatively few, brilliant yellow, unspotted flowers each 3–4 cm across. Leaves flat, oblong or ovate-lance-shaped, the lowest short-stalked, the upper stalkless, more or less covered with star-shaped hairs. Heaths and degraded woodlands. Northern half of Iberian peninsula, W. and C. France. Pl. 29, Page 103. 797c. *H. lasianthum* (Lam.) Spach Like 797b but leaves more acute; flower stalks and sepals with long silvery hairs; sepals often with purple bristles. Flowers 2–3 cm, and petals spotted or unspotted. Cork oak woods. S. Portugal, S–W. Spain. Subsp. *formosum* (Curtis).Heywood Flowers larger 4–6 cm, petals spotted well above base. Algarve. Pl. 29.

797d. *H. ocymoides* (Lam.) Willk. A much-branched shrub to 1 m, readily recognized by two forms of leaves: those on the sterile shoots, obovate, stalked, silvery-white, those on flowering shoots much longer, lance-shaped, stalkless. Flowers brilliant yellow, usually dark-spotted, borne in terminal lax long-stalked clusters. Petals 1–1½ cm; sepals 3, often hairless. Heaths, pinewoods, sands. W. and C. Iberian peninsula. Page 111.

TUBERARIA | Usually distinguished by the presence of 5 sepals with the outer 2 usually smaller than the inner 3. Flowers yellow; style absent. Basal leaf rosette present.

1. Perenns. with woody stock and persistent
   basal rosettes of lvs.; flowering stems with
   small bract-like lvs. only.
   2. Rosette lvs. gradually narrowed at the
      base; petals unspotted.   *lignosa*
   2. Rosette lvs. abruptly narrowed into a
      distinct stalk; petals with a dark spot at
      base.

3. Veins of lv. distinctly forking; bracts
   lance-shaped.   *globularifolia*
3. Veins of lv. not forking; bracts broadly
   egg-shaped, blunt. (S. Portugal)   799b. *T. major* (Willk.) P. Silva & Rozeira
1. Anns., with basal rosette often dead at
   flowering time; fl. stems leafy.
   4. Fls. almost stalkless, in dense curved, one-
      sided clusters.   *echioides*
   4. Fls. distinctly stalked.
      5. Outer sepals much smaller than the inner,
         not enlarging.
         6. Lvs. obovate to lance-shaped or oblong
            (the uppermost linear), shaggy-haired,
            flat (or the uppermost with the margins
            more or less inrolled).   *guttata*
         6. Lvs. lance-shaped or linear, at least the
            upper becoming hairless, the margins
            distinctly inrolled. (S. Portugal, S–W.
            Spain)   798a. *T. bupleurifolia* (Lam.) Willk.
      5. Outer sepals more or less equalling the
         inner, all enlarging in fr. (E–C. Portugal,
         S–W. Spain)   798b. *T. macrosepala* (Cosson) Willk.

799. *T. lignosa* (Sweet) Samp. A handsome perennial with hairy plantain-like rosettes of
elliptic basal leaves. Stems erect with pairs of narrow, hairless leaves, and quite hairless,
branched clusters of pale yellow flowers. Flowers 2–3 cm across. Stony and sandy places,
scrub, pine woods. Widespread except in the north. Page 272.   799a. *T. globularifolia*
(Lam.) Willk. Like 799 but flowers larger 3–5 cm across, with a purple-brown blotch at the
base. Rosette leaves abruptly narrowed to a stalk which is about as long as the blade.
Heaths and scrub on sands. N–W. part of Iberian peninsula. Page 103.

**798. *T. guttata* (L.) Fourr. SPOTTED ROCKROSE. A distinctive slender annual with small
yellow flowers with a striking large, dark purple-brown central blotch, but very variable
in flower size and degree of blotching. Leaf rosette persisting at flowering; flowering stems
slender, hairy; fruit stalks reflexed. Open sandy ground, pine woods. Widespread. Page 163.
798c. *T. echioides* (Lam.) Willk. A rather robust, downy-white annual, readily distinguished
by its densely clustered head of rather inconspicuous yellow flowers, with large triangular
outer sepals. Leaves oblong-lance-shaped, soon falling and plant becoming leafless in fruit.
Coastal sands, waste places. S–W. Spain.

HELIANTHEMUM | Rockrose 27 species occur in our area including about 10 which
are restricted to the Iberian peninsula. Some species are very variable and have distinctive
subspecies, making a key to their identification unsatisfactory.

Shrubby plants with distinctive, branched inflorescences
801a. *H. lavandulifolium* Miller Readily distinguished by its forked inflorescence, which
is at first dense and coiled and later spreading, with numerous drooping fruits, placed
regularly along the lower side of the branches. Flowers yellow, about 1½ cm across, erect;
calyx shorter than the flower stalks. Leaves linear-lance-shaped, inrolled, greyish-green
above, white-woolly beneath. Limestone rocks, dry hills, pine woods. Med. region. Pl. 32,
Page 162.   801b. *H. squamatum* (L.) Pers. Recognized by the silvery, plate-like scales which
cover the leaves and stems. Inflorescence irregularly forked, bearing numerous small
yellow flowers densely clustered along the branches. Petals only slightly longer than sepals

which are 4 mm. Leaves silvery, lance-shaped, fleshy, flat. On gypsum soils. S., C., and E. Spain. Page 272. 801c. *H. caput-felis* Boiss. Distinctive because the whole plant is white-felted and the inflorescence is compact and few-flowered. Calyx covered with long, spreading, white silky hairs giving a cat-like appearance before and after flowering. Petals yellow 9–12 mm, longer than the sepals. Coastal limestone cliffs and dry places inland. S–E. Spain (Alicante province), Balearic Islands. 801d. *H. cinereum* (Cav.) Pers. Inflorescence lax, branched, nearly leafless, bearing numerous small yellow flowers. Flowers about 8 mm across; calyx densely covered with white bristly hairs. Basal leaves elliptic, grey-woolly beneath, green and often hairless, or grey-woolly above, but very variable. Arid sandy and stony places, limestone hills. Med. region.

Straggling, woody-based shrublets with unbranched inflorescences
(a) All leaves with stipules
802b. *H. hirtum* (L.) Miller All leaves with stipules, and lower surface of leaf with dense star-shaped hairs. Flowers 1½ cm across, white or yellow; sepals with bristly hairs on the ribs and margin and star-shaped hairs between the ribs. Scrub, woods and dry places. Widespread except in north. 802c. *H. asperum* Dunal Similar to 802b with long bristly hairs on the margin and ribs of the sepals, but sepals inflated and the spaces between the ribs hairless and shining. Flowers white, to 2 cm across. Leaves with star-shaped hairs on upper surface; grey-woolly beneath. Dry and rocky places on calcareous soil. S., C. and E. Spain. 802d. *H. viscarium* Boiss. & Reuter Distinguished from 802c by being more or less glandular-hairy all over. Leaves hairless beneath, with inrolled margin, dark green above. Flowers white, about 1½ cm. Sepals strongly ribbed, with few bristles, hairless between the ribs. Stony ground. S–E. Spain.

**802. *H. nummularium* (L.) Miller COMMON ROCKROSE. An extremely variable species which can be distinguished by the following combination of characters. Leaves all with rather leaf-like stipules longer than the leaf stalk; leaf blades either green on both sides, or grey or white-woolly below or on both sides. Inflorescence simple, one-sided; flowers golden-yellow, pale yellow, creamy white, orange or pink; sepals hairy or hairless between the ribs. Capsule equalling the sepals. Dry stony, or grassy places. Widespread. 802a *H. croceum* (Desf.) Pers. Distinguished from 802 by its leaves which are usually densely covered with star-shaped hairs on both surfaces, and are somewhat fleshy. Flowers to 2 cm, orange-yellow, bright yellow or white; sepals with star-shaped hairs between the ribs. Scrub, woods, mountain slopes. Widespread. Pl. 32.

**803. *H. apenninum* (L.) Miller WHITE ROCKROSE. Like 802a but leaves green, grey or white and densely woolly-haired above, with dense star-shaped hairs below and with inrolled margin. Stipules not leaf-like, linear. Flowers white with a yellow base (a form with pink flowers occurs in the Balearic Islands); calyx shortly hairy all over. Dry hills. Scattered throughout. Page 76.

803a. *H. pilosum* (L.) Pers. A tufted, much-branched shrublet with branches and leaves covered with dense white felt-like hairs. Flowers small about 1 cm across, white with a yellow base; sepals 5–6 mm in fruit, hairless or with star-shaped hairs on the ribs. Scrub, clearings in forests. Scattered throughout except in the north. 803b. *H. almeriense* Pau Like some forms of 803a but plant quite hairless; flowers white with a yellow base. Spain (Almeria province). Page 91.

(b) Lower leaves without stipules
801. *H. canum* (L.) Baumg. HOARY ROCKROSE. A very variable species distinguished by the absence of stipules to all leaves except on the upper flowering shoots. Flowers small, petals 4–8 mm, yellow. Leaves all grey-woolly beneath, green to grey above, with or without star-shaped hairs, elliptic to linear with wedge-shaped base. Dry sandy and calcareous places.

Spain, S. France.   801e. *H. marifolium* (L.) Miller Distinguished from 801 by its egg-shaped leaves which have a heart-shaped base, hairless or greyish-woolly above and grey- or white-woolly beneath. Flowers 1–1½ cm across, in simple or branched clusters; calyx with bristly hairs. Stony places. Southern part of our area.   801f. *H. origanifolium* (Lam.) Pers. Leaves broadly egg-shaped like 801e, but green on both surfaces with sparse or dense star-shaped hairs. Flowers yellow; petals 3–6 mm. Sandy and rocky places on limestone. S. Portugal, S. and E. Spain.   801g. *H. oelandicum* (L.) DC. Distinguished from 801f by its linear to elliptic leaves which are green and hairless on both surfaces, or with simple or clustered, not star-shaped hairs. A very variable species with yellow flowers. Rocks, dry sandy places from the lowlands to the mountains. Widespread.

FUMANA | Distinguished from *Helianthemum* by the upper leaves which are alternate and more or less linear, and the outer stamens which are sterile. Small heath-like shrubs of dry rocky and sandy places, and low scrub very similar to each other in general appearance.

1. Lvs. more or less equally spaced on stems, not or scarcely reduced above.
  2. Plant procumbent, fr. stalks as long as or shorter than adjacent lvs., recurved from the base.   *procumbens*
  2. Plant erect or straggling-upwards; fr. stalks much longer than adjacent lvs., spreading, with deflexed apex. (South)   (805) *F. ericoides* (Cav.) Gand.
1. Lvs. unequally spaced on stem, more or less abruptly reduced above to form small bracts in the inflorescence.
  3. Lvs. opposite, with stipules.   *thymifolia*
  3. Lvs. alternate, usually without stipules.
    4. Lvs. linear and bristle-like, more or less cylindrical, with stipules. (South)   805b. *F. laevipes* (L.) Spach
    4. Lvs. linear to linear-lance-shaped, more or less three-angled, without stipules.
      5. Fl. stalks and calyx not glandular. (S–E. Spain)   805c. *F. paradoxa* Heywood
      5. Fl. stalks and usually calyx glandular.   *scoparia*

804. *F. thymifolia* (L.) Webb A small twiggy shrub with heath-like leaves, and lax few-flowered clusters of small yellow flowers, each about 1 cm across. Leaves opposite, with shorter clusters of leaves in their axils; bracts very small. Portugal, Med. region.   **805. *F. procumbens* (Dunal) Gren. & Godron. Very like 804 but flowers usually solitary, axillary or terminal; leaves all alternate. Widespread, except in north.   805a. *F. scoparia* Pomel Like 805 with alternate leaves, but flowers in a terminal, lax cluster of 2–5, with glandular hairy flower stalks and calyx, and bracts much smaller than the leaves. Med. Spain. Page 272.

TAMARICACEAE | Tamarisk Family

TAMARIX | Tamarisk

1. Sepals and petals 4. (S–E. Spain)   807a. *T. boveana* Bunge
1. Sepals and petals 5.
  2. Petals 2–3 mm, at least some of them persistent.   *africana*
  2. Petals 1½–2 mm, falling early.

3. Bracts equalling or exceeding calyx; plant covered with
small swellings. (Portugal, Med. region)   807b. *T. canariensis* Willd.
3. Bracts not extending beyond the middle of the calyx;
plant entirely hairless. (Coasts throughout)   806. *T. gallica* L.

**807. *T. africana* Poiret A handsome, feathery shrub or small tree with dark bark, small bright green scale-like leaves and cylindrical clusters of white or pale pink flowers forming a plume-like inflorescence. Coastal marshes, river banks. S. and C. Portugal, S. Spain, S. France.

MYRICARIA | **808. *M. germanica* (L.) Desv. GERMAN TAMARISK. A feathery, glaucous, heath-like shrub to 2½ m, of streamsides, and river gravels, with slender terminal, tassel-like clusters of numerous pink flowers. Leaves numerous, scale-like. Seeds with a plume of hairs. S–E., E. Spain, Pyrenees, France.

FRANKENIACEAE | Sea Heath Family

FRANKENIA | 6 species occur in our area, in salt-rich and maritime places: 810a. *F. corymbosa* Desf. A much-branched, dense, greyish thyme-like shrublet to 30 cm, with numerous tiny, pale pink flowers in dense flat-topped or domed clusters. Buds pink; petals 4–6 mm; calyx 2–3 mm. Leaves linear 2–6 mm, in apparent whorls, partially covered with a white crust or powder. S. Spain. Pl. 32.   810b. *F. boissieri* Boiss. Like 810a, but flowers larger; petals 5–7 mm; calyx 4–6 mm. Leaves without a white crust. S. Portugal, S–W. Spain. 810c. *F. thymifolia* Desf. Distinguished by the flowers which are in long one-sided, terminal spikes. Petals about 7 mm; calyx 2½–4 mm. Leaves completely covered with a white crust. S., C., and E. Spain. Page 296.   810. *F. laevis* L. SEA HEATH. A mat-forming perennial with pink, violet or whitish flowers about ½ cm across. Coasts throughout.

CUCURBITACEAE | Gourd Family

The following are often cultivated in Southern Europe for food, particularly in the rich *huertas* of the coastal regions: 812. *Citrullus lanatus* (Thunb.) Mansfeld WATER MELON; 813. *Cucurbita maxima* Duchesne PUMPKIN;   814. *Cucurbita pepo* L. VEGETABLE MARROW; 816. *Cucumis sativus* L. CUCUMBER;   817. *Cucumis melo* L. MELON.

ECBALLIUM | **811. *E. elaterium* (L.) A. Richard SQUIRTING CUCUMBER. A stout spreading, somewhat fleshy, rough-leaved perennial with yellow bell-shaped flowers and long-stalked, inclined, sausage-shaped fruits, which explode violently when touched. Roadsides, waste places, by the sea. Portugal, Med. region.

BRYONIA **815. *B. cretica T.* WHITE BRYONY. Hedges. Widespread.

CITRULLUS | **(812) *C. colocynthis* (L.) Schrader BITTER APPLE, BITTER CUCUMBER. A spreading perennial with deeply once or twice pinnately lobed leaves, and solitary greenish-yellow flowers. Fruit globular, to about 8 cm, pale yellow. Sandy places by the sea. E. Spain.

CUCUMIS | 816a. *C. myriocarpus* Naudin A rough-haired annual with deep yellow flowers 4–5 mm long; distinguished by its small globular fruits, which are sparsely covered with soft spines. Leaves deeply five-lobed. Native of S. Africa; locally naturalized in C. Portugal and Spain.

CACTACEAE | Cactus Family

OPUNTIA | ****818.** *O. ficus-indica* (L.) Miller PRICKLY PEAR, BARBARY FIG. Quite unmistakable with its large thick, spineless and spiny, racket-shaped joints forming a large tree-like growth. Flowers bright yellow, with numerous petals. Fruit large, fleshy and edible, red, yellow or purple. Widely naturalized in Portugal and Med. region.   818a. *O. tuna* (L.) Miller Differs from 818 in being a smaller plant to 1 m and having oblong, readily detachable joints, and pale yellow spines up to 5 cm long in groups of 3–5. Flowers yellow, red-tinged. Fruit about 3 cm, red. Occasionally naturalized in Portugal and Med. region.

LYTHRACEAE | Loosestrife Family

LYTHRUM | About 11 species occur in our area, mostly plants of small stature and with insignificant flowers, except the following: ****820.** *L. salicaria* L. PURPLE LOOSESTRIFE. Lakes, riversides, marshes. Widespread, particularly in the north.   822a. *L. junceum* Banks & Solander A slender, straggling perennial with long unbranched stems, bearing regularly placed, linear-oblong leaves, each with a solitary pink flower in its axil. Petals 5–6 mm, sometimes cream or white; 'calyx' tubular, spotted with red near the base, with pointed lobes alternating with blunt scales and shorter than the 12 stamens. Leaves of sterile shoots broader. Wet places. Portugal, Med. region. Pl. 32.   822b. *L. acutangulum* Lag. Very like 822a, but an annual with pink flowers with a distinctive, clearly marked white centre. 'Calyx' not red-spotted, but with narrow papery ribs running down from the pointed calyx-lobes, which are twice as long as the alternating scales. Wet places. S. and C. Spain, S. France.   ****822.** *L. hyssopifolia* L. GRASS POLY Like 822a but petals usually 6, 2–3 mm, and stamens usually 4–6, and included in the 'calyx'. An annual with erect or ascending branches, linear-oblong leaves. Damp places, streamsides. Scattered throughout.

MYRTACEAE | Myrtle Family

MYRTUS | ****824.** *M. communis* L. MYRTLE. A dense evergreen shrub with shining leaves and solitary, axillary, sweet-scented white flowers 2–3 cm across, with numerous projecting stamens. Leaves 2–3 cm, oval-lance-shaped, gland-dotted, aromatic. Fruit bluish-black. Bushy places, scrub, damp ground. Portugal, Med. region. Page 161.

EUCALYPTUS | Australian trees well suited to the Mediterranean climate. The following are sometimes extensively planted in the south and centre of our area:

1. Fls. solitary; fr. more than 1 cm.   *globulus*
1. Fls. in umbels; fr. less than 1 cm.
  2. Fr. stalkless or nearly so.
    3. Umbels three-flowered; inflorescence stalk rounded. (826) *E. viminalis* Labill.
    3. Umbels five- to ten-flowered; inflorescence stalk flattened.
      4. Lvs. more than 18 cm; bark smooth; fr. glaucous. 825a. *E. maidenii* F. Mueller
      4. Lvs. less than 18 cm; bark fibrous; fr. not glaucous.

5. Inflorescence stalk 7–10 mm; fr. 7–9 mm, cylindrical.   825b. *E. botryoides* Sm.
5. Inflorescence stalk 2½–3½ cm; fr. 13–20 mm, bell-
   shaped.   825c. *E. gomphocephalus* DC.
2. Fr. distinctly stalked.
  6. Inflorescence stalk flattened, angular or strap-shaped.
    7. Fr. 5–8 mm; ovoid to hemi-spherical.   825d. *E. resinifer* Sm.
    7. Fr. 12–15 mm; cylindrical to urn-shaped.   825e. *E. robustus* Sm.
  6. Inflorescence stalk more or less rounded.
    8. Inflorescence stalk 1–1½ cm; fr. 7–8 mm by 5–6 mm,
      hemispherical.   *camaldulensis*
    8. Inflorescence stalk 5–12 mm; fr. 6–9 mm by 8–10 mm,
      broadly top-shaped.   825g. *E. tereticornis* Sm.

825. *E. globulus* Labill. BLUE GUM. The most commonly planted Gum tree, distinguished by its large solitary flowers and fruits. Very large trees forming impressive road avenues are often the dominant feature of an otherwise treeless landscape. Pl. 33.   825f. *E. camaldulensis* Dehnh. Widely planted in Andalusia as a roadside tree. It is a spreading tree with smooth dull white bark, with umbels of 5–10 white flowers and distinctly stalked fruits. Leaves lance-shaped, pointed, 12–22 cm.

PUNICACEAE | Pomegranate Family

PUNICA | **827. *P. granatum* L. POMEGRANATE. Unlike any other shrub, with large scarlet flowers with crumpled petals, and red, fleshy calyx. Leaves deciduous, oblong-lance-shaped, shining. Fruit reddish-brown, 5–8 cm, with a tough rind and fleshy edible centre. Widely naturalized. Portugal, Med. region.

ONAGRACEAE | Willow-Herb Family

CIRCAEA | **829. *C. lutetiana* L. ENCHANTER'S NIGHTSHADE. Woods and shady places. N. Portugal, N. Spain, France.

OENOTHERA | Evening Primrose   Natives of America usually with large attractive yellow flowers. About 6 species have become naturalized on waste ground and by waysides in our area.

EPILOBIUM | Willow-Herb   16 species occur in our area, the great majority of them widespread European species. They are mostly plants of the cooler temperate climates. The following large-flowered species are easily recognized:

**835. *E. angustifolium* L. ROSEBAY WILLOW-HERB, FIREWEED. Well known for its long cylindrical, terminal spikes of bright rosy-purple flowers and neat, spirally-arranged, lance-shaped, glaucous leaves. An erect, often massed, perennial to 1 m or more. Woods and shady places in mountains. S., C., and N. Spain, France.   **837. *E. hirsutum* L. GREAT HAIRY WILLOW-HERB. A tall perennial to 1½ m with large rosy-purple flowers 1½–2½ cm across, with broad, shallowly notched petals. Leaves half-clasping stem, sharply toothed. Riversides, lakes and marshes. Widespread.   837a. *E. duriaei* Godron A nearly hairless perennial with creeping yellowish stolons, with medium-sized pink flowers 1–2 cm across. Petals 6–10 mm;

279

stigmas four-lobed. Leaves oval, long-pointed, 1½–3½ cm, more or less stalkless, with strongly and irregularly toothed margins. Seeds narrowed to a prominent beak. Mountains. C. and E. Spain, Pyrenees, France.

CORNACEAE | Dogwood Family

CORNUS | **846.** *C. sanguinea* L. DOGWOOD. Hedges, bushy places, hills. N. Portugal, Spain and France except the Med. region.

ARALIACEAE | Ivy Family

HEDERA | **848.** *H. helix* L. IVY. Hedges, walls, rocks, woods. Widespread.

UMBELLIFERAE | Umbellifer Family

A large and quite important family in our area with 77 genera and approximately 210 species. Considerable experience is required to identify many genera and species, and space only permits the description of a very few of the most striking species.

ASTRANTIA | 2 species.   **852.** *A. major* L. GREAT MASTERWORT. Woods and meadows in mountains. C., N., and N–E. Spain, France.

ERYNGIUM | 14 species, including 8 species restricted to our area in Europe. Bracteoles are the small spiny bracts subtending each group of flowers in the flower cluster; bracts surround the whole flower cluster.

853. *E. maritimum* L. SEA HOLLY. Coasts throughout.   853a. *E. ilicifolium* Lam. A small annual 2–15 cm with short, spreading branches, and small ovoid blue flower heads 1–1½ cm, surrounded by narrow spiny bracts up to about twice as long. Bracteoles with 3 spiny teeth. Basal leaves leathery, obovate, with spiny teeth. Dry places. S. Spain.   853b. *E. aquifolium* Cav. A rather robust perennial to 50 cm with bluish, more or less flat-topped inflorescence with up to 12 globular, stalked, blue flower heads. Flower heads 1–2 cm; bracts 5–7, lance-shaped with spines and spiny teeth; some bracteoles with 3 spiny teeth. Leaves broadly oblance-shaped, coarsely spiny toothed. Dry places. S. Spain.

855. *E. campestre* L. FIELD ERYNGO. Dry places. Widespread

855a. *E. bourgatii* Gouan An erect perennial to 45 cm, with a blue-flushed inflorescence with up to 7 stalked globular blue flower heads each with 10–15 much longer narrow, almost spineless, spreading bracts. Flower heads 1½–2½ cm; bracts 2–5 cm, entire or with 1–2 pairs of spiny teeth; bracteoles entire or three-toothed. Leaves rounded in outline, three times cut in narrow spiny-tipped segments, basal leaves with leaf stalk two to four times as long as blade. Dry stony places in mountains. Spain, Pyrenees. Pl. 32.   855b. *E. dilatatum* Lam. Distinguished from 855a by the smaller flower heads ½–1½ cm and shorter, more spiny bracts 1–3 cm. Basal leaves obovate, narrowed to a short, winged, indistinct leaf stalk. Dry places. Portugal, Spain.

855c. *E. tricuspidatum* L. An erect perennial to ¾ m, with a narrow greenish inflorescence with 2–8 small globular flower heads each about 1 cm, and very much longer narrow bracts. Bracts 5–7, with 5–8 pairs of spines, rarely spineless. Basal leaves rounded, heart-shaped, toothed, undivided or three-lobed, long-stalked; stem leaves with short broadly winged leaf stalk and blades cut to the base into many slender spine-tipped segments. Dry places. S–W. Spain. 845a. *E. glaciale* Boiss. An alpine perennial growing above 2500 m in the Sierra Nevada, S. Spain. Distinguished by its bluish inflorescence with usually 3 rounded flower heads 1–1½ cm surrounded by 7–8 long narrow bracts. Bracts 3–5 cm, spine-tipped with 1–2 pairs of lateral spines. Basal leaves leathery, deeply three-lobed with large spiny lobes, upper part of leaf stalk with spiny wing.

ECHINOPHORA | **857. *E. spinosa* L. A much-branched, domed, very spiny shrub with umbels of white flowers about 3 cm across. Calyx, bracts and bracteoles spine-tipped. Maritime sands. Med. region. Pl. 32.

CRITHMUM | **871. *C. maritimum* L. ROCK SAMPHIRE. Coasts throughout.

KUNDMANNIA | Leaves one- to two-pinnate; segments egg-shaped. Petals yellow, broadly egg-shaped, with inturned apex. Fruit nearly cylindrical, ridges slender but prominent. 875a. *K. sicula* (L.) DC. A hairless perennial to 70 cm, with large yellow umbels with 5–30 rays and numerous linear reflexed bracts and bracteoles. Basal leaves twice-pinnate with egg-shaped, toothed segments, the stem leaves once-pinnate, segments more deeply cut. Fruit 6–10 mm. Dry fields, hills. S. Portugal, S. Spain.

CACHRYS | Distinguished by its ovoid fruit with broad thick ribs, or undulate wings. 3 species. 880a. *C. sicula* L. A robust perennial 30–150 cm with numerous yellow umbels in a domed or somewhat flat-topped compound inflorescence and many times cut feathery leaves with linear pointed lobes 1½–5 cm. Rays 20–30. Bracts of central umbel usually twice-cut; bracteoles entire. Fruit stout swollen, with wide rounded prominent ridges with teeth or a crest. Dry hills, rocks and sands on the littoral. S. and C. Portugal, S. and C. Spain. 880b. *C. libanotis* L. Like 880a but usually smaller and stouter. Leaves less divided lobes ½–1 cm, rigid. Rays 8–15. Ridges of fruit smooth or with short flattened, often appressed papillae. Pine woods. S. Portugal, S. Spain.

MAGYDARIS | Flowers white, petals with shaggy hairs beneath. Fruit ovoid, ridges wide and rounded. 880c. *M. panacifolia* (Vahl) Lange A tall rather slender perennial 1–2½ m smelling strongly of new-mown hay (coumarin) and with a large terminal cream-flowered umbel, conspicuously hairy fruits, and leaves with large oblong segments. Rays 10–30; bracts 2–3 cm narrow, deflexed; bracteoles 1–2 cm deflexed. Basal leaves simple or lobed, upper leaves once-cut into 3–5 broad, finely toothed segments 10 cm or more long. Fruit spongy, greyish-haired, cylindrical, ridged with dense woolly hairs. Dry hills, waysides. Portugal, S. and C. Spain.

BUPLEURUM | 20 species, mostly herbaceous.

Shrubby evergreen perennials
1. Lvs. with a well marked midvein and netted lateral veins.
  2. Lvs. crowded near top of woody branches, from the apex
     of which herbaceous flowering stems arise. (S. Spain
     (Gibraltar, near Algeciras))   884a. *B. foliosum* DC.
  2. Lvs. more or less evenly spaced along stems.

3. Primary lateral veins of lvs. reaching lv.-margin; bracts soon falling. — *fruticosum*

3. Primary lateral veins not reaching lv.-margin; bracts persisting. (S. and C. Spain) — 884b. *B. gibraltarium* Lam.

1. Lvs. with several well marked, more or less parallel veins; lateral veins few and inconspicuous.

4. Bracts usually three-veined. (Balearic Islands) — 884c. *B. barceloi* Willk.

4. Bracts one-veined or veinless.

5. Flowering stems and rays becoming hard and spiny, persisting for 2–3 years. (S. and E. Spain) — 884d. *B. spinosum* Gouan

5. Flowering stems and rays not becoming hard and spiny and not persistent. (C. and E. Spain) — 884e. *B. frutescens* L.

884. *B. fruticosum* L. SHRUBBY HARE'S EAR. A bushy evergreen shrub to 2½ m with stalkless shining leaves and yellow umbels 7–10 cm across. Primary rays 5–25, stout; bracts and bracteoles 5–7, soon falling. Leaves 5–8 cm. Fruit 5–8 mm. Sunny hills, walls, rocky places. S. and C. Portugal, Med. region.

Herbaceous perennials

883a. *B. rigidum* L. A tall perennial to 1½ m with numerous spreading or ascending branches and leathery leaves clasping the stem. Umbels yellow, with 2–5 slender rays; bracts 2–4 mm, adpressed to rays. Leaves variable, oblong to linear. Fruit about 4 mm, ellipsoid, ridges prominent. Dry rocky places. Iberian peninsula, S. France.

FERULA | **897. *F. communis* L. GIANT FENNEL. A very robust perennial 1–5 m with a tall inflorescence of many large yellow umbels with 20–40 rays and dark green, much-dissected feathery leaves with thread-like lobes. Bracts absent; bracteoles few. Uppermost leaves reduced to a large sheathing base which at first encircles the young umbels, lower leaves with ultimate lobes up to 5 cm, margins flat, not inrolled. Dry stony hills. C. Spain, Med. region.   897a. *F. tingitana* L. Like 897 but ultimate lobes of leaves not more than 1 cm long, with inrolled margins, partly united to one another. Rocks, hedges, shady damp places. Portugal, S. and S-E. Spain.

ELAEOSELINUM | Leaves three- to five-pinnate. Petals yellowish or white. Fruit orbicular, ovoid or oblong, somewhat compressed with 4 wide lateral wings, and unwinged or narrowly winged dorsal ridges; oil glands solitary in the grooves and in the ridges. 4 species.

904a. *E. asclepium* (L.) Bertol. An almost hairless perennial to 1¼ m with basal leaves much cut into often whorled segments, with ultimate lobes thread-like 2–3 mm. Umbels yellow; rays 8–25; bracteoles few, linear-lance-shaped. Stem leaves reduced to inflated leaf-stalks. Fruit 8–15 mm, with wide whitish, shiny lateral wings extending beyond top of fruit. Sunny hills. S. Spain.

904b. *E. foetidum* (L.) Boiss. Like 904a but ultimate lobes of leaves egg-shaped in outline, usually three-lobed. Bracteoles numerous, bristle-like. Fruit with wings extending to the top. Sunny hills. S. Portugal, S-W. Spain.

THAPSIA | Leaves two- to three-pinnate. Petals yellow, wedge-shaped. Fruit oblong to egg-shaped, compressed, primary ridges slender, inconspicuous, dorsal secondary ridges like the primary ridges and sometimes narrowly winged, the marginal ones broadly winged.

1. Lvs. once-pinnate: lobes 5–12 cm. (S. and E. Portugal, S. and C. Spain) — 906b. *T. maxima* Miller Pl. 33

1. Lvs. two- to four-pinnate; lobes ½–2½ cm.
2. Ultimate lobes of lvs. regularly toothed.   *villosa*
2. Ultimate lobes of lvs. entire, or with 1–2 teeth. (Portu-
   gal, Med. Spain)   906c. *T. garganica* L.

906a. *T. villosa* L. A robust hairy somewhat glaucous usually conspicuous and abundant
perennial to 2 m with large yellow umbels with 9–24 rays. Bracts and bracteoles few or
absent. Lower leaves woolly-haired, three to four times cut into oblong-oval toothed lobes
usually ½–1½ cm, with a spiny point. Upper leaves reduced to inflated sheaths. Fruit 8–15
mm elliptic, with 2 very broad, thin lateral wings. Dry places, hills, sands. Iberian peninsula, S. France. Pl. 33.

PYROLACEAE | Wintergreen Family

PYROLA | **912. *P. minor* L. SMALL WINTERGREEN. Flowers pinkish, globular, in a dense
oval cluster; style 1–2 mm, shorter than petals and stamens. Mountain woods. C. Spain,
Pyrenees, France.

913. *P. rotundifolia* L. ROUND-LEAVED WINTERGREEN. Flowers white, in a lax cluster; petals
spreading, flowers thus not appearing globular; style down-curved, longer than the upcurved stamens. Sepals lance-shaped, 2–3 times shorter than petals. Woods. C. and N.
Spain, Pyrenees, France.   913a. *P. chlorantha* Swartz GREENISH WINTERGREEN. Like 913
but flowers greenish-white and sepals oval-triangular almost as broad as long, 3–4 times
shorter than the petals. Leaves ½–2 cm. Open dry woods, alpine meadows. C. Spain,
Pyrenees, Massif Central.

ORTHILIA | **914. *O. secunda* (L.) House NODDING WINTERGREEN. Mountain woods,
mainly under pine. Pyrenees, Massif Central.

MONESES | **915. *M. uniflora* (L.) A. Gray ONE-FLOWERED WINTERGREEN. Mountain
woods. Pyrenees, Massif Central.

MONOTROPA | **Bird's Nest   **917. *M. hypopitys* L. YELLOW BIRD'S NEST. A fleshy, waxy-
looking, yellowish or brownish saprophytic plant without chlorophyll, with a terminal
cluster of drooping yellowish flowers. Leaves scale-like. Damp woods in mountains. C.
Portugal, C. and N. Spain, Pyrenees, France.

ERICACEAE | Heath Family

RHODODENDRON | **919. *R. ferrugineum* L. ALPENROSE. A small evergreen mountain
shrub with dark green shining leaves 2–4 cm, which are rust-brown beneath, and terminal
clusters of bright pinkish-red, funnel-shaped flowers. Mountain pastures. C. Spain,
Pyrenees.   **920. *R. ponticum* L. RHODODENDRON. A large shrub with dark laurel-like,
evergreen leaves 7–14 cm, and rounded clusters of pale purple flowers. Flowers 5–7 cm
across, hairy within at the base; stamens and style longer than funnel-shaped corolla.
River and streamsides in mountains. S. Portugal, S–W. and W–C. Spain.

LOISELEURIA | **921. *L. procumbens* (L.) Desv. CREEPING AZALEA. Alpine pastures.
Pyrenees.

DABOECIA | **922. *D. cantabrica* (Hudson) C. Koch ST DABEOC'S HEATH. Unmistakable with its conspicuous, nodding, flask-shaped, rosy-purple flowers 1 cm long or more, borne in a lax terminal cluster. A straggling, heath-like plant, with narrow dark green leaves ½–1 cm, white-hairy beneath. Lowlands and montane heaths. N–W. Portugal, C., W., and N. Spain, W. France. Pl. 33, Page 124.

ANDROMEDA | **923. *A. polifolia* L. MARSH ANDROMEDA. Bogs. Pyrenees, Massif Central.

ARBUTUS | **924. *A. unedo* L. STRAWBERRY TREE. A small evergreen tree or tall shrub with large dark green, elliptic, toothed leaves 4–10 cm, and many-flowered clusters of nearly globular, cream-coloured flowers. Fruit unmistakable: globular, red and strawberry-like when ripe, and covered with warts. Common in *matorral* and cork oak woods in the lowlands and hills. Widespread. Page 63.

ARCTOSTAPHYLOS | **925. *A. uva-ursi* (L.) Sprengel BEARBERRY. A spreading, mat-forming shrub with shiny, leathery, oval leaves, and short clusters of nearly globular pink flowers. Corolla about ½ cm. Leaves 1–2 cm, paler beneath. Fruit globular, shining red. Scrub, open woods, moors in mountains. N. Portugal, S., C., and N. Spain, France. Page 117.
**(925) *A. alpinus* (L.) Sprengel BLACK BEARBERRY. Pyrenees.

CALLUNA | 926. *C. vulgaris* (L.) Hull LING, HEATHER. Lowland and montane heaths. Widespread.

ERICA | Heath, Heather   An important genus in S–W Europe with 13 species, 5 of which occur only in our area in Europe. They are important members of the *matorral* of the Iberian peninsula.

1. Corolla not more than 3 mm, green (sometimes tinged red).   *scoparia*
1. Corolla 3 mm or more, purple, pink or white.
  2. At least some anthers longer than the corolla.
    3. Fl. stalk about as long as sepals.
      4. Anthers with basal appendages; bracteoles more than half as long as
         sepals.   *australis*
      4. Anthers without appendages; bracteoles less than one quarter as long
         as sepals.   *erigena*
    3. Fl. stalk 2 or more times as long as sepals.
      5. Fls. in terminal umbels; fl. stalks downy.   *umbellata*
      5. Fls. in axillary clusters or lateral umbels, usually aggregated into long
         clusters; fl. stalks hairless.
        6. Corolla 4½–5½ mm; anthers 1½ mm, the 2 lobes closely contiguous and
           parallel.   *multiflora*
        6. Corolla 3–4 mm; anthers not more than 1 mm, the 2 lobes widely diver-
           gent.   *vagans*
  2. Anthers all included in corolla.
    7. Corolla 7–10 mm; anthers without basal appendages.   *ciliaris*
    7. Corolla 3–9 mm; anthers with basal appendages.
      8. Fl. stalks hairless, with bracteoles confined to lower half and not over-
         lapping the sepals.
        9. Corolla 2–3 mm; lvs. not more than ½ cm long; anther appendages ¼
           length of anther.   *arborea*

9. Corolla 4–5 mm; lvs. up to 1 cm long; anther appendages as long as or
  longer than anthers.   *lusitanica*
8. Fl. stalks downy, with some of the bracteoles on the upper half and overlapping the sepals.
  10. Ovary hairless.
   11. Lvs. linear, not ciliate, green beneath.   *cinerea*
   11. Lvs. oblong, ciliate, white beneath.   *mackaiana*
  10. Ovary hairy.
   12. Sepals with a conspicuous keel; lvs. with inrolled margins which
     completely conceal the lower surface.   *australis*
   12. Sepals without conspicuous keel; lower surface of lv. visible at least
     near base.
    13. Sepals and fl. stalks densely downy to shaggy-haired; sepals usually
      with long stout cilia.   *tetralix*
    13. Sepals and fl. stalk hairless to minutely downy; sepals without long
      stout cilia.   *terminalis*

**927. *E. arborea* L. TREE HEATH. A tall, feathery-looking shrub 1–4 m, or rarely more, with
very numerous, very tiny leaves and flowers. Flowers 2–3 mm, bell-shaped, white or pinkish,
sweet-scented, in dense pyramidal clusters. Anthers brownish, contrasting with and included within corolla; anther appendages flat, ciliate. Young twigs white-hairy; leaves ½
cm. Scrub, heaths, woods, lowland and montane. Widespread except in the north. Page 287.
**(927) *E. lusitanica* Rudolphi LUSITANIAN HEATH. Distinguished from 927 by the longer
cylindrical flowers 4–5 mm, often tinged with pink, and leaves half as thick and twice as
long. Damp heaths, wood margins. Portugal, N–W. Spain, S–W. France. Pl. 34, Page 287.

928. *E. tetralix* L. BOG HEATHER, CROSS-LEAVED HEATH. A dwarf straggling shrub of bogs,
wet heaths, and pine woods, characterised by the four-whorled, bristly-haired leaves and
dense terminal clusters of rose-pink flowers. Flowers 5–9 mm, pitcher-shaped, with very
short spreading or recurved lobes. Calyx downy and with bristly hairs. Twigs softly
downy; leaves 4–6 mm, with glandular bristles and downy hairs. C. and N. Portugal, C.
and N. Spain, Pyrenees, France. 928a. *E. mackaiana* Bab. Like 928 but leaves not downy
but with bristle-like hairs on the margin, and only partially inrolled, leaving much of the
white undersurface exposed. Calyx hairless, except for bristle-like hairs on the margin.
Twigs soon hairless. Mountain heaths. C. and N–W. Spain. Page 287.

**929. *E. ciliaris* L. DORSET HEATH. Easily recognised by its large deep pink, pitcher-shaped
flowers 8–13 mm, which are somewhat curved at the throat and inflated at the base. Flowers
in one-sided terminal clusters; calyx ciliate. Leaves in whorls of 3, glandular and bristly-
haired, 'like insect's legs'. Bogs, heaths, scrub, open woods. Widespread. Page 287.

930. *E. cinerea* L. BELL HEATHER. Distinguished by its bright reddish-purple, urn-shaped
flowers 4–6 mm, borne in short clusters on the upper axillary branches. Calyx hairless;
anthers included in corolla. Leaves in whorls of 3, linear, dark green, hairless. Widespread
outside the Med. region. 930a. *E. terminalis* Salisb. Like 930 but leaves in whorls of 4
and branches erect. Flowers rosy-pink, 5 mm, in a terminal umbel of 3–8. Flower stalk with
3 narrow bracteoles usually placed above the middle; sepals finely downy; anthers with
triangular, basal appendages. Damp shady places, river banks, wooded ravines in the hills.
S–W. Spain. Page 287.

931. *E. scoparia* L. GREEN HEATHER. A slender shrub to 1 m, easily recognized by its very
numerous, tiny greenish flowers in long spike-like clusters. Flowers 2–3 mm, broadly
bell-shaped, anthers included in corolla. Leaves in whorls of 3–4, linear, 3–5 mm, hairless.
Heaths and woods. Widespread. Page 287.

932. *E. australis* L. (*E. aragonensis*) SPANISH HEATH. A slender, erect shrub to 2 m, with striking reddish-pink flowers in umbels of 4–8 borne at the ends of the branches and forming conspicuous branched clusters. Corolla 6–9 mm, cylindrical to narrow bell-shaped, with triangular out-curved lobes and anthers usually partially projecting. Leaves in whorls of 4, linear, hairless, shining. Heaths and scrub in lowlands and hills. Portugal, S., C., and N. Spain. Pl. 34, Page 160, 287.   932a *E. umbellata* L. A striking dwarf shrub with small bright rosy-purple flowers with contrasting dark purple projecting anthers. Corolla nearly globular to broadly bell-shaped, 3–6 mm, with erect lobes; anthers without appendages. Leaves 3–4 mm, in whorls of 3, at first glandular hairy, later hairless. Heaths and scrub in lowlands and hills. Western half of Iberian peninsula. Pl. 34, Page 287.   (933) *E. erigena* R. Ross (*E. hibernica*; *E. mediterranea* auct.). An erect shrub to nearly 2 m, with pale pink or white flowers in short terminal, leafy, spike-like clusters. Flowers cylindrical 5–7 mm, with erect lobes, and with contrasting dark brown anthers usually only partly projecting and without appendages. Twigs finely hairy, ribbed from the projecting leaf bases. N. and C. Portugal, S–W., W. and N. Spain,? France. Page 287.

**934. *E. multiflora* L. An erect, branched shrub to 1 m, with dense rounded or elongate, terminal clusters of pale pink flowers. Corolla 4–5½ mm, cylindrical to narrowly bell-shaped, with erect or spreading lobes; anthers completely projecting and without appendages. Leaves in whorls of 4–5, linear, 6–11 mm. Rocky hills, thickets. Med. region. Page 287. (934) *E. vagans* L. CORNISH HEATH. Very like 934 and easily confused with it, but flowers lilac or white, smaller 3–4 m, widely bell-shaped with erect lobes, and flower clusters usually terminating a tuft of leaves. Heaths, woods. C. and N. Spain, Pyrenees, W. and C. France. Pl. 34. Page 287.

VACCINIUM | Cowberry, Wortleberry, Cranberry

1. Corolla divided nearly to the base and lobes turned back. (Massif Central)	**937. *V. oxycoccos* L.
1. Corolla with small lobes.	
2. Corolla bell-shaped; lvs. persistent, dark green and glossy; fls. in terminal clusters. (Pyrenees, France)	**935. *V. vitis-idaea* L.
2. Corolla pitcher-shaped; lvs. deciduous; fls. 1–4, axillary.	
3. Lvs. finely toothed, bright green; twigs angled. (Widespread except south)	**936. *V. myrtillus* L.
3. Lvs. entire, obtuse, blue-green; twigs cylindrical. (N–E. Spain, France)	(936) *V. uliginosum* L.

EMPETRACEAE | Crowberry Family

EMPETRUM | **938. *E. nigrum* L. CROWBERRY. Heaths and rocky places in mountains. Pyrenees, France.

COREMA | 938a. *C. album* (L.) D. Don A low, erect heather-like shrub with numerous stiff, crowded, whorled linear leaves, and usually terminal clusters of tiny pink flowers. Petals 3, neatly fringed; stamens 3, longer than petals. Leaves ½–1 cm, grooved; twigs downy. Fruit a white or red berry. Maritime sands and dunes on the Atlantic littoral. Western part of Iberian peninsula. Page 296.

1. *Erica multiflora* 934
4. *E. vagans* (934)
7. *E. australis* 932
10. *E. erigena* (933)

2. *E. terminalis* 930a
5. *E. scoparia* 931
8. *E. mackaiana* 928a
11. *E. umbellata* 932a

3. *E. arborea* 927
6. *E. ciliaris* 929
9. *E. lusitanica* (927)

PRIMULACEAE | Primrose Family

PRIMULA | Primrose, Cowslip

1. Fls. yellow.
  2. Fls. in a stalked umbel.
    3. Calyx lobes long-pointed; mature fr. equalling or longer than calyx.   *elatior*
    3. Calyx lobes not long-pointed; mature fr. shorter than calyx. (Widespread except in south)   939.   *P. veris* L.
  2. Fls. solitary, long-stalked, arising directly from the stock.   *vulgaris*
1. Fls. lilac, purple, pink or red.
  4. Lvs. with white powdery flour on lower surface.   *farinosa*
  4. Lvs. without powdery flour.
    5. Upper surface of lvs. hairless or with scattered hairs.
      6. Glandular hairs on margin with reddish tips. (N. Spain)   946a. *P. pedemontana* Gaudin
      6. Glandular hairs on lv. margin with pale tips.   *integrifolia*
    5. Upper surface of lvs. with rather dense glandular hairs.
      7. Fl. stem longer than lvs.; fls. violet-purple without pale centre.   *latifolia*
      7. Fl. stem shorter than lvs.; fls. purple with a white centre.   *hirsuta*

**940. *P. elatior* (L.) Hill OXLIP. Flowers 1½–2 cm, pale yellow without a darker yellow throat, petals wide-spreading, not cup-shaped; calyx constricted at the throat. Pastures and grassy places in mountains. C., N., and E. Spain, France.   **942. *P. vulgaris* Hudson PRIMROSE. A white-flowered very fragrant subspecies is found in the Balearic Islands. Damp meadows, hedges, banks, mountains in south. Widespread except in south.

**943. *P. farinosa* L. BIRD'S-EYE PRIMROSE. Distinguished by its neat rosettes of obovaloblong leaves which are powdery white beneath. Flowers rose-lilac with a pale eye, borne in an umbel of 2 to many flowers; corolla tube twice as long as powdery calyx. Mountain pastures. C. and N–E. Spain, Pyrenees.   946. *P. integrifolia* L. A tiny plant, distinguished by its reddish-lilac flowers in an umbel of 2–3, with deeply notched petals and with the tube of the corolla twice as long as the glandular calyx. Bracts of umbel longer than flower stalks. Leaves with ciliate margin, not sticky. Damp places in alpine pastures. C. and E. Pyrenees.

**945. *P. latifolia* Lapeyr. (*P. viscosa*) Distinguished by its glandular-hairy leaves, with colourless glandular hairs. Flowers 1½ cm across, violet-purple throughout, fragrant, in a one-sided umbel of 3–20 flowers. Alpine rocks. E. Pyrenees.   **(945) *P. hirsuta* All. RED ALPINE PRIMROSE. Usually has yellow or red glandular hairs on the leaves, and 1–3 bright purple flowers with a white eye and throat. Alpine rocks. Pyrenees. Page 148.

VITALIANA | **947. *V. primuliflora* Bertol. Alpine rocks, dry pastures. S. Spain, Pyrenees.

ANDROSACE | 11 species in our area, including 3 rare species not included in the key.

1. Anns. without non-flowering rosettes.   *maxima*
1. Perenns. usually tufted, with non-flowering rosettes.
  2. Fls. in umbels.
    3. Plants sparsely hairy with star-shaped hairs.   *carnea*
    3. Plants shaggy-haired.   *villosa*
  2. Fls. solitary.
    4. Lvs. persistent after withering, forming columnar shoots; plants forming deep cushions.
      5. Fls. distinctly stalked, with a stalk about twice as long as the lvs. and with 2–3 bracts below calyx (endemic Pyrenees)   948a. *A. pyrenaica* Lam.
      5. Fls. stalkless or very short-stalked, without bracts below calyx.
        6. Lvs. covered with white matted hairs.   *argentea*
        6. Lvs. downy with soft, short hairs (endemic Pyrenees).   948b. *A. cylindrica* DC.
    4. Lvs. not persistent after withering, forming short rosettes at the ends of the branches; plants forming low cushions.
      7. Fl. stalks $\frac{1}{2}$–$1\frac{1}{2}$ cm; fls. pink or violet (endemic Pyrenees).   948c. *A. ciliata* DC.
      7. Fl. stalks less than $\frac{1}{2}$ cm; fls. usually white. (Pyrenees)   948d. *A. pubescens* DC.

950. *A. maxima* L. An unmistakable annual with a neat basal rosette of leaves, and leafless stems bearing umbels of tiny pink or white flowers, with much longer green calyx and conspicuous leafy involucres. Fields, dry stony places, in lowlands and mountains. S., C., and E. Spain, France.

**953. *A. carnea* L. A variable perennial with dense umbels of white or pink flowers borne on finely downy stems. Rosette of leaves often nearly hairless, or ciliate, leaves linear-pointed, $\frac{1}{2}$–2 cm. Damp screes, grassy places. Pyrenees. Page 149. **(954) *A. villosa* L. Distinguished from 953 by the long spreading hairs on the stem and the silvery, shaggy-haired rosettes 5–8 mm across of oblong-blunt leaves. Plants usually densely tufted, with numerous rosettes. Flowers white or pink. Alpine rocks and screes, dry turf. N. and E. Spain, Pyrenees. Pl. 34, Page 124. (948) *A vandelli* (Turra) Chiov. (*A. argentea*) It has the appearance of silvery tufts and buttons seaming the cracks in the alpine cliffs. Flowers solitary, almost stalkless, white with a yellow then a purple throat. Cushions about 2–4 cm across, individual rosettes about $\frac{1}{2}$ cm across. Alpine rocks and cliffs. S. Spain (Sierra Nevada), Pyrenees.

SOLDANELLA | 956. *S. alpina* L. ALPINE SNOWBELL. Unmistakable with its delicate, nodding, bluish-violet, deeply fringed tubular flowers set above tiny dark green, shiny, kidney-shaped leaves. Alpine meadows. C. Pyrenees, Corbières, Auvergne. 956a. *S. villosa* Darracq A rare endemic of the Basses-Pyrénées, distinguished by its long stalked glandular hairs on the leaf stalk, which persist to maturity (in 956 glandular hairs stalkless, mostly not persisting). Corolla deeply fringed to four-fifths of its length. Clearings in damp conifer woods. S–W. France.

CYCLAMEN | **960. *C. repandum* Sm. REPAND CYCLAMEN. A spring-flowering species with bright rose-coloured or rarely white flowers, with slightly twisted, reflexed petals 2–3 cm. Leaves triangular-heart-shaped, with shallow angular lobes. Woods and scrub. C. Spain, E. Pyrenees, Med. France. 960a. *C. balearicum* Willk. Distinguished by its small white flowers about $1\frac{1}{2}$ cm with style not projecting. Leaves shallowly toothed, marbled with silver above, reddish beneath. Shady, rocky places. Balearic Islands, S. France. Pl. 36.

289

LYSIMACHIA | Loosestrife

1. Small creeping or spreading plants.
  2. Lvs. acute; calyx teeth awl-shaped. (Widespread)   **962. *L. nemorum* L.
  2. Lvs. obtuse; calyx teeth egg-shaped. (Spain, France)   **961. *L. nummularia* L.
1. Erect plants 25 cm or more.
  3. Fls. solitary in axils of upper lvs., pinkish at the base and greenish at the apex of the lobes (endemic Minorca).   963b. *L. minoricensis* Rodr.
  3. Fls. in spike-like or branched clusters.
    4. Fls. with sepals and petals 7, in axillary spike-like clusters. (France)   **964. *L. thyrsiflora* L.
    4. Fls. with sepals and petals 5, in a terminal branched cluster.
      5. Fls. yellow. (Widespread except south)   963. *L. vulgaris* L.
      5. Fls. whitish to pale pinkish.   *ephemerum*

963a. *L. ephemerum* L. A tall erect, unbranched perennial to 1 m, with a long spike-like inflorescence of white to pale pinkish flowers with conspicuous reddish stamens, and numerous opposite, linear-lance-shaped, glaucous leaves. Flowers about 1 cm across; petals oval, 5; calyx blunt, white-bordered. Riversides, damp grassy places, marshes. W. Portugal, S., C., and E. Spain, Pyrenees, Corbières. Pl. 36.

ANAGALLIS | Pimpernel

1. Plants only 1–4 cm; petals minute, much smaller than sepals. (Scattered throughout)   (966) *A. minima* (L.) E. H. L. Krause
1. Plants 5–50 cm; petals as long as or longer than sepals.
  2. Prostrate plants rooting at the nodes; fls. white or pink.
    3. Fl. stalks shorter than subtending lvs.; fls. white or cream.   *crassifolia*
    3. Fl. stalks longer than subtending lvs.; fls. pink, rarely white.   *tenella*
  2. Ascending or spreading plants, not rooting at the nodes; fls. red or blue (rarely otherwise).
    4. Anns.; fls. 2–7 mm, blue or red. (Widespread)   967. *A. arvensis* L.
    4. Perenns.; fls. 1–2 cm, blue.   *monelli*

966. *A. tenella* (L.) L. BOG PIMPERNEL. Bogs and damp peaty places. Widespread except in the south.   966a. *A. crassifolia* Thore Differs from 966 in having smaller white flowers with petals little longer than the calyx, and flower stalks shorter than the leaves. Leaves alternate, packed tight in 2 overlapping rows, rounded, shining, succulent. Damp, often sandy places. Portugal, S. Spain, S–W. France. Page 296.

**(967) *A. monelli* L. (*A. linifolia*) SHRUBBY PIMPERNEL. A low growing plant with very striking, usually brilliant blue flowers with purple centres, 1–2 cm across, 'the glory of the sand dunes of the Atlantic coast of Spain and S. Portugal'—but other forms have smaller flowers and are more like 967. A woody-based perennial; leaves linear-lance-shaped. Pine woods, vineyards, hedges, waysides, maritime sands. Portugal, S., S–E., and C. Spain. Pl. 35, Page 52.

GLAUX | **968. *G. maritima* L. SEA MILKWORT. A creeping, spreading perennial of salt marshes, with four-ranked, narrow fleshy leaves and small stalkless pinkish flowers in their axils. Sometimes also on salt-rich areas inland. Atlantic coasts, W. Med.

SAMOLUS | **969. *S. valerandi* L. BROOKWEED. Distinguished by its lax branched cluster of tiny white flowers borne above a rosette of pale green leaves. Peaty places, ditches, marshes, particularly near the sea. Widespread.

CORIS | **970. *C. monspeliensis* L. A stiff, branched, thyme-like annual with linear leaves and dense terminal clusters of two-lipped, rose-lilac flowers. Petals 5, 3 upper and 2 lower; calyx with 5, often black-spotted, triangular lobes and 6–21 spiny teeth. Stony hills, maritime sands. Spain, Med. France. Pl. 34, Page 90.

970a. *C. hispanica* Lange A rare biennial species of S–E. Spain (Almeria and Murcia provinces), distinguished by its white to pale pink flowers, and spines on calyx absent or usually with 2–3 developed on the back of the calyx and equal in length; calyx lobes with a large black blotch. Branchlets white. Clayey soils.

PLUMBAGINACEAE | Sea Lavender Family

PLUMBAGO | **971. *P. europaea* L. EUROPEAN PLUMBAGO. An erect branched perennial with dense clusters of violet or pink, veined flowers, and conspicuously bristly, glandular-hairy calyx. Corolla tube twice as long as calyx. Leaves rough, glandular-toothed, the lower oval stalked, the upper linear clasping. Dry rocks and hills. S. and N. Portugal, Med. region.

LIMONIUM | **Sea Lavender** About 46 species occur in our area, with 26 restricted to the Iberian peninsula. The identification of many species is difficult and combinations of small botanical characters of the calyx, corolla, bracts, inflorescence and leaves separate the species. The following are among the more distinctive, or more widespread species:

Stems winged below the inflorescence; leaves lobed
**972. *L. sinuatum* (L.) Miller WINGED SEA LAVENDER. A striking perennial with a rosette of deeply lobed leaves, winged stems, and compact flat-topped clusters of bright blue-mauve, papery 'everlasting' calyx and tiny yellowish corolla. Rocks and sands on the littoral. S–E. Portugal, Med. region. Pl. 36.   972a. *L. thouinii* (Viv.) O. Kuntze An annual with a rosette of lobed leaves and erect stems which are unwinged below but suddenly expanded into 3 broad, pointed wings below the compact inflorescence. Calyx conspicuous, white or pale blue, papery with 5 pointed lobes, calyx tube with curved hooks; corolla yellowish. Dry places near the sea. S. Spain. Pl. 35.

Inflorescences with many sterile branches; leaves entire, usually absent at flowering
972b. *L. ferulaceum* (L.) O. Kuntze An unusual-looking bushy perennial with brush-like tufts of numerous sterile thread-like branches with brown scales. Flowering branches differing, regularly alternate, with small one-flowered spikelets clustered at the ends of the branches. Corolla yellow, 5–6 mm; calyx surrounded by long-pointed bracts. Maritime marshes. S. Portugal, Med. region. Page 293.   972c. *L. insigne* (Cosson) O. Kuntze A very attractive, tufted plant with much-branched, silvery-grey stems without leaves, and branched clusters of beautiful bright rosy-violet flowers 8–9 mm across. Spikelets one- to two-flowered; calyx limb papery with pink ribs; corolla tube 1 cm, a little longer than calyx. Leaves all basal, spathulate. Dry places. S–E. Spain (Almeria and Granada provinces). Page 90.   972d. *L. caesium* (Girard) O. Kuntze A tall graceful plant, the most beautiful

in the genus, with a pyramidal inflorescence up to 1 m covered with a lavender-blue waxy bloom with numerous small pink flowers. Leaves small, spoon-shaped, leathery covered with limey encrustations. Locally abundant in sandy places, salt-rich ground. S–E. Spain.
**974. *L. bellidifolium* (Gouan) Dumort. MATTED SEA LAVENDER. A rather small plant with a hemispherical, much-branched, zig-zag inflorescence with both sterile and flowering branches present. Flowers violet, densely clustered at the ends of the branches and contrasting with the white papery calyx and bracts. Basal leaves withering at flowering. Coastal marshes. E. Spain, Med. France.   **(974) *L. oleifolium* Miller (incl. *Statice virgata*) Like 974 but flowers larger, 8–9 mm, in long, lax, unilateral clusters and branches usually wide-spreading. Bracts conspicuous, brown; calyx lobes curved, with broad white papery margin. Maritime cliffs. Coasts throughout.

Sterile branches not usually present; leaves entire, present at flowering
974a. *L. caprariense* (Font Quer & Marcos) Pignatti A distinctive plant of maritime cliffs of the Balearic Islands, with cushion-like rosettes of spathulate leaves densely clustered on woody stems, and short, branched clusters of pale purple flowers with darker purple midveins. Leaves with a puckered surface and covered with fine glandular hairs.   974b. *L. emarginatum* (Willd.) O. Kuntze A very handsome plant of rocks near the sea on Gibraltar, with large violet flowers about 1 cm across, pale brown papery bracts and dull white, deeply five-lobed calyx. Flower clusters 2–3 cm long; inflorescence wide-branched above. Basal leaves fleshy.

974c. *L. ovalifolium* (Poiret) O. Kuntze Distinguished by its broad oval, glaucous leaves arranged in compact basal rosettes and its numerous tiny violet flowers in flat-topped clusters forming a dense branched inflorescence. Spikelets 4 mm, in dense short spikes about 1 cm long. Leaves with a fine apical point. Atlantic rocks and cliffs, steppes. C. Portugal, S. and E. Spain, France. Page 293.

974d. *L. echioides* (L.) Miller A slender annual with a basal rosette of leaves, and with widely spaced, curved spikelets arranged along one side of the forked inflorescence. Flowers pale pink; calyx with 5–10 curved red spines; bracts with minute swellings. Sands and rocks. S. Portugal, Med. region. Page 293.

ARMERIA | Thrift   About 30 species occur in our area; 7 are restricted to Portugal, 8 to Spain, and a further 8 to the Iberian peninsula in Europe. A very difficult genus.

Leaves 3–8 mm broad
976. *A. alliacea* (Cav.) Hoffmanns. & Link (*A. arenaria*) JERSEY THRIFT, PLANTAIN-LEAVED THRIFT. A rather robust perennial with linear-lance-shaped leaves 3–8 mm wide, and tall stems 20–50 cm, bearing deep pink to white flower heads 1–2½ cm across. Involucral bracts conspicuous, brownish or reddish, pointed, the outer usually longer than the flower head; sheath below flower head 2–6 cm; calyx 5–9 mm. Dry meadows and pastures from lowlands to mountains. Widespread.   976a. *A. villosa* Girard Leaves like 976, but finely hairy, and flower heads often white. Involucral bracts straw-coloured to brownish, the outer lance-shaped, abruptly narrowed to a fine spine-like point; sheath below flower head 1½–2½ cm; calyx 7–12 mm. Stony places in mountains. S. Spain. Page 293.

Leaves more than 1 cm broad
976b. *A. pseudarmeria* (Murray) Mansfeld (*A. latifolia*) A very handsome, robust shrubby plant with stout stems up to 60 cm or more, bearing very large, conspicuous heads of white or very pale pink flowers, 3–4 cm across. Leaves to 20 cm or more and up to 2 cm wide, lance-shaped to spathulate, with a fine point. Involucral bracts straw-coloured with wide papery margins, long-pointed; bracts as long as calyx which is not spurred; sheath loose, conspicuous, 5–8 cm. Maritime pastures. W. Portugal (Cabo da Roca). Pl. 35.   976c. *A.*

1. *Armeria villosa* 976a
3. *Limonium echioides* 974d
5. *Limonium ovalifolium* 974c

2. *Limonium ferulaceum* 972b
4. *Limoniastrum monopetalum* 976e

gaditana Boiss. A similar, very robust but non-shrubby species, differing from 976b in having usually pink flowers and by the absence of bracts or bracteoles among the flowers. Calyx spurred. Sandy and grassy places near the sea. S. Portugal, S–W. Spain.

Leaves usually less than 3 mm broad

975a. *A. filicaulis* (Boiss.) Boiss. A slender plant with short tufts of numerous thread-like leaves and small rounded heads, about 1 cm across, of white or pink flowers. Involucral bracts very unequal, the outer one third as long as the inner; sheath ½–1½ cm. Leaves of two kinds, the outer broader, flat, outspreading, the inner narrower, channelled and very slender, erect. Mountain pastures, scrub. S–E. Spain. Page 84.   975b. *A. juniperifolia* (Vahl) Hoffmanns. & Link (*A. cespitosa*) A tiny, densely tufted, plant with rosettes of linear leaves 4–15 mm by about 1 mm, with white spiny tips and ciliate margins. Flower heads 1–1½ cm across, pale purplish or pink; outer involucral bracts shorter than inner, pointed, with narrow papery margin, inner with wide papery margin; sheath 3–9 mm. Mountain pastures, rock crevices. C. Spain. Pl. 35. Page 111.   975c. *A. splendens* (Lag. & Rodr.) Webb Like 975b. but outer involucral bracts with broad shining papery margins. Leaf tips not spiny-pointed, hairless. Mountain pastures, rock crevices. S–W. and N. Spain.

**(976) *A. pungens* (Link) Hoffmanns. & Link (*A. fasciculata*) SPINY THRIFT. Distinguished by its rigid, glaucous, grooved and spine-tipped leaves in dense clusters, and by its rather large, pale rose-coloured flower heads 1½–3 cm across. Outer involucral bracts leathery, purplish-brown, often not pointed, the inner longer blunt; calyx tube densely hairy; sheath rusty-brown, 1½–3 cm. Maritime dunes. S. Portugal, S–W., N–W. Spain. Pl. 35, p. 44.   976d. *A. macrophylla* Boiss. & Reuter Flower heads large, pink, with involucral bracts with a broad scaly margin, hairless and blunt-tipped. Calyx apex hairy. Leaves narrowly linear, stiff, 12–25 cm. Sandy soils on the littoral. S. Portugal, S–W. Spain.

**975. *A. maritima* (Miller) Willd. THRIFT, SEA PINK. A densely tufted perennial with soft, linear, one-veined leaves and globular heads of pink or purple flowers 1–3 cm across. Outer involucral bracts often shining green with narrow papery margins, the inner papery; sheath ½–1½ cm. A very variable species occupying a wide range of habitats from the coasts to the high mountains. Throughout.

LIMONIASTRUM | Shrubby plants with fleshy leaves; spikelets alternate, pressed to the branches, each two-flowered and with 3 bracts, easily breaking off when dry.

976e. *L. monopetalum* (L.) Boiss. A distinctive, much-branched, silvery glaucous shrub ½–1 m, with narrow fleshy leaves and bright pink flowers 1½ cm across. Corolla funnel-shaped, with rounded lobes; calyx with 5 acute teeth. Leaves covered with white scales, linear-lance-shaped, blunt, gradually narrowed to a fleshy encircling base, 2–4 cm. Salt marshes and sands on the coasts. S. Portugal, S. and E. Spain, S–W. France. Page 293.

OLEACEAE | Olive Family

FRAXINUS | 978. *F. excelsior* L. ASH. Shady woods from the lowlands to mountains. Widespread.   **979. *F. ornus* L. MANNA or FLOWERING ASH. Differs from 978 in having white, sweet-scented flowers appearing with the leaves. Leaves with 5–9 leaflets, at least some leaflets stalked. Mountains. S–E. Spain, Balearic Islands. Cultivated elsewhere.

SYRINGA | 980. *S. vulgaris* L. LILAC. Often planted for ornament, particularly in cooler areas, and sometimes naturalized.

PHILLYREA | (981) *P. latifolia* L. A small evergreen tree with leathery, dark green oval, toothed leaves 1½–4½ cm, and small rounded axillary clusters, about 1 cm across, of green-ish-yellow fragrant flowers. Calyx with triangular lobes. Leaves of two forms: the lower broader with a heart-shaped base, the upper narrower with a rounded base. Fruit without a pointed apex. Open woods and scrub, rocky slopes. Portugal, Med. region. 981. *P. angustifolia* L. Distinguished from (981) by its narrow lance-shaped leaves ½–1 cm across, with 4–6 pairs of indistinct veins. Calyx with short rounded lobes; fruit with an apical point. Portugal, Med. region.

LIGUSTRUM | 982. *L. vulgare* L. PRIVET. Hedges, bushy places. Widespread except in the south.

JASMINUM | **983. *J. fruticans* L. WILD JASMINE. A dense semi-evergreen shrub with green angular stems and small, few-flowered clusters of tubular unscented, yellow flowers 1½ cm across. Leaves oblong, shining, with 1 or 3 leaflets. Fruit black, shining. Bushy places. Portugal, Med. region. **984. *J. officinale* L. JASMINE. A climbing plant with pinnate leaves and clusters of white, very sweet-scented flowers. Grown for ornament and sometimes naturalized in hedges, in the south.

OLEA | **985. *O. europaea* L. OLIVE. The wild olive is a low spiny bush with four-angled stems and small opposite, oval or elliptic leaves, and fruit 8–12 mm long, and thus appearing to differ markedly from the cultivated plant. It is common in *matorral* and bushy places, and can be distinguished from similar-looking *Rhamnus* species by the opposite leaves. The cultivated Olive is a very important oil-bearing tree and is widely grown in much of the southern and central part of our area.

GENTIANACEAE | Gentian Family

CENTAURIUM | **Centaury** A difficult genus with frequent hybridization and much parallel variation within the species. 12 species.

**986. *C. erythraea* Rafn COMMON CENTAURY. A widespread, very variable species, including subsp. *majus* (Hoffmanns. & Link) Melderis, which has relatively large, bright pink flowers with acute corolla lobes 8–10 mm, which are longer than the corolla tube. Calyx ⅔–¾ as long as corolla tube. South of our area. Subsp. *grandiflorum* (Biv.) Melderis has large flowers but blunt corolla lobes and calyx half as long as the corolla tube. Portugal, Med. region. Pl. 35. **987. *C. maritimum* (L.) Fritsch YELLOW CENTAURY. Coasts. Portugal, Med. region.

BLACKSTONIA | **988. *B. perfoliata* (L.) Hudson YELLOW-WORT. A very variable plant. Damp shady places, pine woods. Widespread. Subsp. *imperfoliata* (L. fil.) Franco & Rocha Afonso Stem leaves not or scarcely joined together by their bases round the stem. Inland and maritime damp sands. Scattered throughout except in north.

GENTIANA | Gentian

1. Corolla lobes much longer than the corolla tube. (Fls. yellow.)   *lutea*
1. Corolla lobes not more than half as long as the corolla tube.

1. *Convolvulus lineatus* 1037
2. *Convolvulus boissieri* 1036b
3. *Anagallis crassifolia* 966a
4. *Corema album* 938a
5. *Periploca laevigata* 1009a
6. *Frankenia thymifolia* 810c

2. Fls. crowded in a terminal head and sometimes also in
   axillary clusters; lvs. more than 10 cm.
  3. Corolla blue, unspotted.   *cruciata*
  3. Corolla yellow, usually with dark spots.   *burseri*
2. Fls. solitary or few, not crowded in a terminal head;
   lvs. less than 10 cm.
  4. Corolla appearing ten-lobed due to additional lobes
     more than half as long, between the corolla lobes.
    5. Lvs. linear-lance-shaped; corolla 2–3 cm across.   *pyrenaica*
    5. Lvs. egg-shaped to rounded; corolla 8–10 mm across. *boryi*
  4. Corolla five-lobed with additional lobes much smaller
     than corolla lobes.
    6. Anns., without non-flowering shoots. (Pyrenees)   990. *G. nivalis* L.
    6. Perenns., with non-flowering shoots.
     7. Lower lvs. not forming a well marked rosette.
      8. Lvs. distinctly three- to five-veined; fls. arranged
        along one side of the stem. (E. Spain)   **994. *G. asclepiadea* L.
      8. Lvs. one-veined; fls. erect.   *pneumonanthe*
     7. Lower lvs. in a rosette; stem lvs. few.
      9. Corolla usually 5–7 cm across; tube obconical;
        lobes ascending.   *acaulis* group
      9. Corolla 1½–3 cm across, tube almost cylindrical;
        lobes spreading outwards.
       10. Longest rosette lvs. about twice as long as stem
          lvs.   *verna*
       10. Longest rosette lvs. not or little longer than
          stem lvs.
        11. Rosette lvs. rhombic; calyx teeth one third as
           long as calyx tube. (Pyrenees, S. Nevada)   991a. *G. brachyphylla* Vill.
        11. Rosette lvs. linear-lance-shaped; calyx teeth
           one half to three-quarters as long as calyx tube.
           (E. Pyrenees)   991b. *G. pumila* Jacq.

**989. *G. cruciata* L. CROSS GENTIAN. A robust, leafy perennial with large oval to lance-shaped, three-veined, sheathing leaves, and few large, dull blue flowers 2–2½ cm across, in terminal and axillary clusters. Corolla barrel-shaped, with 4 short spreading lobes. Woods and pastures in mountains. E. Spain, France.   **993. *G. pneumonanthe* L. MARSH GENTIAN. A lowland and mountain plant of wet places and heaths distinguished by its linear to oblong, one-veined, upper leaves and its few large, blue flowers 2½–5 cm long, with 5 green stripes on the outside. Scattered throughout, except in Med. region.

**996. *G. lutea* L. GREAT YELLOW GENTIAN. Readily distinguished from other yellow gentians by the flowers, which are not tubular but have 5–9 narrow spreading lobes. Very robust, with large, strongly veined leaves and terminal and axillary clusters of numerous flowers. Marshes and pastures in the hills and mountains. Much of our area, except Portugal and the Med. region. Page 148.   996a. *G. burseri* Lapeyr. A robust yellow-flowered plant like 996, but flowers bell-shaped, with more or less erect lobes much shorter than the corolla tube. Corolla to 4 cm, often brown-spotted; calyx papery, split down one side. Alpine meadows. Pyrenees. Pl. 37.

995. *G. pyrenaica* L. PYRENEAN GENTIAN. Readily distinguished by its blue corolla which has 10 nearly equal spreading lobes. Flowers solitary; corolla 2–3 cm across, greenish towards the base. A cushion-forming perennial with dense rosettes of rather leathery, pointed leaves. Mountain flushes and bogs. E. Pyrenees.   995a. *G. boryi* Boiss. A tiny tufted perennial of damp alpine pastures endemic to the Cantabrian mountains and the Sierra Nevada, with small, solitary blue flowers with white throats. Corolla 8–10 mm across, funnel-shaped, ten-lobed, copper-green and with white pleats outside. Leaves minute, rounded fleshy. N. and S. Spain.

**991. *G. verna* L. SPRING GENTIAN. A rosette-forming perennial distinguished by its brilliant deep blue flowers often with whitish pleats and its angled and winged calyx. Corolla $1\frac{1}{2}$–$2\frac{1}{2}$ cm across, greenish-blue on the outside. Damp grassy places in mountains. S. Spain (Sierra Nevada), Pyrenees, Massif Central.

G. acaulis group

1. Mature rosette lvs. scarcely longer than wide.	*alpina*
1. Mature rosette lvs. at least $1\frac{1}{2}$ times as long as wide.	
2. Calyx teeth triangular, widest at base, usually more than half as long as the calyx tube (Cévennes)	**(992) *G. clusii* Perr. & Song.
2. Calyx teeth lance-shaped to egg-shaped, narrowed to the base.	
3. Corolla with olive-green spots in the throat; calyx teeth usually less than half as long as the calyx tube.	*acaulis*
3. Corolla nearly or quite without green spots in the throat; calyx teeth usually at least half as long as the calyx tube.	*occidentalis*

**992. *G. acaulis* L. (*G. kochiana*) TRUMPET GENTIAN. Flowers with olive-green spots in the throat and corolla lobes acute or tapering to a rigid point. Leaves usually lance-shaped or elliptical. Mountain pastures. N–E. Spain, E. Pyrenees. Page 149.   992a. *G. alpina* Vill. Flowers with olive-green spots in the throat but corolla lobes blunt, usually rounded. Leaves usually rounded about 1 cm. Mountain pastures. S. Spain (Sierra Nevada), C. Pyrenees. Pl. 37.   992b. *G. occidentalis* Jakowatz Flowers with few or no green spots in the throat; lobes of corolla acute to obtuse, narrowed to a fine point. Leaves elliptical to oblong-lance-shaped. Mountains, N. Spain, W. Pyrenees. Pl. 37.

GENTIANELLA | Distinguished from *Gentiana* by the long hairs in the throat of the corolla tube or by the corolla lobes which are fringed with hairs; usually annuals or biennials.

1. Fls. blue (rarely white or yellowish).	
2. Fls. large $2\frac{1}{2}$–5 cm across; corolla lobes fringed with hairs. (E. Spain, Pyrenees)	1001. *G. ciliata* (L.) Borkh.
2. Fls. small 8–12 mm across (Pyrenees)	1000. *G. tenella* (Rottb.) Börner
1. Fls. bluish-lilac, or white; outer calyx lobes encircling inner lobes. (N–E. Spain, Pyrenees, France)	999. *G. campestris* (L.) Börner

SWERTIA | **1002. *S. perennis* L. MARSH FELWORT. A gentian-like plant, distinguished by its dull violet-purple spotted flowers with 5 wide-spreading petals, each with 2 dark shining nectaries at their base. Leaves pale greenish-yellow, lance-shaped, the upper clasping the stem. Grassy bogs in mountains. N. Spain, Pyrenees, Corbières, Massif Central.

MENYANTHACEAE | Bogbean Family

MENYANTHES | **1003. *M. trifoliata* L. BOGBEAN. A submerged aquatic plant or wet-bog plant, with large trifoliate leaves and pyramidal clusters of pinkish-white flowers borne above water. Petals 5, spreading, with conspicuous white hairs on the upper surface; buds pink. C. and N. Portugal, C., N., and E. Spain, France.

NYMPHOIDES | **1004. *N. peltata* (S. G. Gmelin) O. Kuntze FRINGED WATER-LILY. Readily distinguished by its golden-yellow flowers about 3 cm across, with conspicuously fringed petals, borne above the surface of the water. Floating leaves 3–10 cm, rounded, often blotched. Lakes, slow rivers. N. and C. Portugal, E. Spain, France.

APOCYNACEAE | Dogbane Family

VINCA | Periwinkle

1. Fl. stalks longer than the lvs. and the fls.; corolla 2–3 cm across,
   with blunt lobes. (Widespread)   **1005. *V. minor* L.
1. Fl. stalks shorter than lvs.; corolla 2–5 cm across, lobes more or
   less acute.
   2. Sepals ciliate; lv. margin ciliate. (Scattered throughout)   1006. *V. major* L.
   2. Sepals hairless; lv. margin hairless.   *difformis*

(1005) *V. difformis* Pourret INTERMEDIATE PERIWINKLE. A trailing plant with opposite, oval-lance-shaped leaves and solitary blue, axillary flowers 3–4 cm across. Petals oval, pointed; flower stalk about half as long as the hairless leaves. Hedges, ditches, shady damp places. Portugal, N–W. Spain, Med. region.

NERIUM | **1007. *N. oleander* L. OLEANDER. A handsome, grey-leaved bush with clusters of large pink flowers, often found growing in dry river courses and ravines in the south. Flowers 3–5 cm across, with spreading petals and a frilly throat. Leaves in whorls of 3, linear-lance-shaped, leathery, 10–20 cm. S. Portugal, S. and S–E. Spain.

ASCLEPIADACEAE | Milkweed Family

CYNANCHUM | **1008. *C. acutum* L. STRANGLEWORT. A climbing perennial with opposite, triangular-heart-shaped, glaucous leaves and small axillary umbels of pink or white, sweet-scented flowers. Flowers 5–7 mm across, with 10 projecting scales. Fruit horn-like. Hedges, scrub, waysides. C. Portugal, Med. region.

PERIPLOCA | **1009. *P. graeca* L. SILK-VINE. A climbing perennial with dark shining green, oval to lance-shaped leaves, and long-stalked, lax, axillary clusters of brownish flowers, longer than the leaves. Flowers about 2½ cm across; petals oblong, fringed with hairs. Cultivated and naturalized in S. Spain.   1009a. *P. laevigata* Aiton A thorny, grey-stemmed shrub to 60 cm, with small leathery, oblong leaves 1–1½ cm. Flowers dark brownish-purple and green beneath, 1–1½ cm across. Fruit usually of 2 diverging 'horns'; seeds with long silky hairs. Rocky scrub. S–E. Spain (Almeria and Murcia provinces). Page 296.

VINCETOXICUM | **1010. *V. hirundinaria* Medicus COMMON VINCETOXICUM. A yellow-ish-green perennial with rather leathery, oval-pointed, opposite leaves and terminal and axillary clusters of dull whitish or yellowish flowers. Flowers about ½ cm across, usually hairless. Fruit 4–6 cm, horn-like, lance-shaped, long-pointed. Woods, uncultivated ground, rocks, riversides. Portugal, C., N., and E. Spain, France.   (1010) *V. nigrum* (L.) Moench DARK VINCETOXICUM. Like 1010 but flowers reddish, becoming blackish; petals hairy above. Fruit 6–8 cm. Bushy and stony places, riversides, waste places. C. and N. Portugal, S., C., and E. Spain, Med. France.

CARALLUMA | Stems fleshy, without leaves.

1010a. *C. europaea* (Guss.) N.E. Br. (*Apteranthes gussoneana*) A low tufted succulent plant with fleshy, silvery-glaucous, four-angled stems without leaves; stem about 1½ cm wide, angles toothed. Flowers dusky purple, with a dark blackish-purple centre and violet fringe, 12 mm across, in a dense few-flowered umbel. Arid salt-rich areas. S. Spain (Almeria and Murcia provinces).

GOMPHOCARPUS | **1012. *G. fruticosus* (L.) Aiton fil. BRISTLY-FRUITED SILKWEED. A shrubby perennial with narrow long-pointed leaves, small white flowers in umbels and very distinctive large ovoid-pointed, 'boat-shaped' fruits covered with long bristly hairs. Flowers about ½ cm across, with central projecting horn-like scales. River-courses, damp places, naturalized in the south. Pl. 36.

RUBIACEAE | Madder Family

PUTORIA | **1013. *P. calabrica* (L. fil.) DC. A prostrate shrublet with leathery leaves and dense heads of long-tubed pink flowers. Flowers about 1½ cm long, with 4 spreading lobes; stamens projecting. Leaves opposite, shining, inrolled. Rocks in hills and mountains. S. Spain. Pl. 38.

SHERARDIA | **1014. *S. arvensis* L. FIELD MADDER. Widespread.

CRUCIANELLA | 4 species.   **1016. *C. maritima* L. A spreading woody-stemmed shrub-let with leathery, spiny-tipped, white-margined, overlapping leaves and dense compact terminal spikes of yellow flowers. Flowers about 1 cm; bracts with papery margin. Littoral sands. Coasts of Iberia, Med. France.

ASPERULA | About 15 species.   1017. *A. cynanchica* L. SQUINANCY WORT. Portugal, Spain, France.   1018. *A. arvensis* L. BLUE WOODRUFF. A slender, square-stemmed annual with narrow leaves in whorls, and terminal leafy clusters of tiny blue, slender-tubed flowers 5–6 mm long. Leaves in whorls of 4 in lower part of stem, in whorls of 6–8 above, rough-haired. Cultivations. Widespread.   1017a. *A. hirta* Ramond A low spreading peren-nial, with many slender quadrangular stems with stiff leaves ½–1½ cm in whorls of 6, and dense terminal clusters of pink flowers. Corolla tubular with 4 spreading lobes shorter than the tube; style much longer than the corolla. Leaves linear-lance-shaped, bristly-haired, usually longer than the internodes. Alpine rocks. Endemic in Pyrenees.   1017b. *A. hirsuta* Desf. Distinguished from 1017a by its linear leaves with inrolled margin with the lower and younger leaves densely shaggy-haired, upper and adult leaves rough, almost hairless. Flowers dark pink in terminal heads of 9–15 flowers; styles included in corolla. Sunny rocks, screes, waysides. S. Portugal, S., W., and S–E. Spain.

GALIUM | Bedstraw   About 52 species occur in our area; a very difficult genus. The following are widespread: 1023. *G. aparine* L. GOOSEGRASS;   1024. *G. verum* L. LADY'S BEDSTRAW;   1023a. *G. parisiense* L. Widespread except in south;   (1022) *G. palustre* L. MARSH BEDSTRAW;   1022. *G. uliginosum* L. FEN BEDSTRAW;   1026. *G. saxatile* L. HEATH BEDSTRAW.

CRUCIATA | 3 species.   ****1029.** *C. laevipes* Opiz (*Galium cruciata*) CROSSWORT. Scattered throughout except in south.

VALANTIA | 2 species occurring in the south.

RUBIA | 2 species.   ****1031.** *R. peregrina* L. WILD MADDER. A rough, prickly clinging, scrambling perennial with stiff leaves in whorls of 4–6 and axillary branched clusters of greenish-yellow flowers. Flowers about ½ cm; corolla lobes 5. Stems quadrangular with recurved prickles; leaf margins likewise with recurved prickles. Bushy places. Widespread.

POLEMONIACEAE | Phlox Family

POLEMONIUM | **1032. *P. caeruleum* L. JACOB'S LADDER. Damp woods and pastures in mountains. C. Pyrenees, Massif Central.

CONVOLVULACEAE | Convolvulus Family

IPOMOEA | 1033a. *I. sagittata* Poiret A climbing perennial with showy purple or pink flowers 4–7 cm long, and bright shining green, lance-shaped leaves with a heart-shaped base, the upper arrow-shaped. Ditches and hedges on the littoral, climbing on *Arundo*, and on margins of rice fields. Native of N. America. S. Spain.   1033b. *I. batatas* (L.) Lam. (*Batatas edulis*) SWEET POTATO. A trailing perennial with leaves varying on the same plant from heart-shaped to deeply five-lobed, and with few-flowered axillary clusters of purple, funnel-shaped flowers 3½–5 cm long. Tubers oblong, purple or white. Grown as a vegetable on the *vegas* of Andalusia, S. Spain.

CONVOLVULUS |

1. Dwarf cushion plants of S. Spain.	*boissieri* (*lineatus*)
1. Not cushion plants.	
2. Lvs. very abruptly narrowed into a distinct leaf stalk.	
3. Lvs. with a wedge-shaped or cut-off base.	
4. Plants entirely herbaceous; fls. 7–12 mm.	*siculus*
4. Plants with woody stock; fls. 1½–3 cm. (S–E. Spain)	1034a. *C. valentinus* Cav.
3. Lvs. with a heart-shaped base or with triangular pointed basal lobes.	
5. Upper lvs. and bracts very deeply divided.	*althaeoides*
5. Upper lvs. and bracts not divided except for the basal lobes.	
6. Fls. blue.	*siculus*
6. Fls. white to pink.	

7. Lvs. toothed; sepals narrowed to a long point or fine-
pointed. (Introd. Portugal (Lisbon))   1039a. *C. farinosus* L.
7. Lvs. entire; sepals blunt or notched. (Widespread)   1038.   *C. arvensis* L.
2. Lvs. often much narrowed at base but never abruptly so.
    8. Anns. or short-lived perenns. entirely herbaceous.
        9. Fl. stems shorter than or about as long as calyx; fls.
        violet. (S–E. Portugal, S. and C. Spain)   1035b. *C. humilis* Jacq.
        9. Fl. stems several times longer than calyx.
            10. Sepals with 2 zones, the upper hairy and leafy; fr.
            hairy.   *tricolor*
            10. Sepals without 2 zones, hairless or sparsely hairy with
            papery margins; fr. hairless.
                11. Sepals lance-shaped to narrowly egg-shaped, acute to
                shortly fine-pointed; fls. 14–22 mm long.   *meonanthus*
                11. Sepals elliptic-egg-shaped, blunt and with a fine
                point; fls. 7–10 mm long.   *pentapetaloides*
    8. Perenns.; shoots woody below, herbaceous above.
        12. Inflorescence with all fls. on each branch crowded into
        a compact head.
            13. Fls. pink.   *lanuginosus*
            13. Fls. usually white. (S–E. Spain)   (1037)   *C. cneorum* L.
        12. Inflorescence lax, with at least some branches distinct.
            14. At least lower parts of stems with mostly spreading
            hairs.   *cantabrica*
            14. Shoots with adpressed silvery hairs, sometimes with
            some spreading hairs.   *lineatus*

1034. *C. siculus* L. SMALL BLUE CONVOLVULUS. A hairy, non-climbing annual with small,
five-lobed, funnel-shaped blue flowers with a yellowish tube, only 1 cm across. Calyx with
long hairs. Leaves oval-acute, with shallowly heart-shaped base. Dry rocky, sandy places.
S. Portugal, Med. region.   (1034) *C. pentapetaloides* L. Very like 1034 but leaves narrower,
oblong-spathulate, and calyx hairless. Flowers blue with a yellow throat, 7–10 mm across.
Fields, bushy places. S. Portugal, Med. region.   **1035. *C. tricolor* L. DWARF CONVOLVULUS.
A very conspicuous spreading annual at flowering with funnel-shaped flowers 2–5 cm
across, with an orange-yellow throat, white central zone and blue margin. Leaves oval-
lance-shaped. Waste ground, roadsides, vineyards. S. and C. Portugal, Med. region (abundant on Costa del Sol). Page 61.   1035a. *C. meonanthus* Hoffmanns. & Link Very similar to
1035 with tricoloured flowers, but flowers smaller 14–22 mm long and about 1½ cm across,
and calyx and fruit more or less hairless. S. and C. Portugal, S. and E. Spain.

**1036. *C. cantabrica* L. PINK CONVOLVULUS. A non-climbing perennial with spreading
hairs, and with lax clusters of pink flowers on stems longer than the narrow leaves.
Corolla 3–4 cm across, hairy on the outside; calyx with spreading hairs. Leaves lance-
shaped to linear. Uncultivated ground, sandy fields. C., N., and E. Spain, S. France. Page
152.   1037. *C. lineatus* L. SILVERY-LEAVED PINK CONVOLVULUS. Like 1036 but leaves and
calyx with adpressed silvery hairs and flower clusters with stems shorter than the leaves.
Flowers 2–3 cm across. A low, often matted, cushion-forming plant. Arid stony places, dry
fields. Portugal, S., C., and E. Spain, France. Page 296.

1036a. *C. lanuginosus* Desr. Readily distinguished by its tight cluster of numerous pink-
striped flowers surrounded by an involucre of hairy bracts, and borne at the ends of erect
stems. Flowers 1½–2½ cm across; calyx with dense, spreading, silvery hairs. A silvery-grey
or whitish, tufted perennial covered with woolly hairs; leaves linear. Calcareous rocks in

low lands and mountains. S. and E. Spain, S. France. Pl. 38.   1036b. *C. boissieri* Steudel A beautiful shining, silvery-leaved, dense cushion-forming plant of the higher mountains of S. Spain, with large almost stalkless, rosy-white flowers 2–3 cm across. Calyx with woolly spreading hairs. Leaves tiny, oval, folded, densely covered with shining silvery-white hairs. Page 296.

**1039. *C. althaeoides* L. MALLOW-LEAVED BINDWEED. A slender, usually climbing perennial with large deep pink flowers 3–4½ cm across, and upper leaves deeply divided into narrow segments. Flowers axillary, 1–2, on stems longer than the subtending leaves. Cultivated ground, arid waste places. S. and C. Portugal, C. Spain, Med. region. Page 162.

CALYSTEGIA | 1040. *C. sepium* (L.) R.Br. BELLBINE, GREATER BINDWEED. Hedges, field verges, damp places. Widespread.   **1041. *C. soldanella* (L.) R.Br. SEA BINDWEED. A creeping plant of sands and shingles on the littoral, with fleshy kidney-shaped leaves, and large pink flowers 4–5 cm across. Atlantic and Med. coasts.

CRESSA | 1042. *C. cretica* L. A small, dense, grey-leaved shrub of salt marshes, with terminal globular clusters of yellowish or pale pink flowers 3–5 mm across. Corolla divided to about half its length into 5 spreading lobes; stamens exserted, 5; calyx lobes blunt. Leaves small 2–10 mm, lance-shaped to heart-shaped, silky-haired. S. and C. Portugal, Med. region.

CUSCUTA | **Dodder**   Probably 3 species of these parasitic plants occur in our area. They are separated by small botanical differences and are not easily distinguished.

BORAGINACEAE | Borage Family

HELIOTROPIUM | **1045. *H. europaeum* L. HELIOTROPE. A softly hairy, often greyish, erect annual with tiny white or pale lilac flowers in leafless, forked spikes. Corolla 3–4 mm across; calyx divided almost to the base, spreading in fruit. Leaves oval. Cultivated ground, waste places, waysides. Portugal, Spain, Med. France.   (1045) *H. supinum* L. Very like 1045 but calyx teeth short, not spreading and falling with the fruit. Flowers white, smaller, 1–2 mm across. Field verges, drying-out calcareous soils. S. and C. Portugal, Med. region.

OMPHALODES |

1. Perenns.; fls. blue.   *nitida*
1. Anns.; fls. white, rarely blue.
  2. Margin of nutlets strongly incurved, toothed.
    3. Inflorescence without bracts. (S. and C. Portugal, Med. region)   **1048.   *O. linifolia* (L.) Moench.
    3. Inflorescence with bracts. (C. Portugal)   1048a. *O. kuzinskyanae* Willk.
  2. Margin of nutlets erect, entire.
    4. Nutlets hairy. (W. France)   1048b. *O. littoralis* Lehm.
    4. Nutlets hairless. (S. Spain)   1048c. *O. brassicifolia* (Lag.) Sweet. Pl. 38 Page 68.

1047a. *O. nitida* Hoffmanns. & Link (*O. lusitanica* auct.) An erect slender, pale green perennial to 30 cm, with a lax elongated cluster of blue flowers borne on long slender stalks which are recurved in fruit. Leaves lance-shaped, shining above, somewhat leathery, the upper more or less clasping, the lower long-stalked. Damp shady places, thickets. C. and N. Portugal, S., W., and N. Spain. Pl. 39. Page 103.

CYNOGLOSSUM | Hound's-Tongue

1. Ripe nutlets flat or slightly concave outside with a thickened border.	
2. Inflorescence with bracts.	*cheirifolium*
2. Inflorescence without bracts.	
3. Stem lvs. oblong to lance-shaped; fls. dull purple. (Spain, France)	**1049. *C. officinale* L.
3. Stem lvs. linear-lance-shaped; fls. deep blue. (N–E. Spain, S. France)	1049a. *C. dioscoridis* Vill.
1. Ripe nutlets convex outside, without a thickened border.	
4. Corolla lobes shaggy-haired. (S. Iberia)	1049b. *C. clandestinum* Desf.
4. Corolla lobes hairless.	
5. Lvs. hairless above. (N–E. Spain, France)	(1049) *C. germanicum* Jacq.
5. Lvs. densely hairy on both surfaces.	
6. Fls. 7–9 mm, deep blue, with netted veining. (Widespread)	(1049) *C. creticum* Miller
6. Fls. 4–6 mm, reddish-violet, without netted veining. (Med. Spain)	1049c. *C. nebrodense* Guss.

(1049) *C. cheirifolium* L. (including *C. arundanum*) A densely silvery-white, woolly-haired or felted biennial with stalkless, lance-shaped stem leaves and flat-topped clusters of purple flowers which later become violet or deep blue. Corolla about 8 mm, lobes hairless, shorter than corolla tube. Nutlets 5–8 mm across, densely covered with barbed projections on raised margin and face, or with a nearly smooth face. Dry places, rock crevices. Portugal, Med. region. Pl. 39. Page 68.

SOLENANTHUS | Like *Cynoglossum* but corolla with a long funnel-shaped tube and anthers projecting.   1049d. *S. reverchonii* Degen Flowers 6–9 mm; corolla lobes very small, reddish-yellow, hairy outside. Fruit like *Cynoglossum* and densely covered with flattened barbed spines. A silvery-hairy perennial with linear to narrowly lance-shaped leaves. S. Spain.

LAPPULA | 3 species.   1050. *L. squarrosa* (Ret.) Dumort. (*L. myosotis*) BUR FORGET-ME-NOT. Dry places, waysides, vineyards, dunes. C. and E. Spain, France.   (1050) *L. deflexa* (Wahlenb.) Garcke (*Hackelia d.*) E. Spain.

ASPERUGO | **1051. *A. procumbens* L. MADWORT. A coarse, bristly-haired, spreading or scrambling annual with small stalkless, purplish to violet funnel-shaped flowers, 2–4 mm across, in the axils of the lance-shaped leaves. Calyx five-lobed, enlarging in fruit into 2 kidney-shaped toothed lobes. Waste nitrogen-rich ground, fields. S., C., and E. Spain, France.

SYMPHYTUM | **Comfrey** 1052. *S. officinale* L. COMFREY. Damp meadows, marshes, by water. Scattered throughout except in south.   **1053. *S. tuberosum* L. TUBEROUS COMFREY. Woods, damp shady places. Spain, France.

BORAGO | **1054. *B. officinalis* L. BORAGE. Readily distinguished by its bright blue flowers 2–2½ cm across, with spreading lobes and its forward-projecting cone of blackish-purple anthers. A very bristly-haired, stiff, stout-stemmed annual. Waste places, waysides, ditches. Throughout, less common in the north. Page 60.

ANCHUSA (including *LYCOPSIS*) | Alkanet

1. Corolla irregular, the tube curved, the limb oblique,
with 5 unequal lobes. (Widespread)   1059. *A. arvensis* (L.) Bieb.
1. Corolla regular with equal, spreading lobes.
   2. Calyx divided into lobes more or less to the base.
     3. Nutlets longer than wide.   *azurea*
     3. Nutlets wider than long. (Spain, W. France)   **1055. *A. officinalis* L.
   2. Calyx divided into lobes to not more than two-thirds its
     length.
     4. Calyx lobes acute.
      5. Most hairs on lvs. with a swollen base.   *granatensis*
      5. Most hairs on lvs. without a swollen base. (? Spain,
      W. France)   **1055. *A. officinalis* L.
     4. Calyx lobes blunt.
      6. Plants with sparse, conspicuously swollen-based
      bristles only.   *calcarea*
      6. Plants downy with short soft hairs and with bristles
      which are mostly without swollen bases.   *undulata*

**1056. *A. azurea* Miller (*A. italica*) LARGE BLUE ALKANET. A striking, brilliant azure-blue-flowered perennial with a much branched flower cluster forming a terminal pyramid. Flowers 1–2 cm across, with a brush-like tuft of white hairs in the throat. A rough bristly-haired plant; leaves 10–30 cm, lance-shaped. Fields, vineyards, waysides. Portugal, Spain, S. France. Pl. 39, Page 61. (1057) *A. undulata* L. (including *A. hybrida*) UNDULATE ALKANET. A biennial with both short downy hairs and long bristly hairs mostly without conspicuous swellings at the base, and with small blackish-purple flowers fading to deep blue, 3–8 mm across. Tube of corolla 1½ to 2 times as long as the calyx. Leaves often undulate. Sandy and stony fields. Portugal, Med. region. Page 45. 1056a. *A. granatensis* Boiss. Like (1057) but plant with few or no short hairs; calyx lobes acute; corolla tube about 6 mm, as long as or a little longer than the calyx. Richer soils in mountains. S. and C. Portugal, S. Spain. 1056b. *A. calcarea* Boiss. Like (1057) but usually a woody-based perennial with sparse bristles with conspicuously swollen bases, particularly conspicuous on dead leaves. Whole inflorescence often suffused with reddish-purple; calyx divided to only one third; corolla dark blue, tube 6–8 mm, longer than calyx. Sands on the littoral and calcareous sands inland. Portugal, S. Spain. Pl. 40.

PENTAGLOTTIS | **1058. *P. sempervirens* (L.) L. H. Bailey ALKANET. A rather robust, bristly-haired perennial, with egg-shaped acute leaves and branched leafy clusters of bright blue flowers. Flowers 5–7 mm across, throat-scales white. Upper leaves clasping. Damp and shady places. N–E. Portugal, C. and N. Spain, S. and W. France.

NONEA | Distinguished from *Anchusa* by the funnel-shaped corolla (not bell-shaped or with spreading lobes).

1. Calyx teeth equalling or longer than the calyx
tube; fls. brownish-purple.   *vesicaria*

1. Calyx teeth *c.* one third as long as the calyx tube.
  2. Nutlets symmetrical, kidney-shaped; fls. pale yellowish or white. (S. and C. Spain, S. France)   (1060)   *N. ventricosa* (Sm.) Griseb.
  2. Nutlets asymmetrical, obliquely ovoid.
    3. Corolla 10–14 mm long, usually dark brownish-purple. (France)   **1060. *N. pulla* (L.) DC.
    3. Corolla 7–8 mm long, pale blue or yellowish. (S., C., and E. Spain)   1060b. *N. micrantha* Boiss. & Reuter

1060a. *N. vesicaria* (L.) Reichenb. (*N. nigricans*) Distinguished by its small maroon flowers 3–5 mm across with yellowish corolla tubes 6–12 mm long. Calyx much enlarging in fruit, to 1–1½ cm; nutlets wrinkled. A rough, bristly, grey-leaved annual or biennial with rather long leafy clusters of short-stalked flowers. Sandy hills, rocky and arid places. Portugal, S. and C. Spain. Page 307.

ELIZALDIA | Like *Nonea* but with a flange-like ring at mouth of corolla tube (not scales). Stamens exserted, attached to top of tube.
1060c. *E. calycina* (Roemer & Schultes) Maire (*E. nonneoides*) A very bristly-haired annual with lance-shaped leaves, the uppermost tinged purple, and with violet flowers about ½ cm across, with yellow tube, in a lax leafy spike-like cluster. Calyx about 8 mm, enlarging to 1½ cm, becoming ashy-violet. Nutlets transversely ribbed. Sandy places, waste ground. S–W. Spain.

ALKANNA | **1061. *A. tinctoria* (L.) Tausch DYER'S ALKANET. A greyish, usually spreading, bristly-leaved shrubby perennial with small bright blue flowers about ½ cm across, in leafy forked spikes. Corolla without scales or hairs in throat, but with transverse swellings. Upper leaves clasping. Sandy and rocky places. Med. region.   1061a. *A. lutea* DC. An erect or ascending annual with long, branched leafy spikes of small yellow flowers, 5–7 mm long. Plant with glandular and spreading hairs. Dry stony places, maritime sands. Spain, E. Pyrenees, S. France.

PULMONARIA | **Lungwort**   Probably 5 species, only 2 of which spread further west than the Pyrenees.   **1064. *P. longifolia* (Bast.) Boreau NARROW-LEAVED LUNGWORT. Damp shady places. Scattered throughout except in the south.

MYOSOTIS | **Forget-Me-Not, Scorpion-Grass**   About 18 species occur in our area; often not easy to distinguish in the field. The following are widely scattered in our area: (1065) *M. laxa* Lehm. WATER FORGET-ME-NOT;   (1065) *M. secunda* A. Murray WATER FORGET-ME-NOT;   1068. *M. discolor* Pers. YELLOW AND BLUE SCORPION-GRASS;   (1068) *M. ramosissima* Rochel EARLY FORGET-ME-NOT;   1069. *M. arvensis* (L.) Hill COMMON SCORPION-GRASS.

**1067. *M. alpestris* F. W. Schmidt ALPINE FORGET-ME-NOT. A widespread plant of the higher mountains of Europe, excluding Portugal. Distinguished by its bright blue flowers 4–10 mm across, with flat spreading corolla lobes. Calyx densely hairy, with or without hooked bristles, up to 7 mm long and often much enlarged in fruit. Nutlets black, shining. Very variable, from erect lax plants to mat-forming plants.   1067a. *M. alpina* Lapeyr. Closely related to 1067, but always densely mat-forming, with stalkless basal leaves which are hairless beneath. Calyx with numerous hooked bristles at the base; flower stalks to 2 mm. Alpine habitats, pastures and screes. Pyrenees.

1. *Nonea vesicaria* 1060a
2. *Sideritis scordioides* 1115a
3. *Echium lusitanicum* 1081d
4. *Nepeta apuleii* 1116b

1068a. *M. stricta* Roemer & Schultes A widespread annual with minute bright or pale blue flowers little more than 1 mm across. Flower stalks less than 2 mm; calyx with deflexed hooked hairs and adpressed straight hairs at the base, divided halfway into lobes which are closed in fruit and not deciduous. Dry sandy places. C. and E. Spain. 1068b. *M. minutiflora* Boiss. & Reuter Restricted to the mountains of S. Spain (S. Nevada). With minute white flowers, but calyx open, deciduous; fruit stalks to 3 mm. often deflexed. Nutlets egg-shaped. 1068c. *M. refracta* Boiss. Flowers bright blue but calyx lobed to one third its length in fruit, fruit stalks 1 mm, almost always deflexed towards the base of the stem. Nutlets ellipsoid. Mountains. S. Spain. 1068d. *M. persoonii* Rouy Flowers bright yellow about 2 cm across. Like 1068 but a low-growing compact annual with a dense inflorescence with the lowest flowers subtended by bracts. Calyx with stiff, hooked hairs. Iberian peninsula.

LITHOSPERMUM | Gromwell This genus has recently been divided into 4 genera: *LITHODORA, BUGLOSSOIDES, NEATOSTEMA, LITHOSPERMUM*, but in the key below they have been included in the genus *LITHOSPERMUM*:

1. Herbaceous plants; corolla with swellings or folds in throat.
  2. Fls. white, yellow or bluish; corolla 4–8 mm long, little longer than calyx.
    3. Perenns.; nutlets snow-white, smooth, shining; fls. white, tinged yellow. (Widespread)   **1072. *L. officinale* L.
    3. Anns.; nutlets wrinkled.
      4. Fls. white or bluish; corolla hairless in throat. (Widespread)   1071. *L. arvense* L. (*Buglossoides a.*)
      4. Fls. yellow; corolla hairy in throat. (Widespread except north.)   (1071) *L. apulum* (L.) Vahl (*Neatostema a.*)
  2. Fls. blue, purple or violet; corolla 1½–2 cm long.
    5. Lvs. to c. 1½ cm broad, gradually narrowed to base.   *purpurocaeruleum*
    5. Lvs. c. 3 cm broad, abruptly narrowed, clasping.   *gastonii*
1. Woody shrubs or shrublets with peeling bark; corolla with or without hairs in throat.
  6. Lvs. obovate or oblong-egg-shaped, not more than 3 times as long as wide, blunt at apex.   *oleifolium*
  6. Lvs. linear or lance-shaped, usually more than 3 times as long as wide, acute or blunt at apex.
    7. Stamens inserted at unequal levels on the corolla tube; corolla throat more or less densely hairy.   *diffusum*
    7. Stamens all inserted at the same level on the corolla tube; corolla throat hairless.   *fruticosum*

**1073. *L. purpurocaeruleum* L. (*Buglossoides p.*) BLUE GROMWELL. Distinguished by its large, bright blue flowers 1½–2 cm across, which are at first reddish-purple, borne in leafy, few-flowered terminal clusters. Corolla funnel-shaped, hairy outside. Leaves narrow-lance-shaped, the lower narrowed to a stalk. Nutlets white, smooth, shining. Bushy places, thickets, hedges. C., N., and N–E. Spain, France.   1073a. *L. gastonii* Bentham (*Buglossoides g.*) Like 1073 but leaves broader, crowded, half-clasping, the lowest scale-like. Nutlets yellowish, netted. Rocks in mountain woods. W. Pyrenees of France.

**1074. *L. diffusum* Lag. (*Lithodora d.*) SCRAMBLING GROMWELL. A low, erect or spreading undershrub with blunt, bristly-haired leaves and terminal clusters of bright blue, funnel-shaped flowers. Corolla 1–2 cm long, with dense silky hairs outside. Pine woods, scrub and hedges. N. Portugal, S–W., C. and N. Spain, W. France. Pl. 40. Page 160   Subsp. *lusitanica* (Samp.) Silva & Roz. has erect, much-branched stems forming clumps, and leaves adpressed to stems, linear, with inrolled margins. Sandy places. S. and C. Portugal, S. Spain.
1075. *L. fruticosum* L. (*Lithodora f.*) SHRUBBY GROMWELL. Like 1074 with blue flowers smaller about 1 cm across, but corolla tube hairless outside and only sparsely bristly-haired on the outside of the corolla lobes. A dwarf shrub to ½ m, with erect, interwoven branches; leaves linear, whitish with adpressed bristly hairs, margins inrolled. Stony hills, mainly on lime-stone. S., C., and E. Spain, S–W. France. Pl. 38.   1075a. *L. oleifolium* Lapeyr. (*Lithodora o.*) A rare endemic shrublet of rocks in the Eastern Pyrenees of Spain, distinguished by its broader leaves which are dull green above and silvery-white beneath. Flowers pale purplish, then blue; corolla silvery-haired outside, hairless within. Calyx silvery-haired. Nutlets greyish-white, smooth, shining.

ONOSMA | **1076. *O. arenaria* Waldst. & Kit (*O. echioides*) GOLDEN DROP. A stiff, erect, very bristly-haired, variable perennial with terminal clusters of pendulous tubular, pale yellow flowers. Corolla 12–25 mm long; calyx 6–12 mm in flower, up to 18 mm in fruit. Leaves green or grey, usually with simple hairs only. Nutlets 2–4 mm smooth, not horned. Dry rocky and sandy places. S., C., and E. Spain, France. Pl. 40. Page 153.   1076a. *O. tricerosperma* Lag. Very like 1076 and sometimes not easily distinguished from it. Corolla pale yellow, often minutely downy, 17–25 mm; calyx 15–22 in flower, up to 3 cm in fruit. Nutlets 5–9 mm, wrinkled, with 3 horns. S–E. and C. Spain.

CERINTHE | Honeywort

1. Corolla lobes lance-shaped, erect, almost as long as the rest of the corolla. (Balearic Islands, France)   **1078. *C. minor* L.
1. Corolla lobes egg-shaped, sharply recurved, much shorter than the rest of the corolla.
    2. Corolla 5–8 mm wide, more than twice as long as calyx; anns.   *major*
    2. Corolla 3–4 mm wide, not more than twice as long as calyx; perenns. or bienns.   *glabra*

**1079. *C. major* L. HONEYWORT. A hairless, glaucous annual with broad overlapping, clasping leaves and drooping clusters of yellow cylindrical flowers, often with a varying amount of chocolate-brown or red towards the base. Corolla 1½–3 cm long. Leaves often blotched, with rough swellings. Cultivated ground, waysides and stony places. Widespread except in the north. Page 163.   (1079) *C. glabra* Miller SMOOTH HONEYWORT. Distinguished by its smaller yellow flowers with corollas 8–13 mm long, with 5 dark red spots in the throat; calyx lobes blunt, usually hairless. Montane and alpine meadows. Pyrenees, France.

ECHIUM | An important and conspicuous genus with 16 species in our area, 9 of which are restricted in Europe to the Iberian peninsula. Identification is often difficult and the differences between some species are not always clear-cut. The length of the corolla is given in the descriptions below. The following are fairly widely distributed: 1080. *E. parviflorum* Moench SMALL-FLOWERED BUGLOSS (South); (1080) *E. arenarium* Guss. (Med. region); **1082. *E. vulgare* L. VIPERS BUGLOSS (Widespread); **1083. *E. plantagineum* L. (*E. lycopsis*) PURPLE VIPERS BUGLOSS (Widespread). Page 162.

Flowers flesh-coloured, yellowish or bluish-white

1081a. *E. asperrimum* Lam. An erect biennial to 1 m, with a dense whitish-grey covering of spreading, stinging bristles, and flesh-pink flowers in an intricately branched inflorescence. Corolla 13–18 mm, narrowly funnel-shaped, with 4–5 stamens, with carmine filaments, much longer than the corolla. Basal leaves more or less rounded at the base. Roadsides, rough ground. Spain, except north-west, S. France.   **1081. *E. italicum* L. Usually has a solitary erect stem with yellowish- or bluish-white flowers in a dense spike, or inflorescence a much-branched pyramid. Corolla 10–12 mm; 4–5 stamens projecting, filaments pale. Basal leaves narrowed to the base. Fields and rough ground. Spain, France.   1081b. *E. boissieri* Steudel (*E. pomponium*) A very striking biennial usually with extremely long, wand-like inflorescences 1–2½ m tall remaining conspicuous in winter when dead but still standing. Flowers small 16–18 mm, flesh-coloured, with 5 long-projecting stamens with pink-carmine filaments. Roadsides and uncultivated slopes. S. Portugal, S. Spain. Pl. 38.

1081c. *E. flavum* Desf. Like 1081b with flesh-coloured flowers and carmine filaments, but a much less robust plant, less than 1 m tall, with 1 or several equal flowering spikes. Corolla shorter, 11–16 mm; calyx 5–8 mm. Basal leaves 8–20 cm. Calcareous rocks, scrub-covered slopes in mountains. S. and E. Spain. Pl. 38.   1081d. *E. lusitanicum* L. A comparatively softly hairy perennial with rather small, dark grey-blue or bluish-white flowers in a lax spike-like inflorescence to 1 m or more, with 1 or several main stems. Corolla 7–10 mm; all stamens much longer than corolla, filaments carmine; calyx 5–7 mm. Basal leaves usually more than 25 cm. Roadsides and uncultivated ground. N–C. Portugal, W–C. Spain. Page 307.

Flowers blue, reddish-purple or pink-carmine turning blue (see also 1081d).

1083a. *E. albicans* Lag. & Rodr. The most beautiful of all the Iberian species with large pink-carmine to bluish-purple flowers in a dense cylindrical spike-like inflorescence, and with greyish or snowy-white leaves and stems. Corolla 16–21 mm, funnel-shaped; 2 or 3 stamens a little longer than the corolla, filaments violet. Calyx 10–17 mm at flowering, densely shaggy-haired with long white hairs and very sparse bristles. Rosettes large, dense, with numerous white, linear leaves with dense adpressed whitish hairs and sparse bristles. Rock crevices and ledges usually on limestone mountains. S. Spain (S. Nevada, Serranía de Ronda etc.). Pl. 39.

1083b. *E. creticum* L. Flowers large, pink-carmine turning to bluish-purple, or persistently reddish-purple, in a branched inflorescence. Corolla 14–38 mm, rather broadly funnel-shaped, with 1–2 stamens a little longer than the corolla, some or all stamens with sparsely hairy filaments. Leaves greenish or brownish with rather sparse spreading bristles, usually without prominent white basal swellings. A bristly-hairy biennial ½–1 m. Roadsides and grassy slopes. S. Portugal, Med. region.

1083c. *E. tuberculatum* Hoffmanns. & Link Very like 1083b but flowers dark blue-purple, 16–25 mm, narrowly funnel-shaped, with 2–4 unequal stamens longer than the corolla, filaments hairless. Leaves greyish-white, with stiff adpressed bristles with prominent swollen bases. Roadsides, uncultivated ground. S. and C. Portugal.   1083d. *E. sabulicola* Pomel Flowers red-purple, dark blue or bluish-purple, the corolla 12–22 mm, usually with

1 or 2 longer stamens with filaments sometimes sparsely hairy; calyx enlarging to 16 mm in fruit, densely white-bristly. Stems several, 15–50 cm, with an underlayer of forwardly-directed or irregularly spreading hairs. Roadsides, fields, sandy places near the sea. Med. region. 1083e. *E. gaditanum* Boiss. Like 1083c and doubtfully distinct from it but with clear blue to bluish-violet flowers; corolla 16–23 mm, with 3–4 stamens longer; calyx scarcely enlarging in fruit. Leaves with prominent swellings at the base of the bristly hairs. Sandy places, usually near the sea. S. Portugal, S. Spain. Pl. 40.

1083f. *E. rosulatum* Lange A perennial with a lax inflorescence with conspicuous leafy bracts, with pinkish-violet flowers. Corolla 11–20 mm, usually with a narrow tube, and 3–4 stamens longer than the corolla. Meadows and sandy fields. C. and N. Portugal, N–W. Spain. 1083g. *E. humile* Desf. Whitish or greyish perennials with very dense spreading bristles and more or less linear leaves 1–3 mm wide. Flowers bluish-purple, with 3–4 stamens longer than corolla which is about 13 mm long; calyx 6–8 mm at flowering. S–E. and C. Spain.

VERBENACEAE | Verbena Family

VERBENA | 1084. *V. officinalis* L. VERVAIN. Damp places, waste ground, waysides. Widespread. (1084) *V. supina* L. PROCUMBENT VERVAIN. Differs from 1084 in being procumbent and leaves mostly twice pinnately cut into oval segments. Flowering spike unbranched; corolla pale lilac, 3 mm across. Waysides, sandy places, on the littoral, derelict fields. S. and C. Portugal, S., C., and E. Spain.

LIPPIA | **1086. *L. nodiflora* (L.) Michx A creeping, rooting perennial with long-stemmed, dense ovoid heads of white flowers borne from the axils of the opposite leaves. Calyx lobed almost to base; flower heads 5–7 mm across. Leaves elliptic. Wet grassy places, usually near the sea. Med. Spain. 1086a. *L. canescens* Kunth Very like 1086 but flowers lilac and calyx lobed to not more than halfway; flower heads 9–12 mm across. Native of S. America; locally introduced into S–W. Europe. 1086b. *L. triphylla* (L'Hér.) O. Kuntze LEMON VERBENA. A deciduous shrub with lemon-scented, lance-shaped leaves, usually in whorls of 3. Flowers lilac, in long slender pyramidal spikes. Often grown for ornament in Portugal and S. Spain; sometimes naturalized.

VITEX | **1087. *V. agnus-castus* L. CHASTE TREE. A grey-felted, aromatic shrub with terminal spikes of usually pale lilac, or rarely pink flowers, and distinctive palmately divided leaves. Corolla 6–10 mm long, two-lipped; stamens projecting. Leaflets 5–7, lance-shaped, greyish-green above, white-felted beneath. Fruit fleshy, reddish-black. Damp places, on the littoral, streamsides. N–E. Portugal, Med. region.; sometimes naturalized elsewhere.

LABIATAE | Mint Family

AJUGA | Bugle

1. Lvs. with 3 linear lobes which are sometimes
   further three-lobed. (Widespread)   **1094. *A. chamaepitys* (L.) Schreber
1. Lvs. entire, toothed or shallowly lobed.
  2. Fls. yellow or purplish; lvs. 3–6 mm wide,
     linear. (S. and C. Portugal, Med. region)   1093. *A. iva* (L.) Schreber

2. Fls. blue or blue-purple; lvs. 8–40 mm wide, oblong to rounded.

3. Middle and upper part of stem hairy on opposite sides of the stem, alternating at each node. (Widespread) **1089. *A. reptans* L.

3. Middle and upper part of stem more or less evenly hairy on all sides.

4. Stamens distinctly longer than the corolla tube, the filaments hairy. (Pyrenees, France) **1090. *A. genevensis* L. Page 153

4. Stamens only slightly longer than the corolla tube, filaments hairless. (N. Portugal, C., N. and N–E. Spain, France) **1091. *A. pyramidalis* L.

TEUCRIUM | Germander   A large and important genus in S–W. Europe, with 35 species in our region, of which almost 20 are restricted to the Iberian peninsula. The genus is easily recognised by the flowers which have no upper lip and a lower lip with 5 lobes, with the middle lobe largest. Fairly widely dispersed species are: 1095. *T. botrys* L. CUT-LEAVED GERMANDER;   **1098. *T. scordium* L. WATER GERMANDER;   **1099. *T. chamaedrys* L. WALL GERMANDER;   1102. *T. scorodonia* L. WOOD SAGE.

Spiny plants

1098a. *T. spinosum* L. A much-branched downy, glandular-hairy annual to ½ m, with branches becoming spiny and more or less leafless at flowering, with white flowers in leafy whorls. Flowers solitary, or 4–6 in a whorl, turned upside down; calyx with spiny teeth. Leaves oblong, deeply toothed, the upper entire. Cultivated and uncultivated ground, dry sandy places. S. and C. Portugal, S., C., and E. Spain.   1100a. *T. subspinosum* Willd. A small, very spiny hummock-forming shrub 20–50 cm, with stout grey stems ending in stiff spines, and tiny pink flowers about 8 mm. Leaves 1–6 mm, greyish. Dry places. Majorca. Page 314.

Spineless plants

(a) Flowers solitary in axils of leaf-like bracts

**1096. *T. pseudochamaepitys* L. GROUND-PINE GERMANDER. A distinctive woody-based perennial with lax terminal spikes of conspicuous white or pinkish flowers, and with leaves cut into linear segments. Flowers 1–1½ cm, 2 in each whorl; bracts with 3 linear lobes. Dry hills, arid ground, southern part of our area. Pl. 41.   **1101. *T. fruticans* L. TREE GER-MANDER. A handsome shrub with large pale blue or lilac flowers in a lax leafy cluster, with white-felted calyx. Lower lip of corolla about 2 cm long. Twigs white-felted; leaves ever-green, white-felted beneath. Rocks, wooded hills, on the littoral. S. and C. Portugal, S. Spain, S–W. France. Page 60.

(b) Flowers in compact terminal heads

1103a. *T. pyrenaicum* L. A spreading, hairy, shrub-like perennial 10–30 cm, with a compact terminal head of white or yellowish flowers with purple tips. Flowers about 1½ cm; calyx about 1 cm, with spreading hairs. Leaves 1–2½ cm, rounded, with rounded teeth, leaf stalk much shorter; stem with spreading hairs. Limestone rocks. N., N–E Spain, Pyrenees, France. Pl. 40.   1103b. *T. rotundifolium* Schreber (including *T. granatense*) Like 1103a, but smaller in all its parts; flowers white to pale pink with reddish tips, about 1 cm. Leaves velvety-haired, ½–1 cm, with leaf stalks nearly as long as or longer than the blade. Lime-stone rocks and cliffs in mountains. S. Spain. Pl. 40.   1097. *T. polium* L. FELTY GERMANDER. A widespread, dwarf, downy-stemmed, variable shrub of the south of our area with several quite distinctive subspecies. Flowers white or red, in simple or compound globular heads. Leaves downy, with flat or inrolled margins, and with several rounded teeth.   Subsp.

aureum (Schreber) Arcangeli has golden hairs on its stems; flowers in a simple head; lateral lobes of corolla hairy. Mountains. Spain, France. Subsp. *capitatum* (L.) Arcangeli has white or grey hairs on its stems; flowers in compound branched rounded heads. Widespread. Subsp. *polium* has grey or white hairs on its stems; flowers in solitary globular heads; lateral lobes of corolla hairy. S. and C. Portugal, Med. Spain, France. Subsp. *dunense* Sennen (including *T. vincentinum*) has stout stems with dense golden or grey hairs; leaves to 2 cm; flower heads branched; lateral lobes of corolla hairless. Sandy places near the sea. S–W. Portugal, Med. region.   1097a. *T. gnaphalodes* L'Hér. A very pretty shrublet, covered with dense white woolly hairs, and with compact rounded heads of purple flowers with pale brown throats. Leaves 4–15 mm, narrow lance-shaped, rounded-toothed, often with golden hairs. Calyx 4–5 mm, densely woolly-haired, appearing swollen. Dry arid places on limestone mountains. Spain.

1102a. *T. salviastrum* Schreber An endemic dwarf twisted shrub of the mountains of C. Portugal with purplish flowers about 1 cm, in a lax one-sided inflorescence, and with small glandular-hairy leaves. Calyx two-lipped, strongly net-veined, glandular and shaggy-haired, about as long as the corolla tube, calyx teeth spiny-tipped. Leaves to 1½ cm, broadly oblong, wrinkled, with rounded teeth, grey-woolly beneath.

ROSMARINUS | Rosemary

**1105. *R. officinalis* L. ROSEMARY. A dense, aromatic, usually erect but sometimes prostrate evergreen shrub readily distinguished by its narrow dark green leaves which are inrolled and white-felted beneath, and its lilac flowers. Corolla two-lipped, 2 stamens and styles, all curving outwards well beyond the corolla. *Matorral*, rocks and stony ground. Widespread. 1105a. *R. eriocalix* Jordan & Fourr. (*R. tournefortii*) Like 1105, but branches grey and leaves shorter, ½–1½ cm by 1–2 mm, usually greyish-green in the lowlands, hairless in the mountains. Inflorescence densely woolly-haired; flowers violet with violet-blue tips. Leaf stalk, flower stalk, and calyx with star-shaped and long simple glandular hairs. Calcareous rocks. S. Spain (Granada and Almeria provinces). Page 90.

PRASIUM | **1106. *P. majus* L. Rocky and shrubby places. S. Portugal, S., C., and E. Spain.

SCUTELLARIA | Skullcap

Flowers in short dense clusters: mountain plants
**1107. *S. alpina* L. ALPINE SKULLCAP. Flowers blue-violet, purple or rarely white. Rocks and screes. S. Spain, Pyrenees, Massif Central. Page 139.   (1107) *S. orientalis* L. EASTERN SKULLCAP. Flowers yellow, rarely pink. S. Spain.

Flowers in long one-sided leafy clusters
**1108. *S. galericulata* L. SKULLCAP. Damp places, marshes. Iberian peninsula except in South, France.   1108a. *S. balearica* Barc. Distinguished from 1108 by the leaf stalks which are usually 1–2 times as long as the blade (not about a quarter as long). Majorca.   (1108) *S. hastifolia* L. SPEAR-LEAVED SKULLCAP. Flowers 2–2½ cm; calyx glandular. Damp places. N–W. Spain, France.   (1108) *S. minor* Hudson LESSER SKULLCAP. Flowers about 8 mm; calyx not glandular. Damp fields. N. and C. of our area.

LAVANDULA | Lavender   A genus characteristic of the Western Med. region.

1. Lvs. toothed or twice-divided.
2. Lvs. with rounded teeth or deeper comb-like lobes; fls. 4–10 in each whorl.   *dentata*
2. Lvs. mostly twice-divided; fls. 2 in each whorl.   *multifida*

1. *Marrubium supinum* 1113a   2. *Salvia candelabrum* 1143b
3. *Teucrium subspinosum* 1100a   4. *Salvia phlomoides* 1144a

1. Lvs. entire.
  3. Uppermost bracts much longer than the lower bracts and forming a
    terminal tuft to the inflorescence.
    4. Uppermost bracts purple or white; lvs. with short adpressed hairs.   *stoechas*
    4. Uppermost bracts green; lvs. with shaggy spreading hairs.   *viridis*
  3. Bracts all similar, uppermost not forming a tuft.
    5. Bracts broadly egg-shaped, with a narrow point; bracteoles minute or
      absent.   *angustifolia*
    5. Bracts linear or lance-shaped; bracteoles 2–5 mm, linear or bristle-like.
      6. Lvs. very shortly and densely white-felted when young, grey-green and
        less densely felted when mature; calyx thirteen-veined.   *latifolia*
      6. Lvs. densely and persistently white-felted; calyx eight-veined.   *lanata*

**1110. *L. stoechas* L. FRENCH LAVENDER. An attractive, small, grey-leaved shrub with long-stalked, dense oval spikes of dark purple flowers, with a topknot of narrow pale purple or sometimes white bracts. Bracts subtending flowers papery, oval, strongly veined. Dry stony places, sunny hillsides, pine woods. S. and C. of our area. Page 160. Subsp. *pedunculata* (Miller) Samp. has a brighter, often reddish-purple topknot of bracts which are 1½–5 cm long, and a much longer inflorescence stalk, at least twice as long as the flower spike. Iberian peninsula. Pl. 41, Page 111.   1110a. *L. viridis* L'Hér. Like 1110 but readily distinguished by its topknot of large green bracts, and its white flowers which soon turn brown and are subtended by green bracts. Leaves broader, elliptic, green, with shaggy hairs. Dry uncultivated places. S. Portugal, S–W. Spain. Pl. 41, Page 45.

(1110) *L. dentata* L. TOOTHED LAVENDER. Like 1110 with a topknot of purple bracts but they are much shorter, 8–15 mm, and egg-shaped, but readily distinguished from all other species by the leaves which are shallowly toothed to deeply lobed. Spike of flower 1½–5 cm, violet. A grey-felted shrub to 1 m. Arid regions. S. and E. Spain. Page 91.

**(1111) *L. multifida* L. CUT-LEAVED LAVENDER. Readily distinguished by its leaves which are twice-cut into narrow, green, sparsely hairy segments. Spikes of flowers often lax and branched below; corolla blue-violet, to about 12 mm, much longer than the elliptic, pointed bracts. Dry and stony places. S. Portugal, Med. region. Pl. 41.

1111. *L. angustifolia* Miller COMMON LAVENDER. A very aromatic, greyish-leaved, much-branched small shrub with rather dense, cylindrical, long-stalked spikes of blue-purple flowers. Distinguished by the broad, rhomboid-oval bracts subtending the whorls of 6–10 flowers. Leaves entire, at first white-felted, later green. Stony mountain slopes. Med. region, N–E. Spain, E. Pyrenees. (1111) *L. latifolia* Medicus Distinguished from 1111 by its linear or lance-shaped bracts, grey-green and more densely felted leaves, and slender interrupted spikelets of blue-violet flowers. Calyx thirteen-veined. Arid hillsides, preferring limestone. C. Portugal, Med. region; often grown for ornament elsewhere.   1111a. *L. lanata* Boiss. Like (1111) but whole plant covered with a very dense and persistent white pelt of hairs. Flower spike 4–10 cm, interrupted; corolla lilac; bracts linear-lance-shaped; calyx eight-veined. Dry limestone rocks and screes in mountains. S. Spain. Page 77.

MARRUBIUM | Horehound

1. Calyx with 8–10 equal or unequal teeth.(Widespread)   **1112. *M. vulgare* L.
1. Calyx with 5 more or less equal teeth.
  2. Calyx teeth longer than the corolla, usually shorter than
    the calyx tube.   *alysson*
  2. Calyx teeth shorter than the corolla, and shorter than the
    calyx tube.

3. Lvs. fan-shaped with a heart-shaped base; lv. stalk longer
   than blade; fls. pink or lilac.   *supinum*
3. Lvs. oblong, to egg-shaped; lv. stalk shorter than blade;
   fls. white or yellow. (S. France)   **1113. *M. peregrinum* L.

1113a. *M. supinum* L. A densely silvery-haired perennial, with thick, wrinkled, rounded leaves, and dense rounded whorls of violet-purple to pale pink flowers. Flowers two-lipped, the upper lip deeply divided to half its length into 2 spathulate lobes, the lower lip with the middle lobe largest; calyx woolly-haired, with 5 stiff teeth. Leaves toothed, usually grey-green above, white-felted beneath. Fertile places in the mountains. S., C., and S–E. Spain. Page 314.   1113c. *M. alysson* L. A densely white-felted perennial with almost stalkless fan-shaped leaves and dense whorls of small purple flowers. Distinguished by the 5 lance-shaped, long-pointed calyx teeth, which at length become rigid and spread outwards in a star. Leaves at first white-felted, becoming grey-green, all with a wedge-shaped base and toothed margin. Waste places near cultivation, lowlands. S., C., and E. Spain.

SIDERITIS | A difficult genus with about 18 species occurring in our area; hybridization also occurs among Iberian species. The genus can be distinguished by the dense whorls of small yellow or white flowers, closely invested by broad bracts, by the two-lipped corolla with a flat, very erect, rounded and notched upper lip; and by the spiny calyx teeth.

Annuals

1114. *S. romana* L. Distinguished by its leafy whorls of yellow, white or pinkish flowers, and by its conspicuously ribbed calyx with the upper tooth 2–3 times as broad as the 4 lower teeth. Leaves oval to obovate, toothed, shaggy-haired; bracts similar. Dry, sandy and rocky places. S. Portugal, Med. region. Page 90.   **(1114) *S. montana* L. Also an annual but with all calyx teeth more or less equal, egg-shaped, spine-tipped. Flowers yellow, with or without a brownish lip. Derelict fields. S., C., and E. Spain, France.

Shrubby perennials

1115a. *S. scordioides* L. A green or greyish shrublet with widely separated whorls of pale yellow flowers surrounded by broad, spiny-lobed bracts, and with oblong to obovate, toothed leaves. Whorls 3–10; lower bracts 1–2 cm broad, about equalling calyx; calyx 6–9 mm, with a ring of hairs on inside of the tube; corolla 8–10 mm, sometimes with purple markings. Dry hills, principally on limestone. S., C., and E. Spain, S. France. Page 307. (1115) *S. hirsuta* L. Like 1115a but with more or less spreading, straight hairs on the stem (not curled and more or less adpressed as in 1115a). A very variable plant of field, vineyards, dry rocky places, waysides. N–E and S. Portugal, Med. region.   1115b. *S. glacialis* Boiss. Very like 1115a but whorls of pale yellow flowers 1–3; bracts only 5–6 mm broad, shorter than calyx. Leaves ½–1 cm, entire or with a few rounded teeth, densely silvery-haired. Mountains. S. Spain (S. Nevada).   1115c. *S. incana* L. A slender, white-felted, woody-based perennial, recalling lavender in habit, with entire linear to spathulate leaves, and with an elongate spike of 2–10 separate whorls of yellow or pink flowers. Lower bracts 3–10 mm by 2–8 mm, egg-shaped to heart-shaped, with up to 11 teeth, or without teeth, shorter than the calyx, which has no ring of hairs on the inside. Rocks, screes, dry sunny places. S., C., and E. Spain.

**1115. *S. hyssopifolia* L. A variable shaggy-haired or almost hairless small shrubby perennial, with whorls of yellow flowers in a more or less dense spike. Flowers about 1 cm, sometimes purple-tinged; bracts usually spiny-toothed; calyx 6–8 mm. Leaves very variable, toothed or entire. Rocks in hills and mountains. C. Portugal, N. Spain, Pyrenees, France.

NEPETA | Catmint 10 species.

Bracts shorter than calyx

**1116. *N. cataria* L. CATMINT. Waysides, waste ground, hedges. Widespread. (1116) *N. nepetella* L. LESSER CATMINT. Dry sunny places, waste ground, abandoned cultivation. C. and E. Spain, France. 1116c. *N. latifolia* DC. An erect, densely hairy perennial ½–1 m or more, often bluish above, with a spike-like inflorescence of blue flowers and large stalkless leaves. Corolla 8–11 mm; calyx 7–9 mm, teeth often bluish. Leaves 3–7 cm, egg-shaped to heart-shaped, toothed. Hedges, meadows, woods. E. Portugal, C. and S–E. Spain, Pyrenees.

At least the outer bracts as long as or longer than calyx

1116a. *N. tuberosa* L. A robust, unbranched, hairy, sometimes sticky perennial 25–80 cm, with dense or interrupted spike-like inflorescence of purple or violet flowers subtended by often pinkish or pale purplish papery bracts. Corolla 9–13 mm; calyx 8–11 mm. Leaves egg-shaped to oblong, toothed, the upper stalkless with a heart-shaped base. Dry stony places, waysides. S. and C. Portugal, Spain. Pl. 41. 1116b. *N. apuleii* Ucria Distinguished by its whorls of pink flowers subtended by reddish-purple, oval to lance-shaped, very acute bracts. Bracts longer than calyx; teeth of calyx longer than calyx tube; corolla tube much longer than calyx. Upper leaves with heart-shaped base, stalkless, strongly toothed; a nearly hairless erect perennial. Dry grassy fields. S. Spain. Page 307. 1116d. *N. reticulata* Desf. Differs from 1116a in having oblong lance-shaped leaves which have more or less enlarged arrow-shaped lobes, coarsely toothed, and greyish-green hairy on both surfaces. Flowers pale red, spotted, in a long spike which is interrupted below; bracts oval-pointed papery, translucent, whitish with green veins, and pale violet margin. Bushy places, cultivated ground in mountains. S. Spain. Pl. 41.

GLECHOMA | **1118. *G. hederacea* L. GROUND IVY. Woods, shady places. Widespread, rarer in south.

PRUNELLA (BRUNELLA) | Self-Heal

1. Fls. usually more than 18 mm; inflorescence not subtended by lvs. (Widespread)   **(1120) *P. grandiflora* (L.) Scholler. Pl. 103
1. Fls. usually not more than 17 mm, inflorescence usually subtended by lvs.
  2. Fls. yellowish-white, rarely pink or purplish; at least the upper lvs. deeply lobed. (Widespread)   **1120. *P. laciniata* (L.) L.
  2. Fls. violet, rarely whitish; lvs. entire or toothed.
    3. Lvs. egg-shaped, stalked; middle tooth of upper lip of calyx wider than lateral teeth. (Widespread)   1119. *P. vulgaris* L.
    3. Lvs. linear-lance-shaped to elliptic-lance-shaped, stalkless; teeth of upper lip of calyx more or less equal.   *hyssopifolia*

1119a. *P. hyssopifolia* L. Distinguished by its narrow entire, stalkless leaves and violet flowers borne in a dense head subtended by leaves. Corolla 15–17 mm, rarely white; calyx 8–9 mm; bracts about as long as broad. A stiff erect, hairless or bristly-haired perennial. Rocky limestone areas. C. and E. Spain, S. France.

CLEONIA | Like *Prunella* but style shortly four-lobed; bracts deeply dissected. Calyx teeth all bristly-tipped.   1121a. *C. lusitanica* L. An erect annual with a dense ovoid-quadrangular head of conspicuous tubular pale pink flowers subtended by distinctive bracts which are deeply cut into comb-like teeth. Flowers large, 2–2½ cm long; corolla hairy out-side, rarely white; calyx and bracts with white bristly hairs. Leaves oblong, coarsely toothed or lobed. Dry uncultivated places, pine woods, S. and C. Portugal, S. and C. Spain. Pl. 43, Page 85.

MELITTIS | **1121. *M. melissophyllum* L. BASTARD BALM. A strong-smelling, softly hairy perennial with handsome large pink, or white and pink-spotted flowers borne in leafy whorls. Flowers 2–6 in a whorl; corolla 3½–4½ cm long, funnel-shaped. Damp shady places, limestone hills and mountains. Widespread, rare in the south.

MOLUCCELLA | Calyx very much enlarging, becoming papery and netted in fruit, and with long spiny teeth.   1121b. *M. spinosa* L. A hairless, erect annual to 1 m, with distant whorls of white or pale pink velvety flowers subtended by slender spiny bracts. Distinguished from all other plants by the greatly enlarging two-lipped, funnel-shaped, papery calyx with 8–11 strong spines. Leaves oval-heart-shaped, deeply toothed or lobed. Hedges and ditches on the coast. S. Spain (Málaga province).

PHLOMIS |

1. Fls. pink, purple or rarely white.
  2. Bracts awl-shaped; lvs. tomentose beneath.   *herba-venti*
  2. Bracts lance-shaped, elliptic or oblong; lvs. lanate beneath.
    3. Lvs. softly and minutely hairy above. (Portugal, Spain)   *purpurea*
    3. Lvs. woolly above. (Balearic Islands)   1122c. *P. italica* L.
1. Fls. yellow or brownish-yellow.
  4. Bracts egg-shaped to broadly lance-shaped. (Spain)   **1124. *P. fruticosa* L.
  4. Bracts awl-shaped, linear or narrowly lance-shaped.
    5. Lvs. linear to narrowly elliptic, gradually tapering to the
      lv. stalk.   *lychnitis*
    5. Lvs. egg-shaped or lance-shaped, abruptly narrowed to the
      base.   *crinita*

1122. *P. herba-venti* L. A robust perennial with dense whorls of purple flowers, the upper-most with a pair of terminal leaves. Flowers in whorls of 10–20; corolla about 2 cm; calyx 8–15 mm, with stiff bristly-tipped lobes like the bracts. Leaves broadly lance-shaped, toothed, nearly hairless above, thickly hairy and often greyish beneath, heart-shaped or rounded at base; stems greenish. Dry hills, rocks, waysides, field verges. S. Portugal, S., C., and E. Spain. Pl. 42, Page 85. Subsp. *pungens* (Willd.) De Filipps Distinguished by its whitish stems; narrower leaves which are wedge-shaped at base and its fewer-flowered whorls often with 6–8 flowers. Spain.   1122b. *P. purpurea* L. A shrub to 2 m with whorls of purple flowers and lance-shaped, leathery wrinkled leaves which are densely hairy above and white-woolly beneath, with leaf stalks up to 2 cm; young stems very densely white-woolly. Corolla 23–26 mm; calyx 1½–2 cm, grey-woolly, like the bracts. Dry stony places. S. Portugal, S., C., and E. Spain. Pl. 42, Page 161.

1123. *P. lychnitis* L. A small shrub with 4–8 whorls of yellow flowers, each whorl sub-tended by broad oval bracts and contrasting with the much narrower leaves. Corolla 2–3 cm; calyx 1½–2 cm, shaggy-haired, teeth not bristly-tipped. Leaves linear to narrowly elliptic, white-felted. Dry hills, rocks, stony places. S. and C. Portugal, C., S., and E. Spain, France. Pl. 42.   1123a. *P. crinita* Cav. A handsome shrubby plant of the mountains of S.

and E. Spain with a long spike of many whorls of brownish-yellow to yellow flowers, and stems and leaves densely covered with silvery hairs. Corolla 2–2½ cm; calyx 13–17 mm, teeth linear-lance-shaped; bracts linear, shaggy-haired. Leaves egg-shaped to lance-shaped, entire, thick, leathery, abruptly stalked, snowy-white when young. Stony ground. Pl. 43.

GALEOPSIS | Hemp-Nettle

1. Plants with stiff bristly hairs; stems swollen at the nodes. (Widespread)	**1126. *G. tetrahit* L.
1. Plants without stiff bristly hairs; stems not swollen.	
2. Lvs. and calyx densely and softly silky-haired.	
3. Fls. pale yellow, pinkish, rarely purple; lvs. lance-shaped narrowed to the leaf-stalk, apex more or less acute. (Spain, France)	(1125) *G. segetum* Necker
3. Fls. purple; lvs. broadly egg-shaped, abruptly cut-off or shallowly heart-shaped at base, apex very blunt. (Pyrenees)	1125a. *G. pyrenaica* Bartl.
2. Lvs. and calyx finely hairy or almost hairless, never silky.	
4. Calyx green with spreading hairs; lvs. more or less egg-shaped. (Spain, France)	1125b. *G. ladanum* L.
4. Calyx whitish with adpressed hairs; lvs. usually linear or lance-shaped. (Spain, France)	1125. *G. angustifolia* Hoffm.

LAMIUM | Dead-Nettle 7 species in our area. The following are fairly widespread but less common in the south: 1128. *L. purpureum* L. RED DEAD-NETTLE; **1127. *L. amplexicaule* L. HENBIT; (1128) *L. hybridum* Vill. CUT-LEAVED DEAD-NETTLE; **1130. *L. maculatum* L. SPOTTED DEAD-NETTLE; 1129. *L. album* L. WHITE DEAD-NETTLE.

LAMIASTRUM (GALEOBDOLON) | **1132. *L. galeobdolon* (L.) Ehrend. & Polatschek YELLOW ARCHANGEL. Shady woods, hedges, lowland and montane. C., N., and E. Spain, France.

LEONURUS | **1133. *L. cardiaca* L. MOTHERWORT. An erect leafy perennial with a very long, interrupted, leafy spike of numerous small whorls of pinkish or whitish, densely hairy flowers. Calyx bristly-tipped. Leaves three- to seven-lobed, becoming progressively smaller towards apex of the inflorescence. Hedges, walls, waste places. N., N–E. Spain, France.

BALLOTA | Horehound **1134. *B. nigra* L. BLACK HOREHOUND. Hedges, tracksides, waste places. Widespread. 1134a. *B. hirsuta* Bentham (*B. hispanica* auct.) A densely woolly-haired perennial with compact rounded whorls of dull purple flowers with darker purple spots or paler lines. Corolla 14–16 mm; calyx bell-shaped, with wide-spreading tooth-like lobes. Leaves oval or rounded, toothed, with a heart-shaped or cut-off base, covered with glands and simple and star-shaped hairs. Roadsides, waste places. S. and C. Portugal, S., C., and E. Spain. Pl. 41.

STACHYS (incl. BETONICA) | Woundwort   About 17 species occur in our area; many are widespread.

Predominantly pink-flowered species
Fairly widespread are: 1137. *S. arvensis* (L.) L. FIELD WOUNDWORT;   **1138. *S. palustris* L.

MARSH WOUNDWORT; 1139. *S. sylvatica* L. HEDGE WOUNDWORT; **1140. *S. germanica* L.
DOWNY WOUNDWORT; 1141. *S. officinalis* (L.) Trevisan (*Betonica o.*) BETONY.

1136a. *S. circinata* L'Hér. A handsome perennial with softly silvery-haired, heart-shaped
and toothed leaves, and whorls of large pinkish-purple to white flowers in a terminal somewhat lax spike. Corolla 1½–2 cm, with finely hairy upper lip 5–8 mm, lower lip 10–12 mm;
calyx softly hairy 8–10 mm, the teeth about one third as long as the tube. Cliffs and screes.
S. Spain. Page 323. (1139) *S. alpina* L. ALPINE WOUNDWORT. It occurs in shady places in
the Pyrenees and N. Spain and is distinguished from closely related species by its glandular
hairy stem, at least above. Flowers reddish-brown, hairy, 15–22 mm.

Predominantly yellow-flowered species
Fairly widespread are: 1135. *S. annua* (L.) L. ANNUAL YELLOW WOUNDWORT; **1136.
S. recta L. PERENNIAL WOUNDWORT; **1142. *S. alopecurus* (L.) Bentham (*Betonica a.*)
YELLOW BETONY.

(1136) *S. maritima* Gouan Distinguished by its yellow flowers in numerous whorls forming
a short dense spike, and by its white or greyish, woolly-haired stems and leaves, with a
persistent basal rosette of leaves. Calyx and bracts velvety-white; corolla hairy. Maritime
sands; Med. region. 1136b. *S. ocymastrum* (L.) Briq. (*S. hirta*) A robust erect or ascending,
shaggy-haired annual to ½ m, with large white or pale yellowish flowers 12–15 mm, two-
lipped with the upper lip deeply lobed. Calyx 7–10 mm, hairy, the long spiny teeth equal,
and as long as the calyx tube. Leaves heart-shaped, toothed, hairy. Fields, waysides. S.
and C. Portugal, Spain, S. France. Pl. 41.

SALVIA | Sage 16 species.

1. Connective separating the 2 anthers of each
   stamen shorter than or equal to the filament; arms
   of connective more or less equal.
   2. Stems leafy along their length. (Introd. Portugal,
      Med. region)   1143. *S. officinalis* L.
   2. Stems leafy only in the lower third.
      3. Stems hairless; fls. almost stalkless.   *blancoana*
      3. Stems hairy at least below; fls. with stalks ½–1
         cm.
         4. Lvs. narrow-elliptic; fls. 3–4 cm.   *candelabrum*
         4. Lvs. oblong to oblong-linear; fls. 2–2½ cm   *lavandulifolia*
1. Connective separating the 2 anthers of each
   stamen longer than the filament; arms of connective very unequal.
   5. Sterile arm of connective awl-shaped; usually
      15–30 fls. in each whorl. (C., N–E. Spain, France)   1145. *S. verticillata* L.
   5. Sterile arm of connective enlarged or with a
      sterile cell.
      6. Upper lip of calyx straight, not concave in fr.;
         calyx tubular or bell-shaped.
         7. Bracts longer than corolla, lilac or white.
            8. Corolla 14–18 mm, the hood more or less
               straight.   *viridis*
            8. Corolla 2–3 cm, the hood strongly curved.
               (N–E. Portugal, S., C., and E. Spain, S. France)   **1144. *S. sclarea* L.
         7. Bracts shorter than corolla, white or green.

9. Lvs. finely softly hairy or with short rigid
close-pressed bristle-like hairs.
10. Corolla pink or violet.   *viridis*
10. Corolla yellow with reddish-brown markings. (E. Spain, France)   **1146. *S. glutinosa* L.
9. At least the lower lvs. woolly-haired or felted.
11. Bracts equalling or exceeding calyx, greyish-white.   *phlomoides*
11. Bracts shorter than calyx, green.
12. Corolla 1–1½ cm, the hood weakly curved:
stems not glandular. (N–E. Portugal,
Spain, S. France)   **(1144) *S. aethiopis* L. Pl. 43
12. Corolla 1½–3½ cm, the hood strongly curved;
stems glandular. (S. Portugal, S., C., and E.
Spain)   **(1144) *S. argentea* L.
6. Upper lip of calyx concave, two-furrowed in
fr., calyx bell-shaped.
13. Fl. stalks 8–10 mm; lower lip of corolla white,
the upper lip violet blue.   *bicolor*
13. Fl. stalks less than ½ cm; fls. not as above.
14. Bracts purplish, half as long as calyx or more;
overlapping in bud. Fls. *c.* 8 mm, blue-violet.
(E. Spain)   1148a. *S. valentina* Vahl
14. Bracts green less than half as long as calyx,
not overlapping in bud.
15. Stems glandular below, 15–40 cm. Fls.
purplish or violet, 1½–2 cm. (S. and C.
Portugal, S–W. Spain)   1148b. *S. sclareoides* Brot.
15. Stems without glands below, up to 1 m.
16. Lvs. deeply lobed or cut to the midvein; fls.
usually 6–10 mm. (Widespread)   (1148) *S. verbenaca* L. Page 63
16. Lvs. toothed or almost entire; fls. 1½–3 cm.
(Widespread)   **1147. *S. pratensis* L.

1143a. *S. blancoana* Webb & Heldr. Flowers large, pale violet-blue in short-stalked clusters on a slender, branched, leafless inflorescence. Corolla 2½–4 cm; calyx 1–1½ cm, glandular-sticky. Leaves sage-like, oblong-elliptic, wrinkled; stalked, young leaves white-felted. Stony ground S–E. Spain. Pl. 43.   1143b. *S. candelabrum* Boiss. Flowers large, the upper lip white with violet markings and lower lip blue-violet, borne in lax clusters on an erect, little-branched, leafless inflorescence. Flower stalks long; corolla 3–4 cm; calyx glandular-sticky. Leaves simple or with a pair of lateral lobes at the base. Sunny hills. S. Spain. Page 314.   1143c. *S. lavandulifolia* Vahl Like 1143 but leaves linear-lance-shaped, with small rounded teeth, thickly covered with velvety-white hairs at least beneath. Flowers rather large 2–2½ cm, blue or violet, in leafless branched clusters; flower stalks about ½ cm; calyx 8–12 mm, often reddish-purple, finely hairy, gland-spotted. Dry hills and slopes among bushes. S., C., and E. Spain, S–W. France. Page 116.

**1149. *S. viridis* L. (*S. horminum*) RED-TOPPED SAGE. An annual with erect usually simple stems and whorls of pink or violet flowers, and often with a conspicuous crown of bi-coloured bracts which may be violet, pink, white or green. Flowers 14–18 mm; calyx 7–10 mm, finely hairy. Dry sandy places, cultivated ground, waysides. S. Portugal, S. and C. Spain, S. France. 1144a. *S. phlomoides* Asso Distinguished by its stout spike-like inflorescence of whorls of large white, pinkish to purplish flowers, encircled by broad greyish-white bracts which are as long as or longer than the spiny-toothed calyx. Corolla 2½–3½ cm; calyx 1½–2 cm; glandular, shaggy-haired. Leaves silvery-woolly beneath, with cobweb hairs above. Dry hills, disturbed ground, waysides. S., C., and E. Spain. Page 314.

1147a. *S. bicolor* Lam. A rare, very tall attractive biennial or perennial to 1½ m, with long, lax branched inflorescence of long-stalked flowers which have a blue-violet prominently curved upper lip and a white lower lip. Leaves broadly egg-shaped, toothed or lobed, with a heart-shaped base; stems glandular-sticky above. Dry hills, disturbed ground. S–W. Spain. Pl. 43.

ZIZIPHORA | Calyx long tubular, with 13 strong veins, upper lip three-toothed, the lower two-toothed with blunt tips. Corolla tube slender, lips more or less equal, the upper erect, the lower three-lobed. Fertile stamens 2. 4 Species.

1149a. *Z. hispanica* L. A very aromatic small annual recalling a Thyme, with an elongated dense spike of many whorls of tiny, long-tubed pink flowers subtended by leafy bracts. Calyx swollen at the base, with 13 strong, hairy veins, and 5 narrow erect teeth. Leaves oval-oblong, entire, strongly veined and often purplish beneath and very glandular. Fields, sandy hills. S., S–E., and C. Spain. Page 323.

MELISSA | 1150. *M. officinalis* L. BALM. A much-branched perennial with whitish or pinkish flowers in dense, one-sided, terminal and axillary whorls forming leafy spikes. Corolla 8–12 mm, twice as long as calyx. Leaves oval, toothed, sweetly lemon-scented. Shady places, hedges. Probably throughout, but sometimes naturalized from cultivation.

HORMINUM | **1151. *H. pyrenaicum* L. DRAGONMOUTH. A perennial with a stout rootstock and with a basal rosette of leaves, and erect usually unbranched stems bearing a one-sided cluster of drooping blue-violet flowers. Flowers whorled, about 1½ cm; corolla tubular, twice as long as calyx. Leaves oval, rounded-toothed. Mountain pastures. Pyrenees. Page 148.

SATUREJA | Savory

1. Anns.; at least lower calyx teeth distinctly longer than calyx tube. (Med. region)	(1152) *S. hortensis* L.
1. Small shrubs or perenns. with woody base; calyx teeth shorter or slightly longer than calyx tube.	
2. Calyx teeth blunt. (S–W. Spain)	1152a. *S. inodora* Bentham
2. Calyx teeth acute. (Med. region)	**1152. *S. montana* L. *cuneifolia*

1152b. *S. cuneifolia* Ten. Very like 1152 but distinguished by its lower bracts which are 3–10 mm, shorter than or about equalling the whorl; whorls usually distinct. (1152 has lower bracts 1–2 cm, exceeding the whorl; whorls usually crowded.) Flowers ½–1 cm, white, pale pink to bright purple. Stem and leaves without hairs or with very few long hairs; stems 10–50 cm. Med. region. Page 323.

MICROMERIA | 7 species. 1153. *M. graeca* (L.) Reichenb. A dwarf shrub with a lax spike-like inflorescence with many short-stalked spreading whorls of purplish flowers. Corolla usually 6–8 mm; calyx hairy in throat. Leaves 5–12 mm, the upper lance-shaped to linear, margin inrolled. Sunny rocks, dry sandy ground. South of our area.

1. *Thymus cephalotos* 1162a
2. *Thymus membranaceus* 1162d
3. *Thymus longiflorus* 1162c
4. *Thymus villosus* 1162b
5. *Satureja cuneifolia* 1152b
6. *Stachys circinata* 1136a
7. *Ziziphora hispanica* 1149a

CALAMINTHA | Calamint   A difficult genus; 5 species in our area.   (1155) *C. ascendens* Jordan Probably the most widespread species in the south-west.   **1154. *C. grandiflora* (L.) Moench LARGE-FLOWERED CALAMINT. A striking plant with few large, pink, tubular flowers 2½–4 cm long, in a lax leafy inflorescence, and calyx 12–16 mm, ciliate. Shady places in the mountains. N–E. Spain, Massif Central.   **1156. *C. nepeta* (L.) Savi LESSER CALA-MINT. Flowers pale lilac or white, in lax or dense axillary clusters forming an elongated leafy inflorescence. Corolla about 1 cm; calyx 4–6 mm, the upper lobes nearly straight, the lower lobes longer, ciliate, throat hairy. Leaves oval 1–2 cm; plant strong-smelling. Dry stony and bushy places. Widespread. Page 62.

ACINOS | Like *Calamintha* but flowers in whorls; calyx swollen at the base and with a curved tube contracted in the middle.

1. Lvs. very prominently veined beneath, narrowed to
   a fine point. (S. and C. Spain)   1157a. *A. rotundifolius* Pers.
1. Lvs. not prominently veined, blunt or pointed.
   2. Fls. 1–2 cm, usually longer than subtending bracts.
      (N–E. Portugal, Spain, Pyrenees)   (1157) **A. alpinus* (L.) Moench
   2. Fls. 7–10 mm, not longer than subtending bracts.
      (Widespread)   1157. *A. arvensis* (Lam.) Dandy

CLINOPODIUM | **1158. *C. vulgare* L. WILD BASIL. Hedges, woods, thickets. Widespread, less common in the south.

HYSSOPUS | **1159. *H. officinalis* L. HYSSOP. Dry sandy places, limestone hills, in mountains. S., C., and E. Spain, France.

ORIGANUM | Marjoram   6 species.   **1160. *O. vulgare* L. MARJORAM. Sunny hills. N–W. Portugal, S., C., and E. Spain, France.   1160a. *O. virens* Hoffmanns. & Link Like 1160 but bracts subtending flowers pale green, papery, rounded, twice as long as calyx. Flowers white; calyx hairless. Dry places, field verges, hedges. Portugal, Med. Spain. 1160b. *O. compactum* Bentham A small shrub to 35 cm, with pink or white flowers in densely clustered spikelets with purple gland-spotted calyx. Spikelets 1–2½ cm; corolla 8–10 mm; bracts oval-oblong, acute, purplish, twice as long as calyx. Leaves egg-shaped, gland-spotted above and below, 1½–2 cm. Dry hills. S. Spain.

THYMUS | Thyme   31 species in our area of which 24 are restricted to the Iberian peninsula in Europe. Only the most distinctive and commonly encountered species are described below.

Corolla tube shorter than calyx
(a) Calyx two-lipped but upper lobes similar to lower
(1165) *T. mastichina* L. ROUND-HEADED THYME. A strongly aromatic dwarf shrub 20–50 cm, with dense globular clusters, 1–2 cm across, of whitish flowers and very woolly-haired calyx. Calyx lobes rigid, somewhat spiny-tipped, ciliate; corolla shorter than calyx. Leaves at length pale green, hairless, similar to floral bracts. Dry hills. Portugal, S. and C. Spain. Pl. 43.   1165a. *T. tomentosus* Willd. Like (1165) but flower clusters smaller, 6–8 mm across; calyx lobes soft. Leaves densely white-haired; floral bracts larger, broader. Dry places, sandy ground by the sea. S. Portugal, S. Spain.

(b) *Calyx two-lipped, the upper lobes conspicuously differing from the lower lobes*
(i) *Floral bracts larger and broader than leaves*
**1163. *T. vulgaris* L. THYME. A very aromatic dwarf shrub with velvety-white twigs, narrow leaves which are velvety-white beneath, and white or pinkish flowers in rounded or elongate clusters. Leaves with inrolled margin; floral bracts greyish-green, more or less leafy, but both very variable. Dry sunny slopes. S., C., and E. Spain, S. France. 1163a. *T. hyemalis* Lange Like 1163 but flowers pinkish-purple, twice as long as calyx which is coloured and is covered with bristly spreading hairs. Leaves narrow, to 1 mm wide, densely clustered at each node, greyish. Rough ground. S–E. Spain. 1163b. *T. granatensis* Boiss. Distinguished by its bracts and flowers which are bright deep purple, in a compact rounded terminal cluster. Corolla tube shorter or longer than calyx. Bracts much longer than the leaves, oval-acute, conspicuously veined, margin ciliate. Rock crevices, screes, in limestone mountains. S. Spain. Pl. 42.

(ii) *Floral bracts similar to leaves*
1163c. *T. zygis* L. A small erect shrublet with tiny whitish flowers in an elongated, interrupted cluster. Calyx 2–4 mm, woolly-haired, the upper lobes not ciliate; corolla as long or a little longer. Leaves linear, clustered, with brown glands, sparsely ciliate at base, margin inrolled. Dry sunny hills. Portugal, S., C., and E. Spain. Pl. 42.

Corolla tube longer than calyx
1162. *T. capitatus* (L.) Hoffmanns. & Link (*Coridothymus c.*) A compact, stiff, branched shrublet with terminal oblong clusters of pink flowers in cone-like heads with broad, oval, closely overlapping greenish and often reddish-tinged, ciliate bracts. Calyx tube dorsally flattened. Leaves linear, fleshy, gland-spotted, margin flat. Dry hills, heaths. Portugal, Med. Spain. 1162a. *T. cephalotos* L. A distinctive dwarf shrub with oblong conical heads of purple, long-tubed flowers subtended by broad, leathery, purplish bracts which are closely overlapping. Corolla 1½ cm; calyx with upper lobes much narrower than lower; bracts egg-shaped up to 2 cm by 1 cm. Leaves broadly linear, ciliate. Dry sandy ground. S. Portugal, S. and E. Spain. Page 323. 1162b. *T. villosus* L. Like 1162a but bracts usually greenish and toothed or shallowly lobed. Corolla 6–10 mm, the corolla tube variable in length; calyx 4–6 mm. Heaths, pine woods, dry places. C. Portugal, S. Spain.

1162c. *T. longiflorus* Boiss. Distinguished by its dense globular heads about 2½ cm across of broad purplish overlapping bracts and much longer, slender-tubed purple flowers. Corolla about 1½ cm; calyx 5–7 mm, the upper lobes narrow; bracts up to 13 mm by 8 mm, long-pointed, usually ciliate. Leaves linear-lance-shaped, margins inrolled, not or sparsely ciliate. Dry hills, sunny rocks. S. Spain. Pl. 42. Page 323. 1162d. *T. membranaceus* Boiss. Like 1162c but bracts papery, whitish and flowers white. Dry hills. S. Spain. Page 323. 1162e. *T. camphoratus* Hoffmanns. & Link A much-branched dwarf shrub to 40 cm, with narrow egg-shaped leaves which are woolly-haired above, and terminal, globular, whitish flower heads 1–2 cm across. Calyx 4–6 mm, lobes triangular-awl-shaped, mostly ciliate; floral bracts broadly egg-shaped, purplish, woolly-haired. Dry sandy places, bushy places. S. Portugal. 1162f. *T. capitellatus* Hoffmanns. & Link Like 1162e but flower heads smaller 6–8 mm across, and floral bracts greenish; upper lobes of calyx broadly triangular, not ciliate. Coastal regions. S. Portugal.

LYCOPUS | **1166. *L. europaeus* L. GIPSY-WORT. Damp places, watersides. Widespread.

MENTHA | Mint A difficult genus with many hybrids and cultivated forms which often become naturalized; about 11 species. The following are fairly widespread in our area:

**1167. *M. pulegium* L. PENNY-ROYAL;   1168. *M. arvensis* L. CORN-MINT;   **1169. *M. aquatica* L. WATER MINT;   1170. *M. spicata* L. SPEARMINT;   1171. *M. longifolia* (L.) Hudson HORSEMINT;   1172. *M. × rotundifolia* (L.) Hudson APPLE-SCENTED MINT.

1167a. *M. cervina* L. (*Preslia c.*) An almost hairless, very strong-smelling perennial, with erect stems bearing rather few, dense whorls of pink flowers subtended by palmately lobed bracteoles almost as long as the flowers. Calyx tubular with 4 equal lobes, hairless. Leaves linear-lance-shaped, entire or shallowly toothed. Damp places. Portugal, Med. region.

SOLANACEAE | Nightshade Family

LYCIUM |

1. Lvs. usually widest at the middle, at least some 1 cm wide or more; branches pendulous, flexible, with few slender spines; corolla lobes slightly shorter than the corolla tube. (Scattered throughout)   (1173) *L. barbarum* L.
1. Lvs. usually widest above the middle, ½–1 cm wide; branches stiff, not pendulous, very spiny; corolla tube longer than lobes.
  2. Calyx 5–7 mm; corolla 20–22 mm. (S. Spain, France)   1173a. *L. afrum* L.
  2. Calyx 1–4 mm; corolla 8–20 mm.
    3. Lvs. 2–5 cm; calyx 2–3 mm; corolla 11–13 mm. (Med. region)   **1173. *L. europaeum* L.
    3. Lvs. 3–15 mm; calyx 1–2 mm; corolla 1½–2 cm. (C. Portugal, Med. region).   1173b. *L. intricatum* Boiss. Page 91

ATROPA | **1174. *A. bella-donna* L. DEADLY NIGHTSHADE. Woods and thickets, in mountains. C. Portugal, C. and E. Spain, France.   1174a. *A. baetica* Willk. Like 1174 but flowers greenish-yellow, solitary, erect, long-stalked. Corolla twice as long as calyx and stamens longer than corolla. Shady rocks and stony places on limestone mountains. S. Spain.

HYOSCYAMUS | Henbane   **1176. *H. niger* L. HENBANE. Cultivated and uncultivated ground, waysides, hedges. Widespread, rarer in south.   **1177. *H. albus* L. WHITE HENBANE. Dry uncultivated ground, walls, field margins. S. and C. Portugal, Med. region.

PHYSALIS | **1178. *P. alkekengi* L. BLADDER CHERRY. Cultivated ground, vineyards. S., C., and E. Spain, France.   (1178) *P. peruviana* L. Naturalized occasionally; coast of Spain.

CAPSICUM | (1179) *C. annum* L. RED PEPPER, CHILLIES, CAPSICUM. Widely cultivated in the south; sometimes self-seeding.

WITHANIA | Shrubs with entire leaves. Calyx and corolla bell-shaped, five-lobed; calyx enlarging in fruit; stamens 5. Fruit a berry surrounded by an inflated calyx.

1179a. *W. somnifera* (L.) Dunal A shrub to 1 m or more, with dull-green oval leaves, and several or pairs of small yellowish bell-shaped flowers in the leaf axils. Stem covered with white star-shaped hairs; leaves nearly hairless above, with white star-shaped hairs beneath. Fruit a red berry encircled by the calyx, which enlarges to 12–16 mm. Fields, roadsides, waste ground. S. Portugal, S. and E. Spain. Pl. 44.   1179b. *W. frutescens* (L.) Pauguy Like

1179a but branches hairless and leaves more or less heart-shaped, hairless or nearly so. Flowers usually solitary, nodding; calyx at first 4–5 mm, enlarging in fruit to 1½–2½ cm and enclosing a green berry. Hedges along the coast. S. and S–E. Spain, Balearic Islands.

SOLANUM | Widespread species are: **1181. *S. dulcamara* L. BITTERSWEET, WOODY NIGHTSHADE; 1182. *S. nigrum* L. BLACK NIGHTSHADE; (1182) *S. luteum* Miller Alien weeds or locally naturalized species are: **1180. *S. sodomeum* L. Native of Africa; frequent on the littoral of Portugal and Spain; 1180a. *S. sublobatum* Roemer & Schultes; 1180b. *S. pseudocapsicum* L.; 1183. *S. tuberosum* L. POTATO; 1183a. *S. bonariense* L. Page 62.

LYCOPERSICON | 1184. *L. esculentum* Miller TOMATO. Widely cultivated in the south; often occurring as a casual.

MANDRAGORA | Mandrake **1185. *M. autumnalis* Bertol. (*M. officinarum*) MANDRAKE. Flowers violet, appearing in autumn, to 2½ cm across; lobes of corolla broadly triangular. Fruit reddish-yellow, oblong. Fields, waysides and ditches. S. Portugal, S. Spain. Pl. 44.

TRIGUERA | Flowers irregular, more or less two-lipped, corolla tube very short. Fruit a globular berry with membraneous walls. 3 species.

1185a. *T. ambrosiaca* Cav. An erect leafy annual with stalkless, egg-shaped, strongly toothed leaves and axillary pairs of large deep violet flowers with darker throat and contrasting yellow anthers. Corolla 2½–3½ cm across, bell-shaped with 5 unequal lobes; calyx with dense white hairs, five-lobed. Fields and uncultivated ground. S–W. Spain. Page 52.

DATURA | Thorn-Apple

1. Fls. not more than 10 cm long; fr. erect.	
2. Calyx lobes ½–1 cm, unequal; fr. smooth, or with spines not more than 1 cm long. (Widespread)	**1186. *D. stramonium* L.
2. Calyx lobes 3–5 mm, nearly equal; fr. with spines 1–3 cm long. (Naturalized Med. region)	1186a. *D. ferox* L.
1. Fls. 11 cm or more long; fr. nodding.	
3. Stems and lvs. finely hairy; corolla with 10 lobes; fr. with long slender spines. (Naturalized S. Portugal, Med. region)	1186b. *D. innoxia* Miller
3. Stems and lvs. hairless; corolla with 5 lobes; fr. with short spines or swellings. (Naturalized Med. region)	**(1186) *D. metel* L.

CESTRUM | Fetid shrubs, with alternate simple leaves. Calyx tubular, five-lobed, persistent; corolla tubular; stamens 5, attached to the base of the corolla. Fruit a berry with 2 chambers and 2–11 seeds. 4 species cultivated. 1186c. *C. parqui* L'Hér: A very strong-smelling deciduous shrub with erect branches, lance-shaped entire, hairless leaves, and dense terminal leafy clusters of greenish-yellow to yellow night-scented flowers. Flowers 2–2½ cm, the lobes egg-shaped acute. Young branches white; leaves 3–14 cm. Fruit ovoid, blackish, about 1 cm. Native of S. America; frequently cultivated and sometimes naturalized in S. Spain.

NICOTIANA | Tobacco   **1187. *N. glauca* R. C. Graham SHRUB TOBACCO. Planted and becoming naturalized; coasts of S. and C. Portugal, S. and S–E Spain.   **1188. *N. rustica* L. SMALL TOBACCO. Sometimes cultivated in E. Spain and France.   **(1188) *N. tabacum* L. LARGE TOBACCO. More widely cultivated than the previous species in the south of our area, and sometimes naturalized.

SCROPHULARIACEAE | Figwort Family

VERBASCUM | Mullein   20 species occur in our area including 6 which are restricted to the Iberian peninsula in Europe. A difficult genus with similar-looking species. Usually plants of dry, open, sunny plains and hillsides.

Each bract with a single flower in its axil
(a) Flower stalks 2–10 mm
(1192) *V. virgatum* Stokes TWIGGY MULLEIN. Flowers 3–4 cm across in a lax, usually unbranched spike, the upper solitary in axils of oval-long-pointed toothed bracts, the lower bracts often with clusters of 2–5 flowers. Filament hairs purple on lower 2 stamens. Leaves hairless or glandular hairy. Dry sandy or stony ground. Widespread.   1192. *V. blattaria* L. MOTH MULLEIN. Differs from (1192) in that the plant is hairless below but glandular-hairy above. Flowers solitary, flower stalks longer than calyx. Waysides. Spain, France.

**1196. *V. creticum* (L.) Cav. Inflorescence unbranched; flowers large 4–5 cm across, yellow, the 2 upper petals with brownish-purple spots. Stamens 4, without sterile stamens; bracts oval-pointed, toothed. Leaves strongly toothed, or lobed. Med. region, Balearic Islands. 1196a. *V. hervieri* Degen Inflorescence simple, lax, hairless, up to 3 m, shining mahogany-coloured and contrasting with the very large silvery-silky basal rosette leaves. Flowers rather small 2–2½ cm; filament hairs white; bracts 2–3 mm. S–E. Spain (Sierra de Cazorla etc.).

(b) Flower stalks more than 1 cm
1196b. *V. laciniatum* (Poiret) O. Kuntze A handsome plant with a long leafless spike of large yellow flowers, each 3½–5 cm across, with brownish-purple basal blotch and purple filament hairs, and a basal rosette of almost thistle-like leaves. Calyx 6–9 mm; stamens 4; flower stalks 8–16 mm. Lower leaves oblong-egg-shaped, stalked, with large triangular, toothed lobes. Dry regions in mountains. S–W. Spain. Pl. 46, Page 163.   1196c. *V. barnadesii* Vahl Like 1196b but leaves narrowly lance-shaped in outline; flowers smaller 3–3½ cm; calyx about 4 mm. Dry sandy places. C. Portugal, S–W. and C. Spain.

At least the lower bracts with a cluster of several flowers in their axils
(a) Anthers partially or completely elongated down filament
1193. *V. thapsus* L. AARON'S ROD. Banks, waste places, uncultivated ground. Widespread. 1193a. *V. boerhavii* L. Like 1193 but upper stem leaves not decurrent (i.e. leaf blade not running down stem), and filament hairs violet. Inflorescence simple; flowers 22–32 mm across; flower stalks 1–4 mm; bracts 1–2½ cm, linear acute. A white woolly-haired biennial, with tufts of soft hairs on the upper parts; basal leaves stalked, broadly elliptic, toothed. Med. region, Balearic Islands.   1193b. *V. nevadense* Boiss. Distinguished by its bright yellow flowers 3–4 cm across, which have the upper stamens with violet filament hairs and the lower 2 with hairless filaments. Inflorescence simple, lax; bracts linear; flower stalks 4–9 mm. A greyish, woolly and tufted-haired perennial with oblong-lance-shaped, entire, stalked leaves. Mountains. S. and S–E. Spain.   1194. *V. phlomoides* L. Distinguished from 1193b by the upper 3 stamens which have white or yellow filament hairs, and lower 2 hairless. Upper leaves not or very slightly decurrent. Style spathulate; stigma decurrent. Spain, except north, France. See (1192) *V. virgatum* above.

(b) Anthers kidney-shaped, attached in the middle
(i) Leaves white- or grey-woolly on both surfaces
(1195) *V. pulverulentum* Vill. HOARY MULLEIN. Inflorescence much-branched forming a wide-spreading pyramid with scattered yellow flowers with white filament hairs. Whole plant thickly covered with white mealy woolly hairs, which easily rub off; basal leaves oblance-shaped, toothed or nearly entire, stalked or almost stalkless. Widespread. Pl. 46, Page 46. 1191a. *V. rotundifolium* Ten. A white-woolly-haired perennial usually with a simple inflorescence with yellow flowers and stamens with violet filament hairs. Basal leaves rounded to broadly elliptic, shallowly toothed or entire, stalked. Bracts 8–15 mm, lance-shaped. Med. Spain. 1191. *V. sinuatum* L. Flowers with violet filament hairs like 1191a, but basal leaves deeply pinnately lobed, undulate, stalkless. Inflorescence freely branched; bracts heart-shaped; flowers 1½–3 cm across. Leaves greyish-yellow, woolly-haired. Widespread in south.

(ii) Mature leaves green, at least above
**1190. *V. nigrum* L. DARK MULLEIN. Waysides, banks, dry places. N–E. Spain, Balearic Islands, France. 1190a. *V. chaixii* Vill. Flowers with purple filament hairs, like 1190 but basal leaves with blades with a cut-off or shortly wedge-shaped base (not heart-shaped). Inflorescence much-branched, with wide-spreading branches. N–E. Spain, France. Pl. 46. 1195. *V. lychnitis* L. WHITE MULLEIN. Flowers yellow or white with stamens with yellow or white filament hairs. Inflorescence freely branched. Basal leaves oblong to egg-shaped, wedge-shaped at base, coarsely toothed or nearly entire, green above and whitish-woolly beneath. Widespread, but rare in Med. region.

ANTIRRHINUM | Snapdragon 17 species, of which 14 occur only in the Iberian peninsula in Europe; some with very restricted distribution. A difficult genus owing to hybridization and introgression with more widespread species.

Leaves mostly hairless
1197. *A. majus* L. SNAPDRAGON. A very variable hairless or glandular-hairy, widely distributed perennial, usually with pink or purple flowers 3–4½ cm long. Leaves linear to narrow egg-shaped; bracts egg-shaped. Fruit glandular-hairy or hairless. Subsp. *tortuosum* (Bosc) Rouy often has a hairless inflorescence and narrow leaves 9–12 times as long as wide; sometimes climbing. S. Spain, Med. region. Subsp. *majus* has a glandular-hairy inflorescence; leaves 2½–7 times as long as wide and leaves widest about the middle. N–E. Spain, Pyrenees, S–C. France. Subsp. *linkianum* (Boiss. & Reuter) Rothm. Like subsp. *majus* but leaves egg-shaped to narrow lance-shaped widest near the base; sometimes climbing. S. and C. Portugal, S. Spain. Pl. 46.
1197a. *A. barrelieri* Boreau An erect slender-branched, usually climbing perennial with narrow hairless leaves and spikes of rosy-purple flowers with a white or yellow throat-boss. Flowers 2–3 cm; calyx 3–6 mm; flower stalks 1–4 mm; bracts linear or lance-shaped. Leaves linear to narrow lance-shaped; plant usually hairless below. Hedges, rocky places. S. Portugal, S. and E. Spain. Pl. 46.

Leaves hairy
(a) Leaves and stems glandular-hairy
1197b. *A. graniticum* Rothm. A tall erect, sometimes climbing, glandular-hairy perennial with pink or whitish flowers 25–32 mm long with an orange boss. Flower stalks 3–15 mm, longer than bracts. Leaves egg-shaped to oblong-lance-shaped blunt, opposite below, alternate above. Fruit 8–10 mm, glandular-hairy. Rocks, walls and stony hillsides. C. and N. Portugal, S. and C. Spain. Pl. 46. 1197c. *A. australe* Rothm. Like 1197b but stems erect, stout, sparingly branched. Flowers larger 4–4½ cm, bright pinkish-purple; flower stalks

shorter than bracts. Leaves egg-shaped, mostly opposite or in whorls of 3. Fruit 11–14 mm. Calcareous rocks and walls. S. and S–E. Spain.

1197d. *A. hispanicum* Chav. A variable glandular-hairy or shaggy-haired, ascending or spreading dwarf shrub, much branched, non-climbing and with spike-like clusters of medium-sized white or pink flowers. Corolla 2–2½ cm, throat-boss sometimes yellow; calyx 6–8 mm oval-lance-shaped; flower stalks 2–20 mm. Leaves lance-shaped to rounded, opposite below, alternate above or nearly all alternate; bracts similar to leaves. Fruit 9–15 mm, glandular-hairy. Rocks and walls. S–E. Spain. Pl. 45.   **1197h.** *A. grosii* Font Quer. Flowers yellow, 3–3½ cm. Leaves glandular-hairy; stems spreading. Fruit 8–10 mm. Granite cliffs. Endemic. C. Spain (S. de Gredos).

(b) Leaves and stems not glandular-hairy
1197e. *A. sempervirens* Lapeyr. A softly hairy dwarf shrub of shady limestone rocks with white or cream flowers, often veined with pink with an orange-yellow throat-boss. Flowers 1–1½ cm; calyx lobes 5–6 mm, lance-shaped; flower stalks ½–1 cm. Leaves oblong to elliptical, mostly opposite; leaf stalks 3–7 mm. Fruit about 6 mm, glandular-hairy. C. Pyrenees. Page 138.   **1197f.** *A. pulverulentum* Láz-Ibiza Like 1197e but hairs rather longer; leaves oblong to elliptical, rounded at apex, the upper alternate. Flowers pale yellow, or white, or buff with an orange boss and purple lines on the upper lip. Fruit hairy but not glandular-hairy. Calcareous rocks. S–E. Spain (Jaén to Teruel provinces). Pl. 45. Page 116.   **1197g.** *A. charidemi* Lange Like 1197f but flowers white to lilac-pink and flower stalks longer than subtending leaf. Schistose rocks. Endemic. S. Spain (Cabo de Gata).   **1197i.** *A. braun-blanquetii* Rothm. Flowers cream with a yellow throat-boss and with violet lines on upper lip, 3–4 cm. Leaves and stems glandular-hairy; bracts exceeding flower buds. Fruit 13–17 mm. Rocks, walls. N–E. Portugal, N–W. Spain.

MISOPATES | **1199. *M. orontium* (L.) Rafin. (*Antirrhinum o.*) WEASEL'S SNOUT. Cultivated ground. Widespread.

ASARINA | **1200. *A. procumbens* Miller CREEPING SNAPDRAGON. A creeping, brittle-stemmed, glandular-sticky perennial, with large whitish-yellow, axillary flowers 3–3½ cm, often streaked with pink. Leaves kidney-shaped, with rounded lobes, palmately veined. Fruit hairless, opening by 2 pores. Shady rocks in mountains. N–E. Spain, Pyrenees, Massif Central. Page 139.

LINARIA | Toadflax A difficult genus with about 52 species in our area, including 36 which are restricted to the Iberian peninsula in Europe. The seeds are important in determining many species: they are either *globular* in outline, but strongly angled and often quadrangular (not winged, except sometimes at the angles); or *disk-like* and flattened with a broad or narrow wing, and smooth, warted or netted sides.

Seeds more or less globular not disk-like, without a wing (or narrowly winged at the angles)
(a) Inflorescence glandular-hairy
(i) Flowers mainly violet, purple or white
(x) Flowers small 9–17 mm, including the spur which is not more than 7 mm
1201a. *L. clementei* Boiss. An erect glaucous, sparsely leafy perennial 80–150 cm, with a dense terminal cluster of violet flowers, each 13–17 mm. Flower stalks 2–4 mm, glandular-sticky, a little longer than the bracts; spur of corolla about 3 mm. Leaves linear, fleshy, 1–2 cm. Fruit about 3 mm. Dry sandy and calcareous places. S. Spain (Málaga province). **1201b.** *L. nivea* Boiss. & Reuter An erect leafy, glaucous perennial, with a long slender inflorescence of white or pale lílac-veined flowers, with a pale throat-boss covered with

white hairs. Flowers 12–14 mm; spur 4–5 mm. Leaves linear-oblong acute, 4–7 mm wide. Fruit about ½ cm. Mountains. C. Spain.

(xx) Flowers 16–30 mm, including the spur which is at least 8 mm

1201c. *L. incarnata* (Vent.) Sprengel A slender annual with short, rather lax inflorescences of strongly scented flowers, basically violet-coloured, but conspicuously bi-coloured owing to the white throat-boss with a yellow spot, and with widely gaping lips. Corolla 22 mm, upper lip with diverging lobes; spur slender, straight or curved, up to 11 mm; stigma deeply two-lobed. Leaves linear. Cultivated ground, dry grassland. Portugal, W. Spain.

1201d. *L. elegans* Cav. (*L. delphinoides*) A slender erect annual with a long lax glandular-hairy spike of slender lilac or dark violet flowers with long curved spurs longer than 'the rest of the corolla. Spur 10–14 mm; lips of corolla strongly diverging, throat-boss whitish, mouth of tube more or less open; stigma club-shaped. Leaves narrow linear to narrow lance-shaped. Dry open places. N. Portugal, W., C., and N. Spain. Pl. 45. Page 110   1201e. *L. algarviana* Chav. Like 1201d but stems spreading or ascending and inflorescence usually with 1–8 flowers; stigma deeply two-lobed. Flowers 2–2½ cm, violet, spotted with white or yellow on the throat-boss; spur 11–12 mm. Dry sandy places. S–W. Portugal (Algarve). Page 45.

(ii) Flowers mainly yellow

1201f. *L. hirta* (L.) Moench A rather robust erect annual 15–60 cm, with oblong-lance-shaped, semi-clasping leaves, and a dense terminal glandular-hairy spike of rather stout pale yellow flowers. Corolla 2–3 cm or more; spur 1–1½ cm; throat-boss orange. Calyx 7–8 mm, lobes very unequal. Leaves 4–15 mm broad. Fruit 5–6 mm; seeds tetrahedral with slightly winged angles. Cultivated ground. Portugal, S., C., and E. Spain. Page 60.   1201g. *L. viscosa* (L.) Dum.-Courset Like 1201h but inflorescence densely glandular-hairy; flower clusters dense, even in fruit, usually bright yellow but sometimes violet in S. Spain. Calyx up to 6 mm, lobes often with long-pointed apices; flower stalks not more than 8 mm, erect. Waste places, sandy fields. S. Portugal, S. Spain.

(b) Inflorescence hairless

1201h. *L. spartea* (L.) Willd. A slender erect, sparsely leafy annual with narrow linear leaves and bright yellow flowers 18–30 mm, with lips closely pressed together; spur 9–18 mm. Inflorescence lax, very lax in fruit; calyx about 4 mm, lobes with blunt apices. Fruit about 4 mm; seeds black, tetrahedral. Dry open places. Widespread.   **1202. *L. repens* (L.) Miller PALE TOADFLAX. Distinguished by its slender, elongated, dense spikes of small pale violet flowers with conspicuous darker veins. Corolla 8–15 mm; spur 3–5 mm, conical. Leaves numerous, mostly in whorls, linear to narrow-lance-shaped, 1–3 mm wide. Dry places, screes, in mountains. E. Spain, Pyrenees, France. Page 132   **1208. *L. triphylla* (L.) Miller THREE-LEAVED TOADFLAX Distinguished by its tricoloured flowers and broad oval, slightly fleshy, glaucous leaves, usually in whorls of three. Corolla 2–3 cm, yellowish-white, with orange throat-boss and violet spur. Seeds tetrahedral. Dry places, fields. Med. region.

1201i. *L. pedunculata* (L.) Chaz. A small, usually hairless and glaucous annual of maritime sands, with a lax branched, leafy inflorescence of usually conspicuous violet-veined flowers 1–1½ cm, with an orange throat-boss and with a deep violet spur 6–7 mm. Corolla sometimes cream. Leaves 5–12 mm, fleshy, oblong to elliptic, mostly in whorls below. Fruit stalks to 2½ cm, erect; fruit about ½ cm. S. Portugal, S–W. Spain.

Seeds disk-like, with a marginal wing
(a) Flowers mainly violet, reddish-purple or white
**(1204) *L. triornithophora* (L.) Willd. Readily distinguished from all other species by its

large, violet-purple, strongly veined flowers with a yellow throat-boss, and its long, stout, curved, pointed spur longer than the rest of the corolla. Corolla 3½–5½ cm. Leaves oval lance-shaped, mostly in whorls of three. Hedges and thickets, C. and N. Portugal, N–W. and W–C. Spain. Pl. 44, Page 111.

**1204. *L. alpina* (L.) Miller ALPINE TOADFLAX. Mountains, screes, rocky places, river-gravels. C. Spain, Pyrenees. 1204a. *L. tristis* (L.) Miller A glaucous, spreading or ascending perennial usually with a glandular-hairy inflorescence of brownish-purple flowers with a yellow, often conspicuously brown-veined spur. Corolla 21–28 mm; spur stout 11–13 mm. Leaves numerous, linear to narrow-lance-shaped, mostly alternate above, flat. Limestone rocks. C. Portugal, S. Spain.

1204b. *L. aeruginea* (Gouan) Cav. (*L. melanantha*) Like 1204a but leaves with inrolled margins. Flowers variously tinged with brownish-purple, sometimes completely brownish-purple, violet, yellowish or cream; spur 5–11 mm. Dry hills and rocks in mountains. Portugal, S. and E. Spain, Pl. 45, Page 84. 1204c. *L. amethystea* (Lam.) Hoffmanns. & Link Distinguished from most other species by the thickened wing of the seeds. A slender erect, unbranched annual with a lax, often interrupted, inflórescence of rather few bluish-violet, bright yellow, or cream-coloured flowers, variably spotted with purple and with an orange throat-boss. Inflorescence with often purple glandular hairs. Corolla 10–27 mm; spur slender 4–15 mm; flower stalks shorter than bracts. Fruit 3–5 mm; seeds flat, with warty projections on the side. Cultivated ground, open places. Portugal, Spain. Pl. 45. 1204d. *L. anticaria* Boiss. & Reuter Distinguished by its greyish-lilac flowers with bluish-violet veins and deep bluish-violet throat-boss, sometimes with a yellow spot (corolla very rarely entirely yellow). Corolla 18–26 mm; spur 5–12 mm; inflorescence dense; flower stalks 1–2 mm. A hairless, glaucous, spreading or ascending perennial with oblong leaves in whorls of 4–6. Seeds with a broad, paler wing and smooth or warty sides. Shady limestone rocks. S. Spain. 1204e. *L. glacialis* Boiss. Like 1204d but flower clusters stalkless and overlapped by the upper stem leaves; bracts more than 1 cm, at least as long as the stem leaves. Flowers dull violet suffused with yellow; spur 8–15 mm, violet-veined. Alpine rocks and screes. S. Spain (Sierra Nevada). Pl. 45.

(b) *Flowers mainly yellow* (*sometimes violet-spurred or flushed with violet, purple or brown*) 1206. *L. supina* (L.) Chaz. A glaucous annual to perennial, usually with a glandular-hairy, few-flowered inflorescence of pale yellow flowers sometimes tinged with violet. Lobes of lower lip of corolla long; spur 1–1½ cm. Leaves ½–2 cm, linear to narrowly oblance-shaped, whorled below. Seeds grey or blackish, with a broad wing. Sandy places. Scattered throughout. Page 125. 1206a. *L. caesia* (Pers.) Chav. Like 1206 but flowers yellow with reddish-brown veins, and seeds metallic-shiny. Sandy fields, calcareous hills. Western half of Iberian peninsula.

1205. *L. vulgaris* Miller COMMON TOADFLAX. Waste ground, hedges etc. Widespread. 1205a. *L. platycalyx* Boiss. Recalling 1208 with broad oval stem leaves in whorls of three, but flowers yellow with tip of spur tinged purple, and flowers usually in a terminal cluster of 1–3. Calyx segments egg-shaped, 7–11 mm. A somewhat glaucous, almost hairless, annual or perennial. Seeds flat, smooth or warty, wing pale. Shady limestone rocks. S–W. Spain. Page 68. 1206b. *L. saxatilus* (L.) Chaz. (*L. tournefortii*) Usually a densely glandular-hairy perennial with dense terminal clusters of small yellow flowers and numerous crowded, narrow leaves. Corolla 9–17 mm; spur 5–8 mm, slender. Leaves linear to oblong-elliptic, alternate, hairy. Seeds 6–17 mm, wing narrow. Dry sandy or rocky places. C. and N. Portugal, C. and N. Spain. See: 1204b. *L. aeruginea*; 1204c. *L. amethystea*; 1204d *L. anticaria*.

KICKXIA | Fluellen

1. Lvs all rounded or heart-shaped at the base, without lobes or auricles.
   2. Corolla 1–1½ cm; calyx lobes 4–8 mm in fr., oval-heart-shaped. (Widespread)   1209. *K. spuria* (L.) Dumort.
   2. Corolla 8–11 mm; calyx lobes 3 mm in fr., linear-lance-shaped. (Widespread)   1209a. *K. lanigera* (Desf.) Hand.-Mazz.
1. Lvs. at least some spear-shaped or arrow-shaped.
   3. Corolla 4–6 mm; fr. 1–2 mm. (Portugal, Med. region)   1209b. *K. cirrhosa* (L.) Fritsch
   3. Corolla 7–15 mm; fr. at least 2½ mm.
      4. Seeds with knobbly projections; spur strongly curved. (Med. region)   1209c. *K. commutata* (Reichenb.) Fritsch
      4. Seeds with little pits; spur more or less straight. (Widespread)   (1209) *K. elatine* (L.) Dumort.

**CYMBALARIA | **1210. *C. muralis* P. Gaertner, B. Meyer & Scherb. IVY-LEAVED TOAD-FLAX. Damp rocks, walls; naturalized in our area. 1210a. *C. aequitriloba* (Viv.) A. Cheval. A delicate creeping, finely hairy or shaggy-haired perennial with rounded or kidney-shaped, entire to three-lobed, or sometimes five-lobed leaves. Flowers 8–13 mm, lilac to violet, with a yellow throat-boss; calyx about 3 mm. Fruit hairless. Damp shady places. Balearic Islands.

ANARRHINUM | 4 species in our area, 3 restricted to it. **1211. *A. bellidifolium* (L.) Willd. A hairless plant with a basal rosette of egg-shaped to elliptic leaves, and contrasting, deeply cut, linear stem leaves, and a long rather one-sided spike of tiny pale blue or lilac flowers. Corolla 4–5 mm, shortly spurred; calyx lobes narrow, tapering. Dry places, way-sides, etc. Widespread. 1211a. *A. duriminium* (Brot.) Pers. (*A. hirsutum*) Differs from 1211 in having a glandular-hairy inflorescence, pale yellow or cream-coloured flowers, and three-lobed stem leaves with the central lobe broader and larger. Waysides, hedges, fields and walls. N–W. Portugal, N–W. Spain.

CHAENORHINUM | Leaves entire. Corolla cylindrical with a straight basal spur, two-lipped, the upper lip two-lobed, the lower three-lobed, with a throat-boss which does not close the mouth of the corolla tube. 7 species.

Perennials
1211b. *C. origanifolium* (L.) Fourr. A small, spreading or erect perennial with lance-shaped to rounded leaves, and rather lax clusters of tubular violet flowers with a yellow throat-boss. Corolla 1–2 cm; spur 2–6 mm; calyx lobes narrow oblance-shaped. Fruit stalks erect to spreading; fruit globular. Plants very variable, in leaf shape, hairiness, calyx and capsule. Limestone rocks, walls. Scattered throughout. Pl. 44. Page 124. 1211c. *C. villosum* (L.) Lange A sticky, densely shaggy-haired perennial with yellowish translucent hairs, and with pale lilac, or pale yellow flowers with violet veins, 10–18 mm long; spur 2–6 mm. Leaves obovate to rounded, rather thick. Flower stalks up to 22 mm, recurved in fruit; fruit globular, about half as long as calyx. Walls, rocks in mountains. S. Spain, S–W. France. Pl. 44, Page 90. 1211d. *C. macropodum* (Boiss. & Reuter) Lange A relatively robust, glandular, woolly-haired perennial with large lilac flowers with violet veins, with or without a yellow

or white throat-boss. Flowers 1½–2½ cm, in an elongate cluster; flower stalks about 2½ cm, erect in fruit; spur 5–8 mm; calyx 7–9 mm. Fruit ovoid. Dry hills and rocks in mountains. S. Spain. Page 76.

Annuals

1211e. *C. minus* (L.) Lange SMALL TOADFLAX. Fields, sandy places, rocks in mountains. Scattered throughout.   1211f. *C. rubrifolium* (DC.) Fourr. Like 1211b but basal leaves in a rosette; spur of corolla acute, evenly tapering from the base. Corolla usually less than 1½ cm, blue, violet or lilac, with a yellow throat-boss. Leaves egg-shaped, the lower red beneath. Dry places. Med region.   Subsp. *raveyi* (Boiss.) R. Fernandes has yellow flowers 1½–2 cm, tinged with violet on the tube; it occurs in mountains of S. Spain.

SCROPHULARIA | **Figwort** 18 similar-looking species, 10 of which occur only in our area in Europe. The following widespread species occur scattered throughout much of our area: 1212. *S. peregrina* L. NETTLE-LEAVED FIGWORT;   (1214) *S. auriculata* L. WATER BETONY;   1214. *S. nodosa* L. FIGWORT;   **1215. *S. scorodonia* L. BALM-LEAVED FIGWORT.

(1216) *S. sambucifolia* L. A robust, nearly hairless perennial with flowers up to twice as large as most other species, and with once- or twice-cut leaves. Flowers green with brownish-red or pinkish-red upper lip, 12–20 mm, in short-stalked axillary whorls. Wet places. S. and C. Portugal, S. Spain.

1216. *S. canina* L. Distinguished by its numerous tiny blackish-purple flowers borne in a lax, much-branched, leafless pyramidal-cylindrical inflorescence. Flowers usually 4–5 mm. Leaves once- or twice-cut into toothed lobes. Dry sandy and stony places. Widespread. 1216a. *S. frutescens* L. Like 1216 in having tiny blackish-purple flowers, but with thick leathery, rounded to oblong-lance-shaped, entire or toothed leaves. Corolla 4 mm, lateral petals often with white margins; lower bracts leaf-like. A stout, robust, woody-based plant. Maritime sands. Portugal, W. Spain. Page 337.

MIMULUS | **(1217) *M. moschatus* Lindley MUSK. Native of N. America; naturalized in Portugal.

GRATIOLA | **1218. *G. officinalis* L. GRATIOLE. Wet meadows, riversides. Widespread. 1218a. *G. linifolia* Vahl Like 1218 but glandular-hairy above; stems rounded, not four-angled, and flower stalks at least as long as subtending leaves. Wet places, riversides, marshes. Portugal, S–W. Spain.

LIMOSELLA | (1218) *L. aquatica* L. MUDWORT. Scattered throughout.

VERONICA | **Speedwell** 38 species; the following are widely distributed in our area: 1220. *V. serpyllifolia* L. THYME-LEAVED SPEEDWELL;   (1220) *V. arvensis* L. WALL SPEEDWELL;   **1221. *V. triphyllos* L. FINGERED SPEEDWELL;   1222. *V. cymbalaria* Bodard PALE SPEEDWELL;   (1222) *V. hederifolia* L. IVY SPEEDWELL;   **1223. *V. persica* Poiret BUXBAUM'S SPEEDWELL;   (1223) *V. polita* Fries GREY SPEEDWELL;   (1223) *V. agrestis* L. FIELD SPEEDWELL;   **1225. *V. beccabunga* L. BROOKLIME;   (1225) *V. anagallis-aquatica* L. WATER-SPEEDWELL;   1226. *V. scutellata* L. MARSH SPEEDWELL;   1227. *V. officinalis* L. COMMON SPEEDWELL;   (1227) *V. montana* L. WOOD SPEEDWELL;   1228. *V. chamaedrys* L. GERMANDER SPEEDWELL.

Flowers in terminal clusters

1219. *V. spicata* L. SPIKED SPEEDWELL. Distinguished by its very numerous blue flowers in a dense cylindrical, terminal spike, which may be up to 30 cm long. Corolla 4–8 mm across; calyx lobes oval-elliptic, blunt; flower stalks usually less than 1 mm, much shorter than the bracts. Dry grasslands, rock slopes, in mountains. N. and C. Spain, Pyrenees. 1219a. *V. bellidioides* L. Distinguished by its lilac to violet-blue flowers, which are borne in short, rounded, terminal clusters on erect, sparsely leafy stems. Inflorescence elongating in fruit; corolla 9–10 mm across; flower stalks 2–6 mm. A tufted, greyish-haired plant with obovate, mostly toothed basal leaves in a rosette; stock creeping. Dry alpine pastures. Pyrenees.

**(1220) *V. fruticans* Jacq. A woody-based, spreading perennial with numerous pairs of thick, shining fleshy, oblong leaves, and terminal clusters of few, brilliant blue flowers with reddish centres, each 1–1½ cm across. Calyx, flower stalks, and fruit densely covered with crisped hairs. Rocks and stony places in mountains. Pyrenees. 1220a. *V. fruticulosa* L. SHRUBBY SPEEDWELL. Like (1220) but more bushy and erect, and flowers usually pink with darker veining, in longer spike-like clusters. Calyx, leaf stalk, and fruit with some glandular hairs. Mountain rocks. S. and C. Spain, Pyrenees. Pl. 44. 1220b. *V. nummularia* Gouan Like (1220) with a woody stock, but flowers in a dense terminal cluster closely surrounded by numerous overlapping, rounded leaves. Corolla pink or blue, tiny, 6 mm across, lobes narrow; calyx lobes 4, oblong, ciliate. Leaves 4–5 mm; bracts similar. Damp rocks and screes in mountains. Pyrenees.

1220c. *V. ponae* Gouan Recalling 1229a but slender flower cluster terminal, usually solitary. Flowers bluish-lilac about 1 cm across; calyx 3–4 mm, lobes 4, unequal; flower stalk 2–3 times as long as bracts. Fruit 5–6 mm. Damp or shady places in mountains. S. and N. Spain, Pyrenees.

Flowers in axillary clusters
**1229. *V. austriaca* L. (*V. teucrium*) LARGE SPEEDWELL. A very variable perennial, usually distinguished by its 5 very unequal, linear calyx lobes (sometimes only 4 present). Flowers bright blue, conspicuous in long-stalked, slender axillary clusters. Corolla 1–1½ cm across. Leaves stalkless, variable. Dry places, meadows, mountains in the south. Spain, France. Page 139. 1229a. *V. urticifolia* Jacq. Like 1229 but calyx with 4 elliptic-oblong, more or less equal lobes, and flowers pale blue or rose-lilac, smaller, about 7 mm across. Flower stalks longer than bracts. Leaves triangular-egg-shaped, saw-toothed; stem hairy all round. Fruit 3–4 mm; fruit stalk sharply upturned below fruit. Damp rocks in woods in mountains. E. Spain, Pyrenees, Massif Central.

DIGITALIS | Foxglove

1. Middle lobe of lower lip of corolla projecting far beyond other lobes; shrubby plants.	*obscura*
1. Middle lobe of lower lip only slightly longer than other lobes; herbaceous plants.	
2. Corolla a slender cylindrical tube.	
3. Corolla *c.* 12 mm, dark reddish-brown.	*parviflora*
3. Corolla 2–2½ cm, yellowish or greenish. (C. and E. Spain, France)	**1233. *D. lutea* L.
2. Corolla tube stout, bell-shaped.	
4. Corolla yellow. (Massif Central)	**1232. *D. grandiflora* Miller
4. Corolla purple, pink or white.	
5. Whole plant covered with yellowish-glandular-sticky hairs.	*thapsi*
5. Plant greenish to whitish, not or only slightly glandular-sticky.	

6. Lower lvs. long-stalked, conspicuously toothed.
(Widespread except S.)   **1234. *D. purpurea* L.
6. Lower lvs. short-stalked, entire or shallow-
toothed. (Balearic Islands)   1234a. *D. dubia* Rodr. Pl. 47

**(1233) *D. obscura* L. SPANISH RUSTY FOXGLOVE. A very attractive shrubby plant with long narrow, curved, leathery leaves, and lax one-sided spikes of rusty-brown to orange-yellow flowers, with darker markings within. Leaves linear-lance-shaped, entire, or deeply toothed in subsp. *laciniata* (Lindley) Maire. Rocky mountain slopes. S., C., and E. Spain. Page 69.   1233a. *D. parviflora* Jacq. An erect leafy perennial with a long, very dense spike of very numerous, small, dull reddish-brown flowers with a purple-brown lower lip. Inflorescence white-woolly; calyx lobes egg-shaped. Leaves numerous, oblong-lance-shaped, entire or slightly toothed, leathery. Rocks in mountains. N. and N–E. Spain. Pl. 47.

1234b. *D. thapsi* L. Like 1234 but whole plant covered with yellowish glandular-sticky hairs; inflorescence lax, usually branched at the base; flower stalks about twice as long as calyx. Flowers large 4–7 cm, pink, spotted within, finely hairy outside. Rocky slopes, waste places. E. Portugal, W. and C. Spain. Pl. 47, Page 111.

LAFUENTEA | Flowers in a dense terminal spike; corolla with a long tube, somewhat two-lipped, five-lobed; stamens 4, included, paired.   1235a. *L. rotundifolia* Lag. A strongly aromatic, creeping shrublet with densely grey-haired, glandular leaves and dense cylindrical spikes up to 10 cm long of numerous, small, stalkless, white flowers striped with purple. Corolla 7–8 mm; calyx 4–6 mm, with linear lobes. Leaves about 3 cm, rounded or kidney-shaped, irregularly toothed, long-stalked. Fruit elliptic, with persistent style. Rock crevices, overhanging rocks, in lowlands. S. Spain. Page 337.

ERINUS | **1235. *E. alpinus* L. ALPINE ERINUS. Rocks, screes in mountains. S., N., and E. Spain, Pyrenees, France.

LINDERNIA | Hairless annuals with solitary, irregular flowers in the leaf axils. Corolla · two-lipped; the upper lip small, flat, erect, two-lobed, the lower spreading, three-lobed. Stamens 4.   1235b. *L. procumbens* (Krocker) Philcox (*L. pyxidaria* auct.) A small spreading annual, with entire oval-oblong leaves and small solitary, pale violet or pink flowers on slender stalks usually longer than the subtending leaves. Corolla 2–6 mm, sometimes closed, and shorter than calyx. Muddy river banks, wet sands. Spain, France, introd. C. and N. Portugal.

SIBTHORPIA | A creeping perennial, with rounded leaves and solitary stalked axillary flowers. Corolla tubular, with spreading lobes. 3 species.   1235c. *S. europaea* L. CORNISH MONEYWORT. A delicate creeping perennial, rooting at the nodes, with alternate, rounded, stalked leaves and tiny axillary white or cream flowers more or less tinged with pink. Corolla 1–3 mm, usually five-lobed. Leaf blades 1–2½ cm across, with shallow rounded lobes, hairy. Damp, shady places. Portugal, N. and W. Spain, W. France.

BARTSIA | 3 species.   **1236. *B. alpina* L. ALPINE BARTSIA. Damp places. Pyrenees. Page 148.   1236a. *B. aspera* (Brot.) Lange Flowers in a dense spike, yellow changing to reddish-brown, each about 1 cm, with tube longer than the calyx. Bracts linear-lance-shaped, entire, shorter than calyx. Leaves rough-haired, entire. Dry stony ground, heaths. C. Portugal, S–W. Spain.

BELLARDIA | **1237. *B. trixago* (L.) All. Damp stony, sandy or grassy fields, pine woods. Widespread.

1. *Cephalaria leucantha* (1320)
3. *Pterocephalus intermedius* 1324c
5. *Lafuentea rotundifolia* 1235a
2. *Scrophularia frutescens* 1216a
4. *Jasione amethystina* 1355e
6. *Centranthus nevadensis* 1316a

PARENTUCELLIA | **1238. *P. viscosa* (L.) Caruel YELLOW BARTSIA. Damp, grassy or sandy places. Widespread.   **1239. *P. latifolia* (L.) Caruel SOUTHERN RED BARTSIA. Sandy and stony places, Cork oak woods, on the littoral. South of our area.

ODONTITES | 10 species in area.

Flowers yellow
**1240. *O. lutea* (L.) Clairv. Dry grassland and scrub, in mountains. E. Spain, Pyrenees, France.   1240a. *O. longiflora* (Vahl) Webb A glandular-hairy, usually branched annual with a lax inflorescence of few yellow flowers with very slender corolla tubes 1½–2 cm. Calyx lobes linear; anthers with an apical tuft of hairs, included in corolla. Leaves and bracts linear ½–1 cm. Dry places in mountains. Spain.

Flowers pink, purple and white
**1241. *O. verna* (Bellardi) Dumort. RED BARTSIA. Cultivated ground, cornfields, waste places. Widespread.   1241a. *O. purpurea* (Desf.) G. Don fil. Flowers dull purple, 6–7 mm long, in a rather lax inflorescence of 4–9 flowers. Corolla hairless; calyx 3–4 mm; anthers hairless, included in corolla. A non-glandular, sparsely hairy annual to 40 cm. Scrub. S. Spain.   1241b. *O. kaliformis* (Pourret) Pau Like 1241a but corolla finely hairy, purplish, about 6 mm; anthers hairless, longer than corolla; calyx 3 mm. Leaves linear. Rocky ground. S. and E. Spain.

EUPHRASIA | A very difficult genus with about 10 species in our area.   1244. *E. salisburgensis* Funck Rocks, pastures in mountains. Spain, France. Page 139.

RHINANTHUS | Yellow-Rattle   **1247. *R. minor* L. YELLOW-RATTLE. Meadows, waysides. Widespread, except in the south.

PEDICULARIS | Lousewort   13 species.   **(1256) *P. sylvatica* L. COMMON LOUSEWORT. Damp places, woods and damp heaths. Scattered throughout.   1256. *P. palustris* L. RED-RATTLE. Marshes, wet meadows. Pyrenees, France.

Montane and alpine species: 1249. *P. comosa* L. CRESTED LOUSEWORT. Alpine rocky meadows. S. and E. Spain, Pyrenees, Massif Central.   **1252. *P. foliosa* L. LEAFY LOUSE-WORT. Alpine pastures. N. Spain, Pyrenees, Massif Central. Page 125.   1253. *P. verticillata* L. WHORLED LOUSEWORT. Damp alpine pastures. Pyrenees, Massif Central.

1255a. *P. pyrenaica* Gay A pink-flowered species, distinguished by the long cylindrical beak to the upper lip, and its almost hairless calyx, which has both lance-shaped entire and toothed lobes. Corolla to 2 cm. Leaves and bracts hairless or nearly so, the stem with 2 lines of hairs; stem leaves few. Alpine pastures, screes. Pyrenees.   1255b. *P. mixta* Gren. Like 1255a with a long cylindrical beak but stem leaves more numerous; bracts and calyx with long woolly hairs; flowers pink with a deep crimson upper lip. Alpine pastures and screes. N. Spain, Pyrenees. Page 139.

MELAMPYRUM | Cow-Wheat   5 widespread European species occur mostly in the north and east of our area, including: 1260. *M. pratense* L. COMMON COW-WHEAT.   **1257. *M. cristatum* L. CRESTED COW-WHEAT.   **1258. *M. arvense* L. FIELD COW-WHEAT. Fields. (1260) *M. sylvaticum* L. WOOD COW-WHEAT.

TOZZIA | **1261. *T. alpina* L. Damp places in mountains and hills. Pyrenees, Massif Central.

GLOBULARIACEAE | Globularia Family

GLOBULARIA |

1. Shrubs with erect or spreading woody branches.
  2. Erect shrubs; twigs leafy to the fl. heads.   *alypum*
  2. Procumbent shrubs; fl. heads usually arising direct from lv. rosettes.
    3. Rosette lvs. up to 2½ cm by 8 mm, spathulate, blunt or notched at apex. (N–E. Spain, Massif Central)   **1263. *G. cordifolia* L.
    3. Rosette lvs. 1–2 cm by 1–2 mm, folded, mostly acute.   *repens*
1. Herbaceous plants, often with a woody stock.
  4. Flowering stems usually without small lvs. or bracts (sometimes up to 3); lvs. erect or spreading-upwards.
    5. Stolons absent; lvs. 6–12 cm. (N. Spain, Pyrenees)   (1264) *G. nudicaulis* L.
    5. Stolons 1–5 cm; lvs. 4–7 cm. (Pyrenees)   1264a. *G. gracilis* Rouy & J. Richter
  4. Fl. stems with numerous small lvs. Basal lvs. more or less horizontal.
    6. Rosette lvs. with lateral veins distinctly visible above. (N–W. Spain)   1264b. *G. punctata* Lapeyr.
                             (*aphyllanthes* auct.)
    6. Rosette lvs. with lateral veins not or scarcely visible above.
      7. Rosette lvs. with 2 or more conspicuous spreading spiny teeth on each side.   *spinosa*
      7. Rosette lvs. entire, or three-toothed at apex, or rarely with small, not spiny, forward-pointing teeth on margin.   *vulgaris* group

**1262. *G. alypum* L. SHRUBBY GLOBULARIA. A much-branched, bristly-leaved, evergreen shrub 30–100 cm, with globular lilac-blue flower heads 1–2½ cm across. Involucral bracts oval, overlapping, brown-margined, ciliate. Leaves leathery, with a spiny tip and sometimes spiny teeth. Dry bushy places, rocks. S., C., and E. Spain, Med. France. 1263a. *G. repens* Lam. (*G. nana*) A dwarf creeping woody shrub like 1263, but leaves smaller and flower heads about 1 cm across, usually almost stalkless; involucral bracts lance-shaped. Dry pastures, stony places, rock crevices, in mountains. N. and E. Spain, Pyrenees. Pl. 48.

1264c. *G. spinosa* L. A herbaceous perennial to 20 cm, with a stout stock, easily recognized by its oval to spathulate leaves which have a spiny apex and 2 or more long spreading, spiny teeth on each side, and adult leaves covered with chalky secretions. Flower heads blue, 2–2½ cm across; involucral bracts lance-shaped, long-pointed, bristly-haired. Stem leaves lance-shaped. Limestone rocks in mountains. S–E. Spain. Pl. 48.

1264. *G. vulgaris* L. COMMON GLOBULARIA. Flower heads about 2½ cm across, borne on flowering stems up to 20 cm. Involucral bracts lance-shaped, long-pointed. Leaves elliptic to lance-shaped with a flat margin, three-toothed at apex. Dry hills, sunny rocks, in mountains. Portugal, S. Spain, S. France. 1264d. *G. valentina* Willk. Like 1264 but leaves obovate with undulate, margin toothed. Flowering stems 20–30 cm; involucral bracts egg-shaped with a long rigid point. C. and N–E. Portugal, N–E. Spain.   1264e. *G. cambessedesii* Willk. Like 1264 but leaves obovate, margin toothed, and leaf stalk shorter than blade. Flower heads up to 3½ cm across on flowering stems 20–30 cm. Mountains. N–E. Spain.

ACANTHACEAE | Acanthus Family

ACANTHUS | **1266. *A. mollis* L. BEAR'S BREECH. Hedges, damp shady places. Scattered throughout.

MYOPORACEAE | Myoporum Family

Shrubs or trees with mostly alternate, simple leaves. Flowers usually irregular; calyx five-lobed, persistent; corolla five-lobed; stamens 4–5; ovary superior of 2 carpels; style 1; stigma 1. Fruit a drupe.

MYOPORUM |
1266a. *M. tenuifolium* G. Forster (including *M. acuminatum*) A shrub or small tree, native of S. Australia which is grown for ornament and shelter and is sometimes naturalized. Flowers in axillary flat-topped clusters; corolla about 1 cm across, bell-shaped, white with violet spots, hairy within. Leaves bright glossy green above, oval-lance-shaped to narrow lance-shaped, acute or pointed. Fruit violet-purple, ovoid. Portugal, Spain. Pl. 48.

GESNERIACEAE | Gloxinia Family

RAMONDA | **1268. *R. myconi* (L.) Reichenb. A very beautiful crevice-loving plant with large rosettes of oval, wrinkled leaves with rusty brown hairs beneath, and few large blue to violet-purple flowers with yellow centres. Inflorescence 6–12 cm, glandular-hairy, with 1–6 flowers; corolla 2–4 cm across; petals 5, rounded. Shady limestone crevices in mountains. N–E. Spain, Pyrenees. Pl. 48. Page 138.

OROBANCHACEAE | Broomrape Family

LATHRAEA | **1269. *L. squamaria* L. TOOTHWORT. Parasitic on beech, alder, hazel and other trees. E. Spain, France.   **(1269) *L. clandestina* L. PURPLE TOOTHWORT. Parasitic predominantly on willows, poplars, and alder. N. and E. Spain, France. Page 138.

CISTANCHE | Differs from *Orobanche* in having a five-lobed calyx with equal lobes; large flowers; anthers completely woolly-haired; fruit with persistent style. Parasitic on plants of the *Chenopodiaceae*.

1269a. *C. phelypaea* (L.) Coutinho (*Phelypaea tinctoria, P. lusitanica, P. lutea*) A yellowish plant 20–70 cm, with a swollen base, stout erect stems with numerous oval-lance-shaped, papery-margined scales, and a dense oblong-cylindrical spike of large tubular, yellow flowers. Corolla 3–5 cm long, curved, narrowed at base, with a wide five-lobed mouth; calyx bell-shaped, with rounded, toothed lobes. There are two colour forms: one has bright deep yellow flowers and grows in salt marshes; the other has pale yellow flowers with the folds of the lower lip deeper yellow and usually with touches of violet on the reflexed portions of the petals; bracts violet to grey; stigma white. It grows in the steppe country of Almeria province. S. Portugal, S. Spain. Pl. 47. Page 91.

OROBANCHE | Broomrape 28 species in our area. A key to the commoner species is given below; although this will not enable all plants met with to be identified, it will at least indicate the most important distinguishing features. The host plant is a valuable clue, but very careful dissection of root systems is usually necessary to determine the parasitic connection between the two species. (Note measurements in brackets indicate the maximum or minimum dimensions of an organ in exceptional examples.)

1. Each flower subtended by 2 bracteoles (more or less connected to the calyx) as well as by a bract.
   2. Anthers more or less densely hairy. *arenaria*
   2. Anthers hairless or sparsely hairy at the base.
      3. Corolla 10–20 (–22) mm; bracts 6–8 (–10) mm. *ramosa*
      3. Corolla (18–) 20–25 mm; bracts 8–15 mm. *purpurea*
1. Bracteoles absent.
   4. Stigma purple or dark red.
      5. Corolla not shining dark red inside, usually white or pale
         yellow outside, at least towards the base. *minor* group
      5. Corolla shining red inside, mostly dark red or purple
         outside.
         6. Filaments inserted 3–7 mm above base of corolla; corolla
            glandular-hairy outside. *foetida*
         6. Filaments inserted 1½–2 mm above base of corolla; corolla
            hairless outside. *sanguinea*
   4. Stigma yellow or white (rarely purple).
      7. Corolla shining dark red inside, usually dark purplish-red
         or bright yellow outside.
         8. Lower lip of corolla ciliate; fls. fragrant. *gracilis*
         8. Lower lip of corolla not ciliate; fls. fetid. *foetida*
      7. Corolla not shining dark red inside, usually pale yellow,
         white or bluish outside, at least towards the base.
         9. Parasitic on Ivy. (Widespread) (1273) *O. hederae* Duby
         9. Parasitic on other plants.
            10. Corolla predominantly white, cream, bluish or violet.
               11. Corolla conspicuously inflated, papery and shining
                  at base. *cernua*
               11. Corolla not as above.
                  12. Corolla 20–30 mm, with large strongly divergent lips. *crenata*
                  12. Corolla 10–23 mm, with small lips. *minor* group
            10. Corolla predominantly yellow or reddish.
               13. Stamens inserted 7–15 mm above base of corolla;
                  anthers shaggy-haired; bracts more or less hairless. *latisquama*
               13. Stamens inserted not more than 6 mm above base of
                  corolla; anthers hairless; bracts glandular-hairy.
                  14. Corolla at least 2 cm. *rapum-genistae*
                  14. Corolla usually not more than 2 cm.
                     15. Filaments more or less delta-shaped at base; spike
                        not more than 3 times as long as wide. *densiflora*
                     15. Filaments gradually and only slightly widened to-
                        wards the base; spike (2–) 4–8 times as long as wide. *minor* group

1270. *O. ramosa* T. (*Phelypaea r.*) BRANCHED BROOMRAPE. A very variable species, with 3 subspecies in our area, growing on a wide variety of hosts. Flowers blue or violet with prominent white hairy ridges between the lobes of the lower lip, and yellowish-white at the base. Stems 5–30 cm by 1½–4 mm. Widespread and locally abundant. Page 53. 1271. *O. purpurea* Jacq. (*Phelypaea caerulea*) PURPLE BROOMRAPE. Flowers bluish-violet; stems 15–60 cm by 3–8 mm. On *Compositae*. S., C., and E. Spain, France. (1271) *O. arenaria* Borkh. (*Phelypaea a.*) SAND BROOMRAPE. Flowers bluish-violet 25–35 mm. Stems 15–60 cm by 3–6 mm. On *Artemisia* and perhaps other herbs on the littoral. N. and C. Portugal, S., C., and E. Spain, France. 1271a. *O. gracilis* Sm. Flowers yellow outside, usually with red veins and reddish towards the lips, shining dark red inside, 15–25 mm; stigma orange, style purple, exserted. Stems 15–60 cm by 2–7 mm. On *Leguminosae*, rarely on *Cistus*. Widespread. Pl. 48. 1271b. *O. foetida* Poiret Flowers dark purplish-red, 12–23 mm; stigmas usually orange and conspicuous, sometimes purple, exserted for ½ cm from corolla. Stems 15–60 cm by 4–10 mm. On *Leguminosae*. Portugal, Spain. 1271c. *O. sanguinea* C. Presl Flowers dark red or purple, yellow at the base (rarely entirely yellow), 10–15 mm; stigma purple. Stems 10–40 cm by 3–7 mm. On *Lotus*. Portugal, S. and S-E. Spain. **1272 *O. rapum-genistae* Thuill. GREATER BROOMRAPE. Flowers greyed-red inside, greyed-yellow outside, 20–25 mm; stigma yellow. Stems 20–80 cm by 5–20 mm, usually greyed-red in our area. On shrubby *Leguminosae*. Widespread. 1272a. *O. densiflora* Reuter Distinguished by its slender dark bracts contrasting with its lemon-yellow stems and parchment-coloured flowers; stigma bright yellow. Flowers in a dense terminal spike. On *Lotus creticus* and other *Leguminosae*. Sand dunes. C. Portugal, S–W. Spain. **1274. *O. crenata* Forskål A pest of bean fields, where it may be seen in great numbers; it also attacks the garden sweet-pea and other *Leguminosae*. Widespread. 1274a. *O. cernua* Loefl. Flowers violet-blue, 12–20 mm; stigma whitish. Stems up to 40 cm. On *Artemisia*, rarely on other herbs. S., C., and E. Spain, France.

O. minor group

1. Bracts 7–15 mm; stamens inserted 2–3 mm above base of corolla. *minor*
1. Bracts 10–22 mm; stamens inserted 3–5 mm above base of corolla.
 2. Corolla rather sharply curved inwards near the base, the upper lip deeply two-lobed. *amethystea*
 2. Corolla not sharply curved inwards near the base, the upper lip notched or slightly two-lobed. *loricata*

1277. *O. minor* Sm. LESSER BROOMRAPE. Flowers pale yellow usually tinged with violet towards the top, 10–18 mm; stigma purple (rarely yellow). Most frequently on *Trifolium* but also on many other plants. Widespread. 1278. *O. loricata* Reichenb. (*O. picridis*) Flowers white or pale yellow, tinged and veined with violet, 14–22 mm; stigma purple. On *Compositae* and *Umbelliferae*. S. and C. Portugal, S. and S–E. Spain, France. 1278a. *O. amethystea* Thuill. Flowers white or cream, usually tinged with violet, pink or brown towards the top, 15–25 mm; stigma purple or yellow. On various herbs. Widespread. 1278b. *O. latisquama* (F. W. Schultz) Batt. (*O. macrolepis*) Flowers whitish at base, upper half more or less tinged with purple; bracts bi-coloured, pale at base, dark brown above. Calyx teeth fused below; corolla constricted in the middle; stamens inserted 7–15 mm above the base of the corolla; stigma white or yellow. Stems 20–60 cm by 4–9 mm, lower part purple. On shrubs including *Rosmarinus* and *Cistus*. C. Portugal, S., S–E., and E. Spain.

LENTIBULARIACEAE | Butterwort Family

PINGUICULA | Butterwort

1. Fls. white, with 1 or more yellow spots on the throat-boss.
 (Pyrenees) **1279. *P. alpina* L.

1. Fls. violet, lilac, blue or pinkish.
  2. Lobes of lower lip of corolla notched; plant over-wintering
    as a rosette.   *lusitanica*
  2. Lobes of lower lip not notched; plants over-wintering as a
    winter-bud.
    3. Fls. including spur 2½–4 cm; spur 9–24 mm, more than half
      as long as the rest of the corolla.
      4. Lvs. oblong to oblong-egg-shaped; lobes of the lower lip
        of the corolla about as wide as long.   *grandiflora*
      4. Lvs. strap-shaped to linear-lance-shaped; lobes of the
        lower lip of the corolla much longer than wide.
        5. S–E. and E–C. Spain.   *vallisneriifolia*
        5. Pyrenees and Cévennes.   *longifolia*
    3. Fls. including spur 1–3 cm; spur 4–10 mm, usually less than
      half as long as the rest of the corolla.
      6. Lvs. not much longer than wide. (Spain: Sierra Nevada,
        S. de Alfacar)   *nevadensis*
      6. Lvs. distinctly longer than wide. (Widespread except
        south)   **1280. *P. vulgaris* L.

(1279) *P. lusitanica* L. PALE BUTTERWORT. An inconspicuous plant of sphagnum bogs and wet heaths with tiny, pale lilac to pinkish flowers 7–9 mm, with a yellow throat, and with a short blunt reflexed spur 2–4 mm. Rosettes about 2–4 cm across, with rather greyish oblong-egg-shaped leaves with strongly incurled margins. Flower stems very slender, glandular-hairy; calyx lobes egg-shaped. Scattered throughout except in Med. region. 1279a. *P. nevadensis* (Lindb.) Casper Flowers lilac with white lobes, those of upper lip egg-shaped, those of lower lip oblong-obovate overlapping, more or less equal. Corolla 12–16 mm, including the spur of 3–4 mm. Rosettes 3–4 cm across. Mountains. S. Spain.

**(1280) *P. grandiflora* Lam. LARGE-FLOWERED BUTTERWORT. A most handsome plant with large flowers 2½–3½ cm, including a long straight spur of 1 cm or more, and pale green, glistening rosettes of leaves 4–16 cm across. Corolla broader than long, violet to pinkish or pale blue, with a white throat, lower lip with shallow rounded, overlapping lobes. Wet rocks in mountains. N. Spain, Pyrenees. Page 138.   1280a. *P. longifolia* DC. Distinguished from (1280) by its linear-lance-shaped leaves which are 6–13 cm long, and the narrower lobes to the lower lip of the flower which do not overlap. Flowers very large, up to 4 cm, with spur 1–1½ cm. Rare on wet rocks. Pyrenees, Massif Central. Pl. 49.   1280b. *P. vallisneriifolia* Webb A rare endemic plant, looking very like 1280a, of shady rocks under the constant drip and spray of waterfalls. Flowers large 2½–3½ cm, pale lavender with a white centre, the basal area of the lower lip is usually palest yellow with fine lavender streaks; buds darker violet. Spring leaves elliptic, stalkless, the summer leaves very long 10–20 cm, and strap-shaped, undulate, stalked. Mountains of S–E. and E–C. Spain (S. de Cazorla, S. de Segura, S. de Cuenca) Pl. 49, Page 85.

UTRICULARIA | **1281. *U. vulgaris* L. GREATER BLADDERWORT. A submerged aquatic plant with deeply cut leaves, bearing small animal-catching bladders, and carrying above water a cluster of yellow, spurred flowers. Flowers 1½–2 cm, lower lip with reflexed margin and a large throat-boss; spur conical. Still waters. Spain, France.   (1282) *U. minor* L. LESSER BLADDERWORT. Flowers only 6–8 mm and spur very short, sac-like. Bogs, marshes, lake verges. N. Spain, France.

PLANTAGINACEAE | Plantain Family

PLANTAGO | Plantain   About 33 species occur in our area, about one third of which are widespread in Europe. The following Iberian species are distinctive:

1285a. *P. serraria* L. Distinguished by its rosette of lance-shaped, three- to five-veined leaves, which have widely spaced marginal teeth or tiny narrow lobes. Flowering spike dense, cylindrical, 3–13 cm by 4–6 mm, borne on a hairy stem longer than the leaves; bracts as long as or shorter than the hairy calyx. Sandy and waste places, cultivated ground. S. and C. Portugal, S. Spain.

1288a. *P. albicans* L. A cushion-like perennial with branched woody basal stems and terminal rosettes of silvery-white, narrow lance-shaped, often undulate leaves with 3 veins. Flowering spikes silvery-white, long, slender, often interrupted below. Bracts and sepals oval-blunt, ciliate. A variable plant. Sandy places, by the sea, uncultivated ground, waysides. S. Portugal, Spain, S. France.

1289a. *P. nivalis* Boiss. A small tufted perennial with rosettes of white, extremely woolly-haired, oblong-lance-shaped leaves pressed flat to the ground. Flowering spikes globular, about 1 cm, dark brown or blackish; calyx and corolla hairless; bracts hairy in the centre. Leaves with a fine tip, shallowly toothed; flowering stalk woolly-haired. Dry stony ground in high mountains. S. Spain (S. Nevada). Pl. 48.

CAPRIFOLIACEAE | Honeysuckle Family

SAMBUCUS | **1295. *S. ebulus* L. DANEWORT. Roadsides, waste places. Widespread. 1296. *S. nigra* L. ELDER. Hedges. Widespread.   **1297. *S. racemosa* L. ALPINE or RED ELDER. Montane woods. N. and E. Spain, France.

VIBURNUM | **1298. *V. opulus* L. GUELDER ROSE. Damp woods, thickets, hedges, mountains. C. and N. Spain, France.   **1299. *V. lantana* L. WAYFARING TREE. Open woods, scrub, hedges. C., N., and N–E. Spain, France.   **1300. *V. tinus* L. LAURUSTINUS. An evergreen shrub with dark green, oval leathery leaves, and dense flat-topped clusters of white flowers, pink in bud. Fruit metallic blue-black, poisonous. Bushy places, open woods, in the hills. Portugal, S. and C. Spain, S. France.

LONICERA | Honeysuckle

1. Woody climbers.
  2. Fls. in axillary pairs.   *biflora*
  2. Fls. in heads or whorls.
    3. Lvs. below inflorescence free, stalkless or short-
       stalked. (Widespread)   **1304. *L. periclymenum* L.
    3. At least the uppermost lvs. below inflorescence
       joined together at base (perfoliate).
      4. Fl. clusters with a long stalk. (Portugal, Spain,
         France).   **1305. *L. etrusca* Santi
      4. Fl. clusters stalkless.
        5. Lvs. deciduous, broadly elliptic. (Introd. Spain,
           France)   1306. *L. caprifolium* L.

5. Lvs. evergreen, oblong to egg-shaped.
 6. Corolla tube 3–4 times as long as the lobes; inflorescence with usually 2–6 fls. (Portugal, Med. region)   (1306) *L. implexa* Aiton
 6. Corolla tube 2–3 times as long as lobes; inflorescence with up to 30 fls.   *splendida*
1. Erect shrubs.
 7. Corolla regular or nearly so.
 8. Plant more or less hairy; bracteoles fused at the base..(Pyrenees)   **(1303) *L. caerulea* L.
 8. Plant hairless; bracteoles free.   *pyrenaica*
 7. Corolla two-lipped.
 9. Bracts shorter than ovary; fr. bluish-black. (Pyrenees, Massif Central)   1303. *L. nigra* L.
 9. Bracts at least as long as ovary; fr. red or yellowish.
 10. Twigs and lvs. hairless or nearly so. (Pyrenees, Massif Central)   (1303) *L. alpigena* L.
 10. Twigs and lvs. finely hairy at least on lower surface.
 11. Stalk of fl. clusters very short or absent.   *arborea*
 11. Stalk of fl. clusters 1–2 cm. (Spain, France)   1302. *L. xylosteum* L. Page 139

1304a. *L. biflora* Desf. (*L. canescens* Schousboe) A deciduous climber with pairs of axillary yellowish flowers crowded at the ends of the twigs. Corolla 3–4 cm; tube much longer than the lobes. Leaves egg-shaped, dark green, hairless above, greyish-green and finely hairy beneath. Fruit black. Hedges, salt-rich clayey soils by the sea. S–E. Spain (Alicante to Adra).   1306a. *L. splendida* Boiss. An evergreen climber with stalkless clusters of whitish-yellow flowers like (1306) but clusters with more numerous flowers and corolla more glandular; style hairless. Endemic in mountains of S. Spain.

1303b. *L. pyrenaica* L. A hairless deciduous shrub to 1 m with bluish-green somewhat leathery, obovate to oblance-shaped leaves, and paired sweet-scented, white or pale pink tubular flowers. Corolla about 2 cm; stamens included; bracts lance-shaped, leafy, longer than ovary; bracteoles much shorter. Berries red, scarcely fused. Mountains. N–E. Spain, Balearic Islands, Pyrenees. Pl. 49.   1303a. *L. arborea* Boiss. A shrub or small tree to 9 m, with egg-shaped leaves hairless above, grey-hairy beneath, and paired pink flowers 1–2 cm long. Corolla tube half as long as lobes. Young twigs reddish-violet, becoming white. Fruit yellowish. Mountain valleys (1800–2200 m). S. Spain (S. Nevada).

ADOXACEAE | Moschatel Family

ADOXA | **1308. *A. moschatellina* L. MOSCHATEL. Woods, damp shady places in mountains. C. and E. Spain, Pyrenees, France.

VALERIANACEAE | Valerian Family

VALERIANELLA | **Lamb's Lettuce** About 16 small-flowered, pale pink, lilac or bluish-flowered annuals occur in our area.

FEDIA | About 4 species.   **1312. *F. cornucopiae* (L.) Gaertner A hairless annual with dichotomously branched stems and terminal clusters of numerous pink, long-tubed flowers. Corolla tube with a sac-like swèlling at base, two-lipped; stamens 2. Fruiting branches conspicuously swollen; leaves oval, shallowly toothed. Cultivated ground, cornfields. S. and C. Portugal, Med. region. Page 162.   1312a. *F. caput-bovis* Pomel Like 1312 but fruit with a small persistent crown with 2 triangular teeth. A more robust plant with a dense inflorescence. Portugal, Spain.

VALERIANA | Valerian

1. Stems distinctly grooved, at least $\frac{1}{2}$ m tall; fls. in broad flat-topped clusters.
  2. Lvs. all deeply cut into numerous segments.
    3. Lvs. with 7–11 oval or oblong, strongly toothed segments. (Pyrenees, France) — 1313a. *V. excelsa* Poiret
    3. Lvs. with 13–21 lance-shaped, entire or shallowly toothed segments. (N. Portugal, Spain, France) — **1313.   *V. officinalis* L.
  2. Basal lvs. broadly oval or heart-shaped, entire.
    4. Basal lvs. 1–3 cm broad.
      (W. Pyrenees) — 1313b. *V. hispidula* Boiss.
      (E. Spain, France) — *phu*
    4. Basal lvs. 8–20 cm broad. — *pyrenaica*
1. Stems smooth or finely lined, less than $\frac{1}{2}$ m tall; fls. in compact heads, or lax clusters.
  5. Stock not woody, creeping, giving rise to a single fertile stem.
    6. Stock slender, almost scentless; plant of moist places. (N. Portugal, C., N., and E. Spain, France) — 1314.   *V. dioica* L.
    6. Stock thick, tuberous, strong-scented; plant of dry places. (Portugal, Spain, Pyrenees) — (1313)   *V. tuberosa* L.
  5. Stock woody, branching, giving rise to several fertile stems.
    7. Fls. in flat-topped, somewhat lax inflorescences; plants 20–50 cm.
      8. Lvs. ashy-green or somewhat glaucous; stem lvs. three-lobed. (N. and E. Spain, France) — **1315.   *V. tripteris* L.
      8. Lvs. clear shining green; stem lvs. entire or with a few teeth. (N. Portugal, Spain, Pyrenees) — **(1315) *V. montana* L.
    7. Fls. in crowded heads or in whorls; plants 5–20 cm.
      9. Upper lvs. cut into 5–7 narrow linear segments. — *globularifolia*
      9. Upper lvs. entire.
        10. Lvs. lance-shaped. (Pyrenees) — 1315c. *V. saliunca* All.
        10. Lvs. usually egg-shaped to heart-shaped. (Spain, (N. Aragon)). — 1315b. *V. longiflora* Willk.

1313c. *V. pyrenaica* L. A tall, rather robust, broad-leaved perennial to over 1 m, with large, dense, flat-topped clusters about 7–10 cm across, of pale pink flowers. Stem strongly grooved, hairy at the nodes; leaves all large, broadly heart-shaped, strongly toothed, 8–20 cm broad. Woods, shady places. N. Spain, Pyrenees, Corbières.   1313d. *V. phu* L. Lower leaves oblong-egg-shaped 15–30 cm, upper leaves deeply cut into 3–7 oblong segments; stem smooth. Flowers white or pale pink. Perennial $\frac{1}{2}$–2 m. Woods, damp meadows. Native of the Caucasus; cultivated as a medicinal plant and sometimes naturalized. N. and E. Spain,

France.   1315a. *V. globularifolia* DC. A low hairless tufted plant with a branched, woody stock and with rosettes of oblong spathulate leaves and pale pink flowers in a small compact cluster. Flowering stems 10–20 cm; upper leaves deeply cut; bracts linear-lance-shaped, papery. Alpine rocks. Pyrenees.

CENTRANTHUS (KENTRANTHUS) | 5 species.

**1316. *C. ruber* (L.) DC. RED VALERIAN. Rocks, walls, in mountains. Portugal, Spain, France.   **(1316) *C. angustifolius* (Miller) DC. NARROW-LEAVED RED VALERIAN. Distinguished from 1316 by its glaucous linear to linear-lance-shaped leaves only 2–14 mm broad and its dense, rounded, flat-topped cluster of pinkish-red flowers. Spur of corolla as long as ovary. Calcareous rocks and pastures in mountains. N., E., and S–E. Spain, S–W. France. Pl. 49.   1316a. *C. nevadensis* Boiss. Like 1316 but a more slender plant, with a shorter lance-shaped, blunt-tipped leaves and flowers with a spur 2–3 times as long as the ovary. Fruit with minute swellings on the back and five-ribbed beneath. Alpine rock fissures. S. Spain (S. Nevada). Page 337.   1317. *C. calcitrapae* (L.) Dufresne Cultivated ground, sandy places, etc. Scattered throughout.

DIPSACACEAE | Scabious Family

DIPSACUS | Teasel

1. Stem lvs. shortly stalked; involucral bracts and scales of receptacle similar. (C. and E. Spain, France) — 1319. *D. pilosus* L.
1. Stem lvs. with bases fused round stem; involucral bracts and scales of receptacle differing.
  2. Bracts of receptacle equalling fls., with recurved, rigid, spiny tip. (Introd. throughout) — 1318b. *D. sativus* (L.) Honckeny
  2. At least the lower bracts of receptacle longer than the fls., tip straight, flexible.
    3. Fl. heads more or less globular; involucral bracts spreading or recurved. — *ferox*
    3. Fls. heads more or less cylindrical; involucral bracts curving upwards.
      4. Stem lvs. usually toothed. (Widespread) — 1318. *D. fullonum* L.
      4. Stem lvs. deeply cut into lobes. (C. and E. Spain, France) — **(1318) *D. laciniatus* L.

1318a. *D. ferox* Loisel. A very spiny biennial with dense spines on stems, bracts, both surfaces of leaves and leaf margin, distinguished by its more or less globular, pale blue or white flower heads. Involucral bracts spreading, scales of receptacle abruptly long-pointed, the uppermost often much elongated and forming a crown to the flower head. Middle stem leaves deeply pinnately cut. Fields, dry hills, waysides. S. and C. Portugal, C. Spain.

CEPHALARIA | 1320a. *C. syriaca* (L.) Roemer & Schultes A rough-hairy annual with ovoid heads of pale blue flowers in the axils and at the ends of branches. Involucral bracts egg-shaped, suddenly narrowed to a long spiny tip. Leaves lance-shaped, the lower toothed, the upper entire. Involucel with 4 long and 4 short teeth. Cultivated ground. S., C., and E. Spain, S. France.   (1320) *C. leucantha* (L.) Roemer & Schultes A tall slender-branched perennial with conspicuous, long-stalked, globular white-flowered heads 2–3 cm across.

Involucral bracts numerous, papery, oval-blunt. Stem leaves deeply pinnately cut into lance-shaped, entire or toothed segments. Fields, vineyards, dry hills. S. Portugal, Spain, Massif Central. Page 337.

SUCCISA | 1321. *S. pratensis* Moench DEVIL'S-BIT SCABIOUS. Damp places, grassy places. C. and N–W. Portugal, C. and N. Spain, France. 1321a. *S. pinnatifida* Lange Like 1321 but flower heads small 1–1½ cm across, few-flowered, with flowers white at the base and intense violet above. A woolly-haired, slender, ascending perennial 20–80 cm, dichotomously branched above; the uppermost leaves entire, the middle stem leaves deeply pinnately cut, the basal leaves toothed. Rocks, thickets, heaths. C. and N–W. Portugal, N–W. Spain.

SUCCISELLA | Like *Succisa* but calyx four-lobed, without persistent teeth (calyx with 4 or 5 persistent teeth in *Succisa*). 2 species. 1321b. *S. microcephala* (Willk.) G. Beck Flower heads globular 1–1½ cm, yellow or pink, with broadly oval, acute involucral bracts. An erect, slender, simple or little-branched perennial, with basal leaves densely hairy, lance-shaped, narrowed to a winged leaf stalk, the upper leaves few, narrow linear, fused together at the base, the uppermost bract-like. Dry sandy pastures. W–C. Spain.

KNAUTIA | Distinguished from *Scabiosa* by the absence of scales on the receptacle; calyx usually with 8 or more teeth or bristles. A difficult genus with up to about 19 similar-looking species, many of which are very local in our area.   **1322. *K. arvensis* (L.) Coulter FIELD SCABIOUS. Dry fields, pastures, bushy places. Widespread.   1323. *K. sylvatica* (L.) Duby WOOD SCABIOUS. Shady places and woods in mountains. N–E. Portugal, S., C., and E. Spain, France.

PTEROCEPHALUS | Calyx with 6 to many bristles which have plume-like hairs; receptacle usually with scales.

1. Stems creeping, mat-forming.   *spathulatus*
1. Stems erect.
   2. Anns.; involucel with a long flattened curved spine up
      to 1 cm; calyx with 20–24 hairy bristles. (Portugal, C.
      Spain)   1324b. *P. diandrus* (Lag.) Lag.
   2. Perenns.; involucel without a long spine; calyx with
      6–7 hairy bristles.   *intermedius*

1324a. *P. spathulatus* (Lag.) Coulter A dense mat-forming, woolly-white shrubby perennial with spathulate leaves and very short-stalked flower heads of pink spreading flowers. Involucral bracts 8–10, linear lance-shaped, half as long as the flowers, grey-haired; calyx with about 16 hairy bristles. Alpine rocks. S–E. Spain. Page 84.   1324c. *P. intermedius* (Lag.) Coutinho A slender, erect, scabious-like perennial with mauve flower heads about 2½ cm across. Distinguished by the 6–7 long, hairy bristles of the calyx which are about as long as the silvery-haired corolla. Involucre of 8–12, lance-shaped, silvery-haired bracts, nearly as long as the flowers. Leaves twice cut into narrow linear segments, the uppermost entire. Sandy places, pine woods, by the sea. S. and C. Portugal, S. and S–E. Spain. Page 337.

SCABIOSA | **Scabious**   16 species occur in our area. The following are widely scattered over much of our area, or in the south.   (1325) *S. triandra* L. (*S. gramuntia*);   **1326. *S. atropurpurea* L. (*S. maritima*) MOURNFUL WIDOW, SWEET SCABIOUS;   1325. *S. columbaria* L. SMALL SCABIOUS;   (1328) *S. stellata* L.

1325a. *S. tomentosa* Cav. Like 1325 but densely tufted with rosettes of densely white-felted, entire basal leaves. Lower stem and branch leaves once or twice cut into oval, entire or toothed lobes, the terminal blunt segment slightly longer, the uppermost leaves with linear segments. Flower heads reddish-purple, small. Sunny rocks. S., C., and E. Spain. 1325b. *S. cinerea* Lapeyr. (*S. pyrenaica* auct.) Like 1325a with white-felted basal leaves but hairs star-shaped. Non-flowering rosette leaves lance-shaped, toothed, upper stem leaves once or twice deeply cut, the terminal segment much longer than the lateral lance-shaped segments. Flower heads bluish-purple, large. Sunny alpine meadows. Pyrenees.

**1328. *S. graminifolia* L. GRASS-LEAVED SCABIOUS. Distinguished from other species by its grass-like, silky-haired leaves and its solitary, globular, long-stalked, blue-lilac flower heads 3–4½ cm across. A tufted, woody-based perennial; leaves 1–3 mm wide. Limestone rocks, stony places. E. Spain, Med. France.   1328a. *S. pulsatilloides* Boiss. A woody-based, tufted densely silvery-white-haired perennial, with short white-hairy stems bearing 1 or 3 purple or bluish flower heads usually 2½–3½ cm across. Involucral bracts narrowly egg-shaped, the outer often pinnately lobed, half to one third as long as flowers. Calyx bristles 2–3 times as long as the cup-shaped crown of the involucel. Basal leaves narrow-elliptic, once or twice cut into narrow obovate, entire or slightly toothed segments. Mountains and alpine rocks. S. and N–E. Spain.   1328b. *S. crenata* Cyr. Differs from 1328a in having involucral bracts always entire, densely white-felted. Leaves densely hairy or hairless. Flower heads pale reddish-lilac. Med. Spain.   1328c. *S. monspeliensis* Jacq. Like (1328) *S. stellata* but involucral bracts three-lobed or pinnately lobed (not entire) and flower heads half as large, dull pink. Vineyards, uncultivated ground. S. and C. Portugal, S., C., and E. Spain, Med. France.

PYCNOCOMON | Like *Scabiosa* but involucral bracts fused to the middle into a cup, closely surrounding the flowers.   1328d. *P. rutifolium* (Vahl.) Hoffmanns. & Link An erect, wide-branched perennial 30–80 cm, with small globular creamy or pink flower heads with the outer florets somewhat spreading. Involucre of 6–8 unequal lobes; inner florets with 5 long brown bristles, the outer without bristles. Leaves somewhat fleshy, the basal leaves entire, or toothed, the stem leaves deeply cut into linear-blunt lobes. Coastal sands. S. Portugal, S–W. Spain.

CAMPANULACEAE | Bellflower Family

CAMPANULA | **Bellflower**   About 30 species occur in our area. Frequent and widely distributed species, particularly in N., C., and E. Spain, and France are: **1333. *C. glomerata* L. CLUSTERED BELLFLOWER. Page 152;   (1333) *C. cervicaria* L;   1334. *C. rapunculus* L. RAMPION;   **1335. *C. patula* L. SPREADING BELLFLOWER: 1339. *C. rotundifolia* L. HAREBELL;   **1342. *C. rapunculoides* L. CREEPING BELLFLOWER. Plants of C. and E. Spain, Pyrenees and France are: **1340. *C. trachelium* L. BATS-IN-THE-BELFRY;   1341. *C. latifolia* L. LARGE BELLFLOWER;   **1336. *C. persicifolia* L. NARROW-LEAVED BELLFLOWER. Page 152. Widespread in S. and C. of our area;   1337. *C. erinus* L. ANNUAL BELLFLOWER.

Calyx with 5 large and 5 smaller lobes
1331a. *C. speciosa* Pourret PYRENEAN BELLFLOWER. A very handsome biennial with flat rosettes of grey bristly leaves and erect stems, bearing a pyramidal inflorescence of large blue-violet bells 3–5 cm long, on long, erect stalks. Leaves oblong-lance-shaped, unequally toothed. Rocks in mountains. E. Spain, Pyrenees, Massif Central.   1331b. *C. mollis* L. A densely velvety-haired greyish perennial with slender, fragile, spreading branches, small

oval or elliptic leaves, and a lax terminal cluster of blue, narrow bell-shaped flowers 1½–2 cm long. Calyx lobes linear-lance-shaped, half as long as the corolla, the smaller lobes oval, short. Rock fissures in limestone hills and mountains. S. and S–E. Spain. Pl. 50.   1331c. *C. dichotoma* L. An annual with an erect, wide-branched, pyramidal inflorescence with large blue flowers about 3 cm long. Corolla with oval-acute somewhat spreading lobes; calyx lobes triangular-pointed, with smaller alternating lobes. Leaves oblong-oval, shallowly toothed, the upper half-clasping the stem, all softly shaggy-haired. Sandy ground, dry hills in lowlands and mountains. S. and S–E. Spain.

Calyx with only 5 lobes
1339a. *C. hispanica* Willk. A slender erect stoloniferous perennial closely related to *C. rotundifolia* L. with very narrow, almost thread-like upper leaves, and a lax branched cluster of rather small, deep blue flowers 9–15 mm long. Flower stalks rather long, with minute bracts, at first erect, then arched and finally nodding; corolla 3 times as long as the linear-blunt calyx lobes which are closely pressed against the corolla. Basal leaves rounded, kidney-shaped, toothed. Limestone hills, dry sands, rocks, stony ground. C., N. and E. Spain, Pyrenees. Page 139.   1335a. *C. lusitanica* L. (*C. loeflingii*) A small, erect, very variable annual with few or many dichotomous branches each bearing a terminal tubular-bell-shaped, deeply five-lobed, blue flower which is white at the base. Corolla about 1½ cm; flower stalks long, bractless, at first arched; calyx teeth linear, minutely toothed at base. Leaves variable, toothed, stalkless. Sandy, stony and grassy places, woods and waysides from lowlands to mountains. Portugal, Spain. Pl. 49, Page 45.   1339b. *C. herminii* Hoffmanns. & Link A rather undistinguished species recalling *C. rotundifolia*, but fruit splitting by pores towards the apex of the ovary. A lax, cushion-forming perennial with erect, usually unbranched stems bearing 1–4 rather large, broadly funnel-shaped, blue-violet flowers about 2½ cm across. Calyx lobes linear, spreading at flowering. Lower leaves rounded or long-elliptic, stalked, shallowly toothed, the upper lance-shaped or linear, stalkless, all hairless. Damp pastures and stony places in mountains. C. Portugal, S. and C. Spain.

LEGOUSIA | Venus' Looking-Glass

1. Fls. in a long lax spike at least half as long as the stem.

2. Calyx lobes almost as long as the ovary at flowering; plant not rough to the touch. (S. and E. Spain, S. France)   (1346) *L. falcata* (Ten.) Fritsch

2. Calyx lobes ⅓–½ as long as the ovary at flowering; plant rough to the touch. (Iberian peninsula, S. France)   1346a. *L. castellana* (Lange) Samp.

1. Fls. in small branched clusters or small flat-topped clusters.

3. Corolla half as long as the calyx lobes, which are erect in fr. (Portugal, Med. region)   1346. *L. hybrida* (L.) Delarbre

3. Corolla at least as long as the calyx lobes which are spreading or recurved in fr.

4. Calyx lobes almost as long as the ovary at flowering; fr. 1–1½ cm, narrowed at the apex. (C. and E. Spain, France)   **1347. *L. speculum-veneris* (L.) Chaix. Page 152

4. Calyx lobes ⅓–½ as long as the ovary at flowering; fr. 2–3 cm, not narrowed at the apex. (E. Spain)   (1347) *L. pentagonia* (L.) Druce

TRACHELIUM | 1348. *T. caeruleum* L. THROATWORT. A distinctive perennial with a broad flat-topped cluster of numerous, tiny, long-tubed blue or lilac flowers, with long-projecting styles. Corolla tube very slender, 6–8 mm long, with 5 tiny spreading lobes. Leaves egg-shaped, saw-toothed. Walls, shady rocks. S. and C. Portugal, S., W., and C. Spain. Pl. 49.

PHYTEUMA | Rampion.

Flowers in elongated heads, twice as long as broad or more
**1350. *P. spicatum* L. SPIKED RAMPION. Mountains. N., C., and N–E. Spain. Pyrenees. Page 153. **(1350) *P. betonicifolium* Vill. BLUE-SPIKED RAMPION. Pyrenees; 1350a. *P. scorzonerifolium* Vill. Pyrenees; 1351. *P. ovatum* Honckeny DARK RAMPION. Pyrenees, Cévennes. 1350b. *P. pyrenaicum* R. Schulz A rather robust, erect perennial 20–60 cm, distinguished by its leaves; rosette leaves broad-stalked, long-triangular, toothed, lower stem leaves similar but smaller, the uppermost linear, stalkless. Flowers bright blue or slaty blue, in a cylindrical or rounded head 2–7 cm long, 2–3 cm wide; bracts linear to narrowly triangular. Mountain meadows, woods, bushy slopes. Pyrenees.

Flowers in globular heads as broad as or broader than long
1351. *P. orbiculare* L. ROUND-HEADED RAMPION. Mountains. N. and E. Spain, Pyrenees, France; **1353. *P. hemisphaericum* L. Mountains. E. Pyrenees. 1352a. *P. charmelii* Vill. PYRENEAN RAMPION. A slender plant 10–20 cm, with rather large, rounded heads 1–1½ cm across of blue flowers, and with long narrow toothed bracts spreading beyond the flower heads. Basal leaves of 2 forms: heart-shaped to rounded, or oval-oblong acute, all stalked and toothed, uppermost leaves linear. Calcareous rocks. Pyrenees, S–C. France. 1353a. *P. globulariifolium* Sternb. & Hoppe A dwarf, tufted plant with tongue-like leaves 1½ cm, broader towards the apex and usually 'hooded' due to inrolled apex and margin. Flower heads small, 6–10 mm, blue-violet, the bracts rounded-triangular, longer than the head; flower stems 1–5 cm. Rocks, stony pastures. Pyrenees.

WAHLENBERGIA | Like *Campanula* but fruit splitting by 2–5 apical valves, alternating with the persistent calyx lobes. 2 species. 1353b. *W. hederacea* (L.) Reichenb. IVY CAM-PANULA. A tiny slender, pale green, creeping perennial with rounded to heart-shaped shallowy lobed, stalked leaves and tiny long-stalked nodding, pale blue, bell-shaped flowers. Corolla 6–10 mm long, lobes half as long as tube. Fruit top-shaped. Damp peaty places in mountains. C. and N. Portugal, Spain, Pyrenees, Massif Central.

JASIONE | Sheep's Bit.

1. Involucral bracts few; individual floret stalks longer than the calyx. (S. Spain).   1355a. *J. foliosa* Cav.
1. Involucral bracts usually numerous; individual floret stalks shorter than the calyx.
   2. Calyx lobes hairy.
     3. Calyx lobes spathulate, woolly-haired only at apex. (S. Spain)   1355b. *J. penicillata* Boiss.
     3. Calyx lobes not spathulate, woolly-haired at least at the base. (N. Portugal, Spain, S–W. and C. France)   1355c. *J. crispa* (Pourret) Samp. (*J. humilis*)
   2. Calyx lobes hairless.
     4. Outer involucral bracts deeply toothed.

5. Plants with non-flowering shoots; lvs. not undulate. (S. Spain, S. France)	1355d. *J. laevis* Lam.
5. Plants without non-flowering shoots; lvs. undulate. (Widespread)	**1355. *J. montana* L.
4. Outer involucral bracts entire or shallowly toothed.	
6. Perenns.; involucral bracts and calyx lobes bluish-purple. (Mountains Spain)	1355e. *J. amethystina* Lag. & Rodr. Page 337
6. Usually anns.; involucral bracts and calyx lobes green.	
7. Stems usually leafless on upper half; involucral bracts shorter than fls. (Widespread)	**1355. *J. montana* L.
7. Stems leafy almost to the fl. head; involucral bracts at least as long as the fls. (Portugal, S. Spain)	1355f. *J. corymbosa* Poiret

LOBELIA | 1356. *L. urens* L. ACRID LOBELIA. Rough pastures, heaths. West Iberian peninsula, France.   (1356) *L. dortmanna* L. WATER LOBELIA. Acid lakes. W. France.

LAURENTIA | Like *Lobelia* but corolla tubular, not two-lipped.   1356a. *L. gasparrinii* (Tineo) Strobl (*L. michelii*). A slender, pale green annual with tiny lilac flowers with white centres, borne at the ends of long, very slender stems. Corolla 3–6 mm, tubular, five-lobed. Leaves oboval or oblong, short-stalked, rounded-toothed, or nearly entire; stems 3–10 cm. Marshes, damp woods, springs. Portugal, W. and N. Spain.

COMPOSITAE | Daisy Family

A very important family with over 100 genera of native plants in our area, and more than 750 species. Many of these species are widespread in Europe. Only the most distinctive species, including in particular those of the mountains, the Mediterranean region and the dry semi-desert regions of our area, are briefly described or keyed.

Sub-family **CARDUOIDEAE** | Flower heads with disk-florets, with or without ray-florets. Genera not included: *Conyza, Micropus, Filago, Gnaphalium, Buphthalmum, Tripleurosperum, Matricaria, Arctium, Saussurea.*

EUPATORIUM | **1357. *E. cannabinum* L. HEMP AGRIMONY. Damp places, marshes. Widespread.

SOLIDAGO | About 5 species.   **1358. *S. virgaurea* L. GOLDEN-ROD. Rocks, stony places. Widespread.

BELLIS | About 5 species.   1360. *B. perennis* L. DAISY. Meadows, damp grassy places. Widespread.   (1360) *B. sylvestris* Cyr. SOUTHERN DAISY. Grassy, damp, shady places, in lowlands and hills, and the mountains in the south. Widespread. Pl. 50.

1360a. *B. rotundifolia* (Desf.) Boiss. & Reuter   Distinguished by its large flower heads 2–4 cm across, borne on stout stems up to 25 cm. Leaves with rounded to oblong blades, deeply rounded-toothed, more or less heart-shaped at base and long-stalked. Shady rock fissures in mountains. S. Spain.

**1361. *B. annua* L. ANNUAL DAISY. Sandy or grassy places, waysides. South and centre of our area.

BELLIUM | Distinguished from *Bellis* by its fruits which have alternate rows of papery scales and bristles.   1361a. *B. bellidioides* L.   Very like 1361 but flower heads pink, smaller, 8–12 mm across, and leaves all basal, not toothed. Stony places, usually by the sea. E. Spain, Balearic Islands. Pl. 50.

ASTER | About 12 species.   **1363. *A. alpinus* L. ALPINE ASTER. Rocks and dry pastures in mountains. E. Spain, Pyrenees, Massif Central.   **1365. *A. tripolium* L. SEA ASTER. Salt marshes, cliffs by the sea. Coasts of our area except in the south.

CRINITARIA | **1368. *C. linosyris* (L.) Less. (*Aster l.*) GOLDILOCKS. Shady rocks in mountains. S., C. and E. Spain, France.

ERIGERON | About 10 species.   1370. *E. alpinus* L. ALPINE FLEABANE. A small dull green, downy alpine perennial with narrow leaves and usually solitary violet or pinkish flower heads. Ray-florets numerous, in several rows. Mountains, Pyrenees.

1370a. *E. frigidus* DC.   Like 1370 but plant greyish-green, with crisped hairs, and lower leaves, densely tufted. Flower heads solitary about 1½ cm across; ray-florets in 2–3 rows, violet, about twice as long as the yellow disk-florets; involucre densely woolly-haired. Stem leaves few, linear. Pappus twice as long as fruit. By melting snows. S. Spain, Pyrenees. Pl. 50.   (1370) *E. uniflorus* L.   Very like 1370a but leaves pale green, hairless or finely hairy, often ciliate. Ray-florets in 1 row, purple or white, more than twice as long as the yellow disk-florets. Pappus as long as the fruit. Rocks, stony places in mountains. E. Spain, Pyrenees. Page 149.

1372. *E. karvinskianus* DC. (*E. mucronatus*)   A branched and spreading, often woody-based perennial with whitish daisy-like flower heads about 1½ cm across. Ray-florets spreading, at first pale purple, then white and finally pink. Lower leaves coarsely three-lobed, or toothed, upper nearly entire. Sometimes naturalized on walls and rocks. Page 102.

EVAX | About 6 species.   **1374. *E. pygmaea* (L.) Brot. A tiny annual with a rosette of narrow woolly-haired leaves and a stalkless cluster of tiny yellowish flower heads nestling in the rosette. Dry places, near the sea. Portugal, W., C., and N–E. Spain, Med. France.

ANTENNARIA | **1378. *A. dioica* (L.) Gaertner CAT'S FOOT. Dry sunny stony places in high mountains. Spain, France.   (1378) *A. carpatica* (Wahlenb.) Bluff & Fingerh. CARPATHIAN CAT'S-FOOT. Stony pastures in snowy regions. Pyrenees.

LEONTOPODIUM | **1379. *L. alpinum* Cass. EDELWEISS. Alpine meadows, screes. Pyrenees.

HELICHRYSUM | About 8 species.   **1385. *H. stoechas* (L.) DC.   An 'everlasting' shrubby perennial with dense clusters of lemon-yellow flower heads and usually white-felted stems and leaves, but very variable. Flower heads globular 4–6 mm; involucral bracts papery bright shining yellow, usually not glandular. Dry banks, rocks and sands. Portugal, Spain, France.   1385a. *H. serotinum* Boiss. Very like 1385 but flower heads longer than broad, ovoid-oblong, 3–4 mm broad, bright yellow; involucral bracts glandular. Portugal, Med. region.

1385c. *H. foetidum* (L.) Moench. Easily recognized by its relatively large yellow flower heads 1½–2 cm across with spreading, glistening, papery, white or yellow involucral bracts. A robust, strong-smelling perennial to 1 m or more, with oblong clasping stem leaves which are cottony beneath. Native of Cape of Good Hope; naturalized on rocks and cliffs in Portugal and Spain.

1385d. *H. lamarckii* Camb. A robust and handsome cliff-dwelling perennial, endemic to the Balearic Islands. Leaves lance-shaped, snow-white, covered in a dense close felt of hairs, forming lax cushions. Stems robust, to 30 cm, bearing a dense terminal cluster about 4 cm across of bright yellow flower heads, each about 1 cm across. Involucral bracts yellow, shining. Page 358.

PHAGNALON |

1. Fl. heads of 6–7 mm, in clusters of 2–6; lvs. woolly-
    haired on both sides. (Spain, Med. France)   (1386) *P. sordidum* (L.) Reichenb.
1. Fl. heads 8–12 mm, solitary; lvs. green above.
    2. Outer involucral bracts acute, curved outwards,
        margin undulate; stalk of fl. heads with small
        bracts above. (Widespread)   (1386) *P. saxatile* (L.) Cass.
    2. Outer involucral bracts blunt, pressed against
        fl. head; stalk of fl. head without bracts above.
        (Portugal, Med. region)   **1386. *P. rupestre* (L.) DC.

INULA | About 15 species.   **1387. *I. conyza* DC. PLOUGHMAN'S SPIKENARD. Widespread. 1388a. *I. helenioides* DC. A greyish-white, very hairy, erect perennial to ½ m, with clasping upper leaves and a flat-topped cluster of yellow flower heads each about 3–4 cm across. Ray-florets glandular-hairy; involucral bracts linear lance-shaped, acute, densely woolly. Leaves soft, the upper lance-shaped entire with heart-shaped clasping lobes, the lower narrowed to a long stalk. Fruit velvety. Bushy places, arid banks, rocks. C. and E. Spain, Pyrenees, S. France.   1391. *I. montana* L. Flower head solitary, bright yellow, 4–5 cm across, and stems and leaves densely white-woolly. Involucral bracts unequal, woolly-haired. Upper stem leaves stalkless. Rocky slopes, C. and N. Portugal, S. C., and E. Spain, Pyrenees.

**1390. *I. crithmoides* L. GOLDEN SAMPHIRE. An erect, fleshy-leaved perennial of salt marshes and salt-rich areas with a lax flat-topped cluster of golden-yellow flower heads each about 2½ cm across. Leaves linear or oblong, fleshy, with three-toothed or entire apex. Coasts of our area and salt-rich areas inland.

1392. *I. viscosa* (L.) Aiton AROMATIC INULA. A very strongly resinous-smelling, shrubby perennial with terminal leafy clusters of yellow flower heads, each about 1½ cm across. Leaves lance-shaped, sticky, glandular-hairy, entire or toothed, half clasping the stem. Rocky places, pine woods, olive groves. S. and C. Portugal, S. and E. Spain, S. France.

PULICARIA | Fleabane   5 species.   **1393. *P. dysenterica* (L.) Bernh. FLEABANE. Widespread. (1393) *P. odora* (L.) Reichenb. Distinguished from 1393 by the upper stem leaves which are half-clasping (not with clasping heart-shaped lobes). Flower heads few, yellow,

2–4 cm across, on thickened stems. A woolly-haired perennial. Sandy ground, pine woods, bushy places. Throughout. Pl. 50.

CARPESIUM | 1394. *C. cernuum* L. FALSE BUR-MARIGOLD. Woods, shady places. E. Spain, Pyrenees.

PALLENIS | **1395. *P. spinosa* (L.) Cass. Dry uncultivated ground, tracksides. Widespread.

ASTERISCUS | **1398. *A. maritimus* (L.) Less. A tufted, woody-based perennial with many oblong or spathulate leaves, and short erect stems bearing solitary, deep yellow flower heads 3–4 cm across. Ray-florets numerous, spreading, toothed at apex; involucral bracts about as long spine-tipped. Rocks and stony places by the sea. S. Portugal, Med. Spain. Pl. 51. Page 62. **1399. *A. aquaticus* (L.) Less. Recalling 1395 but outer spreading leafy involucral bracts blunt. An annual (unlike 1398), with pale yellow flower heads 1–2 cm across and involucral bracts longer than ray-florets. Damp places, rocks by the sea, waysides. Scattered throughout.

XANTHIUM | **1401. *X. strumarium* L. COCKLEBUR. A non-spiny annual; fruits 12–15 mm. Naturalized in waste places and damp ground. Scattered throughout.

1401a. *X. orientale* L. Distinguished from 1401 by its large glandular hairy fruits to 4 cm, covered spines, and with 2 stout, hooked, terminal spines. Scattered throughout, but less common.

**1402. *X. spinosum* L. SPINY COCKLEBUR. Spiny annual; leaves dark green above, white-felted beneath. Naturalized by waysides, waste places, disturbed ground. Widespread.

BIDENS | About 4 species. **1406. *B. cernua* L. NODDING BUR-MARIGOLD. Leaves undivided; flower heads drooping. Damp places. C., N., and E. Spain, France. (1406) *B. tripartita* L. TRIPARTITE BUR-MARIGOLD. Leaves at least three-lobed; flower heads mostly erect. Marshes, damp places. Widespread.

1406a. *B. aurea* (Aiton) Sherff A slender erect, nearly hairless perennial $\frac{1}{2}$ m or more, with erect yellow flower heads with involucral bracts all similar, never leaf-like. Ray-florets 5–6, 1–3 cm long, yellow with purplish lines. Leaves narrow lance-shaped, toothed or deeply cut into linear lobes. Native of C. America; naturalized in S. Portugal, Med. region. Page 62.

SANTOLINA |

1. Plant sticky-glandular, stems leafy almost to the flower head. (W–C. Spain)	1408a. *S. viscosa* Lag.
1. Plant not sticky-glandular; stems leafless for some distance below flower head.	
2. Plant silky-haired.	*obtusifolia*
2. Plant with thick curled hairs, not silky-haired.	
3. Most lvs. flat; stock slender. (Spain (S. Nevada))	1408b. *S. elegans* DC.

3. Most lvs. more or less cylindrical in
section; stock stout. (Spain, S.
France)   1408. *S. chamaecyparissus* L., *rosmarinifolia*

1408c. *S. obtusifolia* Boiss. Flower heads yellow, rather broad 15–18 mm across. Leaves of
non-flowering shoots oblong-spathulate with comb-like lobes. W–C. Spain. 1408d. *S.
rosmarinifolia* L. (incl. *S. pectinata* and *S. canescens*) Very like 1408 and difficult to distinguish from it but flower heads bright yellow and all bracts of involucre with a papery
apex which is cut into narrow segments. Leaves grey or green with remote teeth or lobes.
Sandy, stony places, dry rocks from lowlands to mountains. ?Widespread. Page 358.

ANTHEMIS | About 14 species. **1409. *A. tinctoria* L. YELLOW CHAMOMILE. Flower
heads solitary, yellow, 2½–4 cm across. Leaves deeply twice cut into narrow toothed
segments. Sunny slopes, rocks, in mountains. S., C., and E. Spain, France. 1409a. *A.
triumfetti* DC. Like 1409 but ray-florets white with a yellow base; flower heads 3–4 cm
across. Leaves twice cut into narrow, pointed, almost spiny segments. Fruit winged, with
a conspicuous crown. Rocks, banks, in mountains. Widespread except in Portugal.
1411a. *A. maritima* L. A woody-based, strongly aromatic, bushy perennial distinguished
by its fleshy leaves dotted with small cavities. Flower heads large, 3–4 cm across; ray-
florets white. Involucral bracts with a broad white papery margin. Maritime sands.
Portugal, Med. region.

CHAMAEMELUM | **1413. *C. nobile* (L.) All. CHAMOMILE. Involucral bracts white-
margined. Cornfields, grassy places, on the littoral. Widespread.

1413a. *C. fuscatum* (Brot.) Tutin. An erect, pale green, usually branched annual with
twice-cut leaves with narrow segments. Flower heads long-stalked, aromatic, daisy-like
2–2½ cm across, with white reflexed rays and yellow disk-florets. Flower stalk swollen,
hollow, arched in bud, at length erect. Involucral bracts black-margined, reflexed in fruit.
Cultivated and uncultivated ground, waysides. Portugal, S. and C. Spain, S. France. Page
63.

1413b. *C. mixtum* (L.) All. Like 1413a but branches wide-spreading, sometimes almost at
right angles; involucral bracts erect in fruit. Cultivations, roadsides. Widespread.

ANACYCLUS |

1. Ray-florets short, erect, not longer than
involucre.
   2. Outer frs. with wide-spreading lobes at top
   of wings. (S., C., and E. Spain, France)   1414a. *A. valentinus* L. Pl. 50.
   2. Outer frs. with erect lobes at top of wings.
   (N. Portugal, Med. region)   1414. *A. clavatus* (Desf.) Pers. Page 364.
                                                        (*A. tomentosus*)
1. Ray-florets long, 1–1½ cm, spreading, longer
than involucre.
   3. Ray-florets yellow; inner involucral bracts
   with a conspicuous papery flap at apex.
   (Portugal, Med. region)   (1414) *A. radiatus* Loisel.
   3. Ray-florets white or purple beneath; inner
   involucral bracts without a conspicuous
   flap at apex.

4. Anns.; ray-florets white; involucral
bracts silky shaggy-haired. (Portugal,
Med. region)   1414. *A. clavatus* (Desf.) Pers.
4. Perenns.; ray-florets purplish beneath;
involucral bracts sparsely hairy. (S–E.
Spain)   1414b. *A. pyrethrum* (L.) Link

CLADANTHUS | Flower heads terminal, or stalkless in the axil of branches, and surrounded by 5–6 floral leaves. Receptacle at length conical and hard, with both scales and with thread-like fibres.   1414c. *C. arabicus* (L.) Cass. A regularly branched annual with 2–6 spreading branches at each division and with large orange-yellow flower heads 4 cm across. Involucral bracts green, in 3–6 rows. Leaves cut into linear or trifid lobes, the upper encircling the flower heads. Cultivated ground. S. Spain. Pl. 56, Page 163.

ACHILLEA | About 12 species; the following are most commonly encountered but are not easily distinguished.

White-flowered species
1417. *A. millefolium* L. YARROW. Widespread;   1415. *A. ptarmica* L. SNEEZEWORT. Spain and France;   (1417) *A. nobilis* L. Spain and France;   1417a. *A. odorata* L. Spain and France.

Yellow-flowered species
1422. *A. ageratum* L. Scattered throughout except in the north;   **1421. *A. tomentosa* L. YELLOW MILFOIL. S., C. and E. Spain, S. France;   1421a. *A. santolinoides* Lag. S. Spain.

OTANTHUS | 1423. *O. maritimus* (L.) Hoffmanns. & Link COTTON-WEED. An unmistakable snowy-white, cottony plant of sand dunes and sandy shores, with terminal clusters of globular yellow flower heads. Ray-florets absent and involucral bracts white-felted. Coasts of our area.

CHRYSANTHEMUM | Yellow-flowered annuals.   1424. *C. segetum* L. CORN MARIGOLD. Upper leaves entire or toothed, lower leaves deeply cut; stems little branched. Cornfields, cultivated ground. Widespread.   **1425. *C. coronarium* L. CROWN DAISY. All leaves cut into linear segments; stems much branched. Cultivated and waste ground. Portugal, Med. region. Page 52.   **(1424) *C. myconis* L. All leaves oblong entire, saw-toothed. Cultivated ground, vineyards, field verges, waste ground. Portugal, Med. region. Page 44.   1424a. *C. macrotus* Durieu Distinguished from (1424) by its fruits which all have a large ear-shaped crown which is at length longer than the florets. Involucral bracts bordered with black and papery on the margin. Cornfields, dry hills. S. Portugal, S. Spain.

LEUCANTHEMUM | Like *Chrysanthemum* but fruits of one kind, ten-ribbed, with secretory canals between the ribs and mucilage cells present in the epidermis; usually with white ray-florets.   **1427. *L. vulgare* Lam. (*Chrysanthemum leucanthemum*) MARGUERITE, MOON-DAISY, OX-EYE DAISY. Grasslands, waysides, in mountains. Widespread.

TANACETUM | Like *Chrysanthemum* but fruits all similar, usually five- to ten-ribbed; ray-florets yellow, or if absent then flower heads with two kinds of florets.

Flower heads solitary
1. Lvs. linear-spathulate, with 3, 5 or 7 apical
lobes.   *pallidum*
1. Lvs. deeply cut into narrow, comb-like
lobes.

1. *Santolina rosmarinifolia* 1408d
2. *Senecio leucophyllus* 1453a
3. *Artemisia granatensis* 1434b
4. *Artemisia assoana* 1434a
5. *Helichrysum lamarckii* 1385d
6. *Senecio petraeus* 1455b

2. Ray-florets white, with base yellow or
   pink.
  3. Plant 4–8 cm; lvs. with 5–7 pairs of lobes.
    (N–E. Spain, Pyrenees)   1428. *T. alpinum* (L.) Sch.-Bip. Page 132
                                                      (*Chrysanthemum a.*)
  3. Plant 8–15 cm; lvs. with 8–14 pairs of
    lobes. (W. Iberia)   1428b. *T. pulverulentum* (Lag.) Sch.-Bip.
2. Ray-florets yellow.
  4. Plants densely tufted. Yellow petals turn-
    ing reddish-orange after flowering.   *radicans*
  4. Plants laxly tufted. Petals remaining
    yellow after flowering. (Portugal, N–W.
    Spain)   1428d. *T. flaveolum* (Hoffmanns. & Link)
                                         Rothm.

1428a. *T. pallidum* (Miller) Maire A woody-based, often silvery-haired perennial with clusters of leaves cut into narrow lobes only at the apex, and erect unbranched stems bearing white or yellow flower heads 2½–3½ cm across. Ray-florets becoming reflexed, three-toothed. Sandy, stony and rocky places, in mountains. Scattered throughout Iberia. Pl. 51, Page 85, 110. 1428c. *T. radicans* (Cav.) Sch.-Bip. A small, tufted, woody-based grey-leaved plant often with leafless stems bearing a solitary pale yellow flower head 1½–2 cm across. Involucral bracts hairless, with a broad red-brown margin. Fruit with 5–6 ribs. S. Spain (S. Nevada). Pl. 51.

Flower heads numerous
**1426. *T. vulgare* L. (*Chrysanthemum v.*) TANSY. Flower heads numerous, 7–12 mm across, without rays. Very aromatic; leaves feathery. Hedges, field margins, bushy places. Throughout.

PROLONGOA | Like *Tanacetum* but receptacle conical. 1428e. *P. pectinata* (L.) Boiss. A small annual with bright yellow, solitary, flower heads about 1½ cm across, and small, mostly basal leaves with tooth-comb lobes. Flower heads nodding before and after flowering; involucral bracts papery, blunt; flower stalk leafless on the upper half. Leaves about ½–1 cm, with about 4 pairs of linear-pointed lobes. Sandy, cultivated ground. S. and C. Spain. Page 69.

PHALACROCARPUM | Flower heads white or pink; leaves opposite. Receptacle convex, without scales.

1428f. *P. anomalum* (Lag.) Rothm. (*P. oppositifolium*) An attractive perennial with flower heads 3–4 cm across and white spreading ray-florets. Involucral bracts with a narrow papery reddish-brown margin. Leaves twice cut into linear segments, with adpressed silvery hairs. Fruit without pappus. Rocks and stony places in mountains. N. and C. Portugal, C. Spain. Page 102. 1428g. *P. sericeum* (Hoffmanns. & Link) Henriq. Like 1423f but leaves lance-shaped toothed, not deeply cut. Flower heads 2–3 cm across. N–E. Portugal, W–C. Spain (León and Zamora provinces).

COTULA | **1433. *C. coronopifolia* L. BRASS BUTTONS. Damp places, marshes, on the littoral. Introd. to Portugal, S. Spain, France. Page 53.

ARTEMISIA | **Wormwood** About 35 species. 1434a. *A. assoana* Willk. (*A. lanata*) A low cushion-forming, woody-based perennial with tiny silvery-white leaves and erect, often unbranched stems bearing a spike of nodding, globular flower heads. Flower heads pale yellow, about ½ cm across; involucral bracts white-woolly. Leaf blades more or less circular in outline, deeply cut into linear lobes, the lower leaves stalked, the upper stalkless. Stony and sandy places in limestone hills. S–E. and E. Spain. Page 358.   1434b. *A. granatensis* Boiss. A compact, silvery-leaved, cushion-forming perennial with tiny fanshaped leaves and erect stems bearing one or several, greenish, erect, globular flower heads about ½ cm across. Leaves all stalked, deeply cut into short lobes; flower stems 2–7 cm. Summit screes. S. Spain (S. Nevada). Page 358.

1437. *A. campestris* L. subsp. *maritima* (Lloyd) Archangeli A quite hairless, woody-based perennial to 80 cm, with rounded flower heads 3–4 mm across, borne in branched clusters. Leaves fleshy, dark green, rounded in outline, cut into narrow segments which are grooved on the upper surface. Sandy coasts. Portugal, Spain.

TUSSILAGO | **1439. *T. farfara* L. COLTSFOOT. Damp stony places, riversides, waysides. N. Portugal?, C. and N. Spain, France.

PETASITES | 4 species.   **1440. *P. hybridus* (L.) P. Gaertner, B. Meyer and Scherb. BUTTERBUR. Riversides. N. Portugal?, N. and E. Spain, France.   **1441. *P. albus* (L.) Gaertner WHITE BUTTERBUR. Mountains. N. Spain, France.   1442. *P. fragrans* (Vill.) C. Presl WINTER HELIOTROPE. Cultivated and naturalized; Portugal, France.

HOMOGYNE | **1443. *H. alpina* (L.) Cass. ALPINE COLTSFOOT. Damp pastures. Pyrenees.

ADENOSTYLES | 2 species.   **1445. *A. alliariae* (Gouan) A. Kerner Mountains, riversides. Pyrenees, Massif Central.

ARNICA | **1446. *A. montana* L. ARNICA. Mountain pastures. Portugal, N. and E. Spain, France.

DORONICUM | **Leopard's-Bane**   About 8 species in our area.   **1447. *D. pardalianches* L. GREAT LEOPARD'S-BANE. Shady woods in mountains. N. Spain, Pyrenees, France. (1447) *D. plantagineum* L. LEOPARD'S-BANE. Grassy, bushy places, in mountains. Scattered throughout.   1447a. *D. carpetanum* Willk. Very like 1447 and replacing it in Iberia but stem and leaf stalks hairy, and lower leaves deeply heart-shaped with widely spread lobes, middle leaves ovate-lance-shaped, the upper stalkless, oblong-heart-shaped. Flower heads large, 4–5 cm across. Meadows, rocks in mountains. C. and N. Portugal, C. Spain, Pyrenees. (1447) *D. austriacum* Jacq. AUSTRIAN LEOPARD'S-BANE. A tall erect branched perennial to 1 m with a branched cluster of large yellow flower heads each up to 7½ cm across. Basal leaves absent at flowering, stem leaves many, leaf stalk winged, enlarged and clasping at base. Mountain woods, streamsides. Pyrenees, Massif Central.

SENECIO | Nearly 60 species occur in our area. The following are widespread: 1449. *S. vulgaris* L. GROUNDSEL;   (1450) *S. sylvaticus* L. WOOD GROUNDSEL;   1451a. *S. gallicus* Vill.; (1451) *S. aquaticus* Hill MARSH RAGWORT.

Some or all leaves deeply lobed
(a) Plants densely white-felted
1453. *S. cineraria* DC. CINERARIA. Rocks and sands on the littoral. C. and N. Portugal, Med. region.   1453a. *S. leucophyllus* DC. Like 1453 but involucral bracts acute, 10–16; rayflorets 5–7, pale yellow (ray-florets 10–12, golden-yellow in 1453). Fruit finely grey-haired. Stony places, rocks in mountains. E. Pyrenees. Page 358.

(b) Plants green

1451b. *S. adonidifolius* Loisel. An erect perennial to 70 cm, with distinctive leaves which are several times cut into numerous thread-like, pointed lobes. Flower heads numerous, in a dense terminal flat-topped cluster, with short spreading yellow ray-florets. Involucral bracts 6–8 mm long, incurved in fruit. Woods, meadows, rocks. Pyrenees, France.   1451c. *S. minutus* (Cav.) DC. A delicate, erect annual with solitary, long-stalked flower heads 2½ cm across, with 8–12 pale yellow ray-florets which are purple beneath. Involucre of one row of papery-margined bracts, additional bracts nil. The lower leaves obovate, coarsely toothed, the middle deeply lobed, the upper entire, linear. Stony and sandy places in lowlands and hills. Portugal, S. and C. Spain.

Leaves entire or toothed
(a) Plant densely white-felted

1453b. *S. boissieri* DC. A dense silvery-grey, cushion-forming perennial with tiny leaves and short, almost leafless stems 3–12 cm, bearing solitary pink flower heads with contrasting yellow anthers. Flower heads about 1 cm across; ray-florets absent; involucral bracts linear-lance-shaped, blunt; additional bracts few. Leaves in a rosette, oval or wedge-shaped, stalked, toothed, the upper leaves linear, all densely silvery-haired. Alpine rocks and stony ground. S. Spain (S. Nevada). Pl. 51.

(b) Plants green

1455a. *S. linifolius* L. A robust, erect perennial with dense, more or less flat-topped clusters of numerous yellow flower heads each about 1½–2 cm across, and linear, pointed leaves. Ray-florets 10–15, strap-shaped; involucral bracts linear, two-keeled; additional bracts very short, linear. Stony and rocky places from the lowlands to the mountains. S. Spain, Balearic Islands.   1455b. *S. petraeus* Boiss. & Reuter An erect, rather robust annual with oblong, clasping, toothed or shallow-lobed upper leaves, and a spreading cluster of few, large flower heads each about 3 cm across. Ray-florets yellow, spreading, at length rolled back; involucral bracts linear-pointed, black-tipped; additional bracts absent. Plant hairless or with few cobweb hairs. Among rocks. S. Spain (S. de Grazalema). Page 358.   1455c. *S. tournefortii* Lapeyr. A robust plant with erect stems bearing many large leathery, pale green, lance-shaped, toothed leaves, and a terminal flat-topped cluster of few large yellow flower heads. Flower heads showy, 3–4 cm across, rarely solitary; ray-florets 12–14; involucral bracts linear-lance-shaped, dark-spotted and hairy at apex. Lower leaves with a winged leaf stalk, the upper stalkless, not clasping. Pastures, damp places, and rocks in mountains. C. Portugal, Spain, Pyrenees. Pl. 51.   1455d. *S. eriopus* Willk. Leaves very leathery, all basal in lax rosette, oblong heart-shaped, rounded-toothed, shining pale green above, woolly-haired beneath. Flower heads showy, 4–5 cm across, solitary or few, nodding; ray-florets 10–20, rounded-tipped. Stems, leaf stalks and involucral bracts densely woolly-haired. Rocks in mountains. S. Spain.

LIGULARIA | 1459. *L. sibirica* (L.) Cass. Marshes. Pyrenees, France.

CALENDULA | **Marigold**   About 6 species, but a difficult genus with most species very variable.

Annuals

1460. *C. arvensis* L. MARIGOLD. An annual with variable-sized flower heads 1–3½ cm across. Leaves flat or undulate, the upper half-clasping the stem. Fruits of 3 forms, the outer strongly curved. Cultivated and uncultivated ground. Widespread.   (1460) *C. officinalis* L. POT MARIGOLD. Cultivated and sometimes naturalized.

Perennials or biennials

1460a. *C. tomentosa* Desf. A white-woolly, woody-based perennial 20–50 cm, with blunt leaves, the lower oval-spathulate, the upper oval-oblong and half-clasping. Flower heads medium to large; florets twice as long as the white-woolly involucral bracts. Outer fruits smooth on back. Rocks, dry places. S–W. Portugal, S. Spain.   1460b. *C. suffruticosa* Vahl. Like 1460a but leaves green, finely hairy, more or less glandular-hairy, often viscid. Flower heads large, yellow. Outer fruits usually spiny on back. A very variable species. Rocks on the littoral, and in the hills. S. and C. Portugal, S. Spain. Pl. 51.

CRYPTOSTEMMA | Fruits hairy, the hairs hiding the scales of the pappus; receptacle honey-combed; involucral bracts in many rows.   1460d. *C. calendula* (L.) Druce (*C. calendulaceum, Arctotheca c.*) A low-growing perennial with a lax rosette of deeply cut leaves, and erect, long-stalked, solitary flower heads with yellow ray- and black disk-florets. Flower heads about 4 cm across; ray-florets pale at the tips, dark on underside. Leaves pale green above, white-hairy beneath; the terminal lobe broadly triangular, much larger than the 2–4 pairs of lateral lobes. Fruit with very long, dense woolly hairs. Native of S. Africa: naturalized in Portugal and S. Spain. Pl. 52.

ECHINOPS | Globe-Thistle

1. Lvs. rough and bristly on upper surface.	*strigosus*
1. Lvs. smooth on upper surface.	
2. Plant glandular-hairy; lvs. with broad, scarcely spiny lobes; fl. heads very pale blue, 4–8 cm across. (N–C. and E. Spain, France)	1461. *E. sphaerocephalus* L.
2. Plant cottony, not glandular; lvs. with narrow spiny lobes; fl. heads rich blue, 3–3½ cm across. (S., C., and E. Spain, France)	**1462. *E. ritro* L.

(1461) *E. strigosus* L. ROUGH-LEAVED GLOBE-THISTLE. An erect, spiny, extremely rough-leaved greyish annual with globular blue flower heads 4–8 cm across. Involucral bracts very unequal. Leaves 2–3 times cut into narrow, inrolled, spine-tipped segments. Dry places, waysides, field verges. S. and C. Portugal, S. and C. Spain. Pl. 52.

XERANTHEMUM |

1. Outer involucral bracts hairless; fr. with 5 spines.	
2. Fl. heads longer than broad, cylindrical; inner coloured bracts little longer than the width of the disk, scarcely spreading. (N–E. Portugal, S. Spain)	(1464) *X. inapertum* (L.) Miller Page 132
2. Fl. heads broader than long, bell-shaped; inner coloured bracts twice as long as width of the disk, wide-spreading. (S. Portugal, E. Spain)	**1464. *X. annuum* L.
1. Outer involucral bracts white-woolly, blunt. (N–E. Spain, W. France)	1465. *X. foetidum* (Cass.) Moench

CARLINA | Carline Thistle

1. Fl. heads usually solitary, 6–14 cm across.
2. Lvs. all stalked, hairless or nearly so; lv. segments narrow. (Pyrenees, Massif Central)   **1467. *C. acaulis* L.
2. The innermost lvs. stalkless, all white-hairy beneath; lv. segments broad. (Pyrenees, France)   1466. *C. acanthifolia* All.
1. Fl. heads usually several, 2–5 cm across.
3. Inner involucral bracts whitish or pale yellow above. (Spain, France)   1468. *C. vulgaris* L.
3. Inner involucral bracts bright yellow or purplish.
4. Inner involucral bracts purplish. (C. Spain, Med. region)   (1469) *C. lanata* L.
4. Inner involucral bracts yellow.
5. Outer involucral bracts leafy, about as long as the fl. head; inner bracts golden-yellow. (Widespread)   **1469. *C. corymbosa* L.
5. Outer involucral bracts leafy, much longer than the fl. head; inner bracts sulphur-yellow. (Portugal, S., W., and C. Spain)   1469a. *C. racemosa* L. Page 364

ATRACTYLIS | 1470. *A. cancellata* L. Dry rocky places, tracksides, hills. S. and C. Portugal, Med. region. Page 52. 1470a. *A. humilis* L. A slender perennial with stems 5–30 cm, bearing purple flower heads 1½–2½ cm across, with the outer involucral bracts longer than the flower heads and similar to the stem leaves. Middle bracts of involucre rounded-egg-shaped, the upper oblong. Stem leaves oblong-lance-shaped, deeply pinnately cut into spiny lobes. Pappus slightly longer than fruit. Dry places, principally on limestone. S. and S–C. Spain, S. France.

1471. *A. gummifera* L. A usually stemless, thistle-like perennial with large purple flower heads 3–7 cm across. Involucral bracts spiny, 3–7 cm, twice-cut, the outer with 3 spreading apical spines much longer than the lateral spines. Field verges, waysides, hedges. S. and C. Portugal, Med. Spain.

STAEHELINA | 1474. *S. dubia* L. A woody-ᴜ ᴙsed, bushy perennial with white-felted stems and undersides of leaves, and slender cylindrical, purple flower heads with distinctive silvery-haired involucral bracts flushed with purple. Flower heads 2–3 cm by 3–5 mm wide; pappus white, much longer than involucre and fruit. Leaves linear, inrolled, toothed or entire, dark green above, 3–5 mm wide. Dry banks, stony places. Portugal, S., C., and E. Spain, S. France. Page 364. 1474a. *S. baetica* DC. Like 1474 but a smaller bushy perennial to 15 cm; flower heads broader, 7–8 mm wide; involucral bracts hairless. Leaves smaller 8–15 mm, toothed or lobed. Shady places. S. Spain (Málaga province).

JURINEA | 3 species.

1476a. *J. humilis* (Desf.) DC. (*Serratula h.*) A small, usually stemless perennial with a rosette of pinnately lobed leaves, and solitary stalkless, pale purple flower heads 2–2½ cm across. Involucral bracts linear-lance-shaped, straight, spreading or recurved, almost hairless or with cobweb hairs. Leaves usually 2–3½ cm, segments oblong, white-woolly beneath, margins inrolled. Stony and rocky places, dry pastures in mountains. C. Portugal, S., C., and E. Spain, S. France. Page 364. 1476b. *J. pinnata* (Lag.) DC. Distinguished from 1476a by the flower heads which are borne on a short, rather leafless stem usually 6–10 cm tall. Flower heads purple, rarely white, 1½–2 cm across; involucral bracts woolly-haired,

1. *Staehelina dubia* 1474
3. *Carlina racemosa* 1469a
5. *Jurinea humilis* 1476a

2. *Anacyclus clavatus* 1414
4. *Rhaponticum cynaroides* 1498a

often purple-tipped. Stony places, rocks on limestone in lowlands and mountains. S. and C. Spain.

CARDUUS | Thistle Distinguished from *Cirsium* by its unbranched, rough pappus. About 25 species occur in our area, including 10 which are restricted to it in Europe.

Flower heads small or medium-sized, up to 2½ cm across
1480a. *C. carlinoides* Gouan PYRENEAN THISTLE. A very prickly, white-downy perennial covered with yellow spines, and dense clusters of rosy-purple or white flower heads each 1½–2 cm across. Involucral bracts linear, spine-tipped, woolly-haired. Leaves very spiny, narrow-oblong in outline, deeply cut into narrow lobes which are reduced almost to yellow spines; stem winged, 20–50 cm. Rocks, screes, dry pastures in mountains. Endemic Pyrenees. Pl. 52.

Flower heads large, 2½ cm or more
1478a. *C. granatensis* Willk. (including *C. platypus*) A rather low-growing, robust, spiny thistle 20–60 cm, with a spiny-winged stem, narrow spiny-lobed leaves and solitary or few pink flower heads 2½–5 cm across. Involucral bracts distinctive, broadly linear with very long recurved spiny pointed tips, the innermost purple-tipped. Stems and leaves at first white with cobweb-hairs, leaves later becoming hairless. Cultivated and uncultivated ground, in lowlands and mountains. Portugal, Spain. Pl. 52.

CHAMAEPEUCE | Like *Cirsium* but fruits almost as broad as long; differs from *Notobasis* in leaves not veined with white, and leaf-spines in groups of 3–5 (not single). 2 species.

1481a. *C. hispanica* (Lam.) DC. A striking, robust, extremely spiny thistle with long yellow spines, white-felted stems and undersides of leaves, and large, very spiny, purple flower heads. Flower heads several, 2½–4 cm across; involucral bracts numerous, spreading in a star, each gradually tapering into a long slender, very sharp yellow spine, the inner bracts purple. Leaves dark green above, spiny, contrasting with undersides; spines to 1½ cm, in threes. Stony places, rocks in mountains. S. Spain. Page 367.

**NOTOBASIS | ****1481. *N. syriaca* (L.) Cass. SYRIAN THISTLE. A tall erect thistle distinguished by the purple-flushed, very spiny uppermost leaves which encircle and spread beyond the clustered purple flower heads. Field margins, wayside ditches. S. and C. Portugal, S. Spain, S. France.

CIRSIUM | Distinguished from *Carduus* by the feathery hairs on the bristles of the pappus. About 30 species occur, with perhaps 12 restricted to our area in Europe.

Flower heads large, usually solitary
1484. *C. erisithales* (Jacq.) Scop. YELLOW MELANCHOLY THISTLE. Flower heads lemon-yellow, usually solitary, nodding, borne on smooth leafless stems, rarely reddish-purple or clustered. An erect perennial to 1½ m. Hills and mountains. Massif Central.

**1485. *C. eriophorum* (L.) Scop. WOOLLY THISTLE. Distinguished by its large, broad, reddish-purple flower heads with spiny-tipped involucral bracts with cobweb hairs running among them. Uncultivated ground, waysides, lowlands and mountains. Spain, France. 1485a. *C. ferox* (L.) DC. Distinguished by the usually white flower heads, which are longer than broad and surrounded by longer upper leaves. Involucral bracts narrowed to a long, scarcely prickly point, with few cobweb hairs. Leaves white-woolly beneath, green and hairy above, very spiny and with a long spiny tip. Biennial to over 1 m. Rocky ground, uncultivated places in lowlands and mountains. S. Spain, S. France. 1485e. *C. echinatum* (Desf.) DC. Differs from 1485a in its ovoid-conical flower head of purple florets, which is over-topped 2–3 cm by the upper leaves, and with the outer involucral bracts curved out-wards and downwards. Arid places, field verges. S. and E. Spain, Pyrenees, S. France.

Flower heads small, several
(a) Leaves lobed, with stout spines
1485b. *C. scabrum* (Poiret) Durande (*C. giganteum*) A very tall, slender perennial 2–4 m, with relatively small pink flower heads clustered in twos and threes at the ends of short branches. Flower heads 2–4 cm by 2 cm, with cobweb hairs. Leaves white-woolly beneath, rough above, with cobweb hairs and with broad spiny-toothed lobes. Hedges. S. Spain. Pl. 53. 1485c. *C. flavispina* DC. An erect, branched perennial with lobed leaves with long robust yellow spines and a branched cluster of many purple, or rarely white flower heads. Involucre ovoid, about 1 cm long; involucral bracts hairless, the outer spine-tipped, the inner spineless, with outcurved tips. Stems white-woolly; leaves completely grey-woolly, or green and nearly hairless above, upper leaves decurrent. Sandy damp places, by water. N–W. Portugal, S., C., and E. Spain. 1485d. *C. nevadense* Willk. Like 1485c but lilac-pink flower heads larger 2–2½ cm, the outer involucral bracts with a dark purple spot at the base of the spine, the inner with a soft dark purple apex. Leaves glaucous, hairless above, with fine cobweb hairs beneath, very spiny with spines 1–1½ cm long. Damp places in mountains. S. Spain (Sierra Nevada).

(b) Leaves entire, with soft marginal bristles
1489a. *C. monspessulanum* (L.) Hill Distinguished by its lax branched, somewhat flat-topped cluster of purple flower heads, and its large lance-shaped leaves which are hairless and spineless except for stiff, yellowish unequal bristles on the margin. Flower heads several, 1½–2 cm across; involucral bracts adpressed, pale green with a blackish swollen tip. Middle stem leaves with blade running down stem as broad wings, uppermost small, not decurrent; stems ½–1½ m, with cobweb hairs above. Damp places, streamsides. S. Portugal, S., C., and E. Spain, S. France. Page 367. 1489b. *C. welwitchii* Cosson Like 1489a but purple flower heads larger, about 3 cm across, one or few borne at the ends of long, nearly leafless branches. Involucral bracts purple, gradually narrowed into a slender, soft tip. Leaves lance-shaped, hairless, shining, bristles on margin more or less equal, scarcely decurrent; uppermost leaves bract-like. Damp places by streams. C. Portugal, S–E. Spain.

CYNARA | Involucral bracts usually with stout spines.

1. Lvs. without spines; involucral bracts with a large tooth-like apex. (Cultivated in South)	**(1491) *C. scolymus* L.
1. Lvs. spiny; involucral bracts, spiny-tipped.	
2. Stems absent or almost so; spiny apex of middle involucral bracts not more than 7 mm, slender. (S. Portugal, S. Spain)	1491b. *C. tournefortii* Boiss. & Reuter
2. Stems present; spiny apex of middle involucral bracts at least 1 cm, stout.	
3. Stem lvs. 1–2 pinnately cut, with linear inrolled segments; fr. winged.	*humilis*
3. Stem lvs. 1–2 pinnately cut with lance-shaped flat segments; fr. not winged.	
4. Lvs. up to 50 cm by 35 cm, with spines 1½–3½ cm clustered at the base of each segment; fr. 6–8 mm.	*cardunculus*
4. Lvs. up to 40 cm by 15 cm with spines 2–20 mm, not in clusters; fr. 3–5 mm.	

1. *Cynara cardunculus* 1491   2. *Chamaepeuce hispanica* 1481a
3. *Cirsium monspessulanum* 1489a

5. Fls. white; lvs. with short adpressed hairs to woolly-haired, becoming hairless, with a distinct network of veins beneath. (S. Spain)   1491c. *C. alba* DC.

5. Fls. purplish-blue; lvs. white-woolly beneath, with an indistinct network of veins. (S. Portugal)   1491d. *C. algarbiensis* Mariz

1491. *C. cardunculus* L. CARDOON. Flower heads blue, lilac or whitish, 4–5½ cm across, with glaucous or purplish involucral bracts narrowed into a stiff spine 1–5 cm long. Leaves large, white-cottony beneath, green above, with lance-shaped, flat segments, spines yellow 1½–3½ cm. Stony places, tracksides, waste ground. S. and C. Portugal, S. and E. Spain, S. France. Page 367.   1491a. *C. humilis* L. Like 1491 but flower heads blue, solitary, and leaves with narrow, inrolled segments ending in a short spine, shining pale green above, white-woolly beneath. Flower heads 2–4½ cm across; involucral bracts purple, with outcurved spiny tips; flower stems white-woolly, 20–80 cm. Dry, waste places. S. and C. Portugal, S. and C. Spain. Pl. 52.

SILYBUM | **1492. *S. marianum* (L.) Gaertner MILK-THISTLE, HOLY THISTLE. Waste ground, waysides, uncultivated places. Widespread.   1492a. *S. eburneum* Cosson & Durieu Like 1492 but basal leaves rough-haired (not glossy and shiny), and with very stout, shining, glossy white, golden-tipped spines; terminal leaf lobe linear and much longer than lateral lobes. Middle involucral bracts with very long straight spines 6–7 cm, outermost bracts spineless. Steppes. E. Spain (Aragon).

GALACTITES | 2 species.   **1493. *G. tomentosa* (L.) Moench Uncultivated ground, waysides, cultivations and field margins. Widespread particularly in Med. region.

TYRIMNUS | Like *Carduus* but filaments of stamens fused, and fruits four-angled.   1493a. *T. leucographus* (L.) Cass. A white-cottony, thistle-like plant with spiny-winged stems, spiny leaves and a long-stemmed solitary purplish-pink flower head. Flower head nearly globular, 1½ cm; involucral bracts numerous, adpressed, weakly spine-tipped. Leaves oblong-lance-shaped, spiny-toothed or lobed, white-spotted above, cottony beneath. Waste places, stony and sandy ground. Med. region.

ONOPORDUM |

1. Stems absent or less than 15 cm; fls. white.   *acaulon*
1. Stems erect 1–3 m or more; fls. bluish or purple, rarely white.
   2. Involucral bracts erect, often adpressed.   *nervosum*
   2. Involucral bracts, at least the outer, spreading horizontally or reflexed.
      3. Involucral bracts glandular-hairy, otherwise hairless. (Introd. S. France)   (1494) *O. tauricum* Willd.
      3. Involucral bracts hairless or hairy, not glandular.
         4. Involucral bracts very numerous, densely crowded, linear-lance-shaped, 2 mm broad at base, gradually narrowed to a golden spine. (Widespread)   **1494. *O. acanthium* L.
         4. Involucral bracts fewer, not densely crowded, more than 2 mm broad at base.

5. Involucral bracts shorter than the fls., somewhat
abruptly narrowed to a short spine. *illyricum*
5. Involucral bracts equal to or longer than the fls.,
narrowed to a long point. *macracanthum*

1494a. *O. acaulon* L. A stemless, thistle-like biennial with a large flattened rosette of greyish or white, oblong, spiny-lobed leaves, and a solitary central, or cluster of almost stalkless large white flower heads. Flower heads globular, 2–4 cm; involucral bracts hairless, spine-tipped. Leaves very spiny, 15–40 cm long, stalked. Dry rich soils, rocky places in lowlands and mountains. S., C., and E. Spain, S. France. Pl. 52.

1494b. *O. nervosum* Boiss. A very robust plant to 3 m with stout stems with very broad spiny wings, large and shallowly lobed, spiny-margined leaves, and pink spiny flower heads. Flower heads oval-conical, 4 cm long, with numerous, stiff erect, spiny-tipped, almost hairless involucral bracts, shorter than the florets. Leaves with a conspicuous network of tough white veins beneath. Stony arid ground, abandoned cultivation. C. Portugal, S., C., and E. Spain. Pl. 53. **1495. *O. illyricum* L. Like 1494b but outer involucral bracts glandular, narrowly oval, abruptly ending in a short recurved spine, inner bracts erect, purplish; florets glandular. Usually a white-woolly biennial. Dry arid places, waysides. S., C., and E. Spain, S. France. **1495a. *O. macracanthum* Schousboe Distinguished by the narrower lance-shaped involucral bracts, the lower reflexed with long stout spines, the inner erect as long as or longer than the hairless florets. A grey or white-woolly biennial. Dry arid places, waysides. S. and C. Portugal, S. Spain. Pl. 53.

CRUPINA | 1496. *C. vulgaris* Cass. FALSE SAW-WORT. Dry grassy and stony ground. Widespread. (1496) *C. crupinastrum* (Moris) Vis. Dry grassy places. S. Spain.

SERRATULA | 10 species occur, 7 are restricted to our area in Europe. **1497. *S. tinctoria* L. SAW-WORT. Damp meadows, marshes, woods and clearings. Widespread. 1497a. *S. boetica* DC. Flower heads solitary, cylindrical, 3–4 cm long with purplish florets and numerous long, rigid involucral bracts which are gradually narrowed to a stout apical spine. Basal leaves entire, or toothed, with a leaf stalk as long as the blade, stem leaves pinnately lobed; upper part of stem leafless or with a few bracts. Mountains. C. Portugal, S–W. Spain.

1497c. *S. barrelieri* Dufour (*S. pinnatifida*) Flower heads 2–3 cm long, pinkish-purple, with involucral bracts narrowing to a fine needle-like spine, or almost spineless. At least the stem leaves pinnately lobed, upper part of stem leafless or with few bracts. Mountains. S. and C. Portugal, N–W., S. and S–C. Spain. Page 378.

RHAPONTICUM | Flower heads large, solitary; involucral bracts with a broad, toothed, papery margin and gradually narrowed to a long papery point; receptacle with scales; pappus of fragile toothed hairs, in several rows. 1498a. *R. cynaroides* (DC.) Less. CARDOON KNAPWEED. A robust leafy perennial to 1 m or more, with deeply cut leaves which are green above and white-woolly beneath, and with a large solitary, rosy-purple flower head 6–7 cm across. Involucral bracts brown, with white papery margins. Leaves large, with lance-shaped, toothed lobes, the upper less divided. Rocks, pastures. Pyrenees. Page 364.

MANTISALCA | Flower head ovoid-conical strongly constricted at the apex, with adpressed leathery involucral bracts with very short apical spines. About 4 species. 1498b. *M. salmantica* (L.) Briq. & Cavillier (*Microlonchus s.*) A slender, branched, knapweed-like perennial with deeply-lobed, woolly-haired, basal leaves, narrow entire hairless stem leaves, and solitary long-stemmed, purple flower heads 2½ cm across. Involucre yellowish, hairless, with broad, closely overlapping bracts with black tips; florets rarely white. Arid places, rocks, vineyards, waysides, cultivated and uncultivated ground. Widespread in the south. Page 375.

CENTAUREA | Knapweed etc.   An important genus with about 90 species in our area, many of which are restricted to the Iberian peninsula. Only a small number of the most conspicuous and unusual-looking species, excluding those already described in *Flowers of Europe*, can be dealt with.

Involucral bracts ending in a spine
(a) Involucral bracts with a broad papery fringe or spiny border extending some way down each side; flower heads yellow
1503a. *C. ornata* Willd. A stiff, erect, branched perennial with pinnately-cut leaves, and rather large, solitary, yellow flower heads with long stiff spreading and recurved yellowish-brown spines. Involucral bracts pale green, oval, with brownish fringed margin, terminal spine 8–35 mm, longer than the flower heads. Leaf lobes linear, ending in a fine spine. Fruit white, silvery-haired. Dry arid sandy or stony places. E. Portugal, S., C., and E. Spain. Page 372.   1503b. *C. collina* L. Like 1503a but terminal spine of involucral bracts shorter, tawny to dark brown, usually 4–5 mm (sometimes up to 8 mm). Fruit at length black and hairless. Dry places. South of our area. Page 373, 375.   1503c. *C. granatensis* DC. A rather robust, softly white-haired perennial with erect usually unbranched, sparsely leafy stems bearing a large, terminal, deep yellow flower head 2½–3 cm across. Involucral bracts oval, woolly, the middle ones with a curved, dark brown, broad-based spine, 4–8 mm long. Basal leaves pinnately lobed, segments lance-shaped to elliptic, fine-pointed, the uppermost leaves shallowly lobed or entire. Stony places and rocks in limestone mountains. S. Spain. Pl. 54, Page 373.

1503d. *C. haenseleri* (DC.) Boiss. & Reuter A stemless perennial with a rosette of woolly-white leaves of two kinds, and 1–3 large orange-yellow flower heads 2–2½ cm across, borne almost stalkless in the centre of the rosette. Outer involucral bracts with a long terminal erect spine and with numerous comb-like bristles at its base and on the margin of the bracts; inner bracts with a rounded, hooded, fringed appendage. Lowest leaves entire with a large rounded blade, most leaves pinnately lobed with unequal lance-shaped lobes and a terminal linear lobe, all lobes spine-tipped. Mountains. S. Spain (Málaga province). Pl. 54, Page 373.

(b) Involucral bracts without a fringed border; terminal spines with comb-like lateral spines at base
(i) Flower heads yellow
1499a. *C. nicaeensis* All. A rather robust, green, branched, very variable biennial or perennial with entire, clasping upper leaves, and terminal bright yellow flower heads about 2 cm across. Involucral bracts oval, hairless, with a long terminal, pale yellow, spreading spine to 2 cm with much smaller brown spiny bristles at its base. Lower leaves pinnately lobed, the upper oblong lance-shaped, the uppermost close beneath the flower heads. Fields. S–E. and E. Spain. Page 372.   1499b. *C. sulphurea* Willd. A handsome, pale green, branched annual with winged branches, narrow leaves, and terminal yellow conical flower heads with long elegant spines. Involucral bracts oval, closely adpressed, ending in a fan of 7–9 slender spines, the middle spine longest, to 2 cm, often dark purple, the lateral spines whitish. Florets glandular. Leaves either all narrow lance-shaped or lower pinnately lobed, the upper decurrent. Fruit large 5 mm by 3 mm, pappus brownish. Uncultivated ground, rocky places, waysides. S. Spain. Page 373, 375.   (1499) *C. melitensis* L. MALTESE STAR THISTLE. Like 1499b but flower heads more numerous, smaller to 1½ cm long. Involucral bracts oblong, ending in a spine and with pinnately arranged lateral spines ranged from

about the middle of the terminal spine to the base (not arranged in a fan). Dry hills, fields, waysides. Widespread. Page 372.

(ii) Flower heads pink or purple

1500a. *C. polyacantha* Willd. The most beautiful of this group, with large pink flower heads spreading 5–7 cm, with very long outer florets 2–3 times as long as the central florets. Involucral bracts woolly-haired, with a conspicuous terminal tuft of 9–13 stiff brown, or yellowish, spreading spines ½–1 cm long. Lower leaves deeply lobed with broad, spiny-toothed lobes, in a lax rosette, the uppermost lance-shaped entire or shallowly lobed. A robust, erect, little-branched perennial to ½ m. Fruit hairless. Sandy and rocky places, pine woods on the littoral. W. Portugal, S–W. Spain. Pl. 54, Page 373. 1500b. *C. sphaero-cephala* L. Differing from 1500a by the involucral bracts which have 5–7 yellowish spines arranged in one plane only, with the median spine somewhat longer. Flower heads smaller 2½–3½ cm across, the outer florets spreading. Leaves with shallow, triangular, toothed lobes. Fruit hairy. Sandy places by the sea. S. Portugal, S. Spain. Page 372.

Involucral bracts not ending in a spine
(a) Involucral bracts ending in comb-like bristles or teeth
(i) Flowers pink, purple or blue

1501a. *C. triumfetti* All. (*C. variegata*) Like *C. montana* but leaf blades not running down stem, and leaves densely white-cottony. Flower heads solitary, showy, 5–6 cm across, blue-violet or purple, pinker towards centre; involucral bracts with a very broad blackish-brown papery margin and a conspicuous fringe of white bristles. Grassy and rocky places, pastures in mountains. N–E. Portugal, S., C., and E. Spain.

1501b. *C. pullata* L. An annual to perennial with large, usually rosy-purple flower heads 4 cm across or more, with longer spreading outer florets, and flower heads sometimes close-ly surrounded by an involucre of narrow upper leaves. Involucral bracts lance-shaped, with a conspicuous black margin and a comb-like apex of 4–5 pairs of slender bristles and a similar terminal bristle. Florets sometimes pale yellow or white. An erect, simple or branched, rough-haired plant with the lower leaves in a rosette shallowly lobed or entire, the upper leaves deeper lobed, the uppermost entire. Fields, waste ground. S. and C. Portugal, S. and S–E. Spain. Pl. 54, Page 372. 1501c. *C. carratracensis* Lange Flower heads 2½ cm across, purple; involucral bracts adpressed, with a broad triangular brown membraneous tip and white teeth. Rocks and stony places. S. Spain.

(ii) Flowers yellow or orange

1504a. *C. toletana* Boiss. & Reuter A stemless, rosette plant with grey-woolly, much-divided leaves and a central cluster of few, almost stalkless yellow flower heads about 3 cm across. Involucre with conspicuous, pale overlapping, triangular comb-like appendages to the outer bracts, inner bracts linear with a rounded, hooded, fringed appendage. Leaves green above, white-hairy beneath, irregularly pinnately lobed with toothed or lobed, lance-shaped, spine-tipped segments. Bushy places in mountains. C. Spain. Pl. 54, Page 373. 1504b. *C. clementei* DC. A very handsome, white-woolly, robust perennial with deeply lobed leaves. Flower heads yellow, very large, 2½–4 cm across; involucral bracts with large dark triangular appendages with numerous long silvery-white comb-like bristles on each side. Leaf lobes oval-triangular, irregularly toothed, densely white-woolly, particularly the young leaves. Limestone cliffs. S–W. Spain. Page 372, 375. 1505a. *C. prolongi* DC. A rather slender, hairless, bright green perennial with almost leafless, flowering stems bearing a solitary, globular, orange or yellow-golden flower head about 2½ cm across. Involucral bracts broadly oval, green with a brown crescentic upper margin with small regular teeth, the innermost narrow, with a brown papery appendage. Basal leaves pinnately cut into unequal lance-shaped, entire lobes, stem leaves narrow lance-

(1499) *C. melitensis*

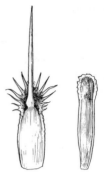

1499a *C. nicaeensis*

1500b *C. sphaerocephala*

1501b *C. pullata*

1503a *C. ornata*

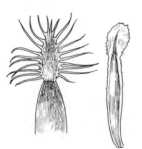

1504a *C. toletana*

1504b *C. clementei*

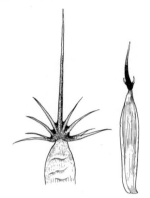

1499b *C. sulphurea*

1500a *C. polyacantha*

1503b *C. collina*

1503c *C. granatensis*

1503d *C. haenseleri*

1505a *C. prolongi*

1505b *C. linaresii*

shaped, entire. Sunny rocks, in hills and limestone mountains. S. Portugal, S–E. Spain. Page 373.

(b) Involucral bracts papery or with an entire papery margin
1505b. *C. linaresii* Láz.-Ibiza Differs from 1505a in having larger yellow flower heads about 4 cm across, with oval involucral bracts, green with brown veins and a papery, entire margin. A taller, more robust plant with lower leaves regularly pinnately cut into oblong spiny-tipped lobes 2½–5 cm long, upper leaves smaller, pinnately lobed. C. Spain. Page 373, 375.   **1507. *C. conifera* L. Distinguished by its purple flower head with numerous, shining brown, oval involucral bracts forming an oval cone-like involucre. Leaves green, with cobweb hairs above, white-woolly beneath. Stems short 2–25 cm. Stony ground. Widespread except in the north. Pl. 53.   1507a *C. rhaponticoides* (Graells) B. D. Jackson A handsome perennial like 1507, but much larger, up to 1½ m. Flower heads purple, globular, 3½–4 cm across. Leaves hairless above, deeply cut, large, 25 cm or more. Woods in mountains. N–E. Portugal, W–C. and C. Spain.

CARTHAMUS | 4 species.   **1508. *C. lanatus* L. A greyish, thistle-like perennial with yellow flower heads 2–3 cm across, and numerous green leafy, spiny involucral bracts surrounding the flower head and as long as it. Waste ground, cultivations. S. and C. Portugal, S., C., and E. Spain, S. France.   (1508) *C. tinctorius* L. SAFFLOWER Leaves entire, oblong to elliptic with marginal bristles. Involucral bracts shorter than the flowers; pappus absent. Sometimes naturalized in the south.   1508a. *C. arborescens* L. A much-branched, glandular-hairy shrub up to 2 m, smelling strongly of goats, with large yellow flower heads up to 3¼ cm across. Inner involucral bracts without an apical appendage. Leaves pale green, wavy-margined with pale spines, half-clasping; stem below flower heads much swollen. Inner fruits finely wrinkled, pappus fragile, feathery, lilac. Hedges and bushy places. S. and S–E. Spain. Pl. 53.

CARDUNCELLUS | Like *Carthamus* but pappus with bristly hairs, and flower heads always bluish. Distinguished from *Cynara* by the involucre bracts in which the outer are more or less leafy, usually spiny, the inner spineless with a semi-circular or egg-shaped fringed appendage. 5 species.

1508b. *C. monspelliensium* All. A stiff thistle-like perennial with solitary often stalkless, blue-violet flower heads with the outer involucral bracts, spiny, and green like the upper leaves. Flower heads 2–4 cm long on stems 2–20 cm. Lower leaves pinnately-cut into 6–9 pairs of linear or oblong spiny-toothed segments; stem leaves not more than 6. Pappus about 4 times as long as fruit. Calcareous soils in hills and mountains. S., C., and E. Spain, Med. France. Pl. 55.   (1508) *C. caeruleus* (L.) C. Presl (*Carthamus c.*) Like 1508b but stems usually more than 20 cm, and stem leaves 10 or more. Flower heads solitary, bluish-purple; involucral bracts shortly glandular, with few or many cobweb hairs. Leaves very variable, simple and toothed to deeply pinnately-cut into 6–10 pairs of spiny lobes, the upper stem leaves narrow egg-shaped. Fruit 6 mm, pappus 1½–2 times as long. Cultivated ground, stony places. S. and C. Portugal, Med. Spain. Pl. 56.

CNICUS | **1509. *C. benedictus* L. BLESSED THISTLE. Cultivated ground, sandy fields. Widespread except in the north.

Sub-family **LIGULIFLORAE** | Flower heads with all florets with strap-shaped corollas (ray or ligulate florets). Many are yellow-flowered and are widespread in Europe. Only the most distinctive or interesting species in this sub-family are included; the following genera are excluded: *Lapsana, Aposeris, Arnoseris, Zazintha, Rhagadiolus, Hedypnois, Hypochoeris, Leontodon, Picris, Podospermum, Taraxacum, Sonchus, Mycelis, Crepis, Hieracium.*

1. *Centaurea sulphurea* 1499b
3. *Centaurea collina* 1503b
5. *Mantisalca salmantica* 1498b

2. *Centaurea clementei* 1504b
4. *Centaurea linaresii* 1505b

SCOLYMUS | 1510. *S. hispanicus* L. SPANISH OYSTER PLANT. Stems narrowly winged; involucral bracts hairless, with narrow papery margin. Pappus with 2–3 hairs. Sandy places, cultivated and uncultivated ground, waysides. Widespread. Pl. 55.   **(1510) *S. maculatus* L. Like 1510 but stems with broad wings 2–5 mm wide at the narrowest part and wings with distinctly thickened white margin; ray-florets with black hairs outside; anthers dark brown. Pappus absent. Cornfields, waysides. S. and C. Portugal, S., C., and E. Spain, S. France. 1510a. *S. grandiflorus* Desf. A perennial to ½ m like 1510 with large terminal and axillary yellow flower heads in a spike-like inflorescence but involucral bracts very hairy. Flower heads with 3 additional very spiny, strongly veined, longer leafy bracts abruptly constricted to a spine; terminal flower heads with 6 such additional bracts; florets up to 3 cm long. Upper part of stem with strong, continuous, spiny wings; stem leaves lance-shaped, deeply cut into spiny, toothed lobes. Pappus with 2–3 hairs. Cultivated ground, waste places. S. Spain, France.

CATANANCHE | **1511. *C. caerulea* L. CUPIDONE. Readily distinguished by its blue, spreading ray-florets and its silvery, papery, inflated involucral bracts with a median brown vein. Dry pastures, bushy places. S., C., and E. Spain, S. France. Pl. 53, Page 139. 1511a. *C. lutea* L. YELLOW CUPIDONE. Distinguished from 1511 by its yellow ray-florets which are shorter than the long, pointed, papery inner involucral bracts. Flower heads about 2½ cm long; outer involucral bracts shining, pale brown. An erect, glaucous, little-branched annual with lance-shaped, rough-haired leaves. Cultivated ground. S. Spain.

HISPIDELLA | Involucral bracts in one row, green, at length encircling the fruits; receptacle honey-combed, without scales; fruit without pappus.

1511b. *H. hispanica* Lam. A small Hawkbit-like plant with solitary, bright yellow flower heads often with a purple centre. Flower heads 2½–5 cm across; ray-florets numerous, often reddish-brown beneath; stalk below flower head swollen. Recalling 1515. *Tolpis barbata* but involucral bracts quite different; they are lance-shaped and shorter than the ray-florets and have intermixed black and white hairs. Leaves oblance-shaped, entire, with long spreading hairs, mostly in a basal rosette. Fields, sandy ground, waysides, principally in mountains. C. and N. Portugal, C. Spain. Pl. 56, Page 110.

CICHORIUM | **Chicory** **1512. *C. intybus* L. CHICORY. Dry fields, waysides, waste places. Widespread.   (1512) *C. endivia* L. ENDIVE. Widely cultivated, sometimes naturalized.   (1512) *C. spinosum* L. SPINY CHICORY. Dry arid places by the sea. S–E. Spain (Almeria province).   1512a. *C. divaricatum* Schousboe A glaucous green perennial with wide-spreading branches, and small blue flower heads in axils of the branches, or at the ends of swollen, hollow branches. Upper leaves small oval, stiff, half-clasping, with a whitish cartilaginous base, lower leaves in a rosette, deeply lobed. Dry fields, derelict cultivation. South of our area.

TOLPIS | **1515. *T. barbata* (L.) Gaertner Readily distinguished by its yellow flower heads with dark reddish-purple centres and its numerous spreading, thread-like, involucral bracts, often spreading beyond the numerous rays. A slender, spreading branched annual, with lance-shaped upper leaves and terminal flower heads 2–3 cm across. Arid, sandy places, uncultivated ground. South of our area. 1515a. *T. umbellata* Bertol. Like 1515 but flower heads smaller 10–18 mm across, yellow throughout, borne in a much-branched, flat-topped inflorescence. A more slender, more hairless annual. Pastures, uncultivated ground, sandy places, etc. Portugal, Spain.

HYOSERIS | 3 species. 1516. *H. radiata* L. A dandelion-like perennial with leaves deeply cut with backward-pointing, toothed, triangular lobes, and longer slender leafless stems bearing a terminal yellow flower head about 3 cm across. Involucral bracts much shorter than the florets, spreading in fruit, the outer about one third as long as the inner; florets showy. Fruit with yellowish pappus, longer than the involucral bracts. Rocks, dry places, cultivated ground. S., C., and E. Spain, S. France. Page 378.

UROSPERMUM | **1526. *U. picroides* (L.) F. W. Schmidt Lower leaves oblong-spathulate, toothed. Cultivated ground, grasslands, rocky places, bushy places. Portugal, S. and E. Spain, S. France. **(1526) *U. dalechampii* (L.) Desf. Lower leaves pinnately lobed. Cultivated ground. Med. region.

TRAGOPOGON | About 7 species. 1527. *T. pratensis* L. GOATSBEARD. Meadows, grassy places, roadsides. Spain, France. **1528. *T. porrifolius* L. SALSIFY. Readily distinguished by its violet-purple, solitary flower heads and its narrow grass-like leaves, often with un-dulate margins. Flower heads variable in size, florets often as long as the involucral bracts and spreading to 4 cm across. Cultivated ground, among herbaceous vegetation. E–C. Portugal, Med. region. (1528) *T. crocifolius* L. Distinguished from 1528 by its contrasting outer violet-purple florets and the yellow central florets, all much shorter than the involucral bracts. Stems not swollen below flower heads; leaves slender, 2–4 mm wide. Stony ground, cultivations. N–E. Portugal, Med. region. Pl. 56. (1528) *T. hybridus* L. (*Geropogon glaber*) An annual 20–60 cm with simple or branched stem, bearing pink or rosy-violet flower heads with 8 much longer narrow acute involucral bracts. Leaves linear, long-pointed. Outer fruits with 5 unbranched hairs, the inner with feathery pappus. Cultivated and uncultivated ground. S. and C. Portugal, S. Spain, S. France.

SCORZONERA | A difficult genus with about 15 species in our area. 1529. *S. purpurea* L. PURPLE VIPERGRASS. Flower heads pale lilac with florets longer than the involucre. Leaves linear, to 3 mm wide; base of stem with old fibrous sheaths. Pastures, rocky ground. Massif Central. Page 152.

1531a. *S. graminifolia* L. An erect perennial 10–45 cm, with numerous glaucous, linear leaves, and large pale yellow flower heads 4–5 cm long and 2–4 cm across. Involucral bracts hairless, or with loose woolly hairs, the outer blunt, shortly triangular-lance-shaped, the inner lance-shaped, narrow-pointed. Leaves entire, 1–3 mm wide, scarcely enlarged in a sheath at the base; stem leafy above. Sandy and stony ground, cultivations, bushy places, hills. Portugal, S., C., and E. Spain. Page 378.

1531. *S. hispanica* L. Cultivated ground, rocks, bushy places. Widespread. Pl. 54. 1531b. *S. baetica* (DC.) Boiss. Like 1531a, but leaves shorter and broader, 4–6 mm wide; flower head smaller 2½ cm long. S. Spain. 1531c. *S. fistulosa* Brot. A hairless, somewhat glaucous plant of marshes and pools with a hollow stem and hollow cylindrical ribbed leaves. Flower heads yellow; rays purplish beneath. Coastal Portugal, S–W. Spain.

ANDRYALA | 6 species. **1533. *A. integrifolia* L. A biennial distinguished by its very woolly foliage and inflorescence which are covered with yellow glandular and star-shaped hairs, and its flat-topped clusters of numerous yellow flower heads. Flower heads about 2 cm across; involucral bracts linear, very woolly-haired and glandular; receptacle with hairs much longer than the fruits. Leaves oblong to linear-lance-shaped. Sands, rocks, tracksides, heaths. Widespread. Pl. 55. 1533a. *A. arenaria* (DC.) Boiss. & Reuter Like 1533 but an annual with upper leaves oval or oval-lance-shaped entire or shallowly lobed, with a rounded or heart-shaped base. Flower heads short-stalked, in a dense flat-topped

1. *Serratula barrelieri* 1497c
2. *Launaea arborescens* 1538c
3. *Hyoseris radiata* 1516
4. *Andryala arenaria* 1533a
5. *Scorzonera graminifolia* 1531a

cluster; florets orange-yellow; receptacle with short hairs. Whole plant covered in very dense, greyish, woolly hairs. Sandy places on the littoral and inland, dry hills. Portugal, S. and C. Spain. Page 378. 1533b. *A. ragusina* L. A tufted, woody-based perennial with a very dense covering of white or yellowish felted hairs, and erect felted stems bearing a lax branched cluster of few conspicuous yellow flower heads with white-felted involucre. Flower heads about 1½–2½ cm across. Leaves lance-shaped, coarsely toothed, or lower leaves pinnately lobed. Stony and sandy places, dry hills, cultivations. S. and C. Portugal, S., C., and E. Spain. Pyrenees.

1533c. *A. agardhii* DC. A shrubby-based, tufted, densely woolly perennial with a close rosette of leaves and erect stems bearing solitary yellow flower heads about 2 cm across. Involucral bracts linear acute, woolly and glandular. Leaves thick, with dense yellowish woolly hairs, lance-shaped to spathulate, stalked; stem leaves few, linear. Rock fissures, stony ground in high mountains. S. Spain (S. Nevada, etc.). Page 76.

CHONDRILLA | 1534. *C. juncea* L. A glaucous, stiff broom-like perennial with green, almost leafless branches and small yellow flower heads in clusters of 2–5 ranged along the upper branches. Flower heads about 1 cm across. Upper leaves linear to lance-shaped. Sandy and stony places, cultivated and uncultivated ground, waysides. Widespread.

REICHARDIA | About 4 species. 1536. *R. picroides* (L.) Roth Fields, waysides, banks, widespread. 1536a. *R. tingitana* (L.) Roth A Hawkbit-like perennial covered with small white projections, with solitary golden-yellow, purplish-centred flower heads 2–2½ cm across. Involucre urn-shaped, bracts with white papery margin, in several rows. Flower stalks leafless, swollen below flower heads. Basal leaves glaucous, in a rosette, cut into finely toothed lobes, stem leaves clasping, entire or toothed. Sandy and grassy places, cultivations, dry hills. S. and E. Spain. Pl. 56. 1536b. *R. gaditana* (Willk.) Coutinho Like 1536a but without small projections. Flower heads large, golden-yellow, 2–4 cm across, rays red beneath; involucral bracts with rust-brown margins, the outer rounded, heart-shaped, with recurved tips. Leaves glaucous with spiny-toothed lobes; stems to ½ m, leafy to the top. Sands and rocks on the littoral. W. Portugal, S. Spain.

CICERBITA | **1537. *C. alpina* (L.) Wallr. BLUE SOW-THISTLE. Flower heads blue-violet, about 2 cm across, in a dense cylindrical cluster; involucre and inflorescence with reddish glandular hairs. A robust leafy perennial ½–2 m. In mountains; damp woods, bushy places, streamsides. Pyrenees, Massif Central. (1537) *C. plumieri* (L.) Kirschleger Differs from 1537 in that the whole plant is hairless. Flower heads blue-violet, in a branched spreading cluster. Acidic mountains; shady places, bogs. Pyrenees, Massif Central. Pl. 56.

LAUNAEA | About 5 species. Like *Lactuca* but woody. 1538a. *L. spinosa* (Forskål) Sch.-Bip. (*L. acanthoclada*) A much-branched, intricately intertwined, shrubby plant to 30 cm, with old branches becoming spiny, and the present year's branches ending in small pale yellow flower heads about 2 cm across. Involucral bracts hairless, with papery margins. Leaves few, linear, coarsely toothed or very shallowly lobed; branches with minute bracts only. Fruit smooth. Dry hills and rocks by the sea. S. Spain. Page 91. 1538b. *L. resedifolia* (L.) O. Kuntze Like 1538a but not spiny and leaves pinnately cut, with widely spaced linear lobes. Flower heads yellow, about 1½ cm across; involucre cylindrical, bracts with white papery margin, the outer egg-shaped, the inner oblong. Fruit with 4 teeth at the base. A variable plant. Dry hills, in lowlands and mountains. S., C., and E. Spain. 1538c. *L. arborescens* (Batt.) Murb. A tiny, intricately branched shrub 4–10 cm, with terminal branches with stiff branched spines. Flower heads yellow, small; involucre cylindrical, about 1 cm long. Leaves linear, shallow-lobed or toothed. Fruit strongly transversely wrinkled. Maritime rocks. Balearic Islands. Page 378.

LACTUCA | About 9, mostly yellow-flowered, species.

**1541. *L. perennis* L. BLUE LETTUCE. Readily distinguished by its beautiful lilac or violet flower heads 3–4 cm across, borne on rather long flower stalks. Leaves deeply pinnately cut into narrow lobes, the upper leaves lance-shaped, clasping the stem, the uppermost bract-like. Fruit black; pappus white. Grassy places, rocks, hills and thickets, principally in mountains. S., C., and E. Spain, S. France.   1541a. *L. tenerrima* Pourret Like 1541 but branches many, slender, forming a tufted growth, and leaves mostly basal. Flower heads violet, smaller, 1½–2 cm across. Fruit brown; pappus yellowish. Rocks, stony ground in the mountains. S., C., and E. Spain, Med. France.

PRENANTHES | **1545. *P. purpurea* L. Shady bushy places in mountains. E. Spain, France.

MONOCOTYLEDONES
ALISMATACEAE | Water-Plantain Family

BALDELLIA | 1558. *B. ranunculoides* (L.) Parl. LESSER WATER-PLANTAIN. Damp places, shallow water. Widespread.

LURONIUM | **1559. *L. natans* (L.) Rafin. FLOATING WATER-PLANTAIN. Lakes, marshes. W. and N. Spain, S. France.

ALISMA | 1560. *A. plantago-aquatica* L. WATER-PLANTAIN. Muddy, shallow water. Widespread.

DAMASONIUM | **1562. *D. alisma* Miller THRUMWORT. Still waters. Very local in Iberia.

SAGITTARIA | 1563. *S. sagittifolia* L. ARROWHEAD. Still and slow-flowing waters. Very local in Iberia.

BUTOMACEAE | Flowering Rush Family

BUTOMUS | **1564. *B. umbellatus* L. FLOWERING RUSH. Still and slow-flowing waters. Very local in Iberia.

HYDROCHARITACEAE | Frog-bit Family

STRATIOTES | **1565. *S. aloides* L. WATER SOLDIER. Still waters. Local in Spain, France.

HYDROCHARIS | **1566. *H. morsus-ranae* L. FROG-BIT. Still waters. Very local in Iberia.

VALLISNERIA | **1567. *V. spiralis* L. Still and slow-flowing waters. Very local in Iberia.

ELODEA | 1568. *E. canadensis* Michx CANADIAN PONDWEED. Still and slow-flowing waters. C. Portugal, Spain, France.

JUNCAGINACEAE (see SCHEUCHZERIACEAE) |
Arrow-grass Family

TRIGLOCHIN | 5 species.   **1569. *T. palustris* L. MARSH ARROW-GRASS. Marshes, damp places. Scattered throughout.   (1569) *T. maritima* L. SEA ARROW-GRASS. Coasts of our area.

POTAMOGETONACEAE

ZANNICHELLIACEAE; NAJADACEAE Widespread species of the following genera occur in our area: *Potamogeton, Groenlandia, Ruppia, Zostera, Cymodocea, Posidonia, Zannichellia, Najas.*

LILIACEAE | Lily Family

TOFIELDIA | **(1583) *T. calyculata* (L.) Wahlenb. Pyrenees.

NARTHECIUM | 1584. *N. ossifragum* (L.) Hudson BOG ASPHODEL. Mountain bogs. N–W. Portugal, N. Spain, Pyrenees.

APHYLLANTHES | **1585. *A. monspeliensis* L. Rush-like, with leafless, ribbed stems and a terminal head of starry blue flowers, each 2½ cm across, encircled by russet-coloured bracts. Dry rocky places. N–E. Portugal, Med. region.

VERATRUM | **1586. *V. album* L. WHITE FALSE HELLEBORINE. A robust plant with whorls of broad, pleated, oval leaves, and a terminal branched cluster of numerous, green, white or yellowish flowers. Leaves finely hairy. Mountain pastures. C. and N. Portugal, C. and N. Spain, France.

MERENDERA | **1587. *M. montana* (L.) Lange Easily recognized in autumnal pastures by its leafless flowers with narrow pinkish-lilac, strap-shaped petals which spread in a star in the turf. Leaves 4–6 mm wide, appearing after flowering. Distinguished from *Colchicum* species by the petals which are not fused together in a tube at the base though closely pressed together. Mountain pastures. C. and N. Portugal, Spain, Pyrenees. Pl. 57. 1587a. *M. filifolia* Camb. Like 1587 but leaves several, very slender, 1–2 mm broad, appearing with the flowers and longer than the fruit stalk. Sandy places S. Portugal (Algarve), Balearic Islands.

BULBOCODIUM | 1589. *B. vernum* L. A spring-flowering crocus-like plant with rosy-lilac, or rarely white flowers. Stamens 6. Leaves 3, lance-shaped, without a white mid-vein, appearing with the flowers. Mountain pastures. Pyrenees. Page 389.

COLCHICUM | Autumn Crocus

1. Spring flowering; lvs. appearing with fls.	*triphyllum*
1. Autumn flowering; lvs. appearing long after fls.	
2. Stigmas arched or slightly curved; stamens all inserted at the same level; fr. elliptical, size of a hazel-nut, usually surrounded by 3 linear-lance-shaped lvs. Corm 2–3 cm across.	*lusitanum*
2. Stigmas curved at the top like a crochet-hook; stamens inserted at two levels; fr. oboval, size of a walnut, usually surrounded by 3 broadly lance-shaped lvs. Corm 3–5 cm across.	*autumnale*

**1588. *C. autumnale* L. MEADOW SAFFRON, AUTUMN CROCUS. Flowers rosy-purple or white, appearing leafless in the autumn from colourless basal sheaths. Petals 3–4½ cm; tube of corolla 5–20 cm. Leaves usually 3, oblong-lance-shaped, appearing in spring. Damp meadows. C. and N. Portugal, C. and N. Spain, France.   1588a. *C. lusitanum* Brot. (*C. bivonae* auct.) Like 1588 but flowers larger 6 cm, and petals broader, pink with network of white veins. Leaves linear-lance-shaped. Stony ground, dry scrub-covered hillsides. C. Portugal, Med. region. Pl. 57.   1588b. *C. triphyllum* Kunze (*C. clementei*) Flowers pink or white, 2 cm across, appearing with the leaves in spring and borne on a distinct stalk above the leaves. Leaves 3, narrow sword-shaped, glandular-ciliate, deeply channelled, curving outwards, as long as the flowers. Wet clay soils in mountains, often near melting snow. S. and C. Spain. Page 77.

ANDROCYMBIUM | Distinguished by its stamens which are attached to the middle of the perianth segments, above yellow glands.   1588c. *A. gramineum* (Cav.) Macbride (*Erythrostictus g.*) A prostrate bulbous plant with a cluster of several white flowers striped with mauve, borne at the centre of a rosette of spreading leaves. Perianth segments lance-shaped acute; stamens arising from a yellow gland in the centre of each segment. Leaves 6–10, linear-lance-shaped, with translucent glands. Sandy and stony places. S. Spain (Cabo de Gata). Pl. 59, Page 91.

ASPHODELUS | Asphodel

1. Roots fibrous, little thickened, without tubers; lvs. rounded in section, less than ½ cm broad.	
2. Anns.; lvs. and stems not hollow. (S. Spain)	1592a. *A. tenuifolius* Cav. Page 90
2. Perenns.; lvs. and stems hollow.	*fistulosus*
1. Roots with oblong, turnip-like tubers; lvs. V-shaped in section, more than ½ cm broad.	
3. Fr. small, 5–8 mm.	
4. Fr. ovoid, longer than wide.	*aestivus*
4. Fr. globular, wider than long. (E–C. Portugal)	1591a. *A. bento-rainhae* A. R. Pinto da Silva
3. Fr. larger, 8–22 mm.	
5. Bracts black or dark brown, longer than the lower fl. stalks.	*albus*
5. Bracts paler not longer than lower fl. stalks.	
6. Fr. ovoid 10–12 mm.	*morisianus*
6. Fr. more or less globular, 13–22 mm.	*ramosus*

1590. *A. albus* Miller WHITE ASPHODEL. A very distinctive species with stout, erect, un-branched stems bearing a dense spike of large white or pinkish flowers each 3–5 cm across, subtended by long, dark bracts. Leaves all basal, V-shaped in section, 1–2 cm broad. Fruit about 1½ cm long. Dry hills, heaths, in mountains. Portugal, Med. Europe. Pl. 57.   1590a. *A. ramosus* L. (*A. cerasifer*) Inflorescence usually with three short lateral clusters and a much longer terminal spike-like cluster, and pale brown papery bracts. Fruit large, cherry-sized, 13–22 mm. Dry or damp places, preferring limestone. South of our area.

**1591. *A. aestivus* Brot. (*A. microcarpus*) ASPHODEL. Distinguished by its tall, freely branched, more or less pyramidal inflorescence, the lower branches often few but wide-spreading. Flowers white with a reddish midvein, 3–4 cm across. Fruit small 5–8 mm, ovoid. Waste ground, heaths, pine woods. Widespread except in the north.   1591b. *A. morisianus* Parl. (*A. lusitanicus*) Like 1591 but a stouter, robuster plant. Inflorescence branched, the terminal spike-like cluster not much longer than the laterals. Fruit 10–12 mm, ovoid. Abundant weed. S. Portugal, S. Spain. Page 61.

**1592. *A. fistulosus* L. HOLLOW-STEMMED ASPHODEL. A rather slender plant 20–60 cm with a lax, sparsely branched inflorescence and pale pinkish flowers about 2 cm across. Leaves 1–3 mm broad, semi-circular, more or less hollow. Fruit globular 4–6 mm, with transverse ridges. Dry sunny places. Widespread except in north. Page 44.

PARADISEA | **1595. *P. liliastrum* (L.) Bertol. ST BRUNO'S LILY. Mountain meadows. N. Spain, Pyrenees, Massif Central. Page 138.

ANTHERICUM | **1596. *A. liliago* L. ST BERNARD'S LILY. Flowers white, 3–5 cm across, in an unbranched inflorescence. Leaves usually as long as the flowering stem, 5–7 mm wide. Wood, scrub, lowlands and in mountains. N–E. Portugal, N–E. and S–E. Spain, France. Page 132.   (1596) *A. ramosum* L. Distinguished from 1596 by the branched inflorescence, smaller flowers 2½ cm across, and leaves 4–5 mm wide, much shorter than the flowering stem. S–W., C., and E. Spain, France.   1596a. *A. baeticum* (Boiss.) Boiss. A smaller plant with an unbranched inflorescence to 30 cm, with 1–10 white flowers 12–24 mm across. Stamens equal in length to perianth segments; style curved downwards. Leaves narrow 2–4 mm, twisted, much shorter than the inflorescence. Damp places in mountains. S. Spain. Page 389.

SIMETHIS | Differs from *Anthericum* in having perianth segments with 5–7 veins; stamens woolly-haired below. Seeds with an aril.   (1596) *S. mattiazzii* (Vandelli) Saccardo A small asphodel-like plant with wide-branched, irregular inflorescence of small white flowers with purplish lower sides to the petals. Petals 8–10 mm, spreading in a star; filaments hairy. Leaves all basal, linear, often twisted, about as long as inflorescence. Fruit globular. Bushy places. Widespread. Page 103.

HEMEROCALLIS | 1597. *H. fulva* (L.) L. DARK DAY-LILY;   **(1597) *H. lilioasphodelus* L. PALE DAY-LILY. Both often grown for ornament and sometimes naturalized in the south.

ALOE | Several species from S. Africa have become naturalized including: 1598. *A. vera* (L.) Burm. fil. Flowers yellow, in an unbranched spike; leaves pale green, teeth few.   1598a. *A. succotrina* All. (*A. purpurascens*) Flowers red; leaves curved upwards, with hooked purple teeth.   1598b. *A. arborescens* Miller Flowers red, in a conical spike; stem 3–4 m; leaves pale green, glaucous, with forward-pointing yellow horny teeth on the edge.

KNIPHOFIA | 1598c. *K. uvaria* (L.) Hooker RED-HOT POKER. Flowers orange-red, borne in an oval-oblong spike-like cluster; stamens longer than corolla. Leaves grey-green, channelled, not spiny, 2–2½ cm wide. Sometimes naturalized from gardens in the south.

GAGEA | About 15 species occur in S–W. Europe, mainly in the mountains. They all appear very similar and are difficult to distinguish. See *Revision monographia della specie di Gagea della Flora Spagnola*, Terraciano A., Bot. Soc. Aragon Cienc. Nat. Zarogoza 1905. *Die Gattung Gagea*, Stroh, G., Salisb. Berh. Bot. Centralblatt 1937; 485–520.

ALLIUM | **Onion, Garlic, Leek**   About 45 species occur in our area with perhaps 12 restricted to the Iberian peninsula in Europe. Space permits treatment of only a few species.

Flowers yellow
**1606. *A. flavum* L. YELLOW ONION. Flowers golden-yellow, bell-shaped, long-stalked and at length drooping; spathes 2, very long and slender. Meadows, woods. N–C. Spain, Pyrenees, Cévennes.   1606a. *A. moly* L. Flowers golden-yellow; bracts 2, short, oval. Leaf glaucous, often solitary, attached towards the base of the stem, 1–2½ cm broad. N–W. Portugal, E. Spain, Pyrenees. Pl. 58.   1606b. *A. stramineum* Boiss. & Reuter Flowers pale yellow; perianth segments acute, 8–10 mm. Leaves 2 or more linear, 3–5 mm wide, all basal; flower stalk angled above. C. and N. Portugal, S–W. Spain. Pl. 58, Page 110.

Flowers white
**1610. *A. triquetrum* L. TRIQUETROUS GARLIC. Distinguished by its one-sided umbel of drooping, white, bell-shaped flowers, with green mid-veins, borne on a sharply three-angled stem. Leaves several, ½–1 cm broad, channelled. Damp places. C. Portugal, S. and E. Spain, Med. France. Page 60.   1608a. *A. subvillosum* J. A. & J. H. Schultes A small plant with many densely clustered white flowers, with blunt spreading perianth segments about 5 mm long. Stamens yellow, as long as perianth. Leaves 3, near base, linear long-pointed, keeled, densely ciliate on margin above. Maritime sands. S–W. Portugal, S. Spain.

1617. *A. victorialis* L. ALPINE LEEK. Distinguished by its globular head of greenish-white bell-shaped flowers which become yellowish. Stamens longer than perianth segments. Leaves broad, oblong-elliptic, short-stalked, sheathed. Woods and rocks in mountains. C. Portugal, N. and E. Spain, France.

Flowers pink
**1611. *A. roseum* L. ROSE GARLIC. Distinguished by its umbel of rather large, erect, attractive, usually bright pink and distinctly bell-shaped flowers. Spathes 2–4, papery, persistent, shorter than the more or less equal flower stalks. Leaves 2–4, keeled. Sandy, stony ground, cultivated. Widespread except in the north. Pl. 58. Page 162.   1611a. *A. montanum* F. W. Schmidt (*A. fallax*) A small thrift-like plant to 30 cm, with a dense rounded head of pink-purple flowers, with stamens longer than the perianth segments which are 6–7 mm. Leaves all basal 3–8, shorter than flowering stem, 1–4 mm broad, flat above, rounded not keeled beneath. Stock stoloniferous, with oval lobes. Stony, rocky places in mountains. C. Portugal, S. and E. Spain, Pyrenees, Massif Central.

1614a. *A. polyanthum* J. A. & J. H. Schultes A robust plant to 1 m or more, distinguished by its large, very dense, globular head of numerous pinkish-purple flowers 4–8 cm across, and short spathe which soon falls. Filaments of stamens with 2 lateral twisted projections; anthers yellow, as long as or a little longer than the perianth. Leaves 3–5, flat, keeled beneath, 1–1½ cm wide. Cultivated ground, in mountains. S. and E. Spain, S. and C. France. Pl. 58.   1604. *A. vineale* L. CROW GARLIC. Flowers usually pink, in lax umbels and usually with stalkless bulbils, or bulbils only present. Filaments of stamens with 2 projections longer than the anthers. Leaves half-cylindrical, hollow, about 2 mm broad. Cultivated ground. Widespread.

LILIUM | ****1618.** *L. martagon* L. MARTAGON LILY. Flowers pink or pale purple, nodding. Woods, meadows, in mountains. N. Portugal, C., N., and E. Spain, Pyrenees, France. ****1619.** *L. pyrenaicum* Gouan YELLOW TURK'S-CAP LILY. Flowers bright yellow, nodding. Woods and meadow, in mountains. N. Spain, Pyrenees, S–W. France.   ****1620.** *L. bulbiferum* L. ORANGE LILY. Flowers erect, bright orange with black spots; leaves with or without bulbils. Mountains pastures. N. Spain, Pyrenees.

FRITILLARIA | **1622. *F. meleagris* L. FRITILLARY. Damp meadows. E. Spain, France.
1623. *F. pyrenaica* L. PYRENEAN SNAKE'S HEAD. Flowers variable, large, bell-shaped, often unpleasantly scented, with deep brown and purple shades prevailing and with a few green markings on the outside, within lustrous green or yellow with slight chequering. Perianth segments 3–4 cm, unequal, the inner broader, oblong, the outer lance-shaped, all outcurved at apex. Leaves 6–10, glaucous; stem leafless below. Woods, pastures. N. Spain, Pyrenees, Cévennes. Pl. 57, Page 148.

1623a. *F. lusitanica* Wikstr. Flowers 2–3 cm long, bell-shaped, dull brick or orange-red, or sometimes purplish outside, yellowish within and becoming sparingly brown-chequered. Leaves glaucous, 7–9, linear-long-pointed, the lower leaves opposite, usually ½–1 cm wide. Subsp. *stenophylla* (Boiss. & Reuter) Coutinho has leaves all linear, the lowest usually 1–4 mm broad, and the fruit more rounded and flat-topped. Stony, bushy places. C. Portugal, S–W. Spain. Pl. 57.   1623b. *F. hispanica* Boiss. & Reuter Flowers bell-shaped, red-brown with a median yellow-green band, but not chequered outside, and with a median yellow band and chequered with brown within. Leaves 7–13, scattered, the lower linear-lance-shaped, the upper linear. Thickets, hills, lowlands to mountains. S., C., and E. Spain. Page 116.

TULIPA / TULIP

1. Lower lvs. oblong-lance-shaped, 3–6 cm broad; fls. red within, usually more than 5 cm long. (N. Spain, S. France) . . . . **1626. *T. oculus-solis* St Amans
1. All lvs. linear to linear-lance-shaped, at most 2 cm broad; fls. yellow or white within, usually less than 5 cm long.
   2. Lvs. 4–5, spreading; bulb woolly under tunic; perianth segments white, the outer pink outside, violet blotched at base. (Introd. C. Portugal, Spain) . . . 1627. *T. clusiana* DC.
   2. Lvs. 2–3, ascending; bulb hairless or nearly so; perianth segments yellow within, never violet-blotched.
      3. Lvs. usually 3; fls. 3–5 cm long; perianth segments hairy within, unequal, the inner broader. (C. Spain, France) . . . . 1624. *T. sylvestris* L.
      3. Lvs. usually 2; fls. 2–3 cm; perianth segments nearly hairless, all nearly equal, lance-shaped. . . . . *australis, celsiana*

**1625. *T. australis* Link Flowers small, with narrow lance-shaped pointed segments which are predominantly yellow, but the outer 3 segments often reddish on the outside; buds nodding or erect. Leaves linear, variable in size, glaucous. Fields, meadows, rocks, in mountains. Portugal, S. and E. Spain, S. France. Pl. 57, Page 84.   1625a. *T. celsiana* DC. Similar coloured to 1625, but differing in possessing stolons. Flowers opening widely in a star. S. Spain.

ERYTHRONIUM | **1628. *E. dens-canis* L. DOG'S TOOTH VIOLET. Alpine pastures. N. Portugal, N. and E. Spain, Pyrenees, Massif Central. Page 133.

URGINEA | **1630. *U. maritima* (L.) Baker SEA SQUILL. Unmistakable with its tall, quite leafless, long-stalked spike 1–1½ m of numerous white flowers arising in the autumn from the bare ground. Leaves broad, persisting through winter to following summer; bulbs very large. Sandy and rocky hillsides, particularly by the coast. S. and C. Portugal, Med. region. Page 62.

SCILLA / SQUILL

1. Bracts paired, about as long as the fl. stalk.
  2. Fr. more or less globular; pollen blue; plant 10–30 cm.   *italica*
  2. Fr. oblong; pollen yellow; plant to *c.* 10 cm (S–W. Portugal)   1634a. *S. vincentina* Hoffmanns. & Link
1. Bracts solitary, or absent.
  3. Bracts much shorter than fl. stalks, or absent.
    4. Bracts 4–7 mm; lv. solitary.   *monophyllos*
    4. Bracts absent or almost so; lvs. more than 1.
      5. Autumn-flowering. (Widespread)   **1636. *S. autumnalis* L.
      5. Spring-flowering.
        6. Lvs. 2 or rarely 3 or more; plant 10–25 cm.   *bifolia*
        6. Lvs. 10–12; plant ½–1 m.   *hyacinthoides*
  3. Bracts half as long as fl. stalks, or longer.
    7. Lvs. 1–5 cm, broad.
      8. Bulb 5–7 cm, with woolly tunic; anthers yellowish.   *peruviana*
      8. Bulb smaller, with hairless tunic; anthers bluish.   *liliohyacinthus*
    7. Lvs. 3–10 mm broad.
      9. Bulb 10–15 mm; fls. few, in a dense flat-topped cluster; lvs. usually shorter than inflorescence. (Widespread except in south)   1633. *S. verna* Hudson
      9. Bulb 18–25 mm; fls. in an elongated cluster; lvs. usually longer than inflorescence.
        10. Fls. blue, scentless, perianth acute; inflorescence with *c.* 30 fls. Lvs. 3–10 mm broad; stem 10–60 cm. (C. and N. Portugal, S. Spain)   1633a. *S. ramburei* Boiss. Page 60, 389
        10. Fls. intense violet-blue, scented; perianth blunt; inflorescence with *c.* 12 fls. Lvs. 2–4 mm broad; stem 6–12 cm. (S. Portugal, S. Spain)   1633b. *S. odorata* Link

1635a. *S. italica* L. (*Hyacinthoides i.*) Flowers blue, rather numerous in a short conical cluster; perianth 5–8 mm; paired bracts unequal. Leaves 3–6, channelled, 4–8 mm broad, usually shorter than the inflorescence. Shady rocks, in the littoral. C. Portugal and Spain? 1636a. *S. monophyllos* Link Flowers blue, few in a lax cluster, each subtended by a bluish bract 3–4 times shorter than the individual flower stalks. Perianth 6–8 mm, segments with a green midvein. Leaf solitary, lance-shaped. Bushy places, sandy ground, pine woods. Portugal, W. Spain. Pl. 59, Page 160.

1631. *S. hyacinthoides* L. A robust plant ½–1 m with a conical-cylindrical spike of very numerous, long-stalked, blue-violet flowers. Flower stalks and anthers violet; bracts very short. Leaves 10–12, 1½–3 cm broad. Fields and rocky places. S. and C. Portugal, S. Spain. **1632. *S. peruviana* L. A striking plant with a broad dense hemispherical cluster of blue or violet-blue flowers with yellowish stamens, borne on a robust stem 20–50 cm. Leaves numerous, very broad 2–6 cm. Bulb large, 5–7 cm. Damp, fertile ground, valleys. S. and C. Portugal, S. Spain.   1632a. *S. liliohyacinthus* L. Distinguished by its lax oval or conical cluster of rather numerous blue flowers each subtended by violet bracts as long as the ascending flower stalks. Perianth segments 1 cm long; anthers bluish. Leaves numerous, 1½–3 cm wide; flowering stems slender. Bulb large, yellowish, with loosely, overlapping scales. Shady fertile places in mountains. N. Spain, Pyrenees, France. Page 133.

**1635. *S. bifolia* L. ALPINE SQUILL. A small plant distinguished by its pair of broad, shining leaves, and its cluster of 2–8 blue flowers without bracts. Perianth segments 6–9, wide-spreading; anthers violet. Shady fertile soils in mountains. C., W., and N. Spain, C. France.

DIPCADI | 1637. *D. serotinum* (L.) Medicus A small bluebell-like plant with a one-sided spike of yellowish, brownish or orange-red flowers, each 12–15 mm long. Perianth segments fused below, the 3 outer with outcurved tips, the 3 inner straight. Leaves narrow, linear, grooved. Sandy and rocky places, scrub, in mountains. Portugal, S., C., and E. Spain, S. France. Pl. 59.

ENDYMION (*Hyacinthoides*) | **1638. *E. non-scriptus* (L.) Garcke BLUEBELL. Inflorescence nodding, one-sided; perianth tubular: anthers yellowish. Woods, thickets, shady places. C. Portugal, C. and N. Spain, France.   **(1638) *E. hispanicus* (Miller) Chouard SPANISH BLUEBELL. Distinguished by its erect, pyramidal inflorescence, and broadly bell-shaped flowers with spreading segments; anthers blue. Bushy places, shady limestone rocks. Portugal, S., C., and W. Spain, S. France. Page 69.

ORNITHOGALUM | Star-of-Bethlehem

1. Fls. stalkless or almost so.
    2. Plant 2–11 cm; lv. usually 1.   *spicatum*
    2. Plant 12–30 cm; lvs. 2–4.   *concinnum*
1. Fls. all stalked.
    3. Stem slender, usually less than 30 cm; lvs. usually narrow; fls. in a flat-topped cluster; fl. stalks unequal.
        4. Lvs. very slender usually 1–2 mm broad, without a well-defined white line; fls. 1–5; fl. stalks 1–2 cm; bulbils absent.
            5. Lvs. 2–4 ascending, equal or slightly longer than inflorescence; fl. stalks erect, the lower longer than the bracts. Fls. 12–15 mm long.   *tenuifolium*
            5. Lvs. more numerous, out-spreading, curved, much longer than inflorescence; fl. stalk spreading, bent after flowering, at first shorter than the bracts. Fls. 15–18 mm long. (S. Spain)   1639c. *O. exscapum* Ten.
        4. Lvs. 2–6 mm broad, with a well-defined white line; fls. 5–15; fl. stalks 3–6 cm; bulbils present. (Widespread)   **1639. *O. umbellatum* L. Page 161.
    3. Stem robust 30–80 cm; lvs broadly linear; fls. in a short or elongated spike-like cluster; fl. stalks equal.

6. Lvs. equal or longer than inflorescence;
   fls. at least 1½ cm long; fr. with 6 ridges.
7. Fls. white within, green outside. (C.
   and E. Spain, C. France)   1642.   *O. nutans* L.
7. Fls. white.
   8. Fls. up to 15 in a short rather flat-
      topped cluster.   *arabicum*
   8. Fls. more numerous in a long spike-
      like cluster.   *reverchonii*
6. Lvs. much shorter than inflorescence;
   fls. 10–12 mm long; fr. with 3 ridges.
   9. Fls. pure white within, not yellowing on
      drying; filaments half as long as peri-
      anth; style longer than stamens. (Wide-
      spread)   1641.   *O. narbonense* L. (*O. pyramidale*)
   9. Fls. greenish-white, yellowing on dry-
      ing; filaments three-quarters as long
      as perianth; style equal to stamens.
      (Widespread)   **1640.   *O. pyrenaicum* L.

1639a. *O. spicatum* Planellas A small plant to 10 cm with a lax spike of few, nearly stalkless white flowers with a green band on the outside. Perianth 1–1½ cm long, bracts about half as long. Leaf usually 1, closely encircling the stem at its base and much longer than it. Sandy heaths, pine woods. Portugal, W. Spain. Pl. 58.   1639b. *O. concinnum* Salisb. Like 1639a but flowers more numerous, 6–15 or more, slightly scented, borne at length in a long spike on a taller stem 12–30 cm. Leaves 2–4. Turf, stony places. C. and N. Portugal. Page 389.

1639d. *O. tenuifolium* Guss. Like 1639 with a flat-topped cluster of few white flowers, but leaves much narrower, usually 1–2 mm wide, without a conspicuous white line. Flowers 1–5, variable, 1½–2½ cm across, green outside; flower stalks erect. Dry hills, low scrub. E. Spain, S. and S–C. France.

1642a. *O. arabicum* L. Flowering stem stout, up to ½ m, bearing a short dense cluster of 12–15 large white fragrant flowers up to 5 cm across, which at length turn yellowish. Perianth segments oval-blunt, without a green band on the outside. Leaves glaucous, without a white line, 1½–2 cm broad. Fruit black. Rock and sandy places on the littoral. C. Portugal, S., W., and E. Spain. Page 53.   1642b. *O. reverchonii* Lange 'Reminiscent of a large white bluebell' growing from the limestone cliffs of the S. de Ronda in S. Spain. Flowers pure white, at length spreading in a star, 3–3½ cm across. Inflorescence 8–10 cm with bracts as long as or longer than the flowers; filaments broad, white; style white. Leaves linear, nearly flat, 2–3 cm wide; stem leafless 35–40 cm. Bulb large. Pl. 58, Page 68.

HYACINTHUS | (1643) *H. amethystinus* L. PYRENEAN HYACINTH. A slender plant to 30 cm, with a lax one-sided cluster of 3–12 bright blue, tubular, nodding flowers, each 7–10 mm long. Perianth segments longer than tube; filaments much shorter than the anthers. Bracts blue. Leaves narrow, 2–4 mm wide, grooved. Meadows, rocks, screes. N–E. Spain, Pyrenees. Pl. 59. Page 148.   1643a. *H. fastigiatus* Bertol A similar plant growing only in the Balearic Islands with a small, flat-topped crowded cluster of 2–6 violet, or sometimes pink, or white flowers. Perianth segments spreading, longer than the tube; filaments twice as long as the anthers. Stony hills, woods.

1. *Biarum carratracense* 1820b
2. *Asparagus stipularis* 1650a
3. *Bulbocodium vernum* 1589
4. *Lapiedra martinezii* 1664a
5. *Anthericum baeticum* 1596a
6. *Scilla ramburei* 1633a
7. *Ornithogalum concinnum* 1639b

B.E.

BELLEVALIA | **1644. *B. romana* (L.) Sweet Grassy places. S–W. France.   1644a. *B. hackelii* Freyn Recalling a small form of 1645, but without a top-knot of bright blue sterile flowers. Flowers in a lax cylindrical cluster, the young flowers bright blue but later turning dark violet. Flower stalks spreading at flowering, shorter than or as long as the perianth. Leaves 2–5, linear, as long as or longer than the inflorescence. Plant 20–25 cm. Scrub-covered slopes. S. Portugal. Pl. 59. Page 45.

MUSCARI | **1645. *M. comosum* (L.) Miller TASSEL HYACINTH. Readily distinguished by its top-knot of bright violet-blue, long-stalked, sterile flowers, contrasting with dark blue and later brownish-green mature flowers. Leaves ½–1½ cm wide, channelled. Rocky ground, fields, vineyards, etc. Widespread. Page 162.   1647. *M. atlanticum* Boiss. & Reuter GRAPE-HYACINTH. Distinguished by its dark blue, plum-scented flowers with ovoid corollas 4–5 mm long, with a pinched-in throat and with white teeth. Sterile flowers stalkless, paler. Leaves 1–3 mm broad, cylindrical. Dry fields, vineyards. Widespread.

**(1647) *M. neglectum* Ten. Very like 1647 but flowers larger 6–8 mm long, more open at the throat, with whitish teeth. Leaves broader 3–5 mm, conspicuously grooved. Fields, vineyards. C. and N. Portugal, E. Spain, S. France.   **1648. *M. botryoides* (L.) Miller SMALL GRAPE-HYACINTH Readily distinguished by its dense conical cluster of very small, almost globular pale blue or violet flowers. Corolla 3–4 mm; teeth white. Leaves broadening towards apex. Meadows, fields, woods. C. Spain, Massif Central.

ASPARAGUS | 7 species, including 3 herbaceous species. The true leaves are reduced to spines or scales and bear in their axils false leaves (*cladodes*) which are solitary or in a cluster and are often needle-like, with or without spiny tips.

Woody species

1. Branches cylindrical, downy. Cladodes 3–6 mm,
   stiff, spiny-tipped. (Widespread except in north)   **1650.   *A. acutifolius* L.
1. Branches angular or ribbed, hairless.
   2. True lvs. reduced to a stout spine. Cladodes in
      clusters, not spiny-tipped. (S. and C. Portugal, S.
      and C. Spain)   (1650)   *A. albus* L.
   2. True lvs. very small or reduced to a spiny spur.
      Cladodes 1 or several, spiny-tipped.
      3. Cladodes in bundles of 3–5. (S. and C. Portugal,
         S. and C. Spain)   (1650)   *A. aphyllus* L.
      3. Cladodes solitary, robust four-angled often 5–7½
         cm long. (C. Portugal, S. and C. Spain)   1650a.   *A. stipularis* Forskål.
                                                              Page 389.

RUSCUS | 1651. *R. aculeatus* L. BUTCHER'S BROOM. Thickets, woods, dry hills, rock crevices. Widespread.   **(1651) *R. hypoglossum* L. LARGE BUTCHER'S BROOM. Shady rocks. E. Spain. 1651a. *R. hypophyllum* L. Like (1651) but flowers in a cluster of 2–6 on the 'undersides' of the 'leaves'; bracts short, papery. S. Spain (Gibraltar).

MAIANTHEMUM | 1652. *M. bifolium* (L.) F. W. Schmidt MAY LILY. Shady woods. C. and E. Spain, C. France.

STREPTOPUS | **1653. *S. amplexifolius* (L.) DC. Woods and damp rocks in mountains. N. and C. Spain, Pyrenees, Massif Central.

POLYGONATUM | ****1654.** *P. odoratum* (Miller) Druce SWEET-SCENTED SOLOMON'S SEAL. Montane woods and rocky places. Widespread. ****1655.** *P. multiflorum* (L.) All. SOLOMON'S SEAL. Montane woods. N. and N–E. Spain, France. 1656. *P. verticillatum* (L.) All. WHORLED SOLOMON'S SEAL. Montane woods. C., N., and N–E. Spain, Pyrenees, Massif Central.

CONVALLARIA | ****1657.** *C. majalis* L. LILY-OF-THE-VALLEY. Montane woods. N. Spain, France.

PARIS | ****1658.** *P. quadrifolia* L. HERB PARIS. Damp woods, C. and E. Spain, France.

SMILAX | 1659. *S. aspera* L. A woody tendril climber with hooked spines on both stems and leaves, small clusters of greenish-yellow flowers, and red fleshy berries. Leaves variable, lance-shaped to triangular heart-shaped, leathery or soft, with or without spines. Berries often black. Scrub, thickets. Widespread. Page 63.

AGAVACEAE | Agave Family

AGAVE |****1660.** *A. americana* L. CENTURY PLANT. Widely naturalized on Mediterranean and S–W. Atlantic coasts for over two centuries.

1660a. *A. sisalana* Perrine SISAL. It is planted both commercially and ornamentally in Portugal and S. Spain. It is distinguished from 1660 by the spear-shaped leaves 1½ m long, without marginal spines but with a dark brown terminal spine. Other similar-looking woody plants often grown for ornament and occasionally self-seeding are: *Agave vivipara* L.; and the liliaceous *Sansevieria zeylanica* Willd. BOWSTRING HEMP; *Dracaena draco* L. DRAGON TREE; *Phormium tenax* J. R. & G. Forster NEW ZEALAND FLAX; *Yucca gloriosa* L. YUCCA.

AMARYLLIDACEAE | Daffodil Family

LEUCOJUM | ****1661.** *L. vernum* L. SPRING SNOWFLAKE. Flowers solitary or rarely 2, 1½–2½ cm long. Damp woods, meadows. N. Spain, Pyrenees. ****1662.** *L. aestivum* L. SUMMER SNOWFLAKE. Like 1661 but flowers in a cluster of 2–8, smaller 1–1½ cm long. Damp places. C. Spain, C. France. 1662a. *L. autumnale* L. A slender, autumn-flowering plant with 2–3 nodding white flowers which are often purplish at the base, and a single papery terminal sheath. Perianth segments 6–10 mm long. Leaves 2–4, thread-like, shorter than flowering stem which is up to 20 cm. Sandy, dry stony places, banks. Portugal, S–W., W., and C. Spain. Page 63. ****(1662)** *L. trichophyllum* Brot. THREE-LEAVED SNOWFLAKE. A very attractive, delicate spring-flowering plant with 1–5 nodding white flowers, often flushed with pink, borne on a slender stem and subtended by 2 papery sheaths. Perianth segments usually 1½ cm long but sometimes up to 2½ cm long. Leaves thread-like, as long as the flowering stem. Sandy places, often by the sea. S. and C. Portugal, S–W. Spain. Pl. 59, Page 61.

GALANTHUS | ****1663.** *G. nivalis* L. SNOWDROP. Damp woods and meadows, in mountains. Pyrenees. W. and C. France.

STERNBERGIA | **1664. *S. lutea* (L.) Sprengel COMMON STERNBERGIA. Marshy and grassy places, rare. S. Spain, Med. France; naturalized in N–E. Portugal.

**(1664) *S. colchiciflora* Waldst. & Kit. SLENDER STERNBERGIA. Stony places on limestone, rare. S. and C. Spain, S. France.

LAPIEDRA | Flowers 3–8; tube of corolla short, lobes spreading in a star, green outside, white within; stamens 6, shorter than perianth, anthers arrow-shaped broader than the filament. 1664a. *L. martinezii* Lag. Easily recognized when not in flower by its dark blue-green leaves with a broad, almost white, stripe down the midrib. Flowers in an umbel, white with a green band or green outside, and white within, lilac-scented, becoming papery, anthers orange. Stem 10–20 cm, compressed; sheaths 2–3. Leaves broadly linear, blunt, spreading. Limestone, rock crevices. S–W. and S. Spain. Page 389.

NARCISSUS | **Daffodil, Narcissus**   A difficult genus, due in part to frequent hybridization, and in part to the great variability within many species. Whilst the majority of plants can be identified in the following keys, individual plants may not conform. Field notes should be made from the living plant of; (a) width and cross-section of leaves and (b) the cross-section of the flower stalk (*scape*). (c) the colour of perianth and in particular the *corona*. In the keys below: the length of perianth tube excludes the ovary; the *corona* is the additional cup or flap of tissue arising from the point where the tube of the perianth flares into the separate perianth segments or 'petals'. About 45 species, including 23 restricted to the Iberian peninsula in Europe. See, *Keys to the Identification of Native and Naturalized Taxa of the genus Narcissus* Fernandes A., Royal Hort. Soc. Daffodil and Tulip Year Book 1968; 37–66.

Keys to sections etc.

1. Tube of perianth (not corona) long and narrow,
     rarely wider at the throat; stamens unequal;
     corona cup-shaped or rudimentary.
   2. Fls. in autumn, appearing without lvs.
      3. Perianth segments pure white; corona
           orange or yellow. (S. Portugal, Med. region)   **1672. *N. serotinus* L. Page 63
      3. Perianth segments and corona green.   *viridiflorus* (see *JONQUILLA*)
   2. Fls. in spring or autumn, appearing with the
        lvs.
      4. Lvs. flat above, bluntly keeled beneath,
           3–20 mm wide; fl. stalk more or less flattened,
           two-edged (sometimes almost rounded).
         5. Fls. in clusters of 4–20; corona uniform in
              texture and colour.   section: *HERMIONE*
         5. Fls. usually 1–3; corona yellow with the
              margin edged with red or becoming papery.   section: *NARCISSUS*
      4. Lvs. rounded in section or nearly so, some-
           times thread-like; finely ribbed or two- to
           four-keeled on the outer surface; fl. stalk
           usually rounded in section.

6. Perianth segments spreading or slightly reflexed; fls. not drooping; pollen golden-yellow.

  7. Lvs. bright green, narrowly linear to round. Fls. 2–6 golden-yellow, or pale yellow, very fragrant, each long-stalked; seeds angular, without a swelling.   section: *JONQUILLA*

  7. Lvs. glaucous, channelled above, two- to four-keeled below. Fls. 1–4, golden-yellow, cream or white, scarcely fragrant; seeds spherical, with a swelling.   section: *APODANTHAE*

6. Perianth segments sharply turned back; fls. drooping; pollen pale yellow.   section: *GANYMEDES*

1. Tube of perianth (not corona) conical, much broader at the throat than at the base; stamens equal or unequal; corona funnel-, bell-, or tube-shaped, usually large.

8. Corona funnel-shaped, shorter than the perianth tube; filaments curved, unequal.   section: *BULBOCODIUM*

8. Corona bell-shaped or tubular, as long as or longer than the perianth tube; filaments straight, equal.   section: *PSEUDONARCISSUS*

Section **PSEUDONARCISSUS**—Trumpet Daffodil*

1. Tube of perianth (not corona) 2–9 mm; stamens attached at base of tube, or almost at the base.

  2. Fls. 1½–2½ cm long, inclined or drooping; perianth tube 5–9 mm long.   *asturiensis*

  2. Fls. 3½–4½ cm long, almost inverted; perianth tube 2–3 mm long.   *cyclamineus*

1. Tube of perianth (not corona) usually longer than 9 mm; stamens attached above base of tube.

  3. Spathe 6–10 cm; tall plants up to 1½ m. (S. Spain (S. de Cazorla))
    1665a. *N. longispathathus* Pugsley Page 85

  3. Spathe shorter.

    4. Fls. large, 4–7 cm long; plants usually tall.

      5. Fls. pale yellow, straw-, cream-coloured or white, one-coloured or two-coloured.   see page 394

      5. Fls. deep yellow, one-coloured or nearly so.

        6. Fl. stalk (between spathe and ovary) 1–3½ cm long, erect but curved.

          7. Fls. 4–6½ cm long, always solitary.

---

*Insufficient knowledge of plants in the field makes the delineation of species in this section tentative.

7. Fls. 3–4 cm long, usually 2 or 3; corona
   golden-yellow, perianth segments yellow.
   (S. Nevada)   1665c. *N. nevadensis* Pugsley Page 76
6. Fl. stalk ½–1½ cm, more or less straight.
8. Lvs. green, 30–35 cm, to 14 mm broad.
   Plant tall, robust. (C. Spain)   1665d. *N. confusus* Pugsley
8. Lvs. glaucous, 8–12 cm or more, *c.* 6 mm
   broad. (N–W. Portugal, Spain)   1665e. *N. portensis* Pugsley
4. Fls. small, 2–3½ cm; plants usually dwarf.
9. Perianth tube (not corona), 11–18 mm long;
   fls. bicoloured.   *minor*
9. Flowers deep yellow, one-coloured or
   nearly so. Perianth tube *c.* 9 mm long.
   (Pyrenees)   1665g. *N. parviflorus* (Jordan) Pugsley

1665b. *N. hispanicus* Gouan (*N. major* Curtis) Distinguished by its large, one-coloured, deep yellow flowers 4–6½ cm long, with twisted perianth segments. Leaves glaucous, spirally twisted, 8–12 mm broad. Damp meadows. S. and N. Spain, Pyrenees, Cévennes. 1665f. *N. minor* L. A dwarf plant 10–15 cm high with pale yellow, lance-shaped perianth segments and a deeper yellow corona. Leaves 7–12 cm long, by 4–6 mm broad. Damp meadows in the lowlands and mountains. Portugal, N. and C. Spain. 1665h. *N. asturiensis* (Jordan) Pugsley A tiny plant up to 10 cm, with tiny inclined, yellow flowers 1½–2½ cm long, with forward-spreading perianth segments as long as the similar-coloured corona. Leaves 2, about ½ cm broad. Alpine meadows. Portugal, N. Spain. Pl. 60, Page 103. 1665i. *N. cyclamineus* DC. Flowers pendulous, bright rich yellow with narrow perianth segments reflexed against the ovary and corona noticeably broadened at the mouth. Leaves narrow, 3–5 mm; stem cylindrical. River margins. N–W. Portugal (near Oporto), N–W. Spain.

Flowers pale yellow, straw-, cream-coloured, or white, one-coloured or two-coloured
(a) Flowers one-coloured
1665j. *N. pallidiflorus* Pugsley PALE LENT LILY. Flowers 4–5½ cm, more or less drooping, pale straw-coloured or primrose-yellow; perianth segments broadly oval, more or less overlapping; corona usually similar-coloured or sometimes slightly darker; perianth tube to 2½ cm, usually orange. Leaves 15–30 cm by ½–1 cm broad. Mountain pastures. N. Spain, Pyrenees. Pl. 60, Page 149. 1665l. *N. alpestris* Pugsley ALPINE LENT LILY. A low plant 10–15 cm, with drooping pure white flowers 3½–4½ cm, with the twisted, narrow perianth segments close round the longer narrow corona and shorter than it. Perianth tube 10–13 mm. Leaves erect, glaucous, channelled. Pyrenees.

(b) Flowers two-coloured
1665. *N. pseudonarcissus* L. LENT LILY, DAFFODIL. Perianth segments oval-lance-shaped, pale yellow, spreading; corona deep yellow, somewhat wider and spreading at the mouth, margin with irregular rounded lobes. Meadows etc. Widespread, except in S. and S–W. Spain. 1665k. *N. nobilis* (Haw.) J. A. & J. H. Schultes PYRENEAN LENT LILY. A magnificent species to ½ m, with very large flowers 5–7 cm, with a golden-yellow corona and contrasting pale yellow or cream perianth segments 4–4½ cm long. Perianth segments overlapping, twisted, elliptic-oblong; corona expanding at the mouth, deeply lobed. Mountain meadows. N. Portugal, N. Spain, C. Pyrenees. Page 124. 1665m. *N. bicolor* L. Like 1665k but flowers smaller and shorter with whitish to cream-coloured, broadly egg-shaped perianth segments and golden yellow corona 2½–3 cm across and dilated at the mouth. Perianth tube 1 cm long. Stamens free almost to base of tube. Leaves wide 12–20 mm. Bulb large. Mountain pastures. Pyrenees. 1665n. *N. abscissus* (Haw.) J. A. & J. H. Schultes Distinguished from 1665m by smaller cut-off corona 1½–2½ cm across and the perianth tube which is obconical (not

slightly hexagonal). Flowers 4½–5 cm; perianth segments pale or sulphur yellow, oval-lance-shaped to narrowly lance-shaped; corona deep golden yellow, cylindrical, not dilated at the mouth. Leaves 10–12 mm. Meadows, hillsides, mountains. Pyrenees, Corbières. Pl. 60.

Section BULBOCODIUM—Hoop Petticoat Daffodil

1. Fls. golden-yellow or yellowish; fl. stalk more than 9 mm long at flowering (rarely shorter).

2. Corona very large and usually contracted at the throat; lvs. prostrate, usually very narrow, 1 mm. (S. and W. Portugal)   1666a. *N. obesus* Salisb.

2. Corona smaller, obconic, not contracted at the throat; lvs. usually ascending, 1½–4 mm broad.   *bulbocodium*

1. Fls. sulphur-yellow, to greenish-white, to pure white; fl. stalk absent or very short at flowering.

3. Fls. pure white or greenish-white; anthers included in corona. (Spain)   **(1666) *N. cantabricus* DC. Page 69

3. Fls. sulphur or pale sulphur or whitish-yellow; anthers usually longer than corona. (S–C. Spain)   1666b. *N. hedraenthus* (Webb & Heldr.) Colmeiro Page 84

**1666. *N. bulbocodium* L. HOOP PETTICOAT DAFFODIL. Easily distinguished by its attractive, widely funnel-shaped flowers with short, narrow, spreading perianth segments. Flowers golden-yellow, but sometimes lemon-yellow to primrose-yellow. (Var. *nivalis* (Graells) Baker has golden-yellow flowers with perianth segments about as long as the corona and stamens, and a particularly long style. A tiny plant of high mountains by melting snow. Page 102.) Spathe conspicuous, papery, partly encircling the perianth tube. Leaves very narrow, semi-cylindrical. Sandy and rocky places in lowlands, hills and mountains. Portugal, Spain, W. France. Pl. 60 (var. *nivalis*), Page 52.

Section JONQUILLA—Jonquils

1. Perianth tube 2–3 cm long, straight.   *jonquilla*

1. Perianth tube 1–2 cm long, curved or straight.

2. Longest fl. stalk usually as long as spathe at flowering and longer than the fl.; plants generally vigorous.
(C. Portugal (Ribatejo))   1667a. *N. fernandesii* G. Pedro
(Portugal, S. Spain)   1667b. *N. willkommii* (Samp.) A. Fernandes

2. Longest fl. stalk usually included in the spathe at flowering, shorter than the fl.; plants slender.

3. Perianth tube straight, 14–18 mm long; lvs. 2–4 mm broad.   *requienii*

3. Perianth tube curved, 8–16 mm long; lvs. thread-like, round or semi-cylindrical.   *gaditanus*

1667. *N. jonquilla* L. JONQUIL. Flowers 2–6, very scented, golden-yellow. Perianth segments egg-shaped, 2–4 mm, similar-coloured to the cup- or saucer-shaped corona. Leaves rush-like, grooved. Meadows. S. Portugal, S. and C. Spain; introd. France.   **1668. *N. requienii* Roemer (*N. juncifolius*) RUSH-LEAVED JONQUIL. Like 1667 but a more slender plant, usually with 1–2 yellow flowers, and with a shorter perianth tube. Corona usually more than half as long as spreading perianth segments. Leaves rush-like, very slender. Rocks in mountains. S. and E. Spain, Pyrenees, S. France. Page 68.   1668a. *N. gaditanus* Boiss. & Reuter (including *N. minutiflorus*) Very like 1668 but easily recognized by the curved perianth tube. Flowers 2–8, pale yellow variable in size and shape; perianth segments usually reflexed, but they may spread in a star. Leaves variable, from thread-like and no longer than the inflorescence, to 65 cm long and 2 mm wide. S. Portugal, S–W. Spain. Page 53.

1668f. *N. viridiflorus* Schousboe Easily recognized by the small green flowers with perianth segments 10–23 mm by 2 mm, and its corona 1 mm long. Autumn-flowering. S. Spain (Algeciras-San Roque). Pl. 60.

Section APODANTHAE (3 species)

1668c. *N. rupicola* Dufour Flowers usually solitary stalkless, semi-erect, up to 2½ cm across, deep yellow. Perianth segments broadly egg-shaped; corona cup-shaped, usually six-lobed; spathe shining, papery. Leaves 2–4, rush-like, grooved, glaucous, erect. Granite rocks in mountains. N. and C. Portugal, S. and C. Spain. Page 111.

Section GANYMEDES

1668d. *N. triandrus* L. ANGEL'S TEARS. Very distinctive, with its cluster of usually 1–3 sulphur-yellow, or whitish drooping flowers with a large cup-shaped corona, and strongly reflexed and twisted perianth segments. Corona 1 cm long or more, sometimes paler than segments. Leaves dark green, 4 mm broad, channelled, keeled or ribbed beneath; stem to 30 cm. Scrub-covered hillsides. C. Portugal, C. and N. Spain, Pyrenees. Pl. 60. Page 125.   1668e. *N. concolor* (Haw.) Link Like 1668a but flowers 1–3, pale golden-yellow; corona cup-shaped to 1 cm long. Leaves bright green. Iberian peninsula.

Section HERMIONE

1. Corona yellow or orange.	*tazetta*
1. Corona white, as well as perianth segments.	
2. Fls. 2–2½ cm across; lvs. flaccid, about as long as the stems. (Portugal, Spain, S. France)	1669a. *N. panizzianus* Parl.
2. Fls. 2½–4 cm across.	
3. Fl. stalk slightly compressed, green like the lvs.; corona usually entire; fls. 2½–3½ cm across. (S. Spain, S. France)	1669b. *N. polyanthos* Loisel.
3. Fl. stalk strongly compressed, somewhat glaucous like the lvs.; corona usually more or less toothed; fls. 3½–4 cm across.	*papyraceus*

1669. *N. tazetta* L. POLYANTHUS NARCISSUS. Readily distinguished by its cluster of 3–18, sweet-scented flowers with wide, spreading perianth segments and golden-yellow, pleated corona. Leaves glaucous, ½–1½ cm broad. Very variable with flowers sometimes all white or all yellow. Damp places, cultivations, on the littoral. Portugal, C., S–E., and E. Spain, S. France.   (1667) *N. papyraceus* Ker-Gawler PAPER-WHITE NARCISSUS. Distinguished by its several, pure white flowers. Perianth tube 2–2½ cm, also pure white, perianth segments oval blunt, 12–15 mm long; corona about 4 mm long. Wet meadows and ditches. Portugal, S. Spain, S–W. France. Page 63.

Section **NARCISSUS**

**1671. *N. poeticus* L. PHEASANT'S-EYE NARCISSUS. Damp meadows in mountains. C. and E. Spain, Pyrenees, France. The hybrid species (1667) *N. × odorus* L.;   1670. *N. × medioluteus* Miller;   (1672) *N. × incomparabilis* Miller, occur scattered in our area.

PANCRATIUM | **1673. *P. maritimum* L. SEA DAFFODIL. Unmistakable with its umbel of large white, sweet-scented, tubular flowers borne on a stout stem, and its thick, glaucous, daffodil-like leaves. Perianth with a long slender tube, wide corona with 12 small teeth, and 6 long, narrow, spreading segments. Summer-flowering. Maritime sands. S–W. Portugal, Med. region. Page 52.   (1673) *P. illyricum* L. Possibly S. Spain.

TAPEINANTHUS | Perianth yellow, with a very short tube and lance-shaped lobes, spreading in a star; corona rudimentary, obscurely twelve-lobed. Style longer than stamens. 1 species.

1673a. *T. humilis* (Cav.) Herbert (*Carregnoa h.*) A small, autumn-flowering, bulbous plant with solitary, pale yellow flowers borne at the end of a very slender, pale green stem 8–15 cm. Spathe 1, equal to flower stalk. Leaves 2, thread-like, much shorter than flowering stem, sometimes absent at flowering. Stony fields, often in dense masses. S–W. Spain Pl.61.

DIOSCOREACEAE | Yam Family

TAMUS | 1674. *T. communis* L. BLACK BRYONY. Hedges, thickets, woods. Widespread.

DIOSCOREA | 1674a. *D. pyrenaica* Gren. A small perennial 5–25 cm, branched above, with slender flexible non-climbing branches, and small clusters of tiny greenish-white male flowers and separate clusters of 1–3 greenish female flowers. Leaves oval-heart-shaped, stalked. Tuber black, with bulbils. A rare Pyrenean endemic of alpine screes.

IRIDACEAE | Iris Family

CROCUS |
Spring-flowering
**1678. *C. albiflorus* Kit. PURPLE CROCUS. Meadows and pastures in hills and mountains. Pyrenees, Massif Central.   1678a. *C. nevadensis* Amo & Campo Endemic in S. Spain (from S. Nevada to S. Bermeja). Flowers pinkish-white with pale lilac veining and greenish-yellow in the throat; stigmas white. Leaves with 2 lateral grooves on the undersides. Corm with a soft brown tunic of parallel fibres. Stony alpine slopes. Pl. 61. Page 76. 1678b. *C. carpetanus* Boiss. & Reuter Flowers usually pale violet with grey-blue veining; stigmas white or violet. Leaves semi-circular in cross-section, with 13 shallow grooves on the curved side. Corm with a tunic of a network of fibres. Mountains. C. and N. Portugal, C. Spain. Pl. 61, Page 110.

Autumn-flowering
(a) *Stigmas scarlet or red, hanging out of perianth, or not*
1677a. *C. sativus* L. SAFFRON CROCUS Flowers reddish-lilac with wide-spreading segments, but readily distinguished from all other species by the long brilliant blood-red stigmas

which are not branched but have broader, toothed apices. Leaves grey-green, appearing before flowers, margin ciliate. Corm with slender netted fibres. Occasionally cultivated and sometimes naturalized in our area.

1677b. *C. cambessedesii* Gay Flowers very small, pale lilac or white, the outer segments straw-coloured and variously feathered with purple; stigmas brilliant scarlet. Leaves slender. Autumn- and early spring-flowering. Heaths and woods. Balearic Islands.

(b) Stigmas orange, erect or arched
(i) Throat of flowers hairless
**1676. *C. nudiflorus* Sm. Distinguished by its solitary or 2–3 purple flowers which appear leafless in the autumn, and in particular by the hairless throat. Leaves appearing in the spring only. Corms with stolons. Grassy places in hills and mountains. Spain, France. Pl. 61.

*(ii) Throat of flowers hairy**
1676a. *C. asturicus* Herbert Like 1676 but violet or purple flowers smaller or more pointed; distinguished by the leaves which, at the time of flowering, are usually just visible above the sheathing bracts. Throat of corolla with pale hairs. Corms without stolons. Hills, mountains. C. and N. Portugal, N. Spain.   1676b. *C. clusii* Gay Flowers deep violet to whitish-violet, white in the throat; perianth segments 18 mm by 13 mm, darker violet towards base. Leaves well-developed at flowering, 1½ mm wide, with 3 prominent ridges on margin. Corm with finely netted tunic, 19–22 mm by 17 mm. Pine woods, dry places. Portugal, S–W. and W. Spain.   1676c. *C. salzmannii* Gay Flowers much longer than in 1676, pale lilac, yellow in the throat; perianth segments 50 mm by 17 mm. Leaves 3½ mm wide scarcely showing at flowering. Corm large, to 2½ cm, with an abundant soft tunic with fine parallel fibres. S. Spain.

ROMULEA | 8 species, including 4 with very restricted distributions in Iberia. A difficult genus.

Spathe of 2 dissimilar segments, the lower outer green, the upper papery
(a) Stigma longer than stamens, divided to the base into 2 linear lobes
1681. *R. bulbocodium* (L.) Sebastiani & Mauri Flowers rose-lilac with lines of purple or yellow, often fading to greenish on the outside and yellow within, rarely wholly white or yellow. Flowers funnel-shaped, 2½–3½ cm long; styles whitish. Leaves rush-like, grooved, 1½–2 mm broad. Rocky and sandy places, scrub. Portugal, Med. region. Page 63.   1681a. *R. clusiana* (Lange) Nyman Like 1681 but flowers larger, 3½–4½ cm long, segments orange-yellow below, white in the middle and violet towards the tips, or rarely one-coloured. Spathe weak, papery, often blue or white. Stigmas one third as long as perianth. Sands by the sea. N–W. Portugal, S–W. Spain. Pl. 61.

(b) Stigma shorter than stamens, two-lobed
(1681) *R. columnae* Sebastiani & Mauri The smallest-flowered species with 1–3 flowers 10–12 mm long, violet, or outer segments greenish with a yellowish base and inner white or bluish with purple veins, throat pale yellow, hairless. Grassy, sandy damp places, cultivations. C. Portugal, Med. region.

Spathe of 2 similar green segments or upper with a narrow papery margin
1681b. *R. gaditana* G. Kunze Flowers 2–3 cm long, the outer segments green and the inner violet outside, all segments violet to purple-violet within, throat pale green with some blue

---

*Differences and distribution of the 3 following species not yet fully known.

wash over the purple in upper throat; tips of segment curl down when fully open. Stigmas white, exserted, glandular hairy; spathe with both segments green. Rough grazing. S–W. Spain (Cadiz province). Page 53.

SISYRINCHIUM | 1682. *S. bermudiana* L. BLUE-EYED GRASS. Naturalized S–W. France.

IRIS | Flowers with 3 outer segments, the *falls*, each with a terminal *blade* and a narrower basal *shaft*; and 3 inner segments, the *standards*, which are usually erect. The falls may have a tuft of hairs, the *beard*, on the upper surface. Style with broad, petal-like arms. The *spathe* is a green or papery bract which encircles the young flower buds.

1. Rootstock a rounded bulb, or corm, covered with fibrous or membraneous scales.
  2. Plant slender, usually less than 30 cm; fls. blue and white.   *sisyrinchium*
  2. Plant robust, usually more than 30 cm; fls. violet, purple or blue, with yellow or orange, or completely yellow.
    3. Bulb with fleshy roots persisting in resting season. Fls. with small spreading standards.   *planifolia*
    3. Bulb without persistent roots. Fls. with large erect standards.
      4. Fls. yellow.   *lusitanica*
      4. Fls. violet, purple or blue.
        5. Perianth tube very short.
          6. Falls with blade much shorter than the fiddle-shaped shaft; fl. stalk long.   *xiphium*
          6. Falls with rounded blade as long as the triangular shaft; fl. stalk short.   *xiphioides*
        5. Perianth tube 12 mm long or more.
          7. Falls bearded. (N–W. Portugal (S. do Gerês))   1685c. *I. boissieri* Henriq. Pl. 62, Page 103
          7. Falls not bearded.
            8. Standards rounded, blunt.   *filifolia*
            8. Standards lance-shaped, acute.   *tingitana*
1. Rootstock a horizontal elongated rhizome.
  9. Falls not bearded.
    10. Stem flattened; fls. smelling of ripe plums. (Pyrenees, S–W. France)   **1686. *I. graminea* L.
    10. Stem rounded.
      11. Standards less than half as long as falls; fls. yellow (Widespread)   **1690. *I. pseudacorus* L.
      11. Standards at least half as long as falls; fls. violet or purple.
        12. Lvs. evergreen; seeds orange-scarlet. (Widespread)   **1689. *I. foetidissima* L.
        12. Lvs. deciduous; seeds not orange-scarlet.   *spuria*
  9. Falls bearded.

13. Stem not branched.
  14. Stems absent or very short.   *pumila*
  14. Stems present.
    15. Perianth tube twice as long as
      ovary.   *chamaeiris*
    15. Perianth tube at least 3 times as
      long as ovary.   *subbiflora*
13. Stem branched.
  16. Spathe wholly green, not dry when
    first fl. opens; fls. white. (Introd.
    Portugal, S. Spain)   1693a. *I. albicans* Lange
  16. Spathe partly green, partly dry when
    first fl. opens; fls. violet. (Cultivated
    and naturalized)   **1693. *I. germanica* L.

**1684. *I. sisyrinchium* L. BARBARY NUT. Flowers very variable in size, 1–3 cm across, with bright blue, rounded and spreading falls with greater or lesser amount of white at the base. Spathes swollen, brownish, papery. Leaves rush-like, grooved, longer than flowering stem. Roadsides, on the littoral, stony ground. S. Portugal, Med. Spain. Page 60.

1684a. *I. planifolia* (Miller) Fiori (*I. alata*) Distinguished by its broad outspread rosette of leaves which are flattened from above and below (not from side to side) and its solitary blue-violet flowers with large broad falls with a broad central yellow to orange patch and narrow outspreading standards. Leaves $1\frac{1}{2}$–$2\frac{1}{2}$ cm broad, sheathing at base. Bulb 2 cm across, with several swollen spindle-shaped roots. Flowering in winter. Grassy places, open hillsides, preferring limestone. S. Portugal, S-W. and S. Spain. Pl. 62, Page 68.

**1685. *I. xiphium* L. (*I. taitii*) SPANISH IRIS. A robust erect plant to $\frac{1}{2}$ m or more with large, usually solitary violet-purple flowers up to 10 cm across, with spreading falls with a conspicuous deep yellow patch running almost to the tip of the blade. Blade of falls oval, shaft purple, much longer; spathe green. Leaves narrow linear, grooved. Damp sandy places in hills and mountains. Portugal, S. Spain, S. France. Pl. 62. **(1685) *I. xiphioides* Ehrh. Distinguished from 1685 by the 2–3 larger, more robust flowers with broad, almost rounded, bright rich blue falls with a large orange central patch. Blades of falls spreading, as long as the shafts and much longer than the erect, bi-lobed standards. Leaves broadly linear, grooved. Damp alpine meadows. N-W. Spain, Pyrenees. Pl. 62. 1685a. *I. tingitana* Boiss. & Reuter Like 1685 but flowers larger and earlier-flowering (about March in S. Andalusia). Distinguished by its longer perianth tube, and narrower, pointed standards. Flowers usually pale bluish-purple, with a pale blue central band on the shaft and a light yellow patch on the blade of the falls which are broad and rounded. S. Spain. 1685b. *I. filifolia* Boiss. A beautiful plant with rich purple flowers, the falls with a conspicuous contrasting yellow streak. Standard and falls nearly equal; blade of falls spathulate, the shaft linear. Leaves long, very slender, bristle- or thread-like, grooved, with inrolled margin. Sandy and rocky places on limestone. S. Spain. Pl. 62. 1685d. *I. lusitanica* Ker-Gawler A robust erect plant to 70 cm with 1–2 yellow flowers with falls with a rounded blade. Leaves broadly linear, channelled. Dry stony places. C. Portugal, E. Spain.

1688. *I. spuria* L. Flowers whitish-lilac, borne on a stem to 30 cm. Falls with a rounded blade and a pale shaft twice as long, not bearded; standards lilac, shorter than falls. Fruit with a long point. Damp pastures, marshes, by the sea. C., E., and S-E. Spain, S. France. Page 132.

**1691. *I. pumila* L. Distinguished by its solitary violet-purple or yellow flowers which are stalkless and nestle among the narrow, glaucous, out-curved, sword-shaped leaves. Falls bearded; perianth tube 4–5 times as long as the ovary. Bushy places. S–E. and E. Spain.
**1692. *I. chamaeiris* Bertol. Differs from 1691 in having a short stem 3–25 cm long, bearing 1 or 2 flowers. Flowers very variable; falls either blue, purple, yellow or white, and tinged or veined with brown; beard orange-yellow. Perianth tube shorter, about 2½ cm. Leaves pale green. Dry stony places. S–E. Spain, S. France.   1692a. *I. subbiflora* Brot. (*I. biflora* auct.) Flowers 1–2, deep purple with a violet beard to the falls, and borne on a stem 20–40 cm. Spathe variable in length, 4–10 cm, green with papery margin. Dry uncultivated ground. C. and N–E. Portugal, S. Spain. Pl. 62.

GLADIOLUS |

1. Anthers as long as or longer than filaments; fls. large, 3 cm or more.
  2. Upper perianth segment separated from,
  longer than, and almost twice as broad
  as lateral segments; fr. nearly globular.
  (Portugal, Med. region)   **1695. *G. segetum* Ker-Gawler Page 45
  2. Upper 3 perianth segments touching and
  almost equal; fr. oval-oblong. (Med.
  France)   **(1695) *G. byzantinus* Miller
1. Anthers shorter than filaments; fls. rarely
3 cm.
  3. Fls. in a one-sided spike; tunic of bulb of
  thick fibres. (Med. France)   1696. *G. communis* L.
  3. Fls. arranged in 2 rows; tunic with very
  fine fibres.
    4. Tunic with mesh at the top; spike with
    3–6 fls.; stigmas swollen in the middle;
    plant slender. (Iberian peninsula, ex-
    cept north)   1697. *G. illyricus* Koch Pl. 61, Page 53
    4. Tunic without mesh at the top; spike
    with 5–8 fls.; stigmas swelling imper-
    ceptibly almost from the base; plant
    robust. (S. & C. Portugal, C. Spain)   1697a. *G. dubius* Guss.

JUNCACEAE | Rush Family

A family of limited importance, particularly in the drier regions of our area. About 36 species of *Juncus*, RUSH occur; the majority are widespread in Europe, apparently only 3–5 species are restricted to our area.   *Luzula*, WOOD RUSH may have up to 20 species, many of which are widespread.

PALMAE | Palm Family

At least 100 species of Palm are grown today in gardens and parks of the Mediterranean region, and they may be planted for shelter and landscaping; few, if any, are naturalized.

PHOENIX | (1718) *P. dactylifera* L. DATE PALM. Distinguished by its tall, rather slender trunk and very large terminal crown of grey-green, pinnate leaves. Fruit in long hanging branches. Cultivated in S. Portugal and S. Spain for its edible fruit. The largest palm grove in Europe is at Elche in Spain. Occasionally naturalized.

**1718. *P. canariensis* Chabaud CANARY PALM. Distinguished from (1718) by its short stout barrel-shaped stem and its leaves which are twice as large, longer and more numerous and arched. Commonly planted for ornament on the Mediterranean littoral.

CHAMAEROPS | **1719. *C. humilis* L. DWARF FAN PALM. The only native palm in Europe, usually found growing as a dwarf bush about ½ m high without a trunk. Readily distinguished by its rounded blades which are deeply divided into 12–15 stiff lance-shaped segments spreading in a fan; leaf stalk spiny. Dry stony and sandy places, near the coast, in mountains and hills. S. Portugal, Med. Spain.

GRAMINEAE | Grass Family

A very important family often dominating large expanses, particularly in the most arid steppe regions of our area, and in the meadows and pastures above the tree line in the mountains. Elsewhere, cutting, grazing and agricultural activities often encourage the development of grass lands. A very large and difficult family comprising over 100 genera and over 450 species in our area, and requiring specialist knowledge for identification. Only a few of the most distinctive species are noted below.

ARUNDO | 1739. *A. donax* L. GIANT REED. A tall, robust, bamboo-like plant 1½–5 m, with terminal plume-like inflorescences which at length become silvery, and numerous broad, maize-like leaves ranged up the stem. Stems about 2–3 cm wide, not brittle; leaves 2–5 cm wide. Damp places, ditches, watersides. South and centre of our area, often planted for shelter.

PHRAGMITES | **1740. *P. australis* (Cav.) Steudel (*P. communis*) COMMON REED. Similar to 1739 but less robust. Leaves 2–4 cm wide; stem usually less than 1½ cm wide, brittle. Watersides, damp places. Widespread.

AMPELODESMA | **1741. *A. mauritanicum* (Poiret) Durand & Schinz (*A. tenax*) A very robust, densely tufted perennial 1–3 m, with a much-branched, interrupted, somewhat one-sided purplish-green, spike-like inflorescence. Leaves very long, tough, rigid and rush-like with inrolled margins. Dry hills. Med. Spain; naturalized in C. Portugal.

CORTADERIA | 1741a. *C. selloana* (J. A. & J. H. Schultes) Ascherson & Graebner (*C. argentea*) PAMPAS GRASS. A very large tufted grass with narrow, rough, arched linear leaves up to 2 m long, and a long spike-like cluster up to 1 m of pale spikelets borne on stems to 4 m. Flower spikes either male or female, the latter broader. Native of S. America; sometimes planted.

AMMOPHILA | **1775. *A. arenaria* (L.) Link MARRAM GRASS. A robust, creeping grass colonizing maritime sands and dunes. Flowering spikes cylindrical, tapering, whitish. Leaves greyish-green, sharp-pointed, rigid, tightly inrolled. Coasts throughout our area.

STIPA | About 10 species.

1787a. *S. tenacissima* L. FALSE ESPARTO. A robust, tufted perennial to 1½ m, with rush-like cylindrical springy-pointed leaves, and a long slender inflorescence up to 35 cm. Glumes large, to 2½ cm; awn robust to 6 cm, spirally twisted and with long white hairs below the angle. Dry sandy calcareous or gypsaceous regions. S. Portugal, S. and C. Spain.   (1787)
S. gigantea Lag. Like 1787a but awns much longer 7–10 cm and without hairs on the lower twisted part. Inflorescence less dense, with florets borne on long very slender stalks. Leaves long 30–90 cm, stout; stems stout 1½–2 m. Sandy places. Portugal, S. and C. Spain.   1787b.
S. lagascae Roemer & Schultes A tufted perennial with stems to 2 m with a long narrow inflorescence with spikelets with very long glumes and extremely long awns. Glumes 5–6 cm; awn of lower glume 25–30 cm, finely hairy on the angles, at length curved. Leaves thread-like, inrolled, finely hairy on inner side. Steppes, dry sandy ground. E–C. Portugal, S. and C. Spain.   1787c. *S. parviflora* Desf. An elegant species with a very lax elongate inflorescence with slender clustered spikelets with very long unequal glumes and thread-like awn. Glumes 15–20 cm; awn 8–10 cm. Arid calcareous regions. S–E. and E. Spain.

ORYZA | 1797. *O. sativa* L. RICE. An aquatic annual with lax, erect or curved, greenish-white inflorescence of numerous long lateral branches bearing many spikelets. Glumes awned or not. Leaves flat, smooth, 1–1½ cm wide. Cultivated in river deltas of Med. and S. Atlantic coasts.

LYGEUM | 1799. *L. spartum* L. ALBARDINE. A tufted rush-like grass, easily distinguished by its broad lance-shaped sheath 4–5 cm enclosing the spikelets. Leaves stiff rush-like, cylindrical. Dry stony steppes, salt-rich ground. S., C., and S–E. Spain.

IMPERATA | **1806. *I. cylindrica* (L.) Beauv. Readily distinguished by its dense cylindrical inflorescence of shining silky-white spikelets, due to the abundance of silky hairs which are much longer than the glumes. A robust plant to 1 m with stiff inrolled glaucous leaves and enlarged sheaths, and creeping rhizomes. Sandy and stony ground. Portugal, S. and E. Spain.

ZEA | 1814. *Z. mays* L. MAIZE. A very robust annual to 5 m with a plume-like cluster of male spikelets and bud-like female spikelets in the axils of the very wide upper leaves. Leaves 5–12 cm wide, numerous, arched. Fruit borne laterally on a swollen axis. Widely cultivated throughout our area.

ARACEAE | Arum Family

COLOCASIA | Leaves peltate. Flowers one-sexed, with female flowers at base of spadix separated by neutral flowers from the upper male flowers; spathe deciduous.   1818a. *C. esculenta* (L.) Schott (*C. antiquorum*) TARO. Leaves very large about 60 cm long, oval-heart-shaped, long-stalked, growing from a short thick rootstock. Spathe pale yellow, 30 cm or more; spadix with a long terminal appendage, but plant rarely flowering. Native of E. Indies; sometimes naturalized in shady places in S. Portugal, S. Spain.

ACORUS | **1816. *A. calamus* L. SWEET FLAG. Still and slow-flowing waters. Pyrenees, France.

ARUM | 1818. *A. maculatum* L. LORDS-AND-LADIES, CUCKOO-PINT. Spathe pale greenish-yellow, edged with purple; spadix purple, rarely yellowish. Leaves appearing in early spring. Shady banks, ditches, woods. N–E. Portugal, N., C., and E. Spain, France.   **(1818) *A. italicum* Miller ITALIAN ARUM. Like 1818 but spathe white or yellowish and spadix almost always yellow. Leaves appearing in autumn, often white-veined. Damp, shady places. S. and C. Portugal, S., C., and E. Spain, S. France.

DRACUNCULUS | **1819. *D. vulgaris* Schott DRAGON ARUM. In scrub, probably introduced and naturalized: N. Portugal, N. Spain, S. France.

HELICODICEROS | 1819a. *H. muscivorus* (L. fil.) Engler (*Arum m.*) Like 1819 with deeply divided leaves and spotted stems and fetid violet-purple 'flowers', but spathe hairy on the inside, much longer than the spadix which is hairy to the tip. Rocky places. Balearic Islands.

BIARUM | **1820. *B. tenuifolium* (L.) Schott A small plant with a long brownish-purple tongue-like spathe and longer cylindrical purple spadix, often appearing direct from the soil and without the leaves. Leaves lance-shaped, usually appearing after flowering. Winter- or spring-flowering. Hedges, waysides, rich soils. S. Portugal, S. Spain. 1820a. *B. arundanum* Boiss. & Reuter Like 1820 with blackish-purple spathe, but spathe with a very short stalk 15–18 mm. Leaves oblong-egg-shaped, narrow to a short sheathing stalk, appearing after summer-flowering. Mountains. S. Spain (S. de Ronda). 1820b. *B. carratracense* (Haenseler) Font Quer (*Ischarum haenseleri*) Spathe purple 10–13 cm, shortly stalked and with a short tube and a flat spreading limb; spadix incurved. Zone of anthers separated from the lower zone of ovaries by a middle zone of awl-shaped swellings; anthers opening by pores. Leaves appearing after flowering, lance-shaped, broadly sheathing at base. Fruit white, knobbly, appearing beside the leaves, not amongst them. Fields, roadsides, banks. S. Spain. Page 389.

ARISARUM | **1821. *A. vulgare* Targ.-Tozz. FRIAR'S COWL. Unmistakable with its striped brown and green, flask-shaped spathe, with a slender forward-curved, brown-purple spadix about the same length. Leaves oval to arrow-shaped. Grassy places, uncultivated ground, cultivations. S. and C. Portugal, Med. region. Page 161. 1821a. *A. proboscideum* (L.) Savi Similar to 1821 but spadix extended into a long thin slender tail up to 10 cm long. Leaves arrow-shaped, with basal lobes distinctly angled outwards from the main part of the blade. Recently rediscovered in marshy ground under Cork oaks. S. Spain (Cadiz province).

LEMNACEAE | Duckweed Family

LEMNA | 4 species. Widespread are: **1823. *L. minor* L. DUCKWEED;   (1823) *L. gibba* L. GIBBOUS DUCKWEED;   1824. *L. trisulca* L. IVY DUCKWEED.

SPARGANIACEAE | Bur-Reed Family

SPARGANIUM | 4 species. Widespread are: **1825. *S. erectum* L. BUR-REED;   1826. *S. emersum* Rehmann UNBRANCHED BUR-REED.

TYPHACEAE | Reedmace Family

TYPHA | 3 species. Widespread are: **1827. *T. latifolia* L. GREAT REEDMACE;   1828. *T. angustifolia* L. LESSER REEDMACE.

CYPERACEAE | Sedge Family

A difficult family requiring special study; the majority of species in our region are widespread in Europe. The following genera with the approximate numbers of species found in our area are: *Cyperus*, 20 species; *Fimbristylis*, 1 species; *Eriophorum*, 4 species; *Scirpus*, 20 species; *Eleocharis*, 4 species; *Schoenus*, 1 species; *Rhynchospora*, 1 species; *Cladium*, 3 species; *Carex*, 90 species.

ORCHIDACEAE | Orchid Family

OPHRYS | There is considerable variation amongst species, and hybridization is not uncommon sometimes making the identification of individual plants difficult. Probably 15 species.

1. Outer 3 perianth segments green or yellowish (sometimes flushed purple).
  2. Lip entire (or very weakly three-lobed), with or without swellings at the base.
    3. Lip dark brown to black with long straight hairs, without an apical appendage.   *atrata*
    3. Lip light to dark brown, with short hairs, usually with an apical appendage.   *sphegodes*
  2. Lip three-lobed.
    4. Inner 2 perianth segments hairless, oblong, blunt; lateral lobes of lip originating from near the tip.
      5. Lip with a brown margin.
        6. Lip with a white band separating brownish reflective patch from brown apex.   *omegaifera*
        6. Lip without a white band separating bluish reflective patch from brown apex.   *fusca*
      5. Lip brown with a conspicuous yellow margin. (Portugal, Med. region)   **1874. *O. lutea* Cav. Page 69
    4. Inner 2 perianth segments with spreading hairs; lateral lobes of lip originating from near the base.
      7. Inner 2 perianth segments linear; lateral lobes of lip narrow, spreading. (France)   **1877. *O. insectifera* L. Page 152

7. Inner 2 perianth segments triangular; lateral lobes of lip rounded not spreading.

  8. Lip *c.* 12 mm, with large bright blue reflective patch.   *speculum*

  8. Lip *c.* 6 mm, with dull brown reflective patch. (Portugal, Med. region)   **1879. *O. bombyliflora* Link Page 60

1. Outer 3 perianth segments pink or white.

  9. Column ending in a long curved beak. (Widespread)   **1878. *O. apifera* Hudson

  9. Column not beaked, or with a short straight beak.

    10. Inner 2 perianth segments linear, 8 mm long; lip usually without basal swellings. (Balearic Islands, France)   **1881. *O. bertolonii* Moretti Pl. 63.

    10. Inner 2 perianth segments broad triangular, less than 8 mm; lip usually with 2 basal swellings.

      11. Inner 2 perianth segments hairless, tongue-like, blunt, with undulate margin; lip without basal swellings. (E. Spain, S. France)   1880b. *O. arachnitiformis* Gren. & Phil.

      11. Inner 2 perianth segments velvety or hairy; lip with 2 basal conical swellings.

        12. Lip more or less quadrangular.   *fuciflora*

        12. Lip spathulate, fan-shaped or more or less cylindrical.

          13. Lip with a broad yellow or greenish-yellow margin   *tenthredinifera*

          13. Lip with a brownish, conspicuously incurved margin.   *scolopax*

**1873. *O. fusca* Link BROWN BEE ORCHID. Flowers with greenish-yellow outer perianth segments and long, dark chocolate-brown, velvet-haired lip with bluish or slaty reflective patches near the base or light blue as in subsp. *atlantica*. Lip notched, about twice as long as broad, with 2 lateral lobes. Rocky, grassy, bushy places. Widespread. Pl. 63.

1873a. *O. omegaifera* Fleischm. Like 1873 in general but readily distinguished by the conspicuous white W–shaped band or line that runs across the middle of the lip (the base of the lip and lower margin are also white giving the impression of the letter omega Ω). Outer perianth segments oval, green, the inner strap-shaped, purplish-green. Spain, Balearic Islands. Pl. 63. 1875. *O. speculum* Link. MIRROR ORCHID. Distinguished by its brilliant metallic-blue reflective patch surrounded by a brown shaggy-haired margin. Lip oval, usually with 2 hairy lumps at the base; outer perianth segments usually greenish, the inner brown-purple. Grassy places. S. and C. Portugal, Med. region. Pl. 63, Page 163.

**1876. *O. scolopax* Cav. WOODCOCK ORCHID. Flowers with a large cylindrical-ovoid to elliptic lip, strongly marked with a variable pattern of white and yellow circles and lines, over a reddish-purple ground, and with a conspicuous yellowish knob-like apex. Lip with 2 hairy knobs or long horns at the base. Outer perianth segments white or purplish, inner usually pink. Grassy places, stony ground. S. and C. Portugal, Med. region. Page 53.

**1882. *O. tenthredinifera* Willd. SAWFLY ORCHID. Distinguished by its often brilliant pink perianth segments and its broad, indented lip which is brown with a broad yellow or greenish-yellow margin. Reflective patch brownish, surrounded by a blue line; basal swellings often present. Stony, grassy places. S. and C. Portugal, Med. region. Pl. 63.

1883. *O. fuciflora* (Crantz) Moench LATE SPIDER ORCHID. A very variable plant with a dark velvety-brown, rather quadrangular lip with a bold symmetrical pattern of yellowish-green lines, often enclosing a blue area. Lip with 2 basal swellings, and a green apical knob; perianth segments pink or white, the inner hairy. Stony, grassy places. S. and C. Portugal, S. Spain, France. 1880. *O. sphegodes* Miller EARLY SPIDER ORCHID. Flowers with an entire velvety brown or blackish-purple ovoid lip, with paler bluish markings either in two parallel lines or in an H-shaped or X-shaped pattern. Perianth segments green, pink or white. Grassy, rocky places. Med. region. Pl. 63. 1880a. *O. atrata* Lindley Very like 1880 and by many considered as a subspecies of it, but distinguished by the indentation at the tip of the lip; the long shaggy hairs on the lip margin and outer part of the basal swellings, and the darker ground colour of the lip. The marking on the lip is H-shaped and much paler. C. Portugal, Med. region.

ORCHIS |

1. Spur very short, sac-like, 1–3 mm. (N. and E. Spain, France)   **1888. *O. ustulata* L. Page 153

1. Spur usually at least half as long as the ovary, linear or conical.

  2. Perianth with 5 segments coming together in a helmet.

   3. Lip not lobed but fan-like and toothed.   *papilionacea*

   3. Lip three- to four-lobed.

    4. Bracts reduced to scales, much shorter than the ovary.

     5. Helmet blackish-purple. (N. Spain, France)   **1892. *O. purpurea* Hudson

     5. Helmet pink or ash-pink.

      6. Lobes of lip spreading, the middle 2 shorter and 2–3 times broader than the lateral lobes. (C., N., and E. Spain, France)   **1891. *O. militaris* L. Page 132

      6. Lobes of lip parallel, recurved in front all approximately similar in width and length.

       7. Tips of lobes conspicuously darker than centre of lip; helmet faintly veined; lvs. with flat margin. (S. and E. Spain, France)   1890. *O. simia* Lam.

       7. Tips of lobes little darker than centre of lip; helmet strongly veined; lvs. with undulate margin.   *italica*

    4. Bracts lance-shaped, nearly as long as or longer than the ovary.

     8. Helmet acute, spur downward-pointing, conical or cylindrical.

9. Fls. pale pink or whitish, almost scent-
less; sepals diverging above; lobes of
lip toothed. *tridentata, lactea*

9. Fls. not pale pink or whitish, strong
smelling; sepals not diverging above;
lobes of lip toothed or not.

  10. Fls. dull reddish-brown, smelling of
bugs. (Portugal, Med. region) **1886.   *O. coriophora* L. Pl. 64, Page 133

  10. Fls. vinous-red, smelling of vanilla.
(Portugal, Med. region) 1886a. *O. fragrans* Pollich

8. Helmet very blunt; spur thick, cut-off,
horizontal or upward-pointing.

  11. Lip scarcely as long as the helmet,
purple-violet and spotted with purple
(lip sometimes pink or white). (Widespread) 1885.   *O. morio* L., Page 45
champagneuxii

  11. Lip longer than the helmet, distinctly
two-coloured, the lateral lobes dark
violet and the central lobe very short,
white and purple-spotted. *longicornu*

2. Perianth segments with 3 central segments
in a helmet and 2 lateral segments spreading.

  12. Bracts (at least the lowest) with 3–7
main veins and with a network of connecting lateral veins.

    13. Lip three-lobed, the middle lobe
shortest; lateral lobes conspicuously
turned down. *laxiflora*

    13. Lip three-lobed, the middle lobe as long
as or longer than the lateral lobes
which are spreading at first. (Spain,
France) (1893) *O. palustris* Jacq.

  12. Bracts with 1–7 main veins, without
smaller connecting veins.

    14. Lip not three-lobed, but oval with
rounded marginal teeth. *saccata*

    14. Lip three-lobed.

     15. Fls. pale yellow or whitish (very rarely
purple and spotted with yellow).

      16. Lvs. oblong, broader towards apex,
not brown-spotted; fls. sweet-scented.
(Pyrenees, France) (1896) *O. pallens* L. Pl. 63.

      16. Lvs. lance-shaped, narrowed towards
apex, brown-spotted; fls. scarcely
scented. *provincialis*

     15. Fls. purple-violet (very rarely pink or
white).

      17. Spur cylindrical, thicker towards the

base and as long as or longer than the
ovary. (Widespread)

17. Spur conical, distinctly shorter than
the ovary.

1894. *O. mascula* (L.) L. Pl. 63.

patens

**1884. *O. papilionacea* L. PINK BUTTERFLY ORCHID. Perhaps the most beautiful of our European orchids with rosy-pink flowers strongly veined with deeper pink, and with a broad fan-shaped lip with a toothed margin quite unlike any other species. Perianth segments pink with darker striped veins. Bracts flushed rosy-purple. Dry grassy places, thickets, etc. S. and C. Portugal, Spain, Med. France. Pl. 64.   **1889. *O. tridentata* Scop. TOOTHED ORCHID. Distinguished by its dense conical, or globular, head of spotted pink and white flowers. Lip four-lobed, comprising a conspicuous large central wedge-shaped lobe which is divided into 2 broad, shallow lobes often with a central tooth and two lateral lobes. Perianth segments usually with darker pink veins; spur stout, nearly as long as the ovary. Bracts small, papery. Grassy places, thickets, woods. S. and C. Portugal, Spain, S–W. France.   1889a. *O. lactea* Poiret Like 1889 but helmet very pale pink or white and sepals more pointed. Lip with central broadly fan-shaped lobe, often not notched. Inflorescence cylindrical, rather lax. Arid ground, pine woods, thickets. Med. region.

1885a. *O. champagneuxii* Barn. Very like 1885. *O. morio* but tubers 3, 2 with long stalks. Lobes of lip folded downwards, centre of lip usually pure white; spur enlarged at the tip, usually two-lobed. Pine woods, thickets and dry places. Portugal, S–E. Spain, Med. France. 1885b. *O. longicornu* Poiret LONG-SPURRED ORCHID. Like 1885 but the lip is distinctly longer than the helmet, three-lobed with the central lobe shorter, the lateral lobes reflexed, dark-purple or blackish-red, the centre of the lip white with dark red dots. Spur long, arched upwards, broader at apex. Dry grassy places, thickets. Portugal, Spain, Balearic Islands.

**(1890) *O. italica* Poiret Distinguished by its dense conical cluster of pink flowers with the pointed helmet conspicuously striped with darker veins, and its fan-shaped spotted lip. Lobes of lip 'arms and legs' little darker than the 'body', central 'tail' often long. Acid ground, bushy places, pine woods. S. and C. Portugal, S., C., and E. Spain. Pl. 64.

1899a. *O. saccata* Ten. Distinguished by its undivided fan-shaped lip, recalling that of 1884, but lip red to dark violet, unspotted or streaked. Spur stout, sac-like, whitish, about half as long as ovary. Perianth greenish-brown to reddish-brown. Bracts purple-violet somewhat longer than ovary. Leaves usually spotted. Arid places, thickets, clearings. S. Spain, S. France. Pl. 64.

1894a. *O. patens* Desf. Like 1894 *O. mascula* but distinguished by the perianth segments which are greenish with purple spots on the inner side, and reddish outside. Lip three-lobed, red-violet, darker red-spotted in the centre, lobes rhomboidal. Spur shorter than the ovary, downward-pointing or more or less horizontal. Leaves sometimes spotted. Open ground, thickets, woods, mountains. S. and C. Spain. Pl. 63, Page 117.

**1893. *O. laxiflora* Lam. JERSEY ORCHID. Flowers large, unspotted, dark purple with pale centres borne in a long lax cluster, the whole inflorescence often purple-flushed. Lip with 2 lobes which are conspicuously turned downwards and almost touching. Spur horizontal or upward-pointing, slender, swollen and notched at apex. Marshy meadows. S. and C. Portugal, Spain, France.

1896. *O. provincialis* Balbis PROVENCE ORCHID. The commonest yellow-flowered orchid of the lowlands. Flowers pale or dark yellow with orange or brownish spots. Lip three-lobed. Spur upward-pointing, as long as or longer than ovary. Bracts pale, about as long as ovary. Bushy and grassy places, pine woods. E. Spain, S. France.

DACTYLORHIZA | Like *Orchis* but tubers lobed; perianth segments not coming together in a helmet. A difficult genus.

1. Fls. yellowish or spotted with yellow; spur very stout, conical, blunt, as long as the ovary. (C. Portugal, N. and C. Spain, France)   **1899. *D. sambucina* (L.) Soó
1. Fls. pink or purple; spur cylindrical, shorter than the ovary.
  2. Stem hollow; lower and middle bracts longer than the fls.
    3. Lip entire, or almost so, lozenge-shaped, blunt, shallowly toothed; lvs. unspotted. (Widespread)   1897. *D. incarnata* (L.) Soó
    3. Lip distinctly three-lobed; lvs. usually spotted. (Widespread)   **1898. *D. majalis* (Richenb.) P. F. Hunt & Summerh. Page 148 *elata*
  2. Stem solid; bracts mostly shorter than the fls.
    4. Lip with 3 narrow, more or less equal lobes. (Widespread)   **(1900) *D. fuchsii* (Druce) Soó, *saccifera*
    4. Lip with 2 broad lateral lobes and a much smaller central lobe. (Widespread)   1900. *D. maculata* (L.) Soó

1898a. *D. elata* (Poiret) Soó Like 1898 but a robust plant to 1 m, with a large broad-heart-shaped lip which is very shallowly three-lobed or entire, red-purple with darker spots and lines. Spur about as long as ovary, down-pointing. Outer perianth segments spreading, inner inflexed at the tip. Bracts longer than the ovary. Leaves unspotted. Marshy places. Iberian peninsula, S–W. France.   1900a. *D. saccifera* (Bory) Soó Like (1900) but spur stout, sac-like, about as long as the ovary. Flowers larger; inflorescence laxer and bracts very long, protruding from the spike. Portugal, Med. region.

TRAUNSTEINERA | 1887. *T. globosa* (L.) Reichenb. (*Orchis g.*) Pastures in mountains, woods. E. Spain, Pyrenees.

NIGRITELLA | **1901. *N. nigra* (L.) Reichenb. fil. BLACK VANILLA ORCHID. Alpine meadows. Pyrenees, Massif Central.

SERAPIAS | **Tongue Orchid**   The lip is divided into a basal *hypochile* which has 2 broad, rounded lateral lobes, and an apical *epichile* which forms the tongue-like lip. Perianth segments forming a helmet.

1. Lip with a solitary dark basal swelling.   *lingua*
1. Lip with 2 parallel or divergent dark basal swellings.
  2. *Epichile* distinctly narrower than *hypochile*.
    3. *Epichile* about twice as long as *hypochile*. (Portugal, S. France)   **1904. *S. vomeracea* (Burm. fil.) Briq.
    3. *Epichile* about 1–1½ times as long as the *hypochile*.

4. Lip bright rusty red.   *parviflora*
4. Lip dull dark red. (S. France)   1904a. *S. olbia* Verguin
2. *Epichile* as broad or almost as broad as the
   hypochile.   *cordigera*

1902. *S. lingua* L. TONGUE ORCHID. Flowers with a long reddish, slightly hairy lip about twice as long as the similar-coloured helmet. The single swelling at the base of the lip (only seen when the lip is removed), distinguishes it from all other species. Cultivated and uncultivated ground, grassy places, pine woods. Portugal, Spain, S. France. Pl. 64, Page 61.
1903. *S. cordigera* L. HEART-FLOWERED SERAPIAS. Distinguished by the basal lobed part, *hypochile*, which, if spread out, is as broad as the apical tongue part, *epichile*, of the lip. Lip black-purple, hairy; helmet reddish-violet or wine-coloured. Bracts pale pinkish, little shorter than the flowers. Damp heaths, sandy places. Portugal, Med. region. Page 161.
(1904) *S. parviflora* Parl. SMALL-FLOWERED SERAPIAS. Distinguished by its small flowers 1½–2 cm, with lip scarcely longer than the helmet. Lip rusty-red, strongly reflexed against ovary; helmet reddish-violet. Grassy places. Portugal, Med. region. Pl. 64.

ACERAS | ****1905.** *A. anthropophorum* (L.) Aiton fil. MAN ORCHID. Grassy and bushy places. S. and C. Portugal, Spain, France. Page 124.

NEOTINEA | 1905a. *N. intacta* (Link) Reichenb. fil. DENSE-FLOWERED ORCHID. A slender plant with a spike of numerous, small, densely clustered pinkish flowers smelling of vanilla. Lip three-lobed, the lateral lobes narrow, the terminal toothed. Spur short, stout. Perianth segments in a helmet, pink or white. Bracts shorter than ovary. Leaves oval, usually spotted. Woods, scrub, stony places, on limestone. C. Portugal, Med. region.

HIMANTOGLOSSUM | ****1906.** *H. longibracteatum* (Biv.) Schlechter GIANT ORCHID. A very robust plant with greenish-yellow flowers more or less flushed with pink or dull purple, in a dense cylindrical spike. Lip large, deeply four-lobed, often wavy-margined. Spur short, stout. Grassy places. S. and C. Portugal, S. and E. Spain, Med. France. ****1907.** *H. hircinum* (L.) Sprengel LIZARD ORCHID. A robust, strong-smelling orchid with a lax cylindrical cluster of greenish-yellow flowers with very long, strap-shaped lips which are irregularly twisted and coiled. Lip brown-spotted, greenish towards tip, and with slender, coiled lateral lobes. Helmet greenish; spur short. Grassy places, banks, etc. S., C., and N. Spain, France. Page 125.

ANACAMPTIS | ****1908.** *A. pyramidalis* (L.) L. C. M. Richard PYRAMIDAL ORCHID. Grassy places. S. and C. Portugal, Spain, France.

HERMINIUM | ****1909.** *H. monorchis* (L.) R.Br. MUSK ORCHID. Dry and damp meadows. C. Spain, Pyrenees, France.

COELOGLOSSUM | 1911. *C. viride* (L.) Hartman FROG ORCHID. Grassy places. Spain, France.

GENNARIA | Differs from *Coeloglossum* and *Gymnadenia* in that the 2 stigmas are raised on very short stalks, and thus resembling *Habenaria*. Leaves 2. Only 1 species.

1911a. *G. diphylla* (Link) Parl. A slender plant 15–30 cm, with a dense one-sided, leafy spike of many tiny, greenish-yellow flowers. Lip three-lobed, with a short rounded spur. Perianth segments oblong, blunt, held closely together. Leaves 2, alternate, heart-shaped acute, the upper smaller. Pine and laurel woods, thickets, screes. S. and C. Portugal, S. and W. Spain.

GYMNADENIA | **1912. *G. conopsea* (L.) R.Br. FRAGRANT ORCHID. Spur nearly twice as long as ovary; lip much broader than long. Grassy places, marshes, woods, in mountains. N–W. Portugal, N. and N–E. Spain, France.   (1912) *G. odoratissima* (L.) L. C. M. Richard SHORT-SPURRED FRAGRANT ORCHID. Like 1912 but spur thicker and shorter than ovary, and lip little broader than long. Flowers rose-violet or whitish-yellow, tiny 5–7 mm, smelling strongly of vanilla. Mountain pastures. N–E. Spain, France.

PSEUDORCHIS | **1913. *P. albida* (L.) Á. & D. Löve (*Leucorchis a.*) SMALL WHITE ORCHID. Mountain pastures. Pyrenees.

PLATANTHERA | **1914. *P. bifolia* (L.) L. C. M. Richard LESSER BUTTERFLY ORCHID. Flowers whitish, 11–18 mm across, with a slender strap-shaped lip and a long curved greenish spur 1½–2 cm. Distinguished by the 2 pale anthers (placed under the helmet) lying parallel to each other. Woods, grassy places, in mountains. C. and N. Portugal, Spain, France.

**1915. *P. chlorantha* (Custer) Reichenb. GREATER BUTTERFLY ORCHID. Like 1914 but flowers larger 18–23 mm across; and 2 yellow anthers conspicuously diverging below. Spur 2–3 cm, down-curved, thickened at tip. Woods, grassy places, in mountains. N. and E. Spain, France.

EPIPACTIS | **Helleborine**   Lip divided into a basal cup-shaped *hypochile* and an apical triangular-heart-shaped *epichile*. The rostellum is a beak-like process formed from the third stigma.

1. *Epichile* attached to *hypochile* by a narrow moveable joint; marsh plant. (C. Portugal, N. and E. Spain, France)	**1916. *E. palustris* (L.) Crantz
1. *Epichile* attached to *hypochile* by 1 or more folds, not movable.	
2. Fls. entirely reddish-purple. (Spain, France)	1918. *E. atrorubens* (Hoffm.) Schultes Pl. 64
2. Fls. not entirely reddish, with varying amounts of green or predominantly green.	
3. Lvs. small 2–2½ cm by 1 cm, shorter than the internodes. (Spain, Balearic Islands, W. France)	1918a. *E. microphylla* Swartz
3. Lvs. larger, longer than or as long as the internodes.	
4. Stem hairless or nearly so; *hypochile* greenish-white inside. (W. France)	1917a. *E. phyllanthes* G. E. Sm.
4. Stem with short hairs; *hypochile* purplish or dark green inside.	
5. *Epichile* longer than broad, arrow-shaped. (France)	1917b. *E. leptochila* (Godf.) Godf.
5. *Epichile* not longer than broad.	
6. Rostellum large, whitish, persistent. (Portugal, S., C., and E. Spain, France)	1917. *E. helleborine* (L.) Crantz
6. Rostellum reduced or absent in bud. (France)	1917c. *E. muelleri* Godf.

CEPHALANTHERA | 1919. *C. damasonium* (Miller) Druce WHITE HELLEBORINE. Flowers white. Leaves oval-lance-shaped; outer perianth segments blunt. Woods, shady, bushy places. N–E. Spain, France.   **(1919) *C. longifolia* (L.) Fritsch LONG-LEAVED HELLEBORINE.

Flowers white. Leaves lance-shaped, the upper linear; outer perianth segments acute. Woods, grassy and shady places. Widespread except in north. **1920. *C. rubra* (L.) L. C. M. Richard RED HELLEBORINE. Flowers bright pink or reddish-violet, with perianth segments pressed close to each other. Lower leaves oblong, the upper lance-shaped, all acute; bracts longer than ovary. Woods, bushy places. N–E. Portugal, S., C., and E. Spain, France. Page 117.

LIMODORUM | **1921. *L. abortivum* (L.) Swartz LIMODORE. Unlike any other orchid in having a long slender spike of large stalkless violet flowers, and violet-flushed scale-like leaves on the stem below. Lip shorter than other segments, with some yellow; spur stout. Pine woods, bushy places, in hills. S. and C. Portugal, Med. region. Pl. 64, Page 116. 1921a. *L. trabutianum* Batt. Like 1921 but spur rudimentary or absent. Woods. C. Portugal, S. Spain.

SPIRANTHES | **1922. *S. spiralis* (L.) Chevall. AUTUMN LADY'S TRESSES. Autumn-flowering. Stem with scales only, leaves all basal, about 2½ cm. Grassy places, heaths. S. Portugal, Spain, France. (1922) *S. aestivalis* (Poiret) L. C. M. Richard SUMMER LADY'S TRESSES. Summer-flowering. Stem leafy, leaves 5–12 cm. Heaths, damp fields. Portugal, C. and N. Spain, France.

LISTERA | 1923. *L. ovata* (L.) R.Br. TWAYBLADE. Plant large 20–60 cm; leaves 5–20 cm. Woods, thickets, in mountains. C., N., and N–E. Spain, France. **(1923) *L. cordata* (L.) R.Br. LESSER TWAYBLADE. Plant small 6–20 cm; leaves 1–2½ cm. Coniferous woods, bogs. Pyrenees, Auvergne.

NEOTTIA | **1924. *N. nidus-avis* (L.) L. C. M. Richard BIRD'S-NEST ORCHID. Shady woods, particularly beech woods in mountains. Widespread, except Portugal (1 locality).

GOODYERA | **1925. *G. repens* (L.) R.Br. CREEPING LADY'S TRESSES. Mossy coniferous woods, in mountains. Pyrenees, France.

CORALLORHIZA | **1926. *C. trifida* Chatel. CORAL-ROOT ORCHID. Woods in mountains. France.

List of Indexes

Index of popular names

GENUS	ENGLISH	SPANISH	PORTUGUESE	FRENCH
Abies	Fir	Abeto	Abeto	Sapin
Acacia	Mimosa, Wattle	Acacia, Mimosa	Acácia, Mimosa	Mimosa
Acanthus	Bear's Breech	Acanto	Acanto, Erva-gigante	Acanthe
Acer campestre	Maple	Arce	Bardo comum	Erable
pseudoplatanus	Sycamore	A. blanco, Falso platano, Sicomoro	Padreiro, Plátano-bastardo	Sycomore
Aceras	Man Orchid		Erva-do-homen-enforcado, Rapazinhos	Homme pendu
Achillea millefolium	Yarrow, Milfoil	Milenrama, Milefolio, Flor de la pluma	Milefólio	Achillée Millefeuille
Aconitum	Monkshood	Acónito	Acónito	Aconit
Actaea	Baneberry, Herb Christopher	Hierba de San Cristobal		Herbe de Saint-Christophe
Adonis (red sp.)	Pheasant's Eye	Gota de sangre	Casadialos, Lágrima-de-sangue	Goutte-de-sang
Aesculus	Horse-chestnut	Castaño de Indias	Castanheiro-da-India	Marronnier d'Inde
Agave	Century Plant	Pita	Piteira	Agave
Agrimonia	Agrimony	Hierbo de San Guillermo	Agrimónia	Aigremoine
Agrostemma	Corn Cockle	Neguillón	Nigela-dos-trigos	Nielle-des-blés
Ailanthus	Tree of Heaven	Árbol del cielo	Ailanto, Espanta-lobos	Ailanthe
Ajuga chamaepitys	Bugle Ground-pine	Bugula Camepiteos		Bugle B.-petit pin
Alcea	Hollyhock	Malva real	Malvaisco	Rose Trémière
Alchemilla vulgaris	Lady's-mantle	Pie de león	Pé de leão (subsp. alpestris)	Alchémille vulgaire
Alisma	Water-plantain	Llantén de agua	Tanchagem da água	Plantain-d'eau

Anthyllis vulneraria	Kidney-vetch	Vulneraria	Vulnerária	Anthyllide Vulnéraire
Antirrhinum	Snapdragon	Boca de dragón	Erva bezarra, Bocas de lobo	Muflier
Apium graveolens	Wild Celery	Apio	Aipo	Céleri
nodiflorum	Fool's Watercress	Berraza	Rabaças	Ache nodiflore
Aquilegia	Columbine	Aquileña	Erva pombinha	Ancolie
Arbutus	Arbutus, Strawberry Tree	Madroño	Medronheiro, Ervodo	Arbousier
Arctium	Burdock	Bardana, Lampazo	Bardana, Pegamaço	Bardane
Arctostaphylos	Bearberry	Gayuba	Medronheiro ursino, Uva-de-urso, uva-ursina	Raisin-d'ours
Arisarum	Friar's Cowl	Rabiacana, Fraillillos	Candeias, Cupuz de fradinho	
Aristolochia baetica	Birthwort	Candiles	Aristoloquia Erva cavalinha	Aristoloche
Armeria	Thrift		Arméria	Arméria
Armoracia	Horse-radish	Rábano rusticano	Armorácia	Raifort
Arnica	Arnica	Árnica	Arnica	Arnica
Artemisia abrotanum	Southernwood	Abrótano macho	Broida, Herba cuquera, Botja	Aurone
absinthium	Wormwood	Ajenjo mayor	Sintro, Losna, Absinto	Absinthe
dracunculus	Tarragon	Dragoncillo, Estragón	Estragão	Estragon
vulgaris	Mugwort	Artemisa, Hierba de San Juan	Artemísia verdadeira	Armoise
Arum	Lords-and-Ladies, Cuckoo-pint	Aro	Jaro, Farro, Erva-da-novidade	Gouet
Arundo	Reed	Caña comun	Cana	Grand roseau
Asarum	Asarabacca	Asaro		Asaret

GENUS	ENGLISH	SPANISH	PORTUGUESE	FRENCH
Asparagus	Asparagus	Esparraguera	Espargo	Asperge
Asperugo	Madwort	Raspilla		Râpette
Asphodelus	Asphodel	Gamón	Abrótea	Asphodele
Aster				
tripolium	Sea Aster		Malmequer-da-praia	
Astragalus	Milk-vetch		Alfavace-dos-montes, Tremoção	Astragale
lusitanica				
Atractylis cancellata	Distaff Thistle			
Atriplex	Orache	Armuelle	Cardo-coroado	Arroche
Atropa	Deadly Nightshade	Belladona	Armoles	Belladone
Avena	Oat	Avena	Beladona	Avoine
Ballota	Black Horehound	Marrubio fétido	Aveão, Aveia	Marrube noir
Barbarea	Winter Cress	Hierba de los carpinteros	Marroio-negro	Barbarée
Bellis	Daisy	Margarita, Chirivita	Erva-de-Santa Bárbara	Pâquerette
Berberis	Barberry	Agracejo, Alvo	Bonina	Epine-vinette
Beta vulgaris	Beet	Acelga salvaje	Bérberis, Uva-espim	Bette, Betterave
Betonica	Betony	Betonica	Acelga brava, Beterraba, Calga	Bétoine
Betula	Birch	Abedul	Betónica	Bouleau
Biscutella	Buckler Mustard	Anteojos de Santa Lucia	Videoiro	Lunetière
Blackstonia	Yellow-wort	Centaura amarilla	Centáurea-meno-perfothada	Centaurée jaune
Borago	Borage	Borraja	Borragem	Bourrache
Brassica				Chou
napus	Rape, Cole, Swede	Nabo, Colza	Colza, Nabo, Nabica	Colza
nigra	Black Mustard	Mostaza negra	Mostarda negra, M. ordinária	Moutarde noire

oleracea	Wild Cabbage	Col	Couve, Bêrça	Chou, Choufleur
rapa	Turnip	Remolacha	Turnepo	Rave, Navet
Briza	Quaking Grass	Cedacillo, Tembladera	Bole-bole	Amourette
Bryonia	Bryony	Nueza blanca, Brionía	Briónia, Norca branca	Bryóne
Butomus	Flowering Rush	Junco florido		Jonc fleuri
Buxus	Box	Boj	Buxo	Buis
Cakile	Sea Rocket	Oruga marítima	Eruca marítima	Roquette-de-mer
Calamintha	Calamint	Calaminta	Néveda, Erva-das-azeitonas	Sarriette
Calendula	Pot-marigold	Maravilla, Caléndula, Flamenquilla	Maravilhas	Souci
Calicotome	Spiny Broom	Retama espinosa		Cytise épineux
Calluna	Ling, Heather	Brecina	Urze, Torga-ordinária	Bruyère
Caltha	Kingcup, Marsh Marigold	Hierba centella	Calta, Malmequer-dos-brejos	Populage
Calystegia soldanella	Large Bindweed Sea Bindweed	Correguela mayor C. marina	Trepadeira, Bons-dias Soldanela, Couve-marinha	Liseron Chou-marin
Campanula lusitanica	Bellflower, Campanula		Rapúncio, Rapôncio	Campanule
Cannabis	Hemp	Cáñamo	Cânhamo	Chanvre
Capparis	Caper	Alcaparra	Alcaparra	Câprier
Capsella	Shepherd's Purse	Bolsa de pastor	Bolsa-de-pastor	Bourse-à-pasteur
Capsicum	Chili, Peppers	Pimiento	Pimentão, P. corni-cabra	Piment
Cardamine	Cuckoo Flower, Bittercress	Mastuerzo de prado		Cardamine
hirsuta	Hairy Bittercress		Agrião menor	
Carduus	Thistle	Cardo	Cardo	Chardon

GENUS	ENGLISH	SPANISH	PORTUGUESE	FRENCH
Carlina	Carline Thistle	Carlina		Carline
acaulis				Chardon argenté
racemosa		Cardo de la uva		
Carpinus	Hornbeam	Carpe, Hojaranzo		Charme, Charmille
Carpobrotus	Hottentot Fig	Flor de cuchillo	Chorões	Mésembrianthème
Carthamus				Chardon béni
lanatus		Azotacristos	Cardo-sanguinho	des Parisiens
tinctorius	Safflower	Alazor, Cártamo	Acafroa	Faux-safran
Castanea	Sweet Chestnut, Spanish Chestnut	Castaño	Castanheiro	Châtaignier
Catananche	Cupidone	Hierba cupido	Sesamóide-menor	Cupidone
Celtis	Nettle Tree	Alméz	Lódão bastardo, Agreira	Micocoulier
Centaurea	Knapweed			Centaurée
calcitrapa	Star Thistle	Cardo estrellado	Cardo estrelado Calcatripa	Chausse-trape
cyanus	Cornflower	Azulejo, Aldiza	Fidalguinhos, Loios-dos-jardins	Bluet
Centaurium	Centaury	Centaura menor, Hiel de tierra	Fel-da-terra, Centáurea-menor	Petite Centaurée
Centranthus	Red Valerian	Valeriana roja	Alfinetes	Centranthe
Ceratonia	Carob, Locust Tree	Algarrobo	Alfarrobeira	Caroubier
Cercis	Judas Tree	Arbol del amor	Olaia, Arvore-de-Juda	Arbre de Judée
Cerinthe	Honeywort	Ceriflor	Chupa-mel, Flor-mel	Mélinet
Chamaemelum	Chamomile	Manzanilla romana		Camomille romaine
Chamaerops	Fan Palm	Palmito	Palmeira anã, Palmeira-das-vassouras	
Cheiranthus	Wallflower	Alhelí amarillo	Goiveiro amarelo	Giroflée, Violier

GENUS	ENGLISH	SPANISH	PORTUGUESE	FRENCH
Citrus				
limon	Lemon	Limonero	Limoeiro,	Citronnier
medica	Citron	Cidrero	Cidreira	Cédratier
sinensis	Orange	Naranjo	Laranjeira-da-China	Oranger doux
Clematis	Clematis	Hierba de los pordioseros	Clematite	Clématite
Vitalba	Travellers' Joy		Vide-branca	Vigne blanche
	Old Man's Beard		Sipó-do-reino	
Clinopodium	Wild Basil	Albahaca silvestre	Clinopódio	Rouette
Cneorum		Olivilla común		
Cnicus	Blessed Thistle	Cardo santo	Cardo-santo	Chardon béni
Colchicum				
autumnale	Meadow Saffron, Autumn Crocus	Cólchico	Cólquico	Colchique
Colutea	Bladder Senna	Espantalobos		Baguenaudier
Consolida	Larkspur	Consuelda, Espuela de caballero	Consólida real	Delphinette
Convallaria	Lily-of-the-Valley	Lirio de los valles	Leirio convale	Muguet
Convolvulus	Bindweed	Corregüela menor	Corriola, Verdeselha Verdisela	Liseron
arvensis	Cornbine			
tricolor		Maravilla	Azuraque, Zuraque	
Corema		Camarina	Camarinheira	
Coriaria		Roldón, Emborachacabras		Corroyère
Cornus				
mas	Cornelian Cherry	Cornejo	Sanguinhe legítimo	Cormier
sanguinea	Dogwood	Cornejo Sanguiñelo	Sanguinho legítimo	Sanguinella
Corydalis	Corydalis	Tijerillas		Corydale
Corylus	Hazel, Cob-nut	Avellano	Aveleira	Noisetier
Cotoneaster		Guillomo		Cotonéaster
Crataegus	Hawthorn	Espino	Pirliteiro, Escalherio	Aubépine, Epine blanche

	English	Spanish	Portuguese	French
azarolus	Mediterranean Medlar	Acerolo	Azaroleira	Azerolier
Crithmum	Rock Samphire	Peregil del mar, Hinojo marino	Funcho marítimo, Perrexil-do-mar	Criste-marine
Crocus	Crocus	Azafrán	Açafrão	Crocus
Cucumis				
melo	Melon	Melón	Melão	Melon
sativus	Cucumber	Pepino	Pepino	Concombre, Cornichon
Cucurbita				
maxima	Pumpkin	Calabaza confitera	Abóbora menina	Courge
pepo	Marrow	Calabaza común	A. porqueira, A. de coroa, Barrete-de-padre	Citrouille
Cupressus	Cypress	Ciprés	Cipreste	Cypres
Cuscuta	Dodder	Cabellos de Venus	Cuscuta, Cabelos	Cuscuta
Cydonia	Quince	Membrillo	Marmeleiro, Gambeiro	Cognassier
Cynanchum	Stranglewort	Escamonea falsa	Escamonéa de Mompelher	Scammonee de Montpellier
Cynara				
cardunculus	Cardoon	Cardo de arrecife, Alcaucil silvestre	Cardo-do-coalho, C. hortense	Cardon
scolymus	Artichoke	Cardo común, C. alcachofero	Alcachofra-de-comer, A.-hortense	Artichaut
Cynoglossum	Hound's-tongue	Cinoglosa	Cinoglossa	Langue-de-chien
Cynomorium		Hongo de Malta, Teticas de doncella	Pútegas, Coalhadas	
Cytinus				Cytinet
Cytisus				
multiflorus		Piorno, Escoba blanca	Giesteira branca	
purgans		Piorno serrano		
scoparius	Broom	Hiniesta	Giesteira das vassouras	Cytise
Daboecia	St. Dabeoc's Heath	Tambarella		Bruyère de Saint-Daboec

	Globe Thistle	Cardo yesquero	Cardo-de-isca	Echinope
Echinops	Globe Thistle	Cardo yesquero	Cardo-de-isca	Echinope
Echinospartum		Piorno fino		
boissieri				
horridum		Erizones		
lusitanicum		Arcebilla		
Echium	Viper's Bugloss	Viborera morada		Vipérine
Elaeagnus	Oleaster	Pangino	Caldoneira	Chalef
Endymion	Bluebell		Jacinto-dos-campos	Jacinthe sauvage
Erica				
arborea	Tree Heath	Brezo, Urce	Urze-branca, Torga Chamiça, Urgeira	Bruyère arborescente
australis	Spanish Heath	Brezo rubio	Lameirinha,	
ciliaris	Dorset Heath	Carroncha	Cordões-de-freira	
cinerea	Bell Heather		Queiró, Torga	
lusitanica	Lusitanian Heath	Argaña	Quiroga, Torga	Bruyère a balai
scoparia	Green Heather	Brezo de escobas	Urze-das-vassouras	
Erinacea	Hedgehog Broom	Piorno azul		Bibacier,
Eriobotrya	Loquat	Níspero de Japón	Nespereira-do-Japão, Magnólio	Néflier du Japon
Erodium	Storksbill	Pico de Cigüeña Relojes	Bico-de-cegonha	Bec-de-grue
Eryngium				Panicaut
campestre	Field Eryngo	Cardo corredor	Cardo-corredor	Chardon roulant
maritimum	Sea Holly		C.-maritimo, C. rolador	
Erythronium	Dog's Tooth Violet	Diente de perro	Dente-de-cáo	Dent-de-chien
Eucalyptus	Gum, Eucalyptus	Eucalipto	Eucalipto	Eucalyptus
Euonymus	Spindle-Tree	Bonetero		Fusain
Eupatorium				
cannabinum	Hemp Agrimony		Eupatório-de-Avicena, Trevo-cervino	Eupatoire-chanvrin

GENUS	ENGLISH	SPANISH	PORTUGUESE	FRENCH
Fagopyrum	Buckwheat	Trigosarraceno Alforfón	Trigo-sarraceno	Sarrasin, Blé noir
Fagus	Beech	Haya		Hêtre, Fayard
Ferula	Giant Fennel	Cañaheja	Canafrecha	Férule
Ficus	Fig	Higuera	Figueira, Baforeira	Figuier
Filipendula ulmaria	Meadow-sweet	Reina de los Prados	Erva-ulmeira	Reine-des-prés
Foeniculum	Fennel	Hinojo	Funcho	Fenouil
Fragaria	Wild Strawberry	Fresa	Moranguerio	Fraisier
Frangula alnus	Alder Buckthorn	Chopera, Arraclán	Sanguinho-da-água, Amieiro-negro	Bourdaine
Frankenia	Sea Heath	Albohol		
Fraxinus	Ash	Fresno	Freixo	Frêne
ornus	Manna A.			Frêne-à-fleurs
Fumaria	Fumitory	Fumaria, Palomilla	Fumaria, Erva-molarinha	Fumeterre
Galanthus	Snowdrop	Campanilla de invierno		Perce-neige
Galega	Goat's Rue, French Lilac	Ruda cabruna	Galega, Caprária	Galéga
Galium	Bedstraw	Cuajaleches		Caille-lait, Gaillet
Genista species	Greenweed, Needle Furze, Petty Whin	Bolina Hiniesta, Aulaga		Genêt
triacanthos			Ranha-lobo, Tojo molar	
Gentiana lutea	Gentian Great Yellow Gentian	Genciana G. mayor, G. amarilla	Genciana G.-das-boticas, Argençana-dos-pastores	Gentiane Grande gentiane
Geranium robertianum	Cranesbill Herb Robert	Hierba de San Roberto	Gernaio Erva-roberta	Géranium Herbe à Robert
Geum urbanum	Herb Bennet, Avens	Carioflada Hierba de San Benito	Erva benta, Carlada, Hiofi, Sanamunda	Benoîte

		Hierba pastel	Pastel-dos-tintureiros	Pastel
Isatis	Woad	Hierba pastel	Pastel-dos-tintureiros	Pastel
Jasminum	Jasmine	Jazmín	Jasmineiro, Giestó	Jasmin
Juglans	Walnut	Nogal	Nogueira	Noyer
Juncus	Rush	Junco	Junco	Jonc
Juniperus	Juniper	Enebro, Sabina	Zimbro, Sabina-da-praia, Oxicedro	Genêvrier
Knautia	Field Scabious	Viuda silvestre		Scabieuse
Laburnum	Laburnum	Codeso de los Alpes	Codeço-bastardo	Cytise, Aubour
Lactuca	Lettuce	Lechuga	Alface	Laitue
Lamium	Dead-nettle		Chuchapitos	Ortie morte, Lamiem
Lapsana	Nipplewort		Labresto, Lapsana	Grageline
Larix	Larch	Alerce		Mélèze
Lathyrus	Vetchling			Gesse
cicera	Red Vetchling	Cicércula	Araca Chícharos-miúdos / Cizirão	Pois Vivace
latifolius	Winged Vetchling	Alverjana	Ervilha-dos-campos	Pois de senteur
ochrus	Sweet Pea	Guisante de olor	Ervilhas-de-cheiro	Gesse commune,
odoratus	Chickling Pea	Guijas	Chícaros	Pois carré
sativa				
Laurus	Laurel	Laurel	Loureiro, Louro / Sempre verde	Laurier
Lavandula	Lavender	Lavanda		Lavande
angustifolia	Common Lavender	Espliego		
lanata		Alhucema		
multifida	Cut-leaved Lavender	Alhucemilla	Alfazema-de-folha-recortada / Alfazema	
spica	French Lavender	Cantueso	Rosmaninho	
stoechas	Green Lavender		Rosmaninho-verde	
viridis				
Lavatera	Tree Mallow	Malva		Mauve royale
Legousia	Venus's Looking-Glass	Espejo de Venus		Miroir de Vénus

	French	Portuguese	Spanish	English
Lysimachia nummularia	Lysimaque, Herbe-aux-écus	Lisimaquia	Hierba de la moneda	Creeping Jenny
Lysimachia vulgaris	Lysimaque		Lysimaquia amarilla	Yellow Loose-strife
Lythrum	Salicaire	Salgueirinha	Lisimaquia roja	Purple Loose-strife
Malcolmia maritima		Goivo-da-praia	Alheli de mahón	Virginia Stock
Malus	Pommier	Maceira, Maçanzeira	Manzano	Apple
Malva	Mauve	Malva	Malva	Mallow
Mandragora	Mandragore	Mandrágora	Mandrágora, Berengenilla	Mandrake
Marrubium	Marrube	Marróio-branco	Marrubio	White Horehound
Matthiola	Violier	Goiveiro-encarnado	Alhelí	Stock
Medicago	Luzerne	Luzerna	Mielga, Carretón	Medick
Medicago lupulina	Minette	Luzerna lupulina		Black Medick
Medicago sativa	Luzerne cultivée	Luzerna	Alfalfa	Lucerne, Alfalfa
Melampyrum	Mélampyre		Trigo vacuno	Cow-wheat
Melia	Margousier	Mélia, Amargoseira, Conteira	Melia, Paraíso	Indian Bead Tree, Persian Lilac
Melilotus	Mélilot	Anafe, Meliloto	Melitoto	Melilot
Melissa	Mélisse	Melissa, Erva-cidreira	Melisa	Balm
Melittis	Mélitte	Melissa-bastarda	Melisa bastarda	Bastard Balm
Mentha	Menthe	Hortela, Mentastro	Menta	Mint
Mentha pulegium	Pouillot	Poejo	Poleo	Penny-royal
Menyanthes	Trèfle-d'eau	Trevo-da-água, Fava-da-agua	Trébol acuático	Bogbean, Buckbean
Mercurialis	Mercuriale	Mercurial, Urtiga morta	Mercurial	Mercury
Merendera		Noselha, Quitamerendas	Quitameriendas	
Mesembryanthemum	Ficoïde glaciale	Erva do orvalho, Choroes	Aguazul, Gazul, Escarchada	
Mespilus	Néflier	Nespereira	Níspero	Medlar
Morus	Mûrier	Amoreira	Moral, Morera	Mulberry
Muscari	Muscari			Grape-hyacinth
Muscari comosum	M.-à-houppe	Jacinto das searas	Jacinto de penacho	Tassel Hyacinth

	English	Spanish	Portuguese	French
Onopordum	Scotch Thistle	Toba, Cardo borriquero	Acanto bastardo	Onoporde
Onosma	Golden Drop	Orcaneta amarilla		Orcanette jaune
Ophrys		Abejera		
apifera	Bee Orchid		Erva-aranha	Ophrys abeille
fusca	Brown Bee Orchid		Moscardo-fusco	Ophrys sombre
lutea	Yellow Bee Orchid		Erva-vespa	Ophrys jaune
scolopax	Woodcock Orchid		Flor-dos-passarinhos	Ophrys bécasse
speculum	Mirror Orchid		Erva abelha	Ophrys miroir
Opuntia	Prickly Pear, Barbary Fig	Chumbera	Figueira-da-India	Figuier de Barbarie
Orchis				
coriophora	Bug Orchid	Sangre de Cristo	Erva perceveja	Orchis punaise
mascula	Early Purple O.		Satirião-macho, Salepeira maior	Orchis mâle
morio	Green-winged O.	Compañón	Testiculo-de-cão, Erva-do-salepo, Fatua	Orchis bouffon
papilionacea	Pink Butterfly O.		Eva borboleta	Orchis papillon
Origanum	Marjoram	Orégano	Ourégão	Origan
virens				
Ornithogalum	Star-of-Bethlehem	Leche de pájaro	Leite-de-galinha	Ornithogale, Dame-d'onze-heures
umbellatum				
Ornithopus	Birdsfoot	Serradella	Serradela	Pied-d'oiseau
sativus				
Orobanche	Broomrape	Guardalobo, Bayón	Brincalheta, Penachos	Orobanche
Osyris		Carrascas de S. Juan, Algodonosa	Cássia-branca	Rouvet
Otanthus	Cotton-weed		Cordeiros-da-praia	
Oxalis	Sorrel	Acederilla	Trevilho, Erva-canária, trevo-azedo	Surelle
pes-caprae				

GENUS	ENGLISH	SPANISH	PORTUGUESE	FRENCH
Paeonia	Peony	Peonia, Celonia	Peonia, Rosa-albardeira, Rosa-de-lobo	Pivoine
Paliurus	Christ's Thorn	Espina de Cristo		Épine du Christ
Pancratium	Sea Daffodil	Nardo marino	Narciso-das-areias	Lis-mathiole
Papaver	Poppy	Amapola, Ababol	Papoila	Pavot, Coquelicot
Parietaria	Pellitory-of-the-Wall	Parietaria	Parietária, Alfavaca-de-cobra	Pariétaire
Paris	Herb Paris	Hierba de Paris, Uva de zorro		Parisette
Parnassia	Grass of Parnassus	Hepática blanca		Parnassie
Pastinaca	Wild Parsnip	Chirivia	Chirivia, Pastinaga	Panais
Peganum		Gamarza		
Pelargonium	Zonal Geranium	Gitanilla	Malva rosa, M. de cheiro, Sardinheira	Géranium
Periploca		Cornical		
Petasites	Winter Heliotrope		Sombreiro	Pétasite
fragrans				Héliotrope-d'hiver
Petroselinum	Parsley	Peregil	Salsa	Persil
Phaseolus	Bean	Judia	Feijoeiro	Haricot
Phillyrea		Labiérnago	Aderno, Cadorno, Lentisco bastardo	
Phlomis		Barbas de macho		
crinita		Aguavientos		
herba-venti		Candilera, Mechera	Salva brava	Herbe-au-vent
lychnitis		Matagallos	Marioila	
purpurea				
Phoenix	Palm	Palmera	Tamareira, Palmeira-das-ingrejas	Palmier
Phragmites	Reed	Carrizo, Cañaborda	Caniço	Roseau-à-balais
Phytolacca	Virginian Pokeweed	Hierba carmin	Bela-sombra Tintureira, Erva-dos-cachos-da-India	Raisin d'Amérique

GENUS	ENGLISH	SPANISH	PORTUGUESE	FRENCH
Prunus				
armeniaca	Apricot	Albaricoquero	Damasqueiro, Alpercheiro, Albricoqueiro	Abricotier
avium	Wild Cherry	Cerezo de monte	Cerdeira, Cerejeire	Merisier
domestica	Plum	Ciruelo	Ameixeira	
dulcis	Almond	Almendro	Amendoeira	Amandier
laurocerasus	Common or Cherry-Laurel	Laurel cerezo	Louro-cerejo	
lusitanicus	Portugal Laurel	Laurel de Portugal	Azereiro	
padus	Bird-cherry	Cerezo aliso	Pado, Azereiro-dos-danados	Merisier-à-grappe
persica	Peach	Melocotonero	Pessegueiro	Pêcher
spinosa	Blackthorn, Sloe	Endrino	Abrunheiro-bravo	Prunellier
Psoralea	Pitch Trefoil	Hierba cabruna	Trevo bituminoso	Herbe-au-bitume
Pulicaria	Fleabane	Hierba pulguera		Pulicaire
vulgaris				
Pulsatilla				
vulgaris	Pasque Flower	Pulsatilla		Pulsatille, Anémone
Punica	Pomegranate	Granado	Romeira, Romanzeira	Grenadier
Putoria		Hedionda		
Pyracantha	Pyracantha	Espino de coral		Buisson ardent
Pyrus	Pear	Peral	Pereira, Maçã craveira, Escambroeiro	Poirier
Quercus	Oak	Roble, Rebollo	Carvalho	Chêne
cerris	Turkey Oak			Chêne chevelu
coccifera	Kermes Oak	Coscoja	Carrasqueiro, Carrasco	Chêne Kermès
faginea	Lusitanian Oak	Quejigo	Carvalho-português	Chêne Zeen
fruticosa			Carvalho anão, Carvalhiça	
ilex	Holm Oak	Encina, Carrasca	Azinheira, Azinho	Yeuse

rotundifolia	Evergreen Oak	Roble	Carvalho cerequeiro	Chêne vert
petraea	Durmast Oak	Melojo		Chêne noir
pyrenaica	Pyrenean Oak		Carvalho cerqinho	Chêne Tauzin
robur	Common Oak	Roble albar	Carvalho commun, Roble	Chêne rouvre
suber	Cork Oak	Alcornoque	Sobreiro, Sobro	Chêne-liège
Ramonda		Hierba de la tos		
Ranunculus	Buttercup, Crowfoot	Ranúnculo		Renoncule
ficaria	Lesser Celandine		Celidónia menor, Ficária	Ficaire
	Pilewort			
Raphanus	Radish	Rábanos	Saramago, Cabresto	Radis
Reseda	Dyer's Rocket,		Lírio-dos-tintureiros,	Réséda
luteola	Mignonette	Gualda	Minhonete	
Rhamnus				
alaternus	Mediterranean Buckthorn	Alaterno, Aladierno	Sanguinho-das-sebes, Aderno bastardo, Sandim	Alaterne
carthaticus	Buckthorn	Espino cerval		Nerprun
Rhinanthus	Yellow Rattle	Cresta de gallo		Cocriste, Rhina
Rhododendron	Rhododendron		Galocrista	Rhododendron
ferrugineum	Alpenrose	Hojaranzo		Rosage, Laurier-rose des Alpes
ponticum	Rhododendron	Bujo	Adeleira, Rododendo	
Rhus	Sumach	Zumaque	Sumagre	Sumac
Ribes				
nigrum	Black Currant	Casis	Groselheira-negra	Cassis
rubrum	Red Currant	Grosellero rojo	G. vermelha	Groseillier à grappes
uva-crispa	Gooseberry	G. espinoso		Groseillier épineux
Ricinus	Castor Oil Plant	Ricino	Bafureira, Carrapêto Mamona, Rícino	Ricin
Robinia	False Acacia	Falsa Acacia, Robinia	Acácia bastarda, Falsa-acácia	Robinier
Roemeria	Violet Horned-poppy	Amapola morada		

GENUS	ENGLISH	SPANISH	PORTUGUESE	FRENCH
Rosa	Rose	Rosa	Roseira, Rosa	Rosier, Eglantier
Rosmarinus	Rosemary	Romero	Alecrim	Romarin
Rubia	Wild Madder	Rubia	Granza brava, Raspa-língua	Garance
Rubus	Blackberry, Bramble		Silva	Ronce
caesius	Dewberry	Zarzamora		Ronce bleuâtre
idaeus	Raspberry	Frambuesa	Framboesa	Framboisier
Rumex	Dock	Acedera, Romaza	Labaça	Patience, Rumex
acetosa	Sorrel		Azedas	Oseille
Ruscus	Butcher's Broom	Brusco, Rusco	Gilbarbeira, Herva-dos-vasculhos	Fragon
Ruta	Rue	Ruda	Arrudão, Ruda, Rudão	Rue
Sagittaria	Arrow-head	Seata de ague, Cola de golondrina	Erva-frecha	Sagittaire, Flèche-d'eau
Salicornia	Glasswort, Marsh Samphire	Sosa		Salicorne
Salix	Willow	Sarga, Sauce	Vimieiro, Salgueiro	Saule
Salsola	Saltwort	Pincho, Barriola, Salicor	Barrilha espinhosa, Soda	Soude
Salvia	Clary, Sage	Salvia	Salva	Sauge
Sambucus				
ebulus	Danewort	Yezgo	Ébulo, Engos, Sabuguei-inho	Hièble
nigra	Elder	Sauco	Sabugueiro	Sureau
Samolus	Brookweed	Pamplina de agua	Alface-dos-rios	Mouron-d'eau
Sanguisorba	Burnet	Pimpinela	Pimpinela	Santuisorbe, Pimprenelle
Sanicula	Sanicle	Sanícula macho	Sanícula	Sanicle
Santolina				
chamaecyparissus	Lavender Cotton	Abrótano hembra, Santolina	Abrótano-femea, Guarda-roupa	Petit cyprès
Saponaria	Soapwort, Bouncing Bet	Jabonera	Erva-saboeira	Saponaire

Sarocapnos				
Satureja	Savory	Ajedrea	Segurelha, Hissopo	Sarriette
Saxifraga	Saxifrage	Saxifraga	Saxifraga	Saxifrage
granulata	Meadow Saxifrage	S. blanca	Quaresmas	Casse-pierre
longifolia				
Schinus	Californian Pepper-tree, Peruvian Mastic-Tree	Corona de rey Pimentero falso	Pimenteira-bastarda	Faux poivrier
Scilla	Squill			Scille
peruviana		Flor de la corona, Jacinto estrellado di Peru		
Scolymus				
hispanicus	Spanish Oyster Plant	Tagarnina	Cardo de oiro, Cangarinha Escólimo malliado	Épine-jaune
maculatus				
Scorpiurus		Lengua de oveja	Cornhilhão	Chenille
Scorzonera	Dwarf Scorzonera	Escorzonera	Escorcioneira	Scorsonère
Scrophularia	Figwort, Water Betony	Escrofularia	Escrofularia	Scrofulaire
Scutellaria	Skull-cap			Scutellaire
galericulata		Tercianaria, Hierba de la celada		
Secale	Rye	Centeno	Centeio	Seigle
Sedum	Stonecrop	Siempreviva		Orpin
acre	Wall-pepper	Uña de gato	Vermicularia	Poivre-de-muraille
album		Siempreviva menor	Uva-de-cão	
telephium	Orpine, Livelong	Hierba callera		Reprise
Sempervivum	Houseleek	Siempreviva mayor	Saião-curto, Sempre-viva-dos-telliados	Joubarbe
tectorum				Artichaut-de-muraille
Senecio	Groundsel	Hierba cana, Suzón	Cardo morto, Tasneirinha	Séneçon
vulgaris				Séneçon vulgaire
Silene	Campion		Alfinetes	Silene
armeria		Colleja		
vulgaris				

Suaeda	Seablite	Sosa	Valverde-dos-sapais	Soude
Succisa pratensis	Scabious, Devil's-bit	Escabiosa mordida	Morso-diabolico, Consolda-maior	Succise
Symphytum	Comfrey	Sínfito	Consoude	Consoude
Syringa	Lilac	Lila	Lilazeiro	Lilas
Tamarix	Tamarisk	Taray, Atarce	Tamarqueira, Tamariz	Tamarin
Tamus	Black Bryony	Nueza negra	Uva-de-cão, Norça-preta	Tamier
Taraxacum officinale	Dandelion	Diente de león, Amargón	Taráxaco, Dente de leão	Dent-de-lion, Pissenlit
Taxus	Yew	Tejo	Teixo	If
Teucrium chamaedrys	Germander, Wall Germander	Camedrios	Carvalhinha	Germandrée, Petit Chêne
Thalictrum	Meadow Rue	Ruibarbo de los pobres	Ruibarba-dos-pobres, Talictro, Vérça-de-cão	Pigamon
Theligonum				
Thymelaea villosa		Bufalaga	Troviso alvar	Passerine
Thymus	Thyme	Tomillo		Thym
cephalotus			Erva-ursa	
serpyllum	Wild Thyme		Serpão, Serpil	
vulgaris	Thyme		Tomilho	
Tilia	Linden, Lime	Tilo	Tília	Tilleul
Tolpis		Flor de viuda	Leituga, Olho de mocho, Viúvas, Flor-de-viúva	Oeil du Christ
Trachelium		Barba cabruna		
Tragopogon porrifolius	Goats-beard, Salsify		Barbas-de-bode	Barbe-de-bouc, Salsifis
Tribulus	Maltese Cross, Small Caltrops	Abrojos	Abrolhos	Croix de Malte
Trifolium	Clover, Trefoil	Trébol	Trevo	Trèfle
Trigonella foenum-graecum	Fenugreek	Alholva	Feno grego, Fenacho, Ervinha alforvas,	Fenugrec

GENUS	ENGLISH	SPANISH	PORTUGUESE	FRENCH
Ziziphus	Jujube	Azufaifo	Açufeifa maior, Anáfega-maie	Jujubier
lotus		Lotus comestible	Açufeifa menor	
Zostera	Eel-grass, Grass-wrack	Sebas de mar, Pelota marina	Limo-de-fita, Limo-seval	Varech, Goémon

Index of place names
listing places of botanical interest

54 *Quercus pubescens* × ½

55 *Quercus pyrenaica* × ¼

(78) *Cytinus ruber* × ⅓

(77) *Aristolochia pistolochia* × ⅓

1

(121) *Mesembryanthemum crystallinum* × ½

78a *Cynomorium coccineum* × ½

140a *Cerastium boissieri* ×½

77a *Aristolochia baetica* ×1

2

(120) *Carpobrotus acinaciformis* ×⅙

126b *Arenaria pungens* × ¾

126c *Arenaria tetraquetra* × ⅘

(128) *Arenaria balearica* × 1

171 *Silene rupestris* × ½

3

163b *Petrocoptis glaucifolia* × ⅓

174a *Silene littorea* × 1

194a *Dianthus lusitanus* × ⅕   194b *Dianthus subacaulis* × ½

185 *Dianthus barbatus* × ⅔   231a *Adonis pyrenaica* × ¼

4

191 *Dianthus seguieri* × 1   257 *Thalictrum flavum subsp. glaucum* × ⅖

55b *Thalictrum tuberosum* × ⅔

216a *Anemone pavoniana* × ⅓

226 *Clematis cirrhosa* × ⅔

239c *Ranunculus demissus* × ⅔

5

248b *Ranunculus acetosellifolius* × ½

245a *Ranunculus abnormis* × ½

258 *Paeonia officinalis subsp. humilis* × ⅕

217 *Anemone palmata* × ⅔

6

(254) *Aquilegia pyrenaica* × ⅓

245 *Ranunculus gramineus* × ½   (199) *Helleborus lividus subsp. lividus* × ½

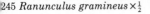

211f *Delphinium verdunense* × 1   (258) *Paeonia broteroi* × ¼

223a *Pulsatilla rubra* ×¼

301b *Matthiola lunata* ×¼

297c *Malcolmia lacera* ×¾

261a *Berberis hispanica* ×⅛

8

279a *Sarcocapnos enneaphylla* ×⅔

279b *Sarcocapnos crassifolia* ×½

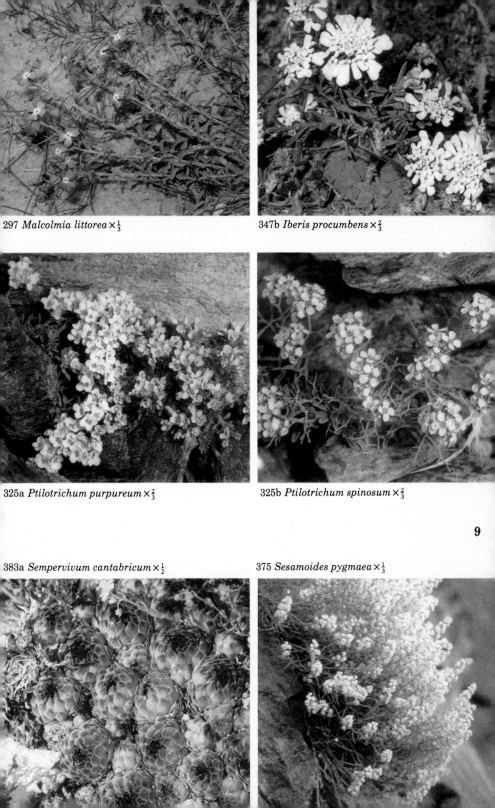

297 *Malcolmia littorea* $\times \frac{1}{3}$

347b *Iberis procumbens* $\times \frac{2}{3}$

325a *Ptilotrichum purpureum* $\times \frac{2}{3}$

325b *Ptilotrichum spinosum* $\times \frac{2}{3}$

9

383a *Sempervivum cantabricum* $\times \frac{1}{2}$

375 *Sesamoides pygmaea* $\times \frac{1}{3}$

361a *Brassica balearica* × ⅓

310a *Cardamine raphanifolia* × ⅔

10

349a *Biscutella vincentina* × ½

347a *Iberis linifolia* × ⅓

365a *Vella spinosa* × ½

295b *Erysimum myriophyllum* × ⅕   295a *Erysimum linifolium* × ¼   (301) *Matthiola fruticulosa* × ⅔

374a *Reseda media* × ⅓

373a *Reseda suffruticosa* × ¼

12

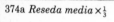

383 *Sempervivum tectorum* × ⅓

394a *Sedum hirsutum* × ⅘

(378) *Drosophyllum lusitanicum* × ¼

379e *Mucizonia sedoides* × ¾

379d *Mucizonia hispida* × ½

379b *Pistorinia hispanica* × ⅘

13

391b *Sedum brevifolium* × ¾

(387) *Sedum sediforme* × ⅘

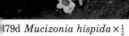

391b *Sedum brevifolium* × ¾

405a Saxifraga spathularis ×¼

404h Saxifraga cuneata ×⅙

411c Saxifraga globulifera ×⅘

(399) Saxifraga longifolia ×¹⁄₁₀

14

411b Saxifraga conifera ×¾

404c Saxifraga pentadactylis ×⅓

410b *Saxifraga aretioides* $\times \frac{1}{2}$

410a *Saxifraga media* $\times \frac{3}{4}$

442a *Geum sylvaticum* $\times \frac{1}{2}$

(404) *Saxifraga aquatica* $\times \frac{1}{10}$

15

476b *Prunus prostrata* $\times \frac{1}{4}$

445a *Potentilla montana* $\times \frac{3}{4}$

491a *Acacia karoo* × ⅕

503 *Cytisus multiflorus* × ⅓

16

489a *Acacia retinodes* × ⅘

498) *Calicotome spinosa* ×1

511a *Genista hirsuta* ×$\frac{1}{8}$

511b *Genista scorpius* ×$\frac{4}{5}$

508b *Genista florida* ×$\frac{1}{2}$

17

511e *Genista pumila* ×$\frac{4}{5}$

511 *Genista germanica* ×$\frac{3}{4}$

513a *Chamaespartium tridentatum* $\times \frac{1}{3}$

513d *Echinospartum lusitanicum* $\times 1$

507b *Teline monspessulana* $\times \frac{1}{3}$

513 *Chamaespartium sagittale* $\times \frac{1}{3}$

18

514a *Lygos sphaerocarpa* $\times \frac{1}{20}$

(515) *Erinacea anthyllis* $\times \frac{1}{4}$

518b *Adenocarpus decorticans* × $\frac{1}{50}$

(517) *Ulex parviflorus* × $\frac{2}{3}$

526 *Astragalus monspessulanus* × $\frac{1}{3}$

519a *Lupinus hispanicus* × $\frac{2}{5}$

19

(528) *Astragalus massiliensis* × $\frac{4}{5}$

519 *Lupinus luteus* × $\frac{1}{3}$

(527) *Astragalus lusitanicus* × ⅘

528d *Astragalus glaux* × 1

20

560b *Lathyrus tingitanus* × 1

565a *Lathyrus laevigatus subsp. occidentalis* × ⅔

533 *Oxytropis campestris* × 1

571f *Ononis striata* × $\frac{4}{5}$

(569) *Ononis fruticosa* × $\frac{1}{2}$

570 *Ononis rotundifolia* × $\frac{2}{5}$

21

(547) *Vicia lutea* × 1 (545) *Vicia villosa* × $\frac{1}{3}$ 636a *Onobrychis saxatilis* × $\frac{1}{2}$

613a *Lotus glareosus* × $\frac{1}{2}$

615 *Lotus creticus* × $\frac{3}{4}$

616 *Tetragonolobus maritimus* × $1\frac{1}{4}$

622 *Anthyllis vulneraria subsp. atlantis* × $\frac{1}{2}$

22

631b *Hippocrepis balearica* × $\frac{1}{12}$

634 *Hedysarum coronarium* × $\frac{1}{3}$

625 *Coronilla valentina subsp. glauca* $\times \frac{1}{20}$

645 *Geranium sylvaticum* $\times \frac{1}{2}$

23

694a *Cneorum tricoccon* $\times \frac{4}{5}$

655b *Fagonia cretica* $\times 1\frac{1}{4}$

653f *Erodium daucoides* × ⅓

653d *Erodium petraeum subsp. glandulosum* × ⅔

653 *Erodium cicutarium* × 1

641a *Geranium cinereum* × ¾

24

661 *Linum narbonense* × ⅔

662 *Linum viscosum* × 1

(658) *Linum campanulatum* $\times \frac{1}{4}$

(664) *Linum suffruticosum* $\times \frac{1}{3}$

694 *Ailanthus altissima* $\times \frac{1}{10}$

707 *Rhus coriaria* $\times \frac{1}{2}$

25

668 *Ricinus communis* $\times \frac{1}{5}$

704 *Pistacia terebinthus* $\times \frac{1}{2}$

688 *Dictamnus albus* $\times\frac{1}{6}$

736a *Malva hispanica* $\times\frac{1}{2}$

665b *Securinega tinctoria* $\times\frac{2}{3}$ 709a *Acer granatense* $\times\frac{1}{2}$

760 *Daphne laureola subsp. philippi* $\times 1$

696b *Polygala microphylla* × ½

701 *Polygala calcarea* × ½

758 *Daphne gnidium* × ¾

756 *Daphne cneorum* × 1¼

27

754b *Thymelaea dioica* × ¾

720b *Rhamnus ludovici-salvatoris* × 1

763a *Hypericum balearicum* × 1

764c *Hypericum caprifolium* × ½

769a *Hypericum nummularium* × ⅔

720 *Rhamnus alaternus (R. myrtifolius)* × ⅘

28

785a *Viola cornuta* × ⅘

781 *Viola arborescens* × ⅘

797c *Halimium lasianthum* × 1

785b *Viola crassiuscula* × $\frac{4}{5}$

797b *Halimium alyssoides* × $\frac{1}{8}$

781a *Viola cazorlensis* × $\frac{1}{2}$

29

795a *Halimium viscosum* × $\frac{1}{2}$

796 *Halimium commutatum* × $\frac{2}{3}$

788 *Cistus albidus* × 1

793 *Cistus ladanifer* × $\frac{1}{2}$

30

793b *Cistus clusii* × $\frac{2}{3}$

789 *Cistus crispus* × $\frac{2}{5}$

793a *Cistus palhinhae* ×⅙

792 *Cistus populifolius* ×⅓

794 *Cistus laurifolius* ×⅓

802a *Helianthemum croceum* × ½

801a *Helianthemum lavandulifolium* × ⅓

822a *Lythrum junceum* × ¾

810a *Frankenia corymbosa* × 1

32

857 *Echinophora spinosa* × 1/10

855a *Eryngium bourgatii* × ½

906a *Thapsia villosa* × $\frac{1}{20}$

906b *Thapsia maxima* × $\frac{1}{15}$

33

25 *Eucalyptus globulus* × $1\frac{1}{3}$

922 *Daboecia cantabrica* × 1

932 *Erica australis* × 1

932a *Erica umbellata* × ¾

(927) *Erica lusitanica* × ⅘

(934) *Erica vagans* × ⅔

34

(954) *Androsace villosa* × ½

970 *Coris monspeliensis* × ½

976b *Armeria pseudarmeria* $\times \frac{1}{15}$

(967) *Anagallis monelli* $\times \frac{3}{4}$

975b *Armeria juniperifolia* $\times \frac{2}{3}$

(976) *Armeria pungens* $\times \frac{1}{3}$

986 *Centaurium erythraea subsp. grandiflorum* $\times 1$

972a *Limonium thouinii* $\times \frac{1}{2}$

960a *Cyclamen balearicum* ×1

963a *Lysimachia ephemerum* ×1

36

972 *Limonium sinuatum*×1

1012 *Gomphocarpus fruticosus* ×⅔

996a *Gentiana burseri* × $\frac{1}{5}$

992a *Gentiana alpina* × 1

992b *Gentiana occidentalis* × 1

1013 *Putoria calabrica* × 1

1036a *Convolvulus lanuginosus* × $\frac{4}{5}$

1048c *Omphalodes brassicifolia* × $\frac{4}{5}$

1081c *Echium flavum* × 1

38

1075 *Lithospermum
fruticosum* × $\frac{2}{3}$

1081b *Echium boissieri* × $\frac{1}{6}$

1083a *Echium albicans* ×⅓

1056 *Anchusa azurea* ×1

39

1047a *Omphalodes nitida* ×1

(1049) *Cynoglossum cheirifolium* ×½

1074 *Lithospermum diffusum* $\times \frac{1}{3}$

1076 *Onosma arenaria* $\times \frac{1}{4}$

1083e *Echium gaditanum* $\times \frac{1}{2}$

1056b *Anchusa calcarea* $\times \frac{1}{10}$

40

1103a *Teucrium pyrenaicum* $\times \frac{2}{3}$

1103b *Teucrium rotundifolium* $\times \frac{1}{2}$

110a *Lavandula viridis* × 1

1110 *Lavandula stoechas subsp. pedunculata* × $\frac{1}{10}$

1111) *Lavandula multifida* × $\frac{1}{3}$

1096 *Teucrium pseudochamaepitys* × 1

1136b *Stachys ocymastrum* × $\frac{1}{2}$

41

134a *Ballota hirsuta* × $\frac{1}{3}$

1116a *Nepeta tuberosa* × $\frac{1}{10}$

1116d *Nepeta reticulata* × $\frac{1}{2}$

1123 *Phlomis lychnitis* $\times \frac{2}{3}$

1122b *Phlomis purpurea* $\times \frac{1}{2}$

1122 *Phlomis herba-venti* $\times \frac{1}{3}$

1163c *Thymus zygis* $\times \frac{1}{2}$

42

1163b *Thymus granatensis* $\times \frac{1}{2}$

1162c *Thymus longiflorus* $\times \frac{4}{5}$

147a *Salvia bicolor* × $\frac{1}{8}$

(1144) *Salvia aethiopis* × $\frac{1}{6}$

(1165) *Thymus mastichina* × 1

43

143a *Salvia blancoana* × $\frac{3}{4}$

1123a *Phlomis crinita* × $\frac{2}{3}$

1121a *Cleonia lusitanica* × $\frac{1}{2}$

(1204) *Linaria triornithophora* × $\frac{3}{4}$

1185b *Mandragora autumnalis* × $\frac{1}{2}$

1179a *Withania somnifera* × 1

1220a *Veronica fruticulosa* × $\frac{2}{3}$

44

1211c *Chaenorhinum villosum* × $\frac{1}{3}$

1211b *Chaenorhinum origanifolium* × $\frac{1}{2}$

201d *Linaria elegans* × ⅓

1204b *Linaria aeruginea* × 1

204c *Linaria amethystea* × ⅔

1204e *Linaria glacialis* × ⅔

45

197f *Antirrhinum pulverulentum* × ⅔

1197d *Antirrhinum hispanicum* × ½

(1195) *Verbascum pulverulentum* $\times \frac{1}{10}$

1190a *Verbascum chaixii* $\times \frac{1}{4}$

1196b *Verbascum laciniatum* $\times \frac{1}{2}$

46

1197a *Antirrhinum barrelieri* $\times \frac{1}{2}$

1197b *Antirrhinum graniticum* $\times \frac{1}{2}$

1197 *Antirrhinum majus* subsp. *linkianum* $\times \frac{1}{3}$

1234a *Digitalis dubia* × $\frac{1}{3}$

1233a *Digitalis parviflora* × $\frac{1}{2}$

47

269a *Cistanche phelypaea* × $\frac{1}{3}$

1234b *Digitalis thapsi* × $\frac{1}{2}$

1268 *Ramonda myconi* × ¼

1271a *Orobanche gracilis* × ¾

1264c *Globularia spinosa* × 1

1263a *Globularia repens* × ⅔

48

1266a *Myoporum tenuifolium* × ½

1289a *Plantago nivalis* × 1

280a *Pinguicula longifolia* × ¼

1280b *Pinguicula vallisneriifolia* × ⅓

303b *Lonicera pyrenaica* × ¾

1348 *Trachelium caeruleum* × ⅓

49

1316) *Centranthus angustifolius* × ⅙

1335a *Campanula lusitanica* × 1

1331b *Campanula mollis* × ½

(1360) *Bellis sylvestris* × ⅓

1361a *Bellium bellidioides* × 1

1370a *Erigeron frigidus* × ⅘

50

1414a *Anacyclus valentinus* × ¾

(1393) *Pulicaria odora* × ¼

398 *Asteriscus maritimus* $\times \frac{3}{4}$

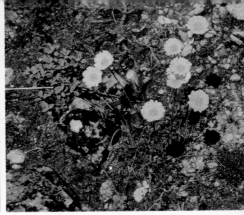

1428a *Tanacetum pallidum* $\times \frac{1}{5}$

1455c *Senecio tournefortii* $\times \frac{1}{2}$

1428c *Tanacetum radicans* $\times \frac{1}{3}$

51

1453b *Senecio boissieri* $\times \frac{2}{3}$

1460b *Calendula suffruticosa* $\times \frac{1}{3}$

1491a *Cynara humilis* × ⅔

1480a *Carduus carlinoides* × ⅓

1478a *Carduus granatensis* × ½

(1461) *Echinops strigosus* × ½

1494a *Onopordum acaulon* × ⅕

1460c *Cryptostemma calendula* × ⅔

1494b *Onopordum nervosum* $\times \frac{1}{10}$

1495a *Onopordum macracanthum* $\times \frac{1}{2}$

1507 *Centaurea conifera* $\times \frac{1}{3}$

1485b *Cirsium scabrum* $\times \frac{1}{20}$

53

1508a *Carthamus arborescens* $\times \frac{1}{3}$

1511 *Catananche caerulea* $\times \frac{3}{4}$

1500a *Centaurea polyacantha* × ⅔

1503c *Centaurea granatensis* × ⅕

1504a *Centaurea toletana* × ½

1503d *Centaurea haenseleri* × ½

54

1501b *Centaurea pullata* × 1

1531 *Scorzonera hispanica* × ⅓

1510 *Scolymus hispanicus* × ⅛

1508b *Carduncellus monspelliensium* × ⅔

1533 *Andryala integrifolia* × ⅓

(1537) *Cicerbita plumieri* × ½

(1508) *Carduncellus caeruleus* × ⅔

1536a *Reichardia tingitana* × ½

1511b *Hispidella hispanica* × 1

56

1414c *Cladanthus arabicus* × ½

(1528) *Tragopogon crocifolius* × ¾

588c *Androcymbium gramineum* × 2/3

(1643) *Hyacinthus amethystinus* × 1

637 *Dipcadi serotinum* × 1¼

1644a *Bellevalia hackelii* × 2/3

59

1622) *Leucojum trichophyllum* × 1⅔

1636a *Scilla monophyllos* × ½

1668d *Narcissus triandrus* × 1

1668f *Narcissus viridiflorus* × $\frac{4}{5}$

1665h *Narcissus asturiensis* × $\frac{1}{2}$

1666 *Narcissus bulbocodium var. nivalis* × $\frac{1}{2}$

60

1665j *Narcissus pallidiflorus* × $\frac{1}{4}$

1665n *Narcissus abscissus* × $\frac{1}{4}$

676 *Crocus nudiflorus* × ¼

1673a *Tapeinanthus humilis* × ¼

678a *Crocus nevadensis* × ½

1678b *Crocus carpetanus* × ⅓

61

681a *Romulea clusiana* × ¾

1697 *Gladiolus illyricus* × ¾

1685b *Iris filifolia* × ⅔

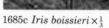

1685c *Iris boissieri* × ⅓

1684a *Iris planifolia* × ⅔

1685 *Iris xiphium* × ¾

62

(1685) *Iris xiphioides* × ⅘

1692a *Iris subbiflora* × ⅓

1880 *Ophrys sphegodes* × ¾

1882 *Ophrys tenthredinifera* × 1⅕

1875 *Ophrys speculum* × 1¼

881 *Ophrys bertolonii* × 1

1873 *Ophrys fusca* subsp. *atlantica* × ⅖

1873a *Ophrys omegaifera* × ½

63

894a *Orchis patens* × ⅔

(1896) *Orchis pallens* × ½

1894 *Orchis mascula* × ½

1899a *Orchis saccata* × ½

1884 *Orchis papilionacea* × ⅓

(1890) *Orchis italica* × ⅔

1886 *Orchis coriophora* × ½

(1904) *Serapias parviflora* × ½

1921 *Limodorum abortivum* × ⅘

1902 *Serapias lingua* × ⅘

1918 *Epipactis atrorubens* × ¾

Index of plants

This index refers to chapter 3 except where the sole reference to a plant occurs in another chapter. Numbers in italic are running numbers; other references are to text page (as 532); colour plate (as pl. 20); and page of line-drawings (as d. 222).

Ragwort
  Marsh, *(1451)*, 360
Ramonda
  myconi, *1268*, 340; pl. 48;
    d. 138
Rampion, *1334*, 349
  Dark, *1351*, 351
  Pyrenean, *1352a*, 351
  Round-Headed, *1351*, 351
  Spiked, *1350*, 351
  Blue-, *(1350)*, 351
Ranunculaceae, 185
Ranunculus
  abnormis, *245a*, 191; pl. 5;
    d. 111
  acris, *239*, 191
  acetosellifolius, *248b*, 190;
    pl. 5; d. 77
  aconitifolius, *246*, 190
  alpestris, *247*, 190
  amplexicaulis, *248a*, 190;
    d. 124
  arvensis, *235*, 191
  bulbosus, *237*, 191
    subsp. *gallecicus*, 114
  bullatus, *234a*, 190; d. 62
  bupleuroides, *245b*, 191
  demissus, *239c*, 191; pl. 5
  ficaria, *233*, 190
  flammula, *243*, 191
  glacialis, *(247)*, 190
  gouanii, 146
  gramineus, *245*, 191; pl. 7;
    d. 116
  gregarius, *239b*, 191
  lingua, *244*, 191
  muricatus, *232*, 190
  nigrescens, 104
  ololeucos, 105
  ophioglossifolius, *(243)*, 191
  parnassifolius, *248*, 190
  parviflorus, *(232)*, 190
  platanifolius, *(246)*, 190
  polyanthemos, *(238)*, 191
  pyrenaeus, *(248)*, 190; d. 149
  repens, *238*, 191
  rupestris, *239a*, 191
  sardous, *(237)*, 191
  sceleratus, *236*, 191
  seguieri, *247a*, 190
  thora, 146
  weyleri, 97
Rape, *360*, 203
Raphanus
  raphanistrum
    subsp. *landra*, 37
Raspberry, *428*, 214
Red-Hot Poker, *1598c*, 383
Redleg, *82*, 176
Red Pepper, *(1179)*, 326
Red-Rattle, *1256*, 338
Reed
  Common, *1740*, 402
  Giant, *1739*, 402
Reedmace
  Great, *1827*, 405
  Lesser, *1828*, 405
Reichardia

gaditana, *1536b*, 379
picroides, *1536*, 379
tingitana, *1536a*, 379; pl. 56
Reseda
  alba, *373*, 204
  glauca, *372b*, 204
  gredensis, *372c*, 204
  lanceolata, *372a*, 204; d. 207
  lutea, *372*, 204
  luteola, *371*, 204
  media, *374a*, 204; pl. 12
  phyteuma, *374*, 204
  suffruticosa, *373a*, 204; pl. 12
  undata, 72
  virgata, *374b*, 204
Resedaceae, 204
Restharrow, *569*, 239
  Large Yellow, *571*, 239
  Round-Leaved, *570*, 239;
    d. 149
  Shrubby, *(569)*, 239; pl. 21
  Spiny, *568*, 239
Retama, see *Lygos*
Rhagadiolus, 374
Rhamnaceae, 259
Rhamnus
  alaternus, *720*, 259
  catharticus, *723*, 260
  frangula, see *724*, 260
  ludovici-salvatoris, *720b*,
    260; pl. 27
  lycioides, *720a*, 259; d. 254
  myrtifolius, 259; pl. 28
Rhaponticum
  cynaroides, *1498a*, 369; d. 364
Rhinanthus
  minor, *1247*, 338
Rhodiola
  rosea, *397*, 209
Rhododendron
  ferrugineum, *919*, 283
  ponticum, *920*, 283
Rhus
  coriaria, *707*, 257; pl. 25
  typhina, *(707)*, 257
Rhynchospora, 405
Ribes
  alpinum, *(414)*, 213
  nigrum, *415*, 213
  petraeum, *(415)*, 213
  rubrum, *414*, 213
  uva-crispa, *416*, 213
Rice, *1797*, 403
Ricinus
  communis, *668*, 255; pl. 25
Robinia
  pseudacacia, *523*, 233
Rock Bramble, *427*, 214
Rocket
  Dyer's, *371*, 204
  Sea, *366*, 204
  Wall, 203
  White, *358*, 203
Rockrose
  Common, *802*, 275
  Hoary, *801*, 275
  Spotted, *798*, 274; d. 163
  White, *803*, 275; d. 76

Rock Samphire, *871*, 281
Roemeria
  hybrida, *271*, 194
Romulea
  bulbocodium, *1681*, 398; d. 63
  clusiana, *1681a*, 398; pl. 61
  columnae, *(1681)*, 398
  gaditana, *1681b*, 398; d. 53
Rosa
  micrantha, 75
  pendulina, *435*, 214
  pimpinellifolia, *434*, 214
  sempervirens, *(430)*, 214
Rosaceae, 214
Rosemary, *1105*, 313
Roseroot, *397*, 209
Rosmarinus
  eriocalix, *1105a*, 313; d. 90
  officinalis, *1105*, 313
  tournefortii, *1105a*, 313; d. 90
Rubia
  peregrina, *1031*, 301
Rubiaceae, 300
Rubus
  caesius, *(429)*, 214
  fruticosus, *429*, 214
  idaeus, *428*, 214
  saxatilis, *427*, 214
Rue, 255
  see also Meadow Rue
Rumex
  acetosella, 105
  bucephalophorus, *95*, 176
Rupicapnos
  africana, *282a*, 195
Ruppia, 381
  spiralis, 159
Ruscus
  aculeatus, *1651*, 390
  hypoglossum, *(1651)*, 390
  hypophyllum, *1651a*, 390
Rush
  Flowering, *1564*, 380
  Wood, 401
Ruta
  angustifolia, *686a*, 255
  chalepensis, *687*, 255
  graveolens, *686*, 255
  montana, *(686)*, 255
Rutaceae, 255

Safflower, *(1508)*, 374
Sage, 320
  Red-Topped, *1149*, 321
Sagina, 178
  saginoides, 155
Sagittaria
  sagittifolia, *1563*, 380
Sainfoin
  False, *(544)*, 235; d. 84
  Italian, *634*, 246; pl. 22; d. 52
  Rock, *636a*, 248; pl. 21
St John's Wort
  Alpine, *772*, 268
  Common, *768*, 267
  Hairy, *765*, 267
  Imperforate, *(767)*, 267
  Mountain, *764*, 267

Bibliography

Reference Floras

COSTE, H., *Flore descriptive et illustrée de la France, de la Corse et des Contrées limitrophes,* Paris, 1900–1906. Every species illustrated by a marginal line drawing. Invaluable for the amateur: in French.

COUTINHO, A. X. P., *Flora de Portugal,* 2nd edn., Lisboa, 1939. The standard flora, but virtually unobtainable; not illustrated; in Portuguese.

FRANCO, J. do A., *Nova Flora de Portugal,* Lisboa, Vol. 1, 1971. The modern flora; not illustrated; in Portuguese.

FOURNIER, P., *Les quatre Flores de la France, Corse comprise,* Poinsin-les-Grancey, 1934–40. Useful pocket manual with many small line drawings of diagnostic characters; in French.

HOFFMANSEGG, I. C., Conte de, & LINK, H. F., *Flore Portugaise,* Berlin, 1809–20. Some of the finest botanical plates ever produced.

LÁZARO E IBIZA, B., *Compendio de la flora Española,* 3rd edn., Madrid, 1920–21. The best flora in Spanish; including lower plants; a few line drawings.

ROUY, G. C. C., et al., *Flore de France,* 14 vols., Asnières, Paris and Rochefort, 1893–1913. The standard detailed flora of France; not illustrated.

TUTIN, T. G., et al., eds., *Flora Europaea,* Cambridge, 1964, 1968, 1972–. The standard flora for Europe, essential for reference; not illustrated; in English.

WILLKOMM, H. M., *Supplementum Prodromi Florae hispanicae,* Stuttgart, 1893.

WILLKOMM, H. M. & LANGE, J., *Prodromus Florae hispanicae,* Stuttgart, 1861–80. The standard flora; not illustrated; in Latin.

Popular Accounts, Guides, Geography

BACON, L., 'A Journey through Spain', *Bull. Alp. Gard. Soc.,* 39, 1971.

CHODAT, R., *Excursions Botaniques en Espagne et au Portugal,* Genève, 1909.

DELVOSALLE, L., & DUVIGNEAUD, P., *Itinéraires Botaniques en Espagne et au Portugal,* Brussels, 1962.

GIUSEPPI, P. L., 'Portugal and its Plants', *Bull. Alp. Gard. Soc.,* 14, 1946.

GORER, R., 'Southern Spain in Spring', *Gard. Chron.,* 25 April, 2 May, 10 May, 1964.
'A Reconaissance in the Montes Universales', *Jour. Roy. Hort. Soc.,* 90, 1965.
'A Jaunt in the Mountains of Central Spain', *Bull. Alp. Gard. Soc.,* 34, 1966.
'Plant hunting in Central Spain', *Gard. Chron.,* 26 April, 3 May, 10 May, 1967.

HAMILTON, A. P., *Gibraltar Walks and Flowers,* Gibraltar, 1970.

HARANT, H. & JARRY, D., *Guide du Naturaliste dans le Midi de la France,* Neuchatel, 1961.

HEYWOOD, V. H., 'Through the Spanish Sierras', *Jour. Roy. Hort. Soc.,* 73, 1948. (Sierras of Urbion, Cazorla, Nevada, Carrascoy, Montserrat.)
'Plant collecting in the mountains of Andalucia', *Jour. Roy. Hort. Soc.,* 75, 1950. (Sierras of Cazorla, Ronda, Nevada, and Murcia-Cartagena area.)

HOUSTON, J. M., *The Western Mediterranean World,* London, 1964.

HUXLEY, A., *Mountain Flowers,* London, 1967.

NAVAL Intelligence Division, *Spain and Portugal,* 1941.

POLUNIN, O., *Flowers of Europe,* London, 1969.

POLUNIN, O., & HUXLEY, A., *Flowers of the Mediterranean,* London, 1965.

RIPLEY, D., 'A Journey through Spain', *Bull. Alp. Gard. Soc.,* 12, 1944. (Sierras of Cazorla, Nevada, Morena, El Torcal de Antequera, Prieta, Cerro Maimon, Baza, Ronda.)

STOCKEN, C. M., *Andalusian Flowers and Countryside,* Kingsbridge, Devon, 1969.

'Sierra Nevada in early Spring', *Bull. Alp. Gard. Soc.,* 30, 1962.

'Plant hunting in Southern Spain', *Jour. Roy. Hort. Soc.,* 89, 1964.

'The Spanish Sierra Nevada', *Bull. Alp. Gard. Soc.,* 33, 1965.

'The Serrania de Ronda', *Bull. Alp. Gard. Soc.,* 34, 1966.

TAYLOR, A. W., *Wild Flowers of the Pyrenees,* London, 1971.

THACKER, T. C., 'A visit to Southern Spain and the Sierra Nevada', *Bull. Alp. Gard. Soc.,* 27, 1959. (Torremolinos, Granada, and Sierra Nevada.)

WALKER, D. S., *The Mediterranean Lands,* London, 1960.

WAY, R., *A Geography of Spain and Portugal,* London, 1962.

WILMOTT, A. J., 'Collecting in Spain from a motor car', *Jour. Bot.* (London), 1927.

Floristic Accounts of Regions, Plant Lists

General

FONT QUER, P., 'La Vegetation', in *Geografía de España y Portugal,* Barcelona, 1954.

LOSA ESPAÑA. M., 'Resumen de un estudio comparativo entre las flores de los Pirineos franco-españoles y los montes cantabroleonses', *Anal. Inst. Bot. Cavanilles,* 13, 1955.

RICKLI, M., *Das Pflanzenkleid der Mittelmeerlander,* Berne, 1948.

RIVAS GODAY, S., 'Los grados de vegetación de la Península Ibérica', *Anal. Inst. Bot. Cavanilles,* 13, 1955.

WILLKOMM, H. M., *Grundzüge der Pflanzenverbreitung auf der iberischen Halbinsel,* Leipzig, 1896.

Portugal

BARRETO, R. R., Dantas, 'Etude phytosociologique des chênaies de la Serra da Peneda', *Agron. Lusit.,* 20, 1958.

BRAUN-BLANQUET, J., SILVA, A. R. Pinto da, & ROZEIRA, A., 'Résultats de deux excursions géobotaniques à travers le Portugal septentrional et moyen', *Agron. Lusit.* 18, 1956; *Agron. Lusit.* 23, 1960.

BRAUN-BLANQUET, J., & FONTES, F. Carvalho, 'Résultats de deux excursions géobotaniques à travers le Portugal septentrional et moyen', (Serra da Estrêla), *Agron. Lusit.,* 14, 1952.

LUISIER, A., 'Apontamentos sobre a flora da região de Setúbal', *Bull. Soc. Broteriana,* 19, 1903.

NATIVIDADE, J. Vieira, *Subériculture,* Nancy, 1956.

PEDRO, J. Gomes, 'Geobotanical study of the Serra da Arrábida', *Agron. Lusit.,* 4, 1942.

PRIMO, S. da Costa, 'Quelques observations sur la végétation de Sagres et du Cap de S. Vicente', *Bull. Soc. Port. Sci. Nat.,* 12, 1936.

ROTHMALER, W., 'Die Pflanzenwelt Portugals', *Nachrichleb.f.d. Deutschen in Portugal,* 3, Lisboa, 1939.

'Vegetationsstudien im Südwestlichen Portugal', *Feddes Repertorium,* 128, Berlin, 1943.

SILVA, A. R. Pinto da, 'Deux herborisaticons', *Revista Agron.,* 28, 1940. (Includes list from São Pedro de Moel (Pinhal de Leiria).)

SILVA, A. R. Pinto da, ROZEIRA, A., & FONTES, F. C., 'Les chênaies de la Serra do Gerês', *Agron. Lusit.*, 12, 1950.

SILVA, A. R. Pinto da, & SOBRINHO, L. G., eds., 'Flora vascular da Serra do Gerês', *Agron. Lusit.*, 12, 1950.

SILVA, A. R. Pinto da, et al., 'First account of the limestone flora and vegetation of North-Western Portugal', *Bull. Soc. Broteriana*, 32, 1958.

TAVARES, C. das Neves, 'Protection de la flore et des groupements vegétaux du Portugal', *Publ. Liga Protec. Nat.*, 16, 1958.

VASCONCELLOS, J. de Carvalho, 'La protection de la flore de la Serra do Gerês', *Agron. Lusit.*, 12, 1950.

Spain

BARRAS, F. de las, 'Datos para la flórula sevillana', *Act. Soc. Esp. Hist. Nat.*, 26, 27, 28, 1897–99.

BOISSIER, E., *Voyage botanique dans le midi de l'Espagne pendant l'année 1837*, Paris, 1839–45. A classic with superb coloured plates.

BORJA CARBONELL, J., 'Una excursión a la Sierra de la Sagra (Granada)', *Anal. Inst. Bot. Cavanilles*, 13, 1965.

CABALLERO, A., 'Apuntes para una flórula de la Serranía de Cuenca', *Anal. Jard. Bot. Madrid*, 2, 1942; 4, 1944; 6, 1946.

CADEVALL y DIARS, J., *Flora de Catalunya*, Barcelona, 1913–37. In Catalan.

CEBALLOS, L., & MARTÍN BOLAÑOS, M., *Estudio sobre la vegetación forestal de la provincia de Cadiz*, Madrid, 1930.

CEBALLOS, L., & VICIOSO, C., *Estudio sobre la vegetación forestal de la provincia de Málaga*, Madrid, 1933.

CUATRECASAS, J., 'Estudios sobre la flora y la vegetación del Macizo de Mágina', *Trab. Mus. Ci. Nat. Barcelona*, 12, 1929.

DRESSER, D. W., 'Notes on the pre-alpine flora of the Picos de Europa, Spain', *Notes Roy. Bot. Gard. Edinburgh*, 33, 1959; 34, 1962.

GALIANO, E. F., 'Anotaciones a la flora de Sierra Morena. Plantas de Aldeaquemada', *Anal. Inst. Bot. Cavanilles*, 12, 1954.

GALIANO, E. F., & HEYWOOD, V. H., *Catálogo de plantas de la provincia de Jaén*, Inst. de Estudios Giennenses, Jaén, 1960.

GUINEA LÓPEZ, E., *Vizcaya y su paisaje vegetal*, Bilbao, 1949.

Geografía botánica de Santander, Santander, 1953.

HEYWOOD, V. H., 'The flora of the Sierra de Cazorla, S.E. Spain, part 1', *Feddes Repertorium*, 64, Berlin, 1961.

KNOCHE, H., *Flora Balearica*, Montpellier, 1921–23.

LAZA PALACIO, M., 'Estudios sobre la flora y la vegetación de las Sierras Tejeda y Almijara', *Anal. Jard. Bot. Madrid*, 1946.

LOSA ESPAÑA, M., *Contribución al estudio de la flora y vegetación de la provincia de Zamora*, Barcelona, 1949.

'Catálogo de las plantas que se encuentran en los montes palentino-leoneses', *Anal. Inst. Bot. Cavanilles*, 15, 1957.

LOSA ESPAÑA, M., & MONTSERRAT, P., 'Aportación al estudio de la flora de los montes cantabricos', *Anal. Inst. Bot. Cavanilles*, 10, 1952.

LOSA ESPAÑA, M. & RIVAS GODAY, S., 'Estudio floristico y geobotánico de la provincia de Almeria', *Arch. Inst. Aclim. Almeria*, 13, 1968.

'Nueva aportación al estudio de la flora de los montes cantabro-leoneses', *Anal. Inst. Bot. Cavanilles*, 11, 1953.

MARCET, A. F., 'Flora Montserratina', *Bol. Soc. Esp. Hist. Nat.*, 1949–54.

MARÈS, P., & VIGINEIX, G., *Plantes vasculaires des Isles Baléares*, Paris, 1880.

MERINO, B., *Flora descriptiva e illustrada de Galicia*, Santiago de Compostella, 1905–1909.

MONTSERRAT, P., 'Plantas de los alrededores de Soria', *Collect. Bot.* (Barcelona), 2, 1949.

El Turbon y su flora, Instituto de Estudios Pirenaicos, Zaragoza, 1953.

Flora de la cordillera litoral Catalana, Mataró, 1968.

PAU, C., 'Herborizaciones por la Sierra de Albarracín', *Bol. Soc. Aragon Ci. Nat.* 9, 1910.

PEREZ LARA, J., 'Flórula gaditana', *Anal. Soc. Esp. Hist. Nat.*, 15, 16, 18, 20, 21, 24, 25, 27, 1886, 1898; and *Mem. R. Soc. Esp. Nat. Hist.*2, 1903.

PRÓSPER, E. R., *Las estepas de España y su Vegetación*, Madrid, 1915.

RIVAS GODAY, S., & BELLOT, F., 'Estudios sobre la vegetación y flora de la comarca Despeñaperros-Santa Elena', *Anal. Jard. Bot. Madrid*, 5, 1945; 6, 1946.

RIVAS GODAY, S., & BORJA, C. J., 'Estudio de Vegetación y Florula del Macizo de Gudar y Jabalambre', *Anal. Inst. Bot. Cavanilles*, 19, 1961.

RIVAS MARTINEZ, S., 'Estudio de la vegetación y flora de las Sierras de Guadarrama y Gredos', *Anal. Jard. Bot. Madrid*, 21, 1963.

SENNEN, F., 'Plantes observées autour de Teruel pendant les mois d'aôut et de septembre 1909', *Bol. Soc. Aragon Ci. Nat.*, 9.

VALVERDE, J. A., 'An ecological sketch of the Coto Doñana', *Brit. Birds*, 51, 1958.

VICIOSO, C., 'Materiales para el estudio de la flora soriana', *Anal. Jard. Bot. Madrid*, 2, 1942.

WOLLEY-DOD, A. H., 'A Flora of Gibraltar and the neighbourhood', *Jour. Bot.* (London), 52, 1914.

Pyrenees and Andorra

BERGERET, J. P., *Flore des Basses-Pyrénées*, Pau, 1909.

BLANCHET, C., *Catalogue des plantes vasculaires du sud-ouest de la France comprenant le département des Landes et celui des Basses-Pyrénées*, Bayonne, 1891.

GAUSSEN, H., *Catalogue-flore des Pyrénées, Monde des Plantes*, 1953 onwards.

GAUTIER, G., *Catalogue raisonné de la flore des Pyrénées-Orientales*, Perpignan, 1898.

LOSA ESPAÑA, M., & MONTSERRAT, P., 'Aportaciones para el conocimiento de la flora del valle de Ordesa', *Collect. Bot.* (Barcelona), I, 1947.

'Aportación al conocimiento de la flora de Andorra', *Primo Congreso Internacional del Pirineo del Inst. Est. Pirenaicos*, 53, 1950.

Various authors: see *Bull. Alp. Gard. Soc.* 1, p. 122; 4, p. 86; p. 96; 7, p. 316; 8, p. 240; 9, p. 98; 16, p. 63; 18, p. 19; 18, p. 24; 19, p. 173; 20, p. 264; 22, p. 351; 24, p. 63; 26, p. 129; 30, p. 299; 33, p. 68; 35, p. 109; 36, p. 233; 36, p. 315; 37, p. 366; 38, p. 141; 38, p. 183; 38, p. 323; 39, p. 236.

France

ACLOQUE, A. N. C., *Flore du sud-ouest de la France et des Pyrénées*, Paris, 1904.

BILLAUGE, A., 'La garrigue de Nîmes', *Soc. Lang. Géo.*, 14, 1943.

BRAUN-BLANQUET, J., *L'origine et le dévelopement des flores dans le Massif Central de France*, Paris/Zurich, 1923.

La végétation alpine des Pyrénées orientales, Barcelona, 1948.

BRAUN-BLANQUET, J., ROUSSINE, N., & NÈGRE, R., *Les groupements végétaux de la France méditerranéenne*, 1951.

CHASSAGNE, M., *Inventaire analytique de la flore d'Auvergne et contrées limitrophes des départements voisins*, Paris, 1956–57.

COSTE, H., 'Florule du Larzac, du Causse Noir et du Causse de Saint-Afrique', *Bul. Soc. Bot. France*, 40, 1894.

DUVIGNEAUD, P., 'Les groupements végétaux de la France méridionale', in *Végétation et faune de la région méditerranéenne française*, Bruxelles, 1954.

FLAHAULT, C., *La distribution géographique des Végétaux dans la Région Méditerranéenne française*, Paris, 1937.

GAUTIER, G., *Catalogue de la Flore des Corbières*, Carcassonne, 1912–13.

LAPEYRÈRE, E., *Flore du département des Landes*, Dax, 1896–1903.

LAUBY, A., *Botanique du Cantal*, Paris, 1903.

MARANNE, I., 'Flore des hauts-plateaux basaltiques du Massif Central', *Bul. Géog. Bot.*, Le Mans 26, 1916.

PAULET, M., 'Les plantes aromatiques des garrigues languedociennes', *Ann. Soc. Hort. Hist. Nat. Hérault*, 98, 1958.

REVOL, J., 'Catalogue des plantes vasculaires du département de l'Ardèche', *Ann. Soc. Linn. Lyon*, 34, 1909.

TALLON, G., 'La végétation de Camargue', *83e Congrès National des Sociétés Savantes Colloqué sur la Camargue*, 1958.